Managing Scotland's Environment

Managing Scotland's Environment

Charles Warren

EDINBURGH
University Press

Edinburgh University Press Ltd
22, George Square, Edinburgh

Typeset in Minion
by Pioneer Associates, Perthshire, and
printed and bound in Great Britain by
The Cromwell Press, Trowbridge, Wilts

A CIP record for this book is available from
the British Library

ISBN 0 7486 1312 9 (hardback)
ISBN 0 7486 1313 7 (paperback)

Contents

Preface

In January 1995 I set about the task of writing a new Honours-level undergraduate course entitled 'Environmental Management in Scotland'. This involved picking up threads that I had laid down in 1987, at the end of an M.Sc. in Natural Resource Management, in order to pursue calving glaciers in remote corners of the globe. Seeking initial advice from John Blyth in my former department at Edinburgh University on the key question of potential course texts, I was told that none existed – and that I should write one myself. This flippantly offered (and – at the time – instantly dismissed) suggestion has now given rise to the present text.

The aims and scope of the book

Why the need for this book? What is it, and what is it not? To my mind, the need for a text such as this is glaringly obvious and urgent. Amazingly, there is no book which introduces and discusses the full sweep of Scottish environmental issues in a contemporary context. It is easy to lay one's hands on material relating to particular sectors – agriculture, water resources or conservation, for instance – and a number of recent edited volumes cover much useful ground,[1] but the whole picture is rarely presented. This is particularly surprising and regrettable given the current emphasis on integration and holistic management. The ground-breaking text edited by Paul Selman (1988a) has now been largely left behind by the speed and extent of subsequent change. More recent popular treatments that do range widely,[2] while lively, interesting and well informed, cannot engage with the literature in very much depth. (What they do offer, of course, is a profusion of superb illustrations with which this contribution cannot compete.)

This book aims to be two things: firstly, a textbook for use in higher education, and secondly an accessible, integrated overview of the Scottish environmental scene for the many organisations, groups and individuals who are involved in these issues. The danger of falling between two stools is clear, but at a time when new environmental policy is being formed at breakneck speed in this 'brave new Scotland', it would seem criminal to write something that would only be read in so-called ivory towers and not in contexts where it might (just possibly) make a difference. In terms of its scope, the book focuses exclusively on the terrestrial and freshwater environments. It stops at the coast, saying nothing about Scotland's rich marine heritage and the challenges of managing it; that would be a book in itself. No attempt to achieve both breadth and depth across the board could hope to succeed to everyone's satisfaction, so doubtless this book underplays, abbreviates or wilfully ignores many people's particular interests. What it does aim to present is an integrated introduction to the key elements and contemporary debates surrounding the demanding challenge of managing Scotland's natural environment.

I have not written another general introduction to environmental management, several of which have appeared recently[3] and whose wider scope complements the tighter geographical focus of this book. Nor have I attempted a campaigning environmental tract. I have tried to be dispassionate, presenting both (or all) sides of the many debates with equal weight and allowing the reader to form her or his own judgement. The book majors on description not prescription: others have ably offered the latter (e.g. Crofts and Holmes, 2000; Holdgate, 2000a). However, while I have not (I think) ground any particular axes, disinterested neutrality and objectivity are impossible when dealing with issues as emotive as treasured landscapes, beloved wildlife and fragile livelihoods. No one stands nowhere on these issues, and doubtless some of my non-neutralities show through.

One motivation for writing the book has been to correct some perceived imbalances and omissions in the literature. The titles of many books appear to promise an overview of environmental management or the rural economy throughout the UK (e.g. Lowe *et al.*, 1986; Allanson and Whitby, 1996; Winter, 1996), but in practice, almost without exception, what they actually offer is a detailed discussion of agricultural issues in England, with occasional allusions to Scotland, Wales and Northern Ireland, and an isolated, token chapter on forestry. I am not alone in being irritated by this. Here I attempt consciously to push the pendulum to the other extreme by focusing exclusively on Scotland, and by giving some prominence to forestry. There are, I believe, sound reasons for this geographical choice and for adopting what might be termed a silvo-centric emphasis. The Scottish context is distinctive in numerous ways – in landscape, climate, history, culture, law, education and politics – and all the more so since the re-establishment of the Scottish Parliament in 1999. Things are different in Scotland, whether one is discussing water resources, conservation, rural development or any other aspect of its multi-faceted natural and human environment. The resulting issues and debates are sometimes unique, invariably complex, and endlessly fascinating, quite apart from being of huge importance for the people and land of Scotland.

In this distinctive environment, forestry plays a key role, covering 16.9 per cent of Scotland's land area as against 14 per cent of Wales, 8.7 per cent of England and just 6.1 per cent of Northern Ireland (FC, 2001). Afforestation comprised the single greatest change in the Scottish landscape during the twentieth century, and has been involved in many of the big conservation controversies. Looking forward, in an era of food surpluses and agricultural restructuring, forestry of a new kind is set to play an expanding role in the new century. A final reason for adopting a woodland emphasis is simply the interest of the author, who rejects Alexander Pope's view of forests as 'collective bodies of straight sticks' in favour of W. H. Auden's dictum that 'a culture is no better than its woods'.

Many of the central terms used in the book are the focus of on-going debates. The precise meaning(s) of words such as nature/natural, rural, environment, community, wilderness, society, and, of course, sustainable development, have been exhaustively picked apart by a myriad of writers.[4] 'Nature', in particular, is 'perhaps the most complex and difficult word in the English language' (Macnaghten and Urry, 1998: 8). These debates are not only interesting but important, because conflicting understandings of these socially constructed terms are woven through many contemporary environmental debates. This text, however, tiptoes gingerly around the edge of these morasses, resisting the temptation to add yet more definitions to the hundreds already in print. Its focus is the practical business of environmental management in Scotland, and most of those involved in this adopt common sense understandings of these slippery terms. As Shucksmith *et al.* (1996) comment, their meaning is obscure only to the clever.

THE STRUCTURE OF THE BOOK

The two chapters in Part One outline the physical and political context within which Scottish environmental management operates. Part Two then introduces the key sectors in turn. I have adopted a measure of structural continuity in these chapters, most of them being organised around the following headings:

- historical background
- management framework
- recent developments
- current issues and debates.

In most cases the historical background is just a thumbnail sketch, partly because of limited space but, more positively, because comprehensive historical reviews exist for most of the sectors addressed here. 'Recent developments' and 'current issues' clearly overlap in many cases, so the themes introduced in the former sections sometimes reappear in the latter. 'Recent' is taken to refer roughly to the last ten years. Part Three explores some of the interactions between the various sectors, and some examples of particular controversies. Part Four might come as a surprise to some. In a book concerned with practical management, some might think it strange to devote space to ethical and philosophical matters, but as the chapter makes clear, these apparently obscure, academic issues are of much greater immediate relevance than many imagine. It is important to plumb the ethical wells from which our practical decisions spring. The concluding chapter identifies some of the common threads in the emerging tapestry of Scottish environmental management, and endeavours to peer a short way forward into the current century. To have attempted far-sighted prediction would have been even more rash than taking on the challenge of discussing the present. My hope is that readers will find as much interest and stimulation in reading the book as I have found in writing it.

NOTES

1. For example, Fenton and Gillmor (1994), Macdonald and Thomas (1997), McDowell and McCormick (1999a) and Holmes and Crofts (2000).
2. For example, Baxter and Thompson (1995), Magnusson and White (1997), Cramb (1998) and McCarthy (1998).
3. Useful texts covering diverse aspects of environmental management include those by Mitchell (1997), Wilson and Bryant (1997), Barrow (1999), O'Riordan (2000a) and Handmer *et al.* (2001).
4. Extended discussions and many references on these debates are provided by, *inter alia*, Simmons (1993, 1997), Pepper (1996), Cronon (1996a), Shucksmith *et al.* (1996), Coates (1998), Macnaghten and Urry (1998) and Phillips and Mighall (2000).

Charles Warren
St Andrews
July 2001

Acknowledgements

It has proved to be both exciting and somewhat foolhardy to write a book during this era of rolling revolution in the Scottish political and rural scene, attempting to keep simultaneous track of multiple moving targets. With so many spinning plates to keep my eye on, doubtless I have overlooked a few. I freely confess to being a generalist, not a specialist, and that in writing this book I have presumptiously pillaged the work of numerous specialists whose knowledge far exceeds my own. The syntheses and emphases are my own, but to the extent that this text interests, informs and stimulates, much of the credit must go to them.

I have racked up considerable debts of gratitude while writing it. Perhaps most importantly, a number of people generously took the time to read drafts of parts of the book, and their helpful and thoughtful suggestions have considerably improved the clarity and accuracy of the text. The following deserve particular thanks: Bill Adams, Andrew Black, John Blyth, Janet Egdell, Felix Fitzroy, Munro Gauld, Debbie Greene, Nick Halfhide, John Harwood, Jo Kerr, David Ledger, John Mackay, Adam Smith, Ian Smith, Chris Smout, Brian Staines, Graeme Swanson, Michael Usher, Graeme Whittington and Andy Wightman. Fraser MacDonald and two anonymous referees made perceptive suggestions on the original book proposal, and I am very grateful to Chris Smout for writing the foreword.

In terms of the production of the book itself, John Davey, my editor, the staff at EUP and the eagle-eyed copy editor, Monica Frisch, have all helped to ensure that my manuscript has actually appeared in book form with a minimum of errors. I am also enormously grateful to Graeme Sandeman for producing all the maps and diagrams to such a high standard. The University of St Andrews granted me one semester of research leave in 1999 to start work on the book. I owe much to Michael & Sue Scott, editors of the incomparable *SCENES* (Scottish Environment News), undoubtedly the finest news digest around; it constantly supplies nuggets of up-to-date information and comment on all aspects of the Scottish environment. Similarly, the Forestry Commission's weekly 'assistance with reading' bulletin has guided me to numerous papers that I might otherwise have missed.

Many individuals within various organisations have assisted me in diverse ways, especially the following:

- the Deer Commission for Scotland: Colin McLean and Helen Pearson
- the Game Conservancy Trust: Adam Smith and Simon Thirgood

- the Forestry Commission: various members of the statistics and public information sections
- the Ramblers Association: Helen Bushnell
- Scottish Executive Environment and Rural Affairs Department: Fiona Currie, Warrick Malcolm, Alasdair Robertson and Venetia Radmore
- Scottish Natural Heritage: Peter Barr, Simon Brooks, Richard Davison, John Gordon, Bob Grant, John Mackay, Kate Munro, Clive Mitchell, Heather Shirra and Peter Pitkin.

Pre-publication material was kindly made available by Bob Aitken, Colin Ballantyne, Andrew Black, John Gordon, Kath Leys, Sir John Lister-Kaye, Hugh Rose, Richard Tipping, Des Thompson and John Phillips. In the 'miscellaneous help of various kinds' category, come the following: Munro Gauld, John Hunt, Rob Lambert, Douglas MacMillan, Alistair McIntosh, Ian Mitchell, Alan Pollock, Rory Putman, Richard Tipping and Andy Wells. I also want to acknowledge the hundreds of students who have taken my Honours course, whose enthusiasm and questions have been tremendously stimulating; their howlers – a selection of which are reproduced in the Appendix – have also been a source of great amusement. Further back in time, Bill Adams, Tim Burt and David Sugden all played key roles in launching me (to my considerable and continuing surprise) down an academic career path, and I want to thank them for their inspiration and guidance.

More personally, I must thank my family. No book can be gestated and brought to birth without impinging on the author's nearest and dearest, and this one is no exception. My wife Sarah has been encouraging and forgiving in equal measure, and my children, Alexander and Iona, have had to learn the hard way that 'soon' is a very elastic concept. My final thanks go to my parents. Their enjoyment of trees and deer, lochs and hills, has rubbed off on me, and their love and prayers have carried me all the way.

List of abbreviations and acronyms

The proliferation and rapid evolution of acronyms is a recipe for instant confusion for anyone unfamiliar with rural land management. Across Scotland, the UK and Europe, the number of organisations that influence aspects of environmental management are legion, and they have a nasty habit of changing names on a regular basis. (Almost every abbreviation used in Selman's (1988a) book is now redundant, for example). Keeping accurate track of these moving targets in a text which ranges across different eras is a recipe for confusion, repetition and longwindedness ('SNH's predecessor body NCC', 'SEERAD, formerly SERAD', etc.). Some of this is unavoidable, but I have chosen to minimise it wherever possible, even though this leads to historical inaccuracies. For example, 'EU' is used throughout the book to refer to what is now the European Union, even when discussing periods when it was actually the EEC or the EC.

ADMG	Association of Deer Management Groups
AEP	Agri-Environment Programme
BAP	Biodiversity Action Plan
CAP	Common Agricultural Policy
CBA	Cost Benefit Analysis
CCC	Cairngorm Chairlift Company (now Cairngorm Mountain Ltd.)
CCS	Countryside Commission for Scotland (now part of SNH)
CEH	Centre for Ecology and Hydrology
CoR	Committee of the Regions
COSLA	Convention of Scottish Local Authorities
CVM	Contingent valuation method
CWP	Cairngorms Working Party
CWS	Community Woodland Supplement
DCS	Deer Commission for Scotland
DEFRA	Department of the Environment, Food and Rural Affairs
DETR	Department of the Environment, Transport & the Regions
DMG	Deer management group
EAP	Environmental Action Plan
ECHR	European Convention of Human Rights
EEC	European Economic Community (now the EU)
ESA	Environmentally Sensitive Area

ESF	European Structural Funds
EU	European Union
FA	Forestry Authority
FC	Forestry Commission
FE	Forest Enterprise
FoE/FoES	Friends of the Earth, Friends of the Earth Scotland
fte	Full-time equivalent
FWAG	Farming and Wildlife Advisory Group
FWPS	Farm Woodland Premium Scheme
GATT	General Agreement on Tarriffs and Trade (now the WTO)
GCT	Game Conservancy Trust
GDP	Gross Domestic Product
GMOs	Genetically Modified Organisms
HAP	Habitat Action Plan
HEI	Habitat Enhancement Initiative
HEP	Hydro-electric power
HGTAC	Home Grown Timber Advisory Committee
HIDB	Highlands and Islands Development Board (now HIE)
HIE	Highlands and Islands Enterprise
HLF	Heritage Lottery Fund
HRC	Highland Regional Council (now Highland Council)
ICM	Integrated catchment management
IFS	Indicative Forest Strategy
IT	Information technology
IUCN	International Union for the Conservation of Nature
JNCC	Joint Nature Conservation Committee
JRS	Joint Raptor Study
LBAP	Local Biodiversity Action Plan
LCA	Landscape Character Assessment
LcA	Life-cycle Analysis
LEADER	Liaisons entre actions de développement de l'économie rurale
LFA	Less Favoured Area
LIFE	L'Instrument Financier pour L'Environment
LLT	Loch Lomond and the Trossachs
LQG	LINK Quarry Group
LRAL	Lafarge Redland Aggregates Ltd
LRPG	Land Reform Policy Group
MCoS	Mountaineering Council of Scotland
MFST	Millennium Forest for Scotland Trust
MLURI	Macaulay Land Use Research Institute
MSP	Member of the Scottish Parliament
NCC	Nature Conservancy Council (now part of SNH)
NFI	Net Farm Income
NFUS	National Farmers Union Scotland

NGF	National Goose Forum
NGO	Non-Governmental Organisation
NHA	Natural Heritage Area
NHZ	Natural Heritage Zone
NIMBY	Not In My Back Yard
NNR	National Nature Reserve
NP	National Park
NSA	National Scenic Area
NTS	National Trust for Scotland
NPPG	National Planning and Policy Guidelines
NVZ	Nitrate Vulnerable Zone
PFAP	Paths for All Partnership
PP	Precautionary principle
RA	Ramblers Association
RDC	Red Deer Commission (now DCS)
RDR	Rural Development Regulation
RIGS	Regionally Important Geological/Geomorphological Site(s)
RPB	River Purification Board (now part of SEPA)
RSFS	Royal Scottish Forestry Society
RSPB	Royal Society for the Protection of Birds
SAP	Species Action Plan
SAC	Special Area of Conservation
SBG	Scottish Biodiversity Group
SCAC	Scottish Countryside Activities Council
SCENES	Scottish Environment News
SCPS	Scottish Countryside Premium Scheme
SCU	Scottish Crofters Union
SD	Sustainable development
SEA	Single European Act
SEERAD	Scottish Executive Environment and Rural Affairs Department
SEL	Scottish Environment LINK
SEPA	Scottish Environment Protection Agency
SERAD	Scottish Executive Rural Affairs Department (now SEERAD)
SLF	Scottish Landowners Federation
SNH	Scottish Natural Heritage
SNW	Scottish Native Woodlands
SOAEFD	Scottish Office Agriculture, Environment and Fisheries Department (now SEERAD)
SODD	Scottish Office Development Department
SOED	Scottish Office Environment Department
SPA	Special Protection Area
SRWS	Scottish Rights of Way Society
SSA	Scottish Sports Association
SPC	Scottish Sports Council

STB	Scottish Tourist Board (now VisitScotland)
SUDS	Sustainable Urban Drainage Systems
SWCL	Scottish Wildlife and Countryside Link (now SEL)
SSSI	Sites of Special Scientific Interest
TCM	Travel Cost Method
TEU	Treaty on European Union
TIBRE	Targeted Inputs for a Better Rural Environment
TIFF	Total Income From Farming
UKRWG	UK Raptor Working Group
UKWAS	UK Woodland Assurance Scheme
UNCED	United Nations Conference on Environment and Development
VMP	Visitor management plan
WCED	World Commission on Environment and Development
WGS	Woodland Grant Scheme
WTA	Willing to accept
WTO	World Trade Organisation
WTP	Willing to pay
WWF/WWFS	World Wide Fund for Nature, World Wide Fund for Nature Scotland

Foreword

The way we treat the environment has become an inescapable central problem of life. On its solution will rest the success of our economy, much of our quality of life, the health of our towns and our rural communities, the viability of farming and fishing and (not least) the fate of other living things with which we share this beautiful land. Yet in Scotland we discuss the environment very little in the public arena, beyond striking postures. The problems are complicated and their solution will need time and consensus, which is of little interest to the media. The issues demand the long-term vision that politicians no longer seem to possess. They ask for thought, concentrated study, open-mindedness and a sense of caring. It is no accident that Charles Warren's book arose from an extremely popular university course directed to those about to live most of their lives in the twenty-first century. Of course the environment matters to them. This is a problem about the future, although it demands knowledge of the past to understand the roots of our problems.

It is easy in environmental politics to take a stance. One side may instinctively treat the land user and the developer as enemies bent on the elimination of the natural world, and those on the other endlessly and shrilly assert that the town does not understand the countryside, which is usually a call to leave them in peace but pay the subsidy cheque.

Passion is understandable when the livelihood of communities or the fortunes of individuals balance against the good of the greater number or the fate of species. But the only way forward to a sensibly managed and sustainable environment must be through negotiation and understanding. This is why *Managing Scotland's Environment* is such a timely book. It is wide ranging, setting the scene in history and explaining current contexts and procedures, discussing agriculture, water and forestry, wildlife management and access, with a wealth of detail ranging from the problems of deer and alien species to flood management and the land reform debate. In case studies, it takes us to the Cairngorm funicular and the Harris superquarry, as well as onto the killing fields where grouse, raptors and gamekeepers interact. Charles Warren touches on the heart of the matter.

> How to strike the right balance? Who judges what is 'right' and by what criteria? How should local, national and international priorities be weighed against each other? How to balance the rights of people against the rights of non-human nature, and the rights of this generation against those of future generations?

He reminds us that at the root of how we treat the environment is a matter of our world-view of right and wrong, and of the rights and responsibilities of *Homo sapiens*: environmental philosophy is 'the unseen air which environmental management breathes'. But this is a book about practicalities, not theories, about options, not judgement. The author often gently suggests that one side of the argument is stronger than the other. Of course, not everyone will be happy with all his conclusions – some would find it hard to countenance culling or translocation of hen harriers while landowners continue so flagrantly to break the law. But that is not the point. His main concern is to put the two sides of the arguments in all their complexities, to indicate where we can find out more, and to leave us to decide the issues for ourselves. Again and again, though, he returns to the point that environmental management today has to be about partnership, and about two sides making serious efforts to understand each other's interests and exploring a common ground. Those who need conflict to make them feel good about themselves will be disappointed. Those who actually wish to help to solve Scotland's environmental problems will find this an invaluable survey and an inspiration. And those who simply want to be better informed could not do better than read it. If every MSP and member of a local planning committee had digested its contents, the cause of a sustainable future for Scotland would have taken a big step forward.

Chris Smout
St Andrews
August 2001

Part One:

The nature and control of the land

CHAPTER ONE

The shaping of Scotland's environment

1.1 Introduction: Natural Scotland?

'A pristine Highland wilderness.' 'The solitude of the rugged coasts and empty glens.' 'An unspoilt natural wonderland.' This is the kind of imagery used by the tourist industry to persuade people to 'savour the grandeur, peace and beauty of the Scottish Highlands'. Clearly there are elements of truth in this appealing picture. With built-up land covering only 4 per cent of the country, Scotland does consist largely of countryside (Fig. 1.1). It is, in fact, one of Europe's most sparsely populated countries, with one of the most urbanised and unevenly distributed populations. Some 80 per cent of its 5.1 million people live in the Midland Valley between the Highlands and the Southern Uplands, and a third live in the four main cities of Edinburgh, Glasgow, Aberdeen and Dundee, leaving large parts of the country with only a scattered human presence. The rural areas which account for 89 per cent of the land mass are inhabited by just one-third of the population; the population density in the Highlands and Islands is just 0.09 per hectare (ha), less than a tenth of the EU average of 1.1 per ha (SERAD, 2000a). It is quintessentially a land of mountains, with over half the country consisting of uplands and 284 peaks classified as 'Munros' (over 3,000 feet or 914m). 'Getting away from it all' is easier in Scotland than in many European countries.

Moreover, Scotland's natural heritage is rich indeed (SNH, 1995a). The land itself provides an unsurpassed backcloth, ranging from magnificent mountains through rolling lowlands to dramatic coastlines (Usher, 1999a) (Fig. 1.2). Abundant, high-quality water resources provide diverse aquatic habitats, energy for hydropower, a challenge for anglers, and beauty and recreation for visitors, not to mention purity for the production of whisky, 'the water of life'. The great variety of terrestrial and marine habitats supports some 90,000 species of animals, plants and microbes, including several which are under threat nationally or internationally (Usher, 1997, 2000). The nation is well endowed with mineral and energy resources. Finally, of course, Scotland's intriguing cultural heritage is rich and ancient, stretching back across nine millennia or more. For these and many other reasons, Scotland's nature can be described as one of the nation's 'top assets, . . . [lying] at the core of what many of us believe makes our country distinct and special' (Scottish Executive, 2001a: 4).

Despite all this, however, the kinds of perceptions of Scotland perpetuated by tourist posters of Highland glens capture very little of the complex diversity of the

FIGURE 1.1 *Map of Scotland.*

nature of Scotland. For a start, Scotland is much more than just the Highlands (Fig. 1.1). But a more fundamental issue is the question of how natural the country's 'natural heritage' actually is. Can any part of Scotland justly be described as an unspoilt wilderness? In Jim Hunter's opinion, to do so is 'to abuse both language and history' and to commit 'the wilderness fallacy' (Hunter, 1986 in McIntosh *et al.*, 1994: 67).

> Highland 'wilderness', as celebrated by today's conservationists, is not so much natural as the wholly artificial result of the nineteenth century landlords having excluded humanity from ecosystems of which people had been part for ten millennia. (Hunter, 2000: 6)
>
> The typical empty Highland glen with its treeless hillsides and its bare moorland is no more natural than a motorway embankment. It is something that has been created by what has been done to it by mankind . . . It has been made the way it is by people maltreating the land, by removing human communities. (Hunter, 1992 in MacAskill, 1999: 44)

What is the basis for such iconoclastic statements? In essence, it is that what we see today is an ancient tapestry woven by constant interaction between humans and the physical environment over many millennia. The Scottish landscape has been dramatically fashioned by both natural and human forces since the disappearance of the last glaciers from these shores some 11,000 years ago (Edwards and Ralston, 1997; Edwards and Smout, 2000). At many times in the past, the speed and extent of natural change has dwarfed any human influence, while at others, especially more recently (and probably in the future), natural processes have been overwhelmed by anthropogenic forces. The next section provides an outline sketch of the ways in which climate change and human activities have combined to fashion Scotland's environment.

1.2 CLIMATIC CHANGE AND HUMAN IMPACT

1.2.1 Climatic and environmental change

Given the brevity of human lifespans, it is easy to imagine that climate and environment simply exist as an unchanging, passive backdrop to human history. The reality, of course, is that neither is ever static; human histories have been played out in concert with complex environmental histories (Smout, 1993a, 2000; Simmons, 2001). Because Scotland is in a climatically sensitive location, situated at the western extremity of Eurasia and adjacent to the North Atlantic, it has experienced extremes of climate during the Quaternary period (the last 2.6 million years). Interludes of temperate climate warmer than or similar to today's alternated with long periods of intense cold during which Scotland was buried under the ice sheets which have fashioned today's mountain and lowland landscapes (Ballantyne and Dawson, 1997). Especially rapid climatic changes resulted from oscillations in the position of the oceanic polar front and in the amount of warm surface water transported northwards across the North Atlantic (Ballantyne, 2002).

FIGURE 1.2 *Landscape contrasts:*
a. The Highlands in late winter: the Fisherfield Hills from Slioch. Photo: © the author.
b. The River Tay winds through rolling lowlands east of Perth. Photo: © the author. (opposite)

Even during the post-glacial Holocene period (the last 11,500 years (Roberts, 1998)), climates and environments throughout the North Atlantic region have been constantly changing (Fig. 1.3).[1] In Scotland the last intense pulse of glaciation (the Loch Lomond Readvance) took place 12,700–11,500 years ago, and was followed by extraordinarily rapid climate change with temperatures reaching today's levels in perhaps as little as a few decades. Between 10,000 and 8,200 years ago, an era traditionally dubbed the Climatic Optimum, Scotland was about 2–3°C warmer than today, and probably drier, with extensive woodland (Walker and Lowe, 1997; Whittington

and Edwards, 1997a). This period was ended by a pronounced but brief cooling event, the last drastic climate fluctuation to affect Scotland (Ballantyne, 2002). Thus when our earliest ancestors arrived in Scotland they encountered a fluctuating climate, and an environment which was adjusting rapidly to the recent retreat of the last Scottish ice sheet (Walker and Lowe, 1997). This adjustment involved rapidly changing coastlines as the seas rose and the land rebounded from its ice-depressed state. It was probably also an era characterised by many large rockslides triggered by frequent severe earthquakes (Ballantyne, 2002). In no sense was Scotland a stable place at this time.

Throughout the early Holocene, progressive melting of the vast mid-latitude ice sheets led to rapidly rising seas, flooding the North Sea basin and eventually severing the land bridge which initially connected Britain with the Eurasian land mass. The British Isles finally became isles no later than 7,800 years ago and perhaps as early as 10,500 years ago (Yalden, 1999). Between the retreat of the continental ice sheet and the isolation of Britain from Europe there was thus a period of just a few thousand years during which plant and animal species could naturally colonise these shores. The significance of this is that many mainland European species did not make this journey, leaving Britain with a relatively impoverished flora and fauna. Those species that arrived naturally are regarded as native, while those subsequently introduced by humans are classed as exotics or aliens (Section 10.1).

Climate changes during the temperate middle and late Holocene were less dramatic than during the immediate aftermath of deglaciation, as demonstrated by the relative stability of Scottish forest cover (Edwards and Whittington, 1997), and their spatial impact became highly variable (Tipping, 2002). Nevertheless, a long-term trend

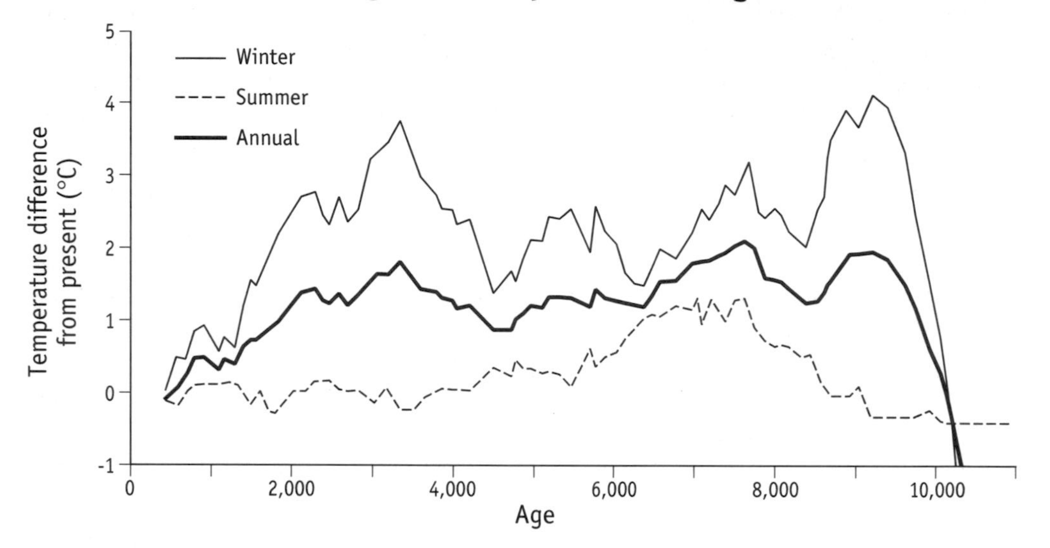

FIGURE 1.3 *Temperature fluctuations in the North Atlantic region during the last 10,000 years recorded in the Greenland Ice Sheet. After Alley* et al. *(1999).*

towards colder and wetter conditions led to an expansion of heathland and peat moors and to a lowering of the treeline from its early Holocene maximum elevation of almost 800m. Superimposed on the long-term trends have been numerous short-term climatic fluctuations during which temperatures have fluctuated by as much as plus or minus 2°C (Ballantyne, 2002). The last millennium, for example, saw temperatures considerably warmer than present during the so-called mediaeval warm period of the twelfth to the fourteenth centuries, followed by the intense cold of the Little Ice Age (seventeenth to nineteenth centuries), the coldest centuries since the last deglaciation. Such changes have led to repeated human occupation and abandonment of marginal land.

An inevitable outcome of on-going climate change has been continuous evolution of environments as geomorphic systems such as rivers, soils and slopes have adjusted, and as floral and faunal species have in turn adapted to the twin influences of climatic and geomorphological change (Ballantyne and Dawson, 1997; Davidson and Carter, 1997; McEwan, 1997). Animal and plant species have come and gone throughout the Holocene, and have existed in ever-changing mixtures within evolving ecosystems (McCormick and Buckland, 1997; Kitchener, 1998; Lusby, 1998). The most intrusive species of them all, *Homo sapiens*, has had an ever greater impact on all aspects of the environment, especially after about 5,800 years ago, an outworking of our progressive transformation from hunters to city-dwellers (Roberts, 1998).

A consequence of this is that from the mid-Holocene onwards it becomes progressively harder to disentangle the various climatic and human influences on environmental changes. For example, the development of the blanket peat which mantles such extensive areas of the uplands could have been triggered by increasing precipitation, by hydrological changes induced by early farming practices, or perhaps by natural soil processes needing no external trigger (Charman, 1994; Whittington and Edwards, 1997a, 1997b). Another well established event for which numerous explanations have been proposed is the elm decline which occurred about 5,800 years ago (Roberts, 1998). In both cases, multi-causal hypotheses (with regional variants) are now seen as being more probable than earlier dichotomous approaches. The key point here is that neither climate nor environment are static. Both are in constant dynamic flux, with or without human influence (Tipping, 2002), a reality which raises both practical and philosophical challenges for environmental conservation.

1.2.2 Human impacts

Scotland has never been the easiest of places for humans to live. Making a living has always required shrewd environmental evaluations and the development of subtle strategies to cope with the vagaries of the physical world (Morrison, 1983). Despite these inherent challenges, there is evidence for human occupation in Scotland throughout post-glacial time, and there was probably a permanent population of hunter-gatherers from at least 9,000 years ago (Ritchie and Ritchie, 1997). The early impacts were small-scale and transient, the tiny population being dominated by and subservient to the natural world. However, as the numbers of people and their grazing animals increased, improving technology combined with the widespread use of fire progressively gave human beings the power to shape their surroundings for their own ends. Cumulatively, over the millennia, human activity has modified most of the country and in places transformed it, the extent depending on the sensitivity of natural systems (Thompson *et al.*, 2001). Examples are numerous, both from prehistory (Smout, 1993a) and in recent times (Crofts, 1995; Smout, 2000a), but three will suffice.

1.2.2.i Forest removal

Amongst the most visually dramatic and spatially extensive of human impacts has been the transformation of Scotland from a dominantly forested country into one of the least wooded in Europe. At their maximum extent in the mid Holocene, forests covered at least 75 per cent of the country (Newton and Humphrey, 1997; Fig. 4.2). In the first century AD enough forest remained in the Highlands for the Romans to name the country 'Caledonia' meaning 'the woods on the heights', but even then most of southern and central Scotland was already deforested (Smout, 1993b, 2000a). By the fifteenth century Scotland was importing timber from Scandinavia to counter serious shortages, and by 1600 woodland occupied just 4 per cent of the country (Smout, 1997a), eliciting Samuel Johnson's famous remark that 'a tree might be a show in Scotland as a horse in Venice'. The subsistence agriculture which replaced the forests was itself replaced by the commercial agriculture which now utilises some 80 per cent of the land. Only in the twentieth century was this long decline in forest cover halted and reversed (Section 4.2).

1.2.2.ii Extinctions, introductions and domestication

Many of the animal and bird species which made up Scotland's native inheritance were driven to extinction by human beings, and many alien species were introduced, intentionally or otherwise (Lambert, 1998; Yalden, 1999) (Tables 1.1 and 1.2). If humans

TABLE 1.1 Extinctions and introductions of mammals in Scotland during the last 10,000 years. 'Introductions' refers to mammals found breeding in the wild today. Reindeer, polecat and muskrat have been reintroduced. 'Present' = 2000. After SNH (1995a), Harris *et al.* (1995), Kitchener (1998) and Yalden (1999).

Extinctions		*Years before present*
Pika	*Ochotona pusilla*	10,000
Arctic fox	*Alopex lagopus*	10,000
Collared lemming	*Dicrostonyx torquatus*	10,000
Northern vole	*Microtus oeconomus*	10,000
Narrow-skulled vole	*Microtus gregalis*	10,000
Horse	*Equus ferus*	9,700
Reindeer	*Rangifer tarandus*	8,300
Aurochs	*Bos primigenius*	3,250
Lynx	*Lynx lynx*	1,770
Moose	*Alces alces*	1,100
Brown bear	*Ursus arctos*	1,000
Beaver	*Castor fiber*	450
Wild boar	*Sus scrofa*	380
Wolf	*Canis lupus*	300
Polecat	*Mustela putorius*	88
Muskrat	*Ondatra zibethicus*	70
Introductions		*Years before present*
Feral goat	*Capra hircus*	8,000
Feral sheep	*Ovis aries*	7,000
Harvest mouse	*Micromys minutus*	5,800
Orkney vole	*Microtus arvalis*	5,500
House mouse	*Mus domesticus*	4,800
Brown hare	*Lepus europaeus*	2,000
Feral cat	*Felis catus*	2,000
Ship rat	*Rattus rattus*	1,800
Fallow deer	*Dama dama*	1,000
Rabbit	*Oryctolagus cuniculus*	800
Brown rat	*Rattus norvegicus*	250
Sika deer	*Cervus nippon*	130
Grey squirrel	*Sciurus carolinensis*	108
Muskrat	*Ondatra zibethicus*	66
American mink	*Mustela vison*	62
Wallaby	*Macropus rufogriseus*	48
Reindeer	*Rangifer tarandus*	48
Polecat	*Mustela putorius*	15
Muntjac	*Muntiacus muntjak*	6

had never reached Scotland, the species composition today would be radically different from what we see around us. In some cases, such as the wolf, species were deliberately hunted to extinction. In others, extinction was an indirect and unintended result of habitat change. Thus, for example, the extinctions of the capercaillie and the bittern were caused respectively by the decline of Caledonian pine forests and by drainage of wetlands (Kitchener, 1998). In still other cases extinctions or near-extinctions were caused by unrestricted and over-enthusiastic game shooting, notably in the eighteenth and nineteenth centuries (Smout, 1993c; Lister-Kaye, 1994).

TABLE 1.2 Extinctions of birds in Scotland during the last 10,000 years. Causes of extinction: K = hunting, H = habitat loss, P = persecution. After Kitchener (1998).

Species		*Date of Extinction*	*Cause of Extinction*
Crane	*Grus grus*	Unknown	K, H
White stork	*Ciconia ciconia*	1416	?H
Great bustard	*Otis tarda*	16th century	?H, ?K
Capercaillie	*Tetrao urogallus*	1785	H, K
Bittern	*Botaurus stellaris*	1830	H
Great auk	*Pinguinus impennis*	1840	K
Great spotted woodpecker	*Dendrocopos major*	1840–50	H
Goshawk	*Accipiter gentilis*	1883	P
Red kite	*Milvus milvus*	1884	P
Spotted crake	*Porzana porzana*	1912	H
Osprey	*Pandeon haliaetus*	1916	P
Sea eagle	*Haliaeetus albicilla*	1918	P

Turning to introductions, some have been deliberate, others accidental. Rabbits, introduced in Norman times for their fur and meat, have bred proverbially and have had a profound impact on the flora and fauna ever since, a story echoed in the twentieth century with the American mink (Section 7.7.2). Uninvited arrivals such as the common rat (*Rattus norvegicus*) hitched rides on ships. Human beings have also had a profound effect on Scotland's animal kingdom through the domestication of sheep, cattle, goats, horses, dogs and cats. Selective breeding has altered their characteristics, and populations have been multiplied to many times their natural size. The breeding of millions of grazing animals has also led to the transformation of the vegetation composition of open land. Sheep, in particular, have been blamed for widespread impoverishment of the soils and flora of the uplands, notably by Fraser Darling whose descriptions of Scotland as a 'devastated landscape' and a 'wet desert' have been much quoted. Although recent reassessments of this sweeping indictment have shown that he was substantially overstating his case (Mather, 1992, 1993a; Smout, 2000a), there is no doubt that the direct and indirect impacts of sheep grazing have significantly modified the landscape (Mather, 1993a). Even from this brief outline it is immediately apparent that natural relationships amongst faunal species and between faunal and floral species have been fundamentally transformed by human activities. Natural predator/prey relationships in particular were convulsed by the progressive loss of all the top predators.

1.2.2.iii Climate change

Looking to the future, a growing body of evidence suggests that human activity has set in train processes which will dramatically alter Scotland's climate during the present century and beyond. Average temperatures may rise by 1.2–2.6°C, more in winter than summer, and annual rainfall by 5–20 per cent, mostly in the autumn and winter, accompanied by changes in seasonality and increasing frequency of extreme events such as storms, floods and droughts (Kerr *et al.*, 1999; Scottish Executive, 2000a). Such changes would, of course, have far-reaching implications for Scotland's natural heritage, especially for biodiversity (Watt *et al.*, 1997) and particularly in the uplands (Hossell *et al.*, 2000; Harrison and Kirkpatrick, 2001). To put the predicted warming in context, an increase of just 1.0°C would be equivalent to moving high mountain habitats 200–300km north or increasing their altitude by 150–200m (SNH, 1995a). The possible responses of floral and faunal species include toleration (with or without adaptation), dispersal, invasion, displacement and local extinction (Watt *et al.*, 1997; SNH, 1999a). Some animals and plants which at present only have a tenuous hold in this country (including snow bunting, ptarmigan, mountain hare and several rare alpine plants) might be driven to extinction through loss or reduction of their habitat and/or through competition from opportunistic incomers. The effects of climate change will be exacerbated by the fragmented character of the landscape which will limit species' options. On the other hand, the predicted changes would be beneficial to a range of species, and would probably work to increase biodiversity in the lowlands (SNH, 1999a). Such scenarios represent great challenges for nature conservation and environmental management (Hossell *et al.*, 2000), not least because of the considerable uncertainties surrounding climate change predictions; managing for uncertainty and accelerating change may be the only sensible option (Harrison and Kirkpatrick, 2001).

The key point brought out in Sections 1.2.2.i and 1.2.2.ii is that anthropogenic environmental change is nothing new. It is ancient and profound. It is also ongoing. For example, the rates and patterns of recent changes in land use and vegetation have been documented in detail by the National Countryside Monitoring Scheme (Tudor *et al.*, 1994; Mackey *et al.*, 1998). In the forty years to 1988 there were profound transformations across much of Scotland, driven by urbanisation, agricultural intensification and afforestation (Fig. 1.4). The area of woodland tripled, most semi-natural habitats declined (e.g. the area of lowland mire decreased by 44 per cent), grass expanded at the expense of heather, and the total length of hedgerows was halved.

The prospect of human-induced climate change exemplifies the extent to which the technological advances of recent centuries have now given us the power to transform our world at a rate and to an extent unprecedented in history. To imagine that we are no longer dependent on natural processes and resources for our survival is, of course, an arrogant and dangerous mistake, but the pendulum has swung a long way from subservience towards mastery during the millennia in which *Homo sapiens* has

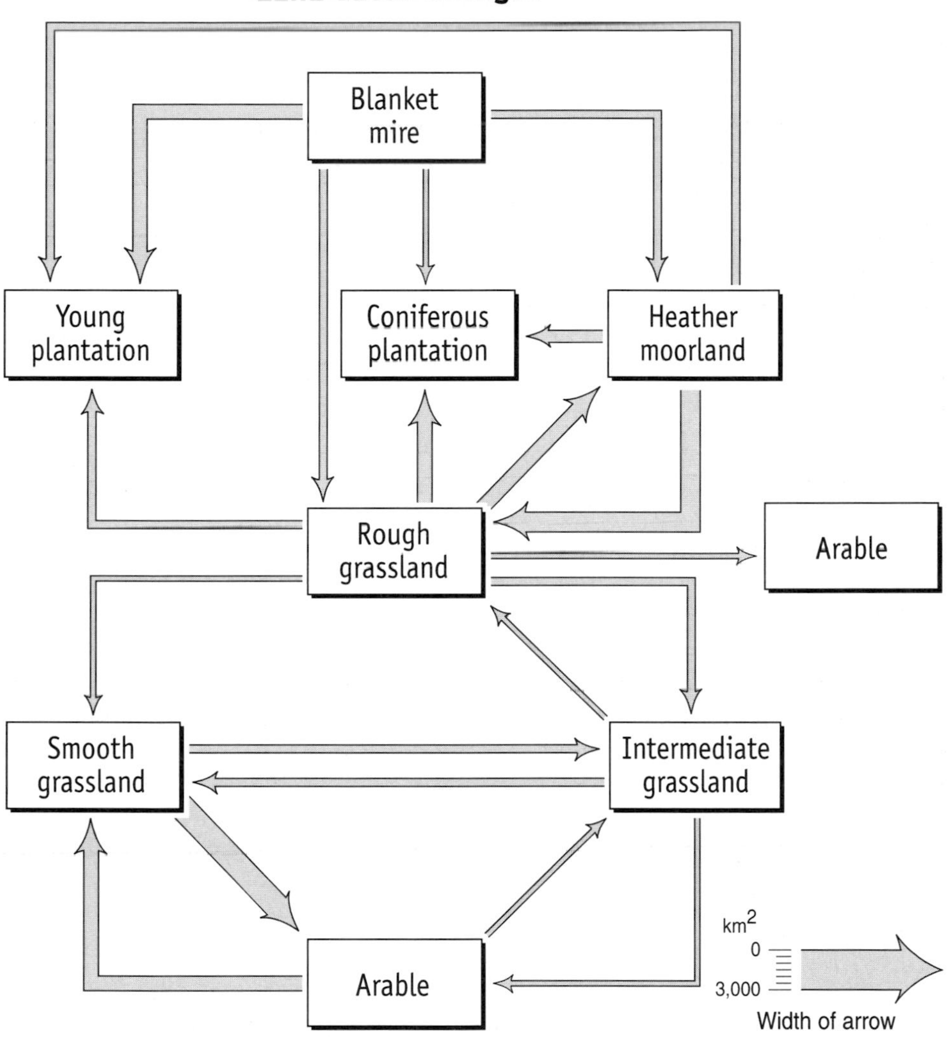

FIGURE 1.4 *Primary changes in land cover in Scotland, 1940s to 1980s. This diagram illustrates changes which affected more than 0.5 per cent of the total land area. After Mackey* et al. *(1998).*

been resident in Scotland. With this much enhanced power comes the obligation to make responsible choices about the ways in which that power is exercised: how should we manage the environment which we have inherited?

1.3 MANAGING THE 'NATURAL' ENVIRONMENT

The length of time during which human and non-human processes have been working in concert to fashion Scotland's landscapes leads some to argue that human beings now belong here as much as any 'natural' element of the environment, and consequently that we should not attempt to turn the clock back in pursuit of some mythic idyll of unspoilt nature. MacAskill (1999: 43) goes so far as to say that 'wilderness is not the natural state of the Highlands and Islands', and Hunter (1995) argues for resettlement of the glens that have been devoid of human communities since the nineteenth century Highland Clearances (Section 3.1). This reflects the tenor of much recent thinking about rural regeneration in Scotland which is predicated on the need to reverse the long-established trend of depopulation.

In stark contrast to this, conservation organisations such as the Scottish Wildlife Trust (SWT) and Trees for Life advocate the exact opposite, the creation of large 'wilderness areas' devoid of people but repopulated with Scotland's extinct fauna, including the top predators (Featherstone, 1997; Cairns, 2000) (Section 10.4). The concept of wilderness, however, is linked in the collective imagination with the Clearances, so such proposals generate storms of protest. SWT's ideas, for example, prompted the headline: 'Clearances fear over proposal to bring back the missing lynx' (*The Scotsman*, 7th January 2000). It is obvious even from this one example that the place of human beings in the world is a controversial topic (Section 13.3.1). The difficulty of striking a balance between human needs and stewardship of the non-human world also lies at the heart of debates about sustainable development (Section 13.2.1).

As the extent of the human transformation of nature – in time, space and severity – becomes ever more apparent, it is logical to ask whether Scotland has a 'natural environment' at all. Much of what appears wild and timeless is actually far from being in an unaltered state; most is, at best, semi-natural as the following quotations bring home:

> There is not much that is natural about the natural heritage. (Magnusson, 1993: xi)
>
> In this small, old country . . . nothing is wilderness. (Smout, 2000a: 172)
>
> The natural heritage of Scotland is a managed heritage. (Johnston, 2000: 9)

Most uncompromising of all is Bennett (1995: 36) who states flatly that 'the whole Scottish landscape is anthropogenic'. In this sense, as Giddens (1999: 27) provocatively asserts, 'our society lives after the end of nature'. Clarion calls to return the land to its natural state are guaranteed a sympathetic hearing in this green age, but the picture sketched above of the perpetual evolution of human and environmental history shows that it is far from easy to ascertain just what that state was or should be. Nor is it simple to select a moment along the continuum of past change when the environment was in a state of nature. Which moment from the past should be canonised as our vision for the future? To select a time before any human influence is to choose the era of climatic and environmental instability immediately after the glaciers retreated, clearly neither a possible nor a desirable baseline. But to choose any subsequent moment along the sliding scale of increasing human influence is unavoidably arbitrary.

To avoid this cul-de-sac, some suggest that the most 'natural' option is the removal or at least the reduction of human interference so that nature is allowed to take its course. But the results of such a 'hands off' approach will still be shot through with human influence because we can only begin from a profoundly altered starting point (Section 8.3.1.ii). Budiansky (1995: 16, 126) provocatively highlights the irony that:

> to have nature be 'natural' requires constant human intrusion . . . Thousands of years of human history have effectively blotted out the very meaning of 'artificial' and 'natural'. We cling to these terms at the cost of endless confusion and muddled thinking.

If 'natural' is taken to mean a complete absence of human influence, then it can be argued with considerable justification that 'natural environment' is a term that has no place in Scotland. Indeed, virtually none of the world's environments are entirely natural in that sense; pristine is history. In common parlance, however, 'natural environment' is usually used as a synonym for the countryside, in contradistinction to the 'built environment'. A more useful term which recognises the interplay of the natural and human processes that have shaped today's landscape is 'natural heritage'. This can be defined as follows (SODD, 1999a: 5):

> The combination and interrelationship of landform, habitat, wildlife and landscape and their capacity to provide enjoyment and inspiration. It therefore encompasses both physical attributes and aesthetic values and, given the long interaction between human communities and the land in Scotland, has important cultural and economic dimensions.

Another term which has come to the fore in recent years to encompass the complex origins of the Scottish environment is 'cultural landscape'.

An acceptance that, in Scotland at least, true wilderness should be allowed to rest in peace in the remote past has certain important contemporary implications. One is that the easy polarity whereby non-human nature is good and humans and all their works are bad is simplistic and unsupportable. Another is that human management of the environment is an unavoidable responsibility, however romantically appealing the ideas of wilderness and non-intervention are in theory. As Brennan (1993: 17) points out, 'non-interference in nature is no more an option for us than for any other mammal'. Cronon (1996b: 80) eloquently drives this point home when he observes that the concept of wilderness represents:

> the false hope of an escape from responsibility, the illusion that we can somehow wipe clean the slate of our past and return to the *tabula rasa* that supposedly existed before we began to leave our marks on the world.

The choice, therefore, is not between interference and non-interference, but about the form and extent of interference (or management). In its current state, nature needs nurture.

Many questions arise from these conclusions, including the following:

- What should the place of human activity be within the natural world? Given that aiming for 'no impact' is unrealistic, how should human well-being and economic development be weighed against conservation and the desire to preserve wild places?
- What should we be trying to conserve, why, and who for?
- What should the aims of enlightened management be? Good management, as with beauty, is largely in the eye of the beholder. The eye of the beholder is inevitably a product of its time and culture, influenced by particular beliefs and values. Objective and subjective are thus inevitably interwoven, and susceptible to changes of fashion.
- Is the world ours to manage?

There are, of course, no easy or clearcut answers to such questions. They are explored later in the book but are raised here because they underlie so much of what follows. If we accept that we have a responsibility to manage our environment, what exactly is the nature of that challenge? The next section outlines some key elements of the theory of environmental management. The rest of the book explores how these are being worked out in practice in the Scottish context.

1.4 ENVIRONMENTAL MANAGEMENT IN THEORY

Environmental management (EM) can be described as the manipulation of physical and human resources to achieve chosen objectives.[2] As both a multi-layered process and a field of study, it touches on innumerable disciplines (Fig. 1.5), requiring an understanding not only of the earth's physical systems in all their interacting diversity, but also of the equally diverse perceptions and values which humans attach to those systems. This essential multidisciplinarity and interdisciplinarity makes it the antithesis of reductionist specialisation. On the contrary, it constitutes a fascinating meeting ground between arts and science, and between physical and human geography. Barrow (1999: 11) describes EM as being 'as much an art as a science'.

Partly because of its multifaceted nature, there is no accepted definition of EM. Both 'environment' and 'management' mean different things to different people, so that innumerable definitions have been proposed (see Wilson and Bryant (1997) for a review). As with geography (with which it has close affinities because of the importance of space and time in environmental issues), EM consists of a (relatively) uncontroversial core surrounded by many peripheral areas which shade imperceptibly into other disciplines. Attempting clear delimitations is neither possible nor profitable. Traditional definitions tend to emphasise the application of science to specific problems in the physical environment, but these overlook the complex political, economic and social spheres within which EM has to operate (Boehmer-Christiansen, 1994).

Environmental management is about far more than biophysical manipulation

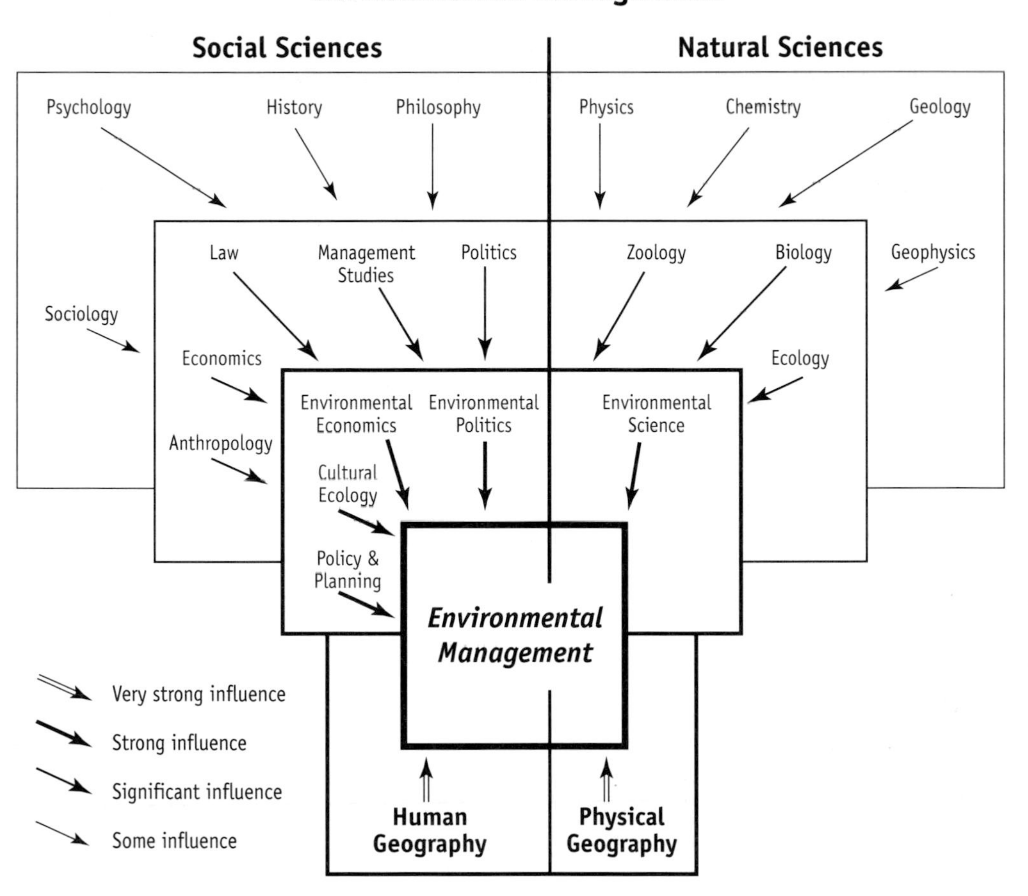

FIGURE 1.5 *Disciplinary influences on environmental management. After Wilson and Bryant (1997).*

and control – it concerns the mutually beneficial management of the humankind-nature interaction to ensure environmental and social quality for future generations . . . [It] is an intensely political process. (Carley and Christie, 1992: 11)

EM has its origins in the rise of the environmental movement in the 1960s and 1970s, and it was an applied and interdisciplinary field from the start. In addition to its close association with geography, it has also drawn on ecology, biology, planning studies, and economics (this last since the rise of environmental economics (Section 13.4) from the early 1980s). Although EM draws heavily on the fruits of environmental science (O'Riordan, 2000a), it differs from it both in its emphasis and its methodologies: EM has a strong social science element, whereas environmental science focuses more on the natural sciences. The former has also been more closely associated with the conservation of nature and landscapes. As Wilson and Bryant (1997: 39) state, 'the enduring tension between exploitation and conservation has always been at the heart of EM'. During the 1990s EM came to be structured (at least in principle) within the over-arching concept of sustainability (Section 13.2.1).

Planners, analysts and politicians like to imagine (all too often) that we live in an orderly, understandable and stable world, but this is dangerously misleading. More realistically, Mitchell (1997) identifies four characteristics of the real world that make EM enormously challenging and that are themselves key elements of EM:

1. **Change**. One of the few certainties in EM is that change is an integral part of the *status quo*. Natural and social systems are continuously evolving, both in time and space, meaning that management objectives and plans must be under constant review.
2. **Complexity**. EM addresses few simple situations. Typically, it has to engage with a wide range of diverse and interacting systems, each a complex system in its own right, all operating on contrasting temporal and spatial scales. In the light of this daunting reality, Kay and Scheider (1994: 33) comment that 'our traditional managerial approaches, which presume a world of simple rules, are wrong-headed and likely to be dangerous'. For EM, complexity arises as much from human behaviour as from physical processes. Contrasting attitudes and perceptions, especially in terms of environmental world views, lead to radically different conclusions about acceptable levels of harm or impact, about what solutions are appropriate, and about who should make those decisions (Boehmer-Christiansen, 1994). Political and market forces (both national and global) add extra layers of complexity. Thus EM is affected – one might say afflicted – by environmental, socio-cultural, political and market factors, usually in combination.
3. **Uncertainty**. Uncertainty and ignorance are the constant companions of environmental managers (Dovers *et al.*, 1996). This is because of the great complexity and continuously changing nature of most systems (both human and physical), combined with the contrast between the long timescales of many natural processes and the shortness of most datasets. As Wilson and Bryant (1997: 7) put it, 'the central predicament of all environmental managers [is] the quest for predictability in a context of increasing social and environmental uncertainty'. Our knowledge of natural systems, and of the interactions between natural and social systems, is (and will always be) partial, and yet decisions have to be made. Consequently, as Mitchell (1997) puts it, we have to start sailing while still building the ship.
4. **Conflict**. Given the complexity of the physical world, it is frequently claimed that the concept of managing the environment is a misnomer. Kay and Scheider (1994: 33) argue that 'we don't manage ecosystems, we manage our interaction with them'. Mitchell (1997) makes this argument one of the central messages of

> his book. If this is so, then much of EM boils down to the management of conflict. Scotland, of course, has no monopoly on environmental conflict; it is a characteristic of environmental issues worldwide. Steering a path through conflict towards a constructive outcome is no easy task, and the difficulty is exacerbated by two factors. The first is that all pluralistic societies consist of groups with divergent values, interests and priorities. The second concerns the benefits and costs involved in environmental conflicts; not only do these tend to be unequally distributed but they typically include intangible elements (aesthetics, emotional attachments, beliefs). For these two reasons, choosing desirable future outcomes would not be straightforward even in a simple and predictable physical world; enviromental complexity makes such choices doubly hard. Consequently, 'obtaining agreement on what constitutes a 'common future' is often a formidable challenge' (Mitchell, 1997: 96).

In the light of all the above, it is clear that environmental problems almost never have single, simple solutions. As a consequence, EM typically involves charting a course through a host of complex dilemmas, a number of which are identified by Barrow (1999):

- Ethical dilemmas: what to prioritise when there is a choice between threatened ecosystems and fragile human communities?
- Efficiency dilemmas: how much damage is acceptable?
- Equity dilemmas: who benefits, who pays and who chooses?
- Liberty dilemmas: how much restriction of people's freedoms is acceptable in order to protect the environment?
- Uncertainty dilemmas: how to make choices without adequate understanding?
- Evaluation dilemmas: how to predict and compare the effects of different courses of action?

Another consequence of the nature of EM is that environmental decision making is a complex business and is often not very formalised, especially in terms of choosing the appropriate approaches and tools (Tonn *et al.*, 2000).

A rational response when faced with a field of such near-infinite complexity is to throw up one's hands in despair and take refuge in theorising (by becoming a university lecturer, for example). More positive responses have been devised. The *ecosystem approach* is one which has received much attention. This is an holistic perspective which stresses the need to consider 'the big picture', the whole system, as an antidote to reductionist approaches based on artificial parcellings of reality. There is impregnable logic to the argument that all relevant factors should be incorporated in the decision making process, but a major problem afflicting any attempt to be all-inclusive is the sheer time and cost involved in gathering and analysing all the necessary data. The danger of drowning in complexity – of 'paralysis by analysis' – becomes very real. By the time any plans or decisions are finalised they are redundant; the world has moved on. A more practical halfway house is the *integrated approach* advocated by Mitchell (1997). This still aims to adopt an holistic perspective but homes in on the key components and linkages of the systems concerned. It capitalises on the fact that in most situations a small number of variables account for a large percentage of the variation. Understanding these critical variables usually permits effective management while avoiding the diminishing returns of attempts to achieve 'full' understanding.

In many ways the integrated approach is an eminently sensible compromise, but it obviously begs the question of what the critical variables are (and who decides what they are). Taking a short cut to the key linkages may not be possible because until a reasonably broad and comprehensive understanding is achieved it will not be apparent just what the critical components are. Even in a so-called ideal world in which all necessary data could be gathered and analysed before decisions were made, uncertainty would persist; many complex systems operate in inherently unpredictable ways, as chaos theory has shown. If to the unknowability of physical systems one then adds the unpredictability of societal change, trying to see any distance into the future becomes a fog-shrouded exercise. There are therefore both practical and theoretical reasons for accepting the uncomfortable reality that environmental managers will always have to operate with incomplete understanding of the present, and with only the haziest of perceptions of possible futures. Change, complexity, uncertainty and conflict combine to make error-free EM a figment of an over-optimistic imagination.

In summary, successful environmental management requires all of the following:

- A knowledge and understanding of resources: the natural sciences.
- A knowledge and understanding of the influences determining resource use: the social sciences.
- Management skills.
- Leadership: hard decisions are unavoidable in the real world. Single issue activists have the luxury of being able to see things in black-and-white, but environmental managers invariably inhabit a complex world of multi-faceted demands and disparate viewpoints. Decision making rarely involves simple choices between obvious goods and bads; more often it requires a delicate act of balance and/or integration involving compromise, uncertainty, and difficult choices between shades of grey.
- A moral/ethical framework: it is impossible to manage in a moral vacuum (Section 13.1).

Managing the unique and irreplaceable environment in which we live and on which we depend (not only for life but for quality of life) is one of the most pressing and testing challenges facing us in the present century.

Notes

1. Some of the sources quoted in this section provide ages based on radiocarbon dates. However, the radiocarbon timescale is somewhat elastic, diverging significantly from the calendrical timescale during parts of the Holocene. To avoid confusion, all ^{14}C dates have been converted to calendrical dates using the conversion table in Roberts (1998: 253).
2. This section lightly sketches some key characteristics of environmental management. Ample flesh is put on these bare bones in several excellent recent texts which this section draws on heavily. See, for example, the detailed discussions provided by Mitchell (1997), Wilson and Bryant (1997), Barrow (1999) and O'Riordan (2000a), and the many references therein. Owen and Unwin (1997) collate a stimulating range of readings and case studies with a global perspective. Owens and Owens (1991) offer an interesting introduction to the field.

CHAPTER TWO

The political and planning context

Living as we do in a global village, no nation is an island. Understanding even very local issues in Scotland is now often impossible without a grasp of the wider political and legal framework within which decisions are made, a framework which has Scottish, UK, European and global dimensions. Starting broad and finishing narrow, this chapter sketches three separate, though closely linked, aspects of the wider political and planning framework, and explores the evolving policy context in which Scottish environmental management operates. It begins with a discussion of the greening of the European Union (EU) and the implications of that transformation for the UK, before examining in turn the relationship of the UK with the EU, the evolving place of Scotland within the UK, and the nature of land use planning in Scotland.

2.1 European environmental policy

2.1.1 Discovering the green

The 1957 Treaty of Rome, the foundation stone of the European Economic Community (EEC), made no reference to the environment. Until 1987, when the Treaty of Rome was amended by the Single European Act (SEA), there was no legal basis for action on the environment. This omission, extraordinary to modern eyes, reflects the fact that the environment only became established as a political category in Western Europe as recently as about 1970, a time when a transformation was occurring in the established view of the relationship between society and nature (Hanf and Jansen, 1998). The new view stressed the interdependence and vulnerability of humankind and the natural world, and was popularised through the use of striking metaphors such as 'spaceship Earth' (Jansen *et al.*, 1998). Prior to that time, issues such as nature conservation, pollution and landscape preservation had been addressed separately, but from then on they came to be redefined and grouped as 'environmental issues'. This rapidly gave birth to the new field of environmental policy, and to government departments to make and enforce it; the UK Department of the Environment was established in 1970. By the late 1980s 'the environment had become popularised, politicised and commodified' (McDowell and McCormick, 1999b: 1), generating enormous public interest and concern. Initially this emergent sphere of interest was the preserve of national governments, but in the space of just thirty years the scope of European environmental policy-making has enlarged to such an extent that environmental policy in any one country (the UK and Scotland included) cannot be understood if divorced from their European context (Lowe and Ward, 1998a). Like an incoming tide, the overall influence of the EU has penetrated ever further (albeit unevenly) into national systems (Jordan, 1998). In particular, the influence of the EU's developing environmental agenda on British environmental policy has grown inexorably, enforcing ever tighter regulatory standards (Lowe and Ward, 1998b).

As its initial name accurately suggested, the EEC's original rationale was sustained economic growth for its member states. The subsequent changes of name, firstly to European Community and then to European Union, reflect the organisation's evolving visions as it has developed and enlarged from six nations to fifteen, with further enlargement to come. From EEC, promoter of economic growth, it has transformed itself into EU, 'a leading promoter of environmental innovation and change' on the world stage (Connelly and Smith, 1999: 217).[1] Prior to the 1970s, environmental legislation was merely a by-product of the desire to create a 'level playing field' for the common market. Thereafter, environmental policy-making took on a life of its own, developing rapidly at EU-level and (in diverse ways) in the member states. By the 1980s it had become one of the most vigorous areas of policy-making in the EU (Lowe and Ward, 1998a), raising the question of whether the simultaneous promotion of economic growth and effective environmental policy is inherently contradictory or whether the two can be reconciled. Put another way, is sustainable development an achievable goal or a mirage? The heads of government who signed the 1997 Treaty of Amsterdam clearly believed the former, committing the EU to the pursuit of 'balanced and sustainable development of economic activities [and] a high level of protection and improvement of the environment' (in Connelly and Smith, 1999: 218).

A rolling programme of five-year Environmental Action Plans (EAPs) was initiated in 1973, the most recent of which, the Fifth EAP, ran from 1993 to 2000. During the 1970s and 1980s these established a range of key principles (Pearce, 1998), including the following:

- the preventive principle: prevention is better than cure
- the environmental protection principle: significant change to the ecological balance should be avoided, and critical natural capital should be protected
- the 'polluter pays' principle
- the integration principle: environmental policy should not be a separate policy area but integral to all policy-making.

The 1992 Maastricht Treaty on European Union (TEU) added a fifth principle to these four, namely the precautionary principle (Section 13.2.2). Prior to 1987, all environmental policy required unanimous decisions, but the SEA and TEU introduced qualified majority voting for most environmental policy, making it easier to push legislation through (Connelly and Smith, 1999). EU Directives have also increasingly incorporated a sixth concept, BATNEEC – the Best Available Techology Not Entailing Excessive Cost. 'BAT' refers to the environmentally cleanest technology available, while 'NEEC' is a reminder that economic considerations inevitably provide a cost ceiling (Pearce, 1998). These principles, together with the need for public participation, have become almost universally adopted as foundational to sustainable development. Within Scotland, they are advocated as the best guiding principles for planning (Raemaekers and Boyack, 1999).

During the 1990s both the EU and national governments made increasingly bold environmental commitments, notably by incorporating sustainable development into statutes. For example, the Fifth EAP, significantly entitled *Towards Sustainability*, made an explicit commitment to sustainability, adopting a long-term, anticipatory, integrated approach. The EU as a whole finally committed itself to the principle of sustainable development in the 1997 Treaty of Amsterdam. Within the UK the phrase 'sustainable development' made its first appearance in statute in the Natural Heritage (Scotland) Act of 1992, and a range of white papers and initiatives have emerged since the 1992 Earth Summit in Rio. While the official aim of incorporating green thinking into all policy making across government departments is still far from being realised, the standing and influence of the Department of the Environment within Whitehall steadily increased during the 1990s, and a commitment to integration was shown by the incorporation of the DoE into the Department of the Environment, Transport and the Regions (DETR) in 1998 (Connelly and Smith, 1999). The growing boldness of EU environmental commitments has been mirrored by a steadily growing environmental budget. For example, the second phase of the EU's LIFE (L'Instrument Financier pour L'Environment) project had a budget of €450 million, and this has been almost doubled for the third phase which began in 2001. Such European funds have been an important facet of environmental projects in Scotland, including, for example, the LIFE Peatlands Project in the Flow Country (Warren, 2000a) and the Duthchas project which is encouraging sustainable development in north-west Scotland (Crofts, 2000).

On the face of it, then, Europe has gone green. However, as so often in politics, the gulf between words and deeds is broad. As Hanf and Jansen (1998: 2) comment,

> . . . a general commitment to [environmental] objectives is 'costless' compared to the difficult choices to be faced in operationlizing them in the form of concrete decisions through which the balance between economic development and environmental quality is to be struck.

Or, in the words of Pearce (1998: 69):

> Politicians are adept at embracing high-sounding objectives – especially when they are so loosely defined as to be consistent with almost any form of action (or inaction).

Most unfortunately, politics and the environment typically operate on very different timescales, to the detriment of the latter. Moreover, in contrast to sectoral policies (which tend to lead to direct and obvious results), environmental policies often yield public benefits which are diffuse and less readily apparent (Lowe and Baldock, 2000). These realities, combined with politicians' submission to economic imperatives, mean that excellent environmental policies which will not bear tangible fruit for a generation or a century are not yet significant vote-winners.

2.1.2 Europe's future: how green?

A number of important issues affecting the future direction and effectiveness of EU environmental policy are identified by Connelly and Smith (1999), all of which will have significant implications for environmental policy and practice in Scotland.

1. **Implementation, monitoring and enforcement.** European environmental legislation is widely acknowledged to be weak in its implementation and in the monitoring of its practical effectiveness. Member states themselves are solely responsible for the former, and largely responsible for the latter. Although the policing of this 'follow through' by member states is being tightened up, there is still plenty of scope for non-compliance, and there are no effective enforcement mechanisms. This 'implementation deficit' causes widespread concern (Philip, 1998), leading some to argue that the European Environment Agency's role should be strengthened.
2. **Eastward enlargement: how to maintain standards?** There are fears that the admission of the former communist bloc countries to the EU will bring with it a pressure to focus on economic development at the expense of environmental standards. This is because the need for economic regeneration is so acute in these nations, and because the attainment of existing EU standards will take years if not decades to achieve. If a level economic playing field is to be maintained across the enlarged EU, there is a danger that environmental standards will be levelled down rather than up. The already vexing problem of how to make uniform environmental policy for countries with widely divergent conditions is set to become more acute.
3. **The economy *versus* the environment.** While economic considerations are no longer the sole force driving the EU, they nevertheless remain the primary *raison d'être*. Despite the growing strength of environmental legislation, and the general assent to the integration principle, most commentators believe that there is always likely to be conflict between economic and environmental objectives (Philip, 1998; Connelly and Smith, 1999). The projected growth in economic activity across the EU is likely to accelerate the loss of biodiversity, involving loss of habitats, loss of species and loss of genetic diversity (Mackey *et al.*, 1998). Halting, let alone reversing, the decline of the continent's landscape and biological diversity remains a daunting challenge. Given that strong interpretations of sustainable development challenge the *status quo* profoundly, the weaker concept of 'ecological modernisation' seems to be coming to the fore (Section 13.2.1). It is noteworthy, for example, that the momentum of EU environmental policy slowed during the 1990s (Carter and Lowe, 1998).
4. **Policy integration.** Philip (1998) regards this as the major problem facing EU policy-makers, citing numerous examples of policy conflicts and contradictions. Obvious cases include EU-supported agriculture and fisheries policies which

create environmental problems (Section 5.2.2), and European Structural Funds (Section 2.2.1) which frequently have adverse impacts on unspoiled environments. Such incoherence means that the EU ends up using funds from one budget to undo damage caused by funds from another. For the EU's environmental objectives to be fully realised, the integration principle enshrined in the EAPs needs to be thoroughly applied. At present the 'yeast' of environmental policy remains largely separate from the 'dough' of other policy areas, and it will remain so unless or until the EU abandons economic integration as its primary objective.

> Torn between a desire to establish a single European market and to adhere to the philosophy of sustainability, the EU maintains a deep ambivalence of policy approaches as . . . good intentions . . . collide with . . . economic imperatives. (Philip, 1998: 273)

5. **Subsidiarity**. This has been high on the European political agenda since 1992, with Britain in the lead, but it is not yet clear whether it will prove to be a force for good or for bad in environmental terms. On the one hand, it should lead to more decision making being carried out close to the affected citizens themselves, and in a participatory manner, which is good from a green perspective. On the other, leaving decisions in the hands of regional and local actors could lead to inaction rather than action.
6. **The role of the regions**. The Committee of the Regions (CoR) was created following the Maastricht Treaty and is likely to become more important as the principle of subsidiarity leads to the devolution of decision making to regional and local level. The CoR was of limited significance for the UK prior to Labour's election in 1997. The subsequent devolution of power to the Scottish Parliament and the Welsh Assembly, and the possibility of regional assemblies in England, demonstrates Labour's commitment to the regionalisation of policy making and implementation, and considerably enhances the importance of CoR funding for the UK.

2.2 Scotland and the UK in Europe

2.2.1 The widening stage

Despite its location on the Atlantic fringe of Eurasia, Scotland has maintained strong links with the countries of mainland Europe for many centuries (Smout, 1986a), most notably through the 'auld alliance' with France, England's historic enemy. In fact, trade across the water considerably predates the existence of Scotland as a nation state. Notwithstanding this long history of interaction with the peoples and countries of Europe, the day to day experience of the Scottish people throughout much of history remained largely unaffected by events further afield except in times of war or invasion. Only during the last three centuries, with the rise of the British Empire, has the overseas world come into sharp focus for ordinary Scots, and only during the last three decades since the UK's accession into the EU has that wider world come to determine many aspects of people's lives.

The accelerating processes of Europeanisation and globalisation are ensuring that fewer and fewer aspects of Scottish life can be understood without reference to economic, political and environmental trends in the global village. In particular, as part of the UK and the EU, Scotland is bound by a wide range of national and international laws, treaties, protocols and agreements, many of which it had little role in creating. Given that 'the maturing of the EU has been at the expense of the nation state' (Danson,1997: 16), the freedom of Scottish and UK politicians to influence and direct their own affairs has been substantially curtailed. The EU increasingly sets the agenda.[2] Thus the spatial frame of reference for management and decision making has widened progressively and ever more rapidly in recent centuries, from the glen, to the region, to the nation, and now to Europe and the world. It can even be argued that the rise of the EU, combined with deindustrialisation and restructuring, has left the Scottish people with 'less control over their own economies and lives than at any time in the last two centuries' (Danson, 1997: 18). The creation of the Scottish Parliament has repatriated some of this control, but Dunion (1999) notes the irony that devolution has come at a time when the ability of individual states to implement distinctive policies is rapidly diminishing.

On the plus side, Scotland has benefitted substantially from the EU, both through the Common Agricultural Policy (CAP) (Section 5.2.1) and through the European Structural Funds (ESF) which aim to reduce regional disparities in social and economic development across the EU. Much of rural Scotland is eligible for 'Objective 2' funding which supports rural areas, and the Highlands and Islands, as a relatively under-developed part of Europe, qualifies for 'Transitional Objective 1' funding (although there are mounting pressures for this lucrative status to be removed) (Fig. 2.1).[3] Consequently, tens of millions of pounds come to Scotland each year through the ESF, much of the funding being spent on infrastructure projects. The LEADER (Liaisons entre actions de développement de l'économie rurale) programmes for rural development have also brought substantial amounts of European money into rural Scotland (Black and Conway, 1996a). It is very apparent that the CAP and the ESF represent major interventions which have had huge direct impacts on land use, although these have often been by default and have had mixed outcomes both socially and environmentally (Reynolds, 1998).

It is, however, unlikely that the ESF and other inputs will ever achieve their stated objective because they represent a rather weak force for convergence when set against the powerful forces exacerbating uneven development such as the globalisation of production and the increased competition for mobile investment (Danson, 1997). Unless the EU chooses to transfer massive resources to Scotland and to other lagging, peripheral regions, the unavoidable realities of economic and time-geographic marginality will ensure that they remain relatively disadvantaged. Structural funds may reduce the disparities and alleviate some of the pain but they are not able to manufacture a level playing field. From long before the UK's accesssion to the EU,

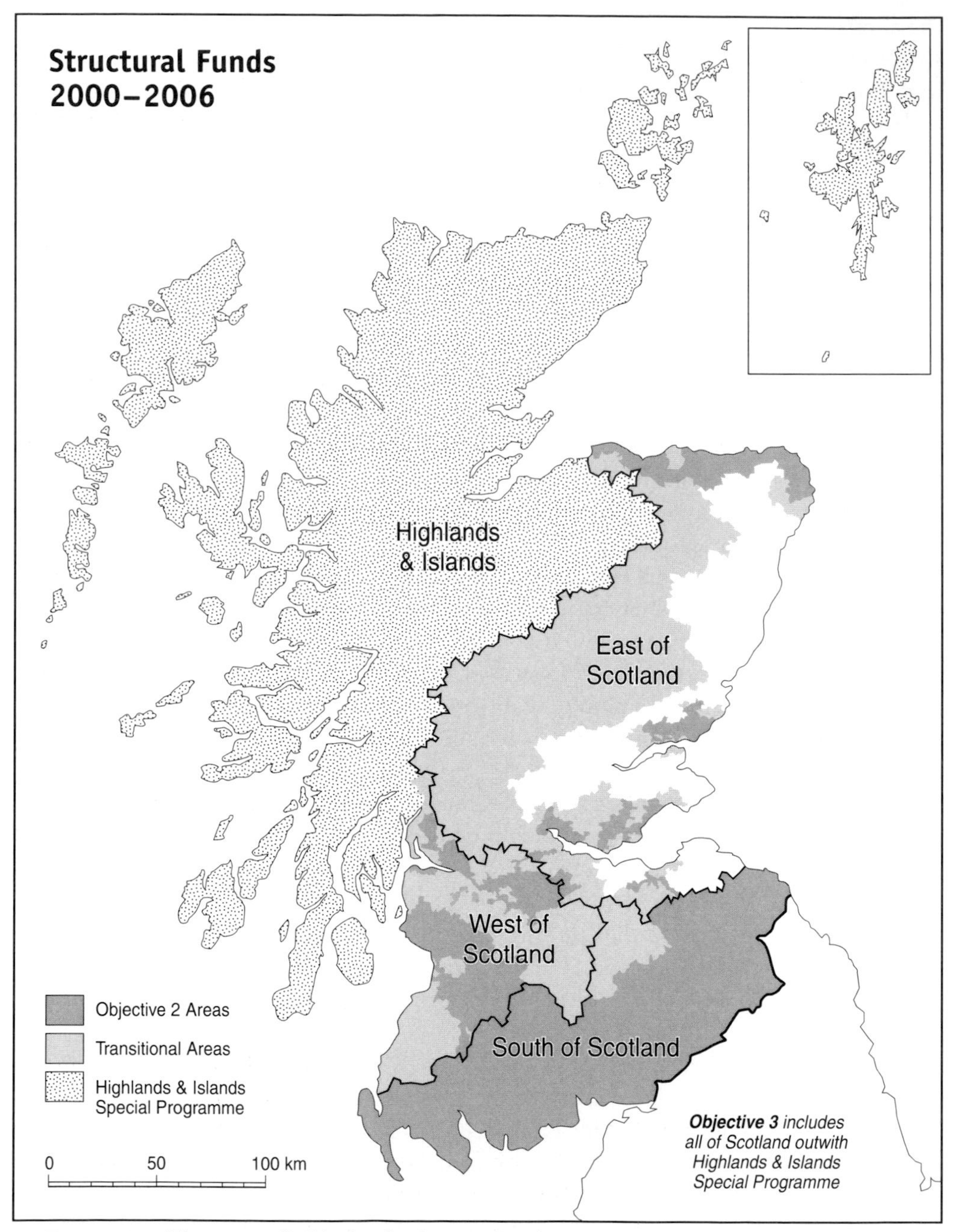

FIGURE 2.1 *The areas eligible for funding under the 2000–6 European Structural Funds programme.*

land use trends in Scotland such as arable reduction and forest expansion were beginning to mirror those across the rest of Europe (Mather, 1994); now the future of Scotland's environment is inextricably bound to a common European and global future.

2.2.2 The UK's environmental record

Britain has the oldest and one of the most elaborate systems of environmental protection of any industrialised nation, built up on an incremental, *ad hoc* basis over 150 years, and had long played a leading role in international conservation politics (Carter and Lowe, 1998; Lowe and Ward, 1998a). On joining the EU, the British perception was that 'they' had plenty to learn from 'us'. No one envisaged that the tables would so soon be turned to such a degree. Indeed, to begin with, the UK had few problems implementing EU environmental directives under existing law. Less than two decades later, however, a major reappraisal of the UK's policies and structures was underway as British politicians ate humble pie and reluctantly adopted continental approaches. This turnaround has received a mixed reception in the UK. Despite the pro-European pronouncements of several British Prime Ministers following their elections, Britain has long had a reputation as a reluctant if not downright recalcitrant EU member, detached by geography and inclination from Europe. The fact that, in many instances, EU law takes precedence over national law has been a long-standing bone of contention between the UK and the EU because of British fears about loss of sovereignty.

Part of the explanation for Britain's reluctance to accept EU environmental directives lies in a clash of philosophies and regulatory styles (Philip, 1998). Concerning pollution control, for example, the EU has followed the continental preference for uniform emission standards backed by law, whereas the UK has always preferred to rely on environmental quality standards implemented *via* a self-regulated, informal voluntary approach (Carter and Lowe, 1998). The UK argues that uniform standards are inefficient, legalistic, and unjustified on scientific grounds (Jordan, 1998), and, moreover, that they are blunt instruments which ignore natural diversity. As an island nation with high rainfall, strong winds and fast-flowing rivers, the UK argues that pollution problems can be effectively tackled with less stringent emission standards than are needed in continental Europe.

This clash between a pragmatic, reactive, case-by-case approach and a codified, principle-led one is more about means than ends, but it has often resulted in Britain being branded as the 'dirty man of Europe'. Is this disparaging epithet justified? The answer to this is complex because Britain's environmental record is in some ways amongst the best and in others amongst the worst in Europe (Hanley, 1998; Connelly and Smith, 1999). Overall, Britain's 'dirty' reputation is probably exaggerated, out of date and partially unjust. Some countries appear to have a better environmental record than the UK simply because their reporting is poorer and because they have a less vigilant voluntary environmental sector; in this sense, the UK is a victim of its own good practice (Hanley, 1998; Connelly and Smith, 1999).

There can be little doubt that the UK is now 'greener' as a result of EU environmental legislation than it would have been if it had remained outwith the aegis of Brussels (Sharp, 1998), especially with regard to water policy (Ward, 1998). Over 80 per cent of British environmental legislation emanates from Brussels and Strasbourg (Lowe and Ward, 1998a), and recent innovations in UK nature conservation have been strongly steered by EU policy (Dixon, 1998), exemplified by the Natura 2000 programme (Section 8.2.1). Part of the content and certainly the timing of several important pieces of UK environmental legislation were largely forced on Westminster by the requirements of EU directives (e.g. the 1981 Wildlife and Countryside Act; the 1990 Environmental Protection Act). Membership of the EU has forced the British Government to crystallise and formalise its procedures and principles. European environmental policy has thus forced the pace, offering a more progressive, far-reaching regulatory regime than the traditional British approach could ever have done (Carter and Lowe, 1998). During the last decade, the influence of the EU has begun to have a significant impact even at local authority level, directly and indirectly affecting planning and land use decisions (Bishop *et al.*, 2000a).

From a standing start just three decades ago, the environment has become an established, mainstream political category throughout Europe. In Britain, however, the shifts in policy have so far lacked real substance because the major political parties do not yet regard the environment as important enough to justify a radical sustainable development strategy (Carter and Lowe, 1998). In turn, the reason that it rarely rises to the top of the political agenda is because it ranks below health, education and employment in public opinion (Crofts, 2000). This gulf between rhetoric and reality is likely to persist because 'at its heart, the British establishment is not very green' (Sharp, 1998: 55) and because the UK is still a reluctant supporter of precautionary measures (Jordan, 1998). In terms of the 'overall greenness' of EU countries, Sharp (1998) characterises Britain as being in a middle group of pragmatic countries which want to strike a balance between environmental concerns and other national interests. Following devolution, Scotland now has some freedom to take a green lead within the UK, but it is too soon to discern whether the brave words of Scotland's new political class will bear sweet green fruit.

2.3 SCOTLAND IN THE UK

For the majority of its long and turbulent association with England, Scotland has had its own parliament. Even during the 292 years between 25th March 1707 and 12th May 1999 when this was not the case, Scotland retained (in varying degrees) a measure of autonomy, together with distinct educational and legal systems more akin to those of continental Europe than of England. Co-existing with this separateness is the sense of a shared place in the world that inevitably springs from long-standing political union and from having had a joint stake in a global empire, with all the common endeavour (both peaceful and military) that that entailed. *Rule Britannia* was written by two Scots. Devine (1999) identifies the British Empire as a linchpin of the unity of the UK, and its demise as one of the factors which has promoted devolution.

Political links between Scotland and England go back over a thousand years. The two countries made common cause against the Vikings in 924 AD and the Normans in 1066 (Parman, 1990), and during the twelfth and thirteenth centuries Scottish kings married English wives and maintained close connections with the Anglo-Norman state to the south (Smout, 1969). It was Edward the First's brutal attempts to assert English overlordship which sundered this relationship, plunging the two countries into 250 years of animosity and war. Old ethnic loyalties within Scotland were buried in this common cause, forging national unity and engendering fierce nationalism and independence. With the abandonment of Catholicism following John Knox's reformation in the sixteenth century, enmity gave way to co-operation, leading to the Unions of 1603 (the Crowns) and 1707 (the Parliaments). The faultlines in the Jacobite rebellions of 1715 and 1745 ran not so much between Scotland and England as between Catholic and Protestant. The subsequent story of the Scottish people is compellingly told by Smout (1969, 1986b).

Given this millennium-long history of interaction within the bounds of a small island, 'pure blooded' Scottish and English individuals may be few in number. On the global stage there is far more that unites than divides the two nations. Nevertheless, Scots were never entirely happy in what was a subordinate or even a subsuming relationship with England (symbolised by the contemporary English habit of using 'England' and 'Britain' interchangeably). The social and political forces that propelled devolution to the top of the agenda in the late twentieth century, too complex to be addressed here, are discussed by Devine (1999). They led initially to an unsuccessful referendum in 1979, and then to the overwhelming 'yes, yes' vote of 1997 which paved the way for the re-establishment of a Scottish Parliament with tax-varying powers in 1999.

Prior to 1999, the country was governed by a Westminster-appointed Secretary of State for Scotland, supported and advised by the civil servants of the Scottish Office. Given that final decisions on matters stretching across the whole spectrum of Scottish life rested with one man, this position was not dissimilar to that of a colonial governor-general. With devolution, that broad responsibility is now distributed amongst the 129 elected Members of the Scottish Parliament. The former Scottish Office has become the Scottish Executive. The division of responsibilities between Holyrood and Westminster (Table 2.1) shows that the Scottish Parliament is responsible for all important decisions concerning land use planning and management, subject, of course, to its European and international commitments. A notable practical source of discontent with the pre-1999 system was the lack of parliamentary time for specifically Scottish legislation amidst the packed legislative schedule at Westminster. Only the most pressing issues were addressed while other important matters were repeatedly postponed. Devolution has now solved this by creating a legislative body which can be much more fleet of foot in dealing with the immediate issues which concern the Scottish people. The post of Secretary of State for Scotland continues for the moment, but it now mainly comprises a liaison role between the two legislatures.

The Scottish Executive has thus far given considerable prominence to rural affairs and to the environment. Its Programme for Government, 'Making It Work Together', made a commitment to support and enhance rural communities and the rural economy, a commitment which has been fulfilled by the appointment of a Minister for Rural Affairs, and the creation of a unified Scottish Executive Environment and Rural Affairs Department (SEERAD). At the time of writing it is too soon to be passing verdicts on the benefits or otherwise of the Parliament for Scotland's environment. The early signs were encouraging (Alexandra, 2000), but there was widespread dismay when, in 2001, the position of Minister for the Environment was dissolved, the portfolio being combined with rural development.[4] This step seems inconsistent with the Executive's avowed commitment to place environmental sustainability at the heart of policy-making, but Ministers argue that distributing the responsibility more widely will put the environment firmly into other portfolios.

TABLE 2.1 The division of responsibilities under Scottish devolution: those that have been devolved to the Scottish Parliament and those that have been retained by the UK Government.

Devolved matters	*Reserved matters*
Health	Constitutional matters
Education and training	UK foreign policy
Local government, social work and housing	UK defence and national security
Economic development and transport	UK fiscal, economic and monetary system
Law and home affairs	Common markets
Environment	Employment legislation
Agriculture, forestry and fishing	Social security
Sport and the arts	Professional regulations
	Transport safety and regulation

2.4 LAND USE PLANNING IN SCOTLAND

2.4.1 The characteristics of the system

Throughout the UK, the planning system is the primary policy machinery responsible for ensuring that the quality of the natural heritage is not compromised by development and land use. It is a system of discretionary development control which is operated by local authorities; they, in turn, act within a framework of indicative policy guidance set by central government (Rowan-Robinson, 1997; Boyack, 1999). Lyddon (1994) characterises land use planning as making arrangements for future change in the use of land. He summarises its purpose as being to indicate:

- which resources (actual or potential) are of national or regional significance (e.g. agricultural land; aggregates)
- where such resources should be safeguarded, promoted, or subject to environmental assessment
- what action should be taken by those involved
- who will take decisions.

Overall, he describes it as an early warning system for anticipating conflict and reducing uncertainty.

Scotland's strategic planning regime has long been separate from the rest of the UK. It is a unique and highly regarded framework which diverged significantly from that in England and Wales during the 1970s and 1980s (Lloyd and Rowan-Robinson, 1992; Rowan-Robinson, 1997).[5] National Planning Guidelines were introduced in

Scotland in 1974. The planning framework was strengthened and updated in 1992 when these were replaced with National Planning Policy Guidelines (NPPGs) which provide statements of policy together with some locational guidance. This moment marked the beginning of a significant increase in intervention by the Scottish Office in planning policy, involving a shift from identifying specific land resources of national importance to issuing more general policy guidance on land use issues (Rowan-Robinson, 1997). Around the same time the concept of indicative strategic planning for future land use emerged with the development of Indicative Forestry Strategies (IFSs) (Section 4.5.2). Several new NPPGs have been issued in recent years, including two which specifically address environmental and rural issues for the first time (Numbers 14 and 15, *Natural Heritage* and *Rural Development* (SODD, 1999a, 1999b)). NPPG 14 suggests that 'a key role of the planning system is to ensure that society's [needs] are met in ways which do not erode environmental capital' (SODD, 1999a: 7).

The planning system is plan-led and comprises four levels (Raemaekers and Boyack, 1999):

1. Guidance on nationally important land use and other planning matters, sometimes supported by a locational framework (NPPGs, supplemented by technical advice in Planning Advice Notes).
2. Strategic forward planning (notably statutory Structure Plans which set the strategic land use framework for a decade or more ahead).
3. Local forward planning (especially statutory Local Plans which look five to ten years ahead).
4. Development control.

Most planning is delivered by local authorities through their Structure Plans and Local Plans, and through development control (the granting of planning permission). However, the Scottish Executive has ultimate control, retaining the power to 'call in' a planning application for a public inquiry in especially controversial cases. Since the planning system was already devolved administratively, it has been minimally affected by devolution, being simply transferred to the Parliament. However, changes may well come in time because devolution opens up possibilities for improving planning integration (Section 2.4.3) and for democratising the planning process (Boyack, 1999).

Twice in the last twenty-five years the map of local authority areas in Scotland has been substantially redrawn. A two-tier hierarchy of Regions and Districts was put in place in 1975, but the earlier system of unitary authorities was reintroduced in 1996. The original county boundaries were not followed, however, even though the county names are still in widespread common use. Recent literature is thus bedevilled with a confusion of overlapping nomenclature. At present there are thirty-two local authorities (Fig. 2.2) and seventeen Structure Plan areas. Six of these cover two or more authority areas, requiring councils to prepare joint structure plans.

Although the Scottish planning system is highly regarded in some respects, its effectiveness can be undermined by delays in preparing plans (Boyack, 1999), and its

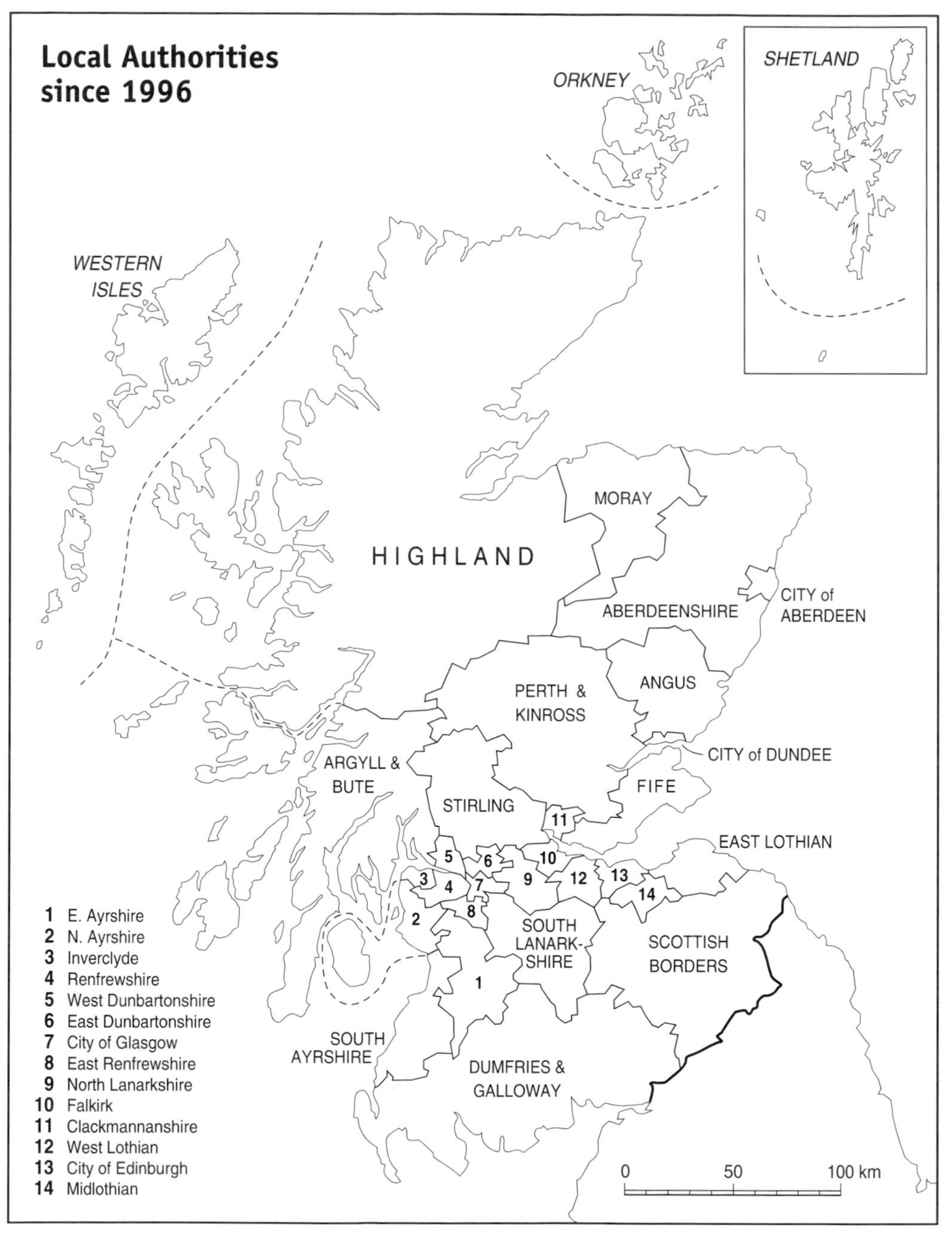

FIGURE 2.2 *The local authority areas in Scotland.*

adversarial nature is frequently criticised. Austin (1997: 32), for example, believes that 'the current adversarial planning system has generated more heat than enlightenment'. When consensus cannot be achieved, the warring parties end up fighting acrimonious and expensive battles through public inquiries or in the courts. This was seen in the

1990s concerning, respectively, the proposal for a superquarry on Harris (Section 12.2) and the Cairngorm funicular railway (Section 12.1). While the recent trend towards management through partnership is not a panacea (Gemmel, 1996; Sections 8.2.2 & 14.1.2), it does allow middle ground to be discovered, and it lessens the chances of public show-downs.

Of all parts of Scotland, rural planning problems are at their most acute in the Highlands and Islands, a region where both natural and human resources are at their most fragile (Selman, 1988b) and which has long been regarded as a problem in development terms (Turnock, 1983). Here, as in many of Europe's peripheral regions, both nature and people tend to need a high level of protection because economic opportunities are often limited and much of the environment is relatively unspoiled. The result is the problematic 'jobs v. nature' dilemma in which planners get ensnared (Section 8.3.2). On the one hand, attempts to 'develop' for the sake of the people are not only severely constrained by the natural environment but run the risk of damaging it. On the other, the need to protect semi-natural habitats leads to a concentration of conservation designations which are frequently regarded by local communities as unwelcome constraints. Baxter and Thompson (1995: 47) comment that 'the balancing act is near-impossible'. Selman (1988b: xii) sees only tourism and recreation as being compatible with the land, yet notes that a reliance on this risks 'reducing the Highlands to a quaint survival of folksy Celtic culture'.

The organisation which, above all others, has had to try to walk this tightrope between communities and the natural heritage is Highlands and Islands Enterprise (HIE, formerly HIDB). HIE and its sister agency Scottish Enterprise (formerly the Scottish Development Agency) were set up in 1965 and 1975 to help the people of their respective regions to improve their economic and social conditions (Lloyd, 1997). Initially the needs of people were given unquestioned priority over environmental concerns. More recently, however, the pendulum has swung decisively away from this single focus to incorporate the needs of the environment, a swing from economic development towards sustainable development; HIE was the first British development agency to embrace sustainable development (Minay, 1997). Traditionally, development has gone ahead unless there has been good reason to refuse it. Now the burden of proof is shifting. Under the Town and Country Planning (Scotland) Act 1997 the presumption in favour of development has been superseded by one in favour of conformity with the statutory plans, although the Act makes no reference to sustainable development *per se* (Raemaekers and Boyack, 1999).

2.4.2 Planning controls on farming and forestry?

The influence of the planning framework on activities in rural Scotland is pervasive but not universal. Many policies which affect aspects of rural life, such as housing and power generation, originate outwith the planning system, and several important aspects of land use change such as road building are specifically excluded. Planning control also ceases at the coast below the low water mark. Most controversially, however, the 1947 Town and Country Planning Act, the foundation stone of post-war planning in the UK, excluded agriculture and forestry (and the conversion of farm land to forestry) from development control. This exclusion is typical throughout Europe (Lyddon, 1994) and is retained under the Town and Country Planning (Scotland) Act 1997 despite the profound impacts of intensive agriculture and commercial afforestation; forest expansion constituted arguably 'the most striking change in land use and landscape in rural Scotland during the twentieth century' (Mather, 1996a: 83). Consequently, there were insistent calls throughout the 1970s and 1980s to bring farming and forestry within the planning framework (Mather, 1991), especially arising out of conservation arguments (CCS, 1986; Lowe *et al.*, 1986).

These calls continue (Ratcliffe, 1998), and the issue was on the agenda of the Land Reform Policy Group (LRPG, 1999). However, relative to the situation in the 1970s when both sectors remained largely unconstrained, farming and forestry practices are now subject to a plethora of controls and safeguards. Subsidies to farmers are becoming less production-orientated and more concerned with environmental and social objectives (Section 5.2.2), and farm building developments are controlled. In forestry, the pressure for more control led to the introduction of consultation procedures in 1974 and to Indicative Forest Strategies during the 1990s (Section 4.5.2). Moreover, forestry grants are now conditional on the implementation of environmental standards and Forestry Commission guidelines. So although neither sector is subject to full planning controls, the potential for largescale negative impacts is now much reduced. The rapid transition towards a multi-objective, environmentally sensitive approach in recent years suggests that there may be no need to bring them within the planning system.

2.4.3 *Sectoral* versus *integrated planning*

For almost thirty years the Scottish rural planning system has been criticised for being too sectoral and piecemeal in its approach, and there have been repeated calls for greater integration in land use planning and practice (e.g. Lowe *et al.*, 1986; Mowle, 1988; CCS, 1990; Mackay, 1995). These arguments stem from the obvious contradiction between, on the one hand, the continual interplay between different sectors 'on the ground', and, on the other, the treatment of these sectors in watertight compartments in the planning process. How much better, it has often been argued, if a more holistic and realistic approach could be adopted. Environmental issues in particular are intrinsically cross-sectoral (Lowe and Baldock, 2000), a fact which was driven home in the late 1980s by the abject failure of the sectoral approach in dealing with the Flow Country controversy (Mather, 1996b; Warren, 2000a). Integration is widely perceived as the key to good resource use; it has even been described as 'a magic word . . . imbued with the power of a panacea' (Alexander *et al.* (undated) in Milton, 1994: 209).

Yet integrated land use is hard to find in practice. Lyddon (1994) identifies a tendency towards reluctance within British administration when it comes to coordinating strategies for the future, the preference being for single-issue policies and a limited timeframe. Only on the large private estates is there a long-standing tradition of integration in Scotland. Notwithstanding the many criticisms of concentrated landownership (Section 3.4.1), estate management at its best provides an exemplar for integrated management (Selman, 1988b). The development of sustainable tourism alongside traditional farming and forestry on Glenlivet Estate during the 1990s (Wells, 1998) is a striking and award-winning example.

While policy coordination makes obvious sense in theory, how practical is it in the rural context, and how wide should the net be cast? Land use in rural Scotland is affected to a greater or lesser degree by a truly immense spectrum of policies and incentives at EU, UK and Scottish levels, stretching from social security, through housing to tourism and transport, quite apart from the more obvious sectors of agriculture, forestry, water resources and conservation. In pointing this out, Lyddon (1994) suggests that any system which attempted to plan and control all aspects of rural change would be simply unworkable. There are two particular dangers. The first is that any advantages of integration would be lost in the unwieldy, bureaucratic machinery of such a 'Ministry of Everything'. In the practical business of government, divisions of responsibility are unavoidable in the interests of efficiency. The second is that all key land use decisions could end up being made behind closed doors. This was the reason why as long ago as 1972 a proposal for a Scottish Office 'Department of Rural Affairs' was rejected (Mackay, 1995).

Although 'it has proved hard to achieve integrated policies that cut across economic sectors' (Mackey *et al.*, 1998: 242), this is not to say that integration and coordination are not worthy goals. Indeed, they are a prerequisite for coherent policy-making, and essential if duplication and conflict between different arms of government are to be avoided. This was officially recognised a decade ago by the Scottish Office (1992: 4) in its blunt declaration that 'tackling rural issues in a sectoral manner does not work'. Overcoming sectoralism is now seen as integral to the realisation of a sustainable future (McDowell and McCormick, 1999b). The need for a more holistic approach was acknowledged in the Rural White Paper (Scottish Office, 1995),[6] and 'joined-up thinking' is now built into rural development policy, with planning authorities being encouraged to work much more closely with Local Enterprise Companies, the private sector, public agencies and local communities (SODD, 1999b).

With this official recognition of the limitations of a sectoral approach has come a steady broadening of the remits of government departments. For example, the original Scottish department of agriculture (DOAS) took on fishing to become DAFS and then SOAFD (the Scottish Office Agriculture and Fisheries Department), before gaining responsibility for the environment as SOAEFD in 1995. In 1999, it was rebranded by the Scottish Parliament as SERAD, the Scottish Executive Rural Affairs Department, and in 2001, with the addition of 'environment', as SEERAD. Its responsibilities now stretch wider than ever before, encompassing not only agriculture, fisheries, forestry and the environment, but all aspects of rural life, from social inclusion to transport to housing. Its creation therefore represents a decisive step away from sectoralism and towards improved integration which may deliver better coordination in practice.

At both national and European level, integration seems to be becoming institutionalised. For example, funding under the EU's Rural Development Regulation (Section 5.2.1.ii) is conditional on the adoption of an integrated approach. The repeated calls for more flexible structures, for indicative strategies, and for multiple objectives appear to have been heeded; Raemaekers and Boyack (1999) identify encouraging signs of 'joined-up thinking' in several of the NPPGs issued in the late 1990s. It has even been suggested that a Chair of Integrated Land Use should be created in a Scottish university to help the effects of this shift trickle down (Lister-Kaye, 1998). Since the mid-1980s there has been a transformation in rural policy. Production-orientated policies have given way to a more broadly based rural policy framework with a strong environmental component, and one result is that the conventional sectoral divisions are becoming progressively blurred. Farmers are planting trees, foresters are managing for wildlife, conservationists are involving local communities, and land managers are learning to welcome walkers. The era of single-issue policies already seems long gone.

Notes

1. For a clear and helpful summary of the ways in which the various EU institutions operate and interact in policy making, see Connelly and Smith (1999: 218–25).
2. Notwithstanding this centrally-set agenda, the actual decisions are taken by the Council of Ministers elected from each member state.
3. Under the 2000–6 European Structural Funds programme, Objective 1 became Transitional Objective 1, and the former Objectives 2 and 5b became Objective 2.
4. This change created the position of Minister for Environment and Rural Development. SERAD also changed its name, becoming the Scottish Executive Environment and Rural Affairs Department (SEERAD).
5. Key aspects of the development of the Scottish planning system and its practice are described by Fledmark (1988) and Lyddon (1994). Descriptions and critiques of the present system are given by Raemaekers and Boyack (1999) and Boyack (1999).
6. For commentary on the Rural White Paper see Bryden and Mather (1996) and Randall (1997).

Part Two:
The pieces of the jigsaw

CHAPTER THREE

The land: who should own Scotland?

Who owns this landscape? –
The millionaire who bought it or
The poacher staggering downhill in the early morning
With a deer on his back?

Who possesses this landscape? –
The man who bought it or
I who am possessed by it?

False question, for
this landscape is
masterless
and intractable in any terms
that are human.

(From *A Man in Assynt* by Norman MacCaig, 1990)

3.1 Historical background

Land is the most basic resource of all and is unlike any other commodity. As the saying goes, 'the thing about land is that they don't make it any more'. It is sacred to a few, useful to some, loved by many, taken for granted by most. All of us, by default, have a stake in the land, but even in rural areas less than 10 per cent of the workforce now work the land (Shucksmith *et al.*, 1996), and even fewer actually own any. Remarkably, Scotland has the most concentrated pattern of private land ownership in the world (Wightman, 2001a). This is a direct result of certain key historical events.[1]

The dearth of human beings in once-populated rural areas is mainly the result of the infamous Highland Clearances of the eighteenth and nineteenth centuries, a time when economic considerations led many landlords to 'clear' people forcibly from the land, replacing paltry rents with profitable sheep, cattle and deer (Richards, 2000). Although the chronic over-population and poverty would probably have forced many to leave of their own volition, as was happening in Norway at the time (Smout, 1997b), and notwithstanding the open warfare that the Clearances continue to generate amongst social historians (Hunter, 2000; Ross, 2000a), the awful injustices perpetrated during that dark time have seared themselves into popular consciousness. In the words of MacLennan (1960 in Hunter, 1995: 25), 'in a deserted Highland glen you feel that everyone who ever mattered is dead and gone'. The Clearances have left such an enduring 'legacy of impotent rage' and anti-landlordism that they retain powerful historical symbolism (Richards, 2000: 12; Cameron, 2001). On Lewis, for example, they remain 'elemental in community awareness' (MacDonald, 1998: 239). They are frequently invoked in response to the actions or proposals of government, landowners, big business or NGOs (Parman, 1990; Mackenzie, 1998; Cairns, 2000), and have re-emerged prominently in the emotions, if not the explicit arguments, of the contemporary land reform debate (Section 3.4; Fig. 3.1). That these historical events remain a live issue was vividly demonstrated in September 2000 when the Scottish Parliament debated a motion expressing 'deepest regrets' for the Clearances.

The later nineteenth century saw the so-called 'Balmoralisation' of Scotland (McIntosh *et al.*, 1994). This flowed from Queen Victoria's love affair with the Highlands and her purchase of Balmoral in 1848, fuelling the romanticisation of upland Scotland which had begun with the writings of William and Dorothy Wordsworth and Sir Walter Scott. This not only initiated tourism (Smith, 1997) but prompted those who could afford it – many of them newly rich industrial magnates from south of the Border – to emulate their queen in buying sporting estates. The collapse of sheep prices in the 1870s released cheap land for sporting use, especially deer stalking and grouse shooting, so that shooting estates comprised almost 60 per cent of Scotland at that time (Orr, 1982). It was in the closing decades of that century

FIGURE 3.1. (above and opposite) *How* The Scotsman*'s cartoonist saw the early moves towards land reform. The Secretary of State for Scotland, the late Donald Dewar, is seen visiting revenge for the Clearances on stereotypical lairds. Published on 6th January, 1999 and 9th July, 1999, and reproduced with the permission of Graham High.*

that much of today's sporting infrastructure of lodges and hill paths was constructed.

Although land ownership in Scotland had been unusually concentrated for centuries, by the end of the nineteenth century much of the indigenous population had been replaced by sheep and then by deer, making a stark situation starker still. In 1873, 118 people owned half of Scotland (Devine, 1999) and the Duke and Duchess of Sutherland held 91 per cent of their county (Clark, 1983). Despite the considerable changes since then, the pattern of land ownership at the start of the third millennium still reflects the historical forces of the early and middle parts of the second. A core of fewer than 1,500 private estates have held most of Scotland for nine centuries, with several of the great houses such as Seafield, Roxburghe, Stair and Hamilton having had hereditary occupation for over thirty generations (Devine, 1999).

Does this matter, and if so, why? Land ownership is of fundamental importance in at least five ways (Wightman, 1996a, 1999):

1. **Economically**. Land ownership determines investment patterns, employment opportunities, and local economic development (MacGregor and Stockdale, 1994).

2. **Environmentally**. Because land use is 'merely the physical expression in the landscape of decisions taken by landowners and occupiers' (Mather, 1995: 127), landowners' motivations and desires strongly affect standards of environmental stewardship (Cramb, 1996). Due to the great size of many holdings, and the historical laird/tenant relationship, landowners have great influence. They have even been described as the *de facto* rural planners (MacGregor, 1988, 1993), although their room for manoeuvre has been progressively curtailed. The days when lairds were 'lord of all they survey' are long gone, but they can, nevertheless, wield great power for good or ill.
3. **Culturally**. Land is central to modern Scottish culture (Cameron, 2001). Down the ages the land has inspired writers, poets, and singers; it is a motif woven through Scottish music as diverse as ancient Celtic ballads and Runrig rock anthems about the Clearances such as 'Rocket to the Moon' (Hunter, 1995). That it still excites the passions of ordinary people today is brought out vividly by MacAskill (1999), and was powerfully illustrated in 2000 by the outspoken reactions to the proposed sale of the entirety of the Black Cuillin of Skye.
4. **Politically**. Land ownership confers not only economic but political power (Bryden, 1997), both in terms of local influence and also through the influential network of 'landed connections' so revealingly investigated by Wightman (1996a). 'Power and landownership have been synonymous in Scotland from time immemorial' (Timperley, 1980 in Wightman, 1999: 47).
5. **Spiritually**. The land has long been a powerful icon and influence in people's beliefs, and continues to be so (Meek, 1987; McIntosh, 2001a). Crofters in Assynt, for example, believe that the transition from being tenants to owners of their land has brought about spiritual change in the community, engendering a sense of freedom (MacAskill, 1999).

Reasons such as these lie behind the Environment Minister's observation in 1997 that 'land is a defining issue for the people of Scotland' (Callander, 1998: 17).

3.2 WHO OWNS SCOTLAND?

Many varied and complex issues surround the question of land ownership. These have been examined in great detail by, amongst others, Wightman (1996a, 1999), Callander (1998), Wightman and Higgins (2000, 2001) and Cameron (2001), and much of this section is based on these sources. The human face behind the statistics and legalities is investigated revealingly by Cramb (1996).

3.2.1 Private ownership

A majority of private rural landowners are farmers, and significant amounts of land are held by private companies, notably forest companies. But the distinctive feature of the pattern of private ownership in Scotland is the areal dominance of sporting estates, especially in the Highlands where estates larger than 2,000ha predominate. There are some 340 sporting estates in the Highlands and Islands, covering 2.1

million hectares and accounting for over half of all the privately owned land (Wightman and Higgins, 2001). The present concentration of ownership in Scotland yields some staggering statistics (Wightman, 1996a, 1999; McCarthy, 1998). Half of the entire country is held by just 608 owners and a mere eighteen owners hold 10 per cent of Scotland. Of Scotland's private land, 30 per cent is held by 103 owners, each with 9,000ha or more, and 50 per cent by 343 owners. A miniscule 0.025 per cent of the population owns 67 per cent of the privately owned rural land. Thirty owners have more than 25,000ha each, while the largest private landowner of them all, the Duke of Buccleuch, has Scottish estates covering some 124,000ha.

Rights to hunt game have been tied to land ownership for centuries, and game sport remains of prime importance on many private estates and a key motivation for their owners (MacGregor, 1988). This is demonstrated by the fact that their capital value is based not on land area but on the current value of available game, an anomaly which can produce 'wildly inflated paper values for land which is amongst the poorest in Europe' (McCarthy, 1998: 101). The capital value of the key game species is put at £22,000 per red deer stag, £2,200 per red deer hind, £3,000 per brace of red grouse and £3,500 per fly-caught salmon (Strutt and Parker, 2000), values which have fluctuated widely in recent years (Fig. 3.2). The quantity and value of game species is also crucial to the economic operation of estates, many relying heavily

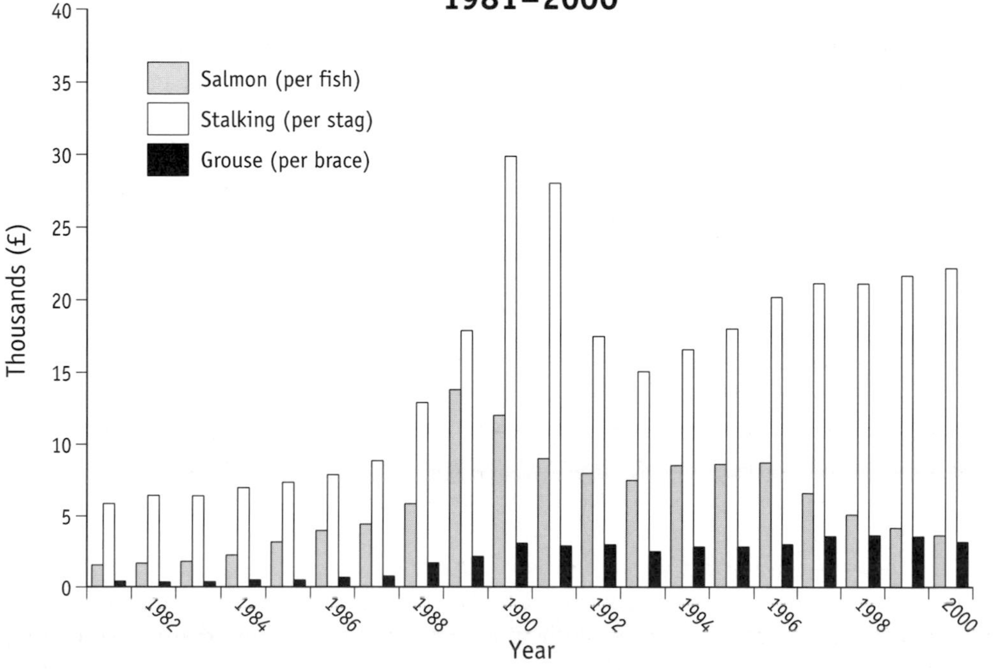

FIGURE 3.2. *Capital values of Scottish sporting estates, 1981–2000. After Strutt and Parker (2000). Used with the permission of Strutt and Parker's Edinburgh Office.*

on income generated from stalking, grouse shooting and fishing. However, most estates, even those worth millions on paper, run at substantial annual losses (MacGregor and Stockdale, 1994). In the early 1990s Mar Lodge Estate was losing £300,000 each year. It seems that most owners aim to minimise losses rather than maximise income (NC/SG, 1998); in economic terms they are satisficers, not profit maximisers (Wightman and Higgins, 2000).

The dire financial condition of many estates is not entirely surprising given that most were originally set up or purchased to display wealth, not to make it. This in turn explains the existence of many of the ostentatious castles and stately homes in rural Scotland. The extraordinarily high sale value of such spectacularly loss-making enterprises is largely due to this 'status symbol' aspect, and to the fact that some owners are absentees who purchased estates for non-monetary reasons. Many estates are luxuries rather than businesses, often representing only a small part of their owners' total assets. For such owners, finances may be only one of several considerations in management decisions, and this blunts the persuasive edge of fiscal and monetary incentives designed to shape land use patterns (Bryden, 1997).

3.2.2 Public ownership

The increase in state ownership of land was a striking feature of the twentieth century. Whereas in 1872 just 0.3 per cent of Scotland was in public ownership, by 1979 public bodies and nationalised industries owned and leased 16.8 per cent of the country (Clark, 1983). Today, some 1.03 million ha, a little over 12 per cent of Scotland's rural land, is owned by public bodies (Scottish Office, 1998a). The single largest owner is the Forestry Commission (FC) with 773,000ha, followed by SEERAD and local authorities. Relatively small areas are also owned by the Ministry of Defence and SNH.

3.2.3 Social ownership

The middle ground between private ownership and state ownership is occupied by the not-for-profit sector, a disparate group of trusts and organisations with diverse management objectives (Dwyer and Hodge, 1996). Though small, its rapid growth and often radical approach has swiftly made it into a significant piece of the jigsaw, as local communities, crofting trusts and conservation organisations have bought land in all parts of Scotland (Boyd and Reid, 2000; Sections 3.3.2 & 3.3.3).

Ownership, of whatever ilk, no longer brings with it the unfettered freedoms that it once did. The options open to land owners and managers are now constrained by local, national and international laws, and their choices are governed by grants, incentives, planning controls and conservation designations, not to mention the fluctuations of international markets. Those taking a keen interest in land use issues now include not only central government and its agencies (FC, SEERAD, SNH) but the EU, international bodies such as the IUCN, and a wide range of special interest groups such as the RSPB, the RA, FoES and WWF (many of them now collaborating within Scottish Environment LINK). Local decisions were once dictated mainly by local considerations, but in recent decades the spatial scale of management perspectives has rapidly expanded.

3.3 RECENT TRENDS

3.3.1 Sale of estates into foreign ownership

Only the super-rich can sustain large financial losses indefinitely. Notwithstanding the substantial capital appreciation in the value of sporting estates (Wightman and Higgins, 2000), during recent decades the large annual running costs have become untenable for increasing numbers of owners, forcing them to put their land on the market. This has affected some of the longest established landowning families in Scotland, such as the Lovat family who broke up and sold their ancestral holdings around Inverness in 1995. In many cases, the buyers have come from overseas, notably from Denmark and the Middle East, so that over 30 per cent of sporting estates are now in foreign ownership. Overseas ownership has increased fourfold since 1970 and now totals almost 6 per cent of Scotland (Wightman, 1996a). In the Highlands, approaching twice as much land is in foreign ownership as is held by not-for-profit organisations.

3.3.2 Purchase by conservation groups

One of the most striking developments in recent years has been the rapid emergence of charitable environmental organisations as major landowners, a phenomenon which is widely perceived as a symptom of failure, of the ineffectiveness of conservation policy. Aitken (1997: 11), for example, interprets the trend as a response to 'our continued chronic failure in Scotland to achieve direct public protection for our best country'. Land purchase by the not-for-profit sector is not a new phenomenon; the National Trust for Scotland (NTS) bought part of Glencoe in 1935, for example, and now owns 5,811ha of the glen (Johnston, 2000). But between 1980 and 1995 the total area owned by this sector rose by 146 per cent to reach 133,500ha (Wightman, 1996a). With the purchase of Mar Lodge Estate in 1995 for £5.5 million, the NTS became the third largest landowner in Scotland with almost 80,000ha (Johnston, 2000), and the RSPB now manages about 51,000ha spread over more than fifty reserves (RSPB, 2000). The John Muir Trust has become a major landowner on Skye, and in 2000 it bought Ben Nevis for £450,000. Conservation organisations as a group now constitute the largest non-public landowner, and if present trends continue the area owned by this sector could easily double in the near future, overtaking public ownership (Boyd, 1999). Aitken (1997), however, doubts whether the growth of NGO ownership can be sustained indefinitely.

The motivations leading such organisations to buy estates, forests, bogs and mountains are diverse, but common threads include the safeguarding of land for future generations, restoration of damaged areas, and the desire to move from advocacy to practical demonstration (Dwyer and Hodge, 1996). Though criticised for being too narrowly focused, there is no doubt that the RSPB (amongst others) has confounded its critics by demonstrating the practical feasibility of many of its ideas, notably at their Abernethy Forest Reserve which now sustains eighty-seven jobs (Beaumont *et al.*, 1995; Taylor, 1995). Organisations like the NTS and the RSPB are also able to explore innovative approaches to land use and management which cash-strapped (and often tradition-bound) private owners are unwilling or unable to risk (Johnston, 2000). Such trail-blazing experimentation is of great value.

At the same time, some of the conservation charities have made themselves unpopular with locals by behaving in arrogant and secretive ways reminiscent of absentee landowners (Mitchell, 1999). There are also some hard questions to which only time will provide answers. For example, in the long term will the money keep coming? It is one thing to raise funds for a one-off purchase, quite another to raise sufficient money for the unglamorous, expensive business of long-term land management. Running Mar Lodge costs the NTS around £250,000 annually. Secondly, there is concern that some of the new conservation owners are more concerned with wildlife and landscape than people (Mitchell, 1999; Cairns, 2000), and that their growing influence may frustrate efforts to repopulate rural areas and give local communities control of their own destinies. Competing agendas could thus undermine moves towards sustainable development (Wightman, 1996a). Given the large areas now in the hands of not-for-profit groups, it is important that these large scale, long-term experiments succeed. Success will be determined not just by the quality of management on the ground but by the willingness of charitable memberships to stay generous down the decades.

3.3.3 Community ownership

Apart from the land reform debate itself (Section 3.4), the most headline-grabbing aspect of land ownership over the last decade has been the unanticipated explosion of interest in community ownership. From a standing start in 1992, community ownership has made a rapid journey from the radical fringe to become the centrepiece of land reform proposals. In 1992, a series of special circumstances came together which enabled the Assynt crofters to purchase the 9,500ha North Lochinver Estate for £300,000 and bring it into community control through a crofting Trust (MacAskill, 1999). The Assynt crofters were not breaking entirely new ground. Community ownership has existed in the Parish of Stornoway on Lewis since 1923, for example, but 1992 marks a decisive turning point.

When various island communities were offered ownership in the 1970s and 1980s, they turned down the opportunity, but since 1992 numerous communities have set out to follow the trail blazed by the Assynt crofters. High profile examples in the 1990s included the purchases of Borve and Annishadder on Skye in 1993, Knoydart in 1999, and the re-crofting of the Orbost estate on Skye, a unique (if troubled) pilot project in rural regeneration (Cameron, 2001). Perhaps best known of all was the long-running saga of the inner Hebridean island of Eigg which was finally bought by residents in 1997 for £1.5 million (McIntosh, 1997; Dressler, 2000). By early 1999 Highlands and Islands Enterprise (HIE) was dealing with over thirty applications from communities interested in purchasing their land (Hunter, 1999), and communities in many areas are now experimenting with diverse forms of social ownership (Boyd and Reid, 2000).

Prior to 1997 community purchases were achieved against heavy odds, but the election of a Labour government with a manifesto promise to examine land issues in Scotland represented a watershed. The establishment in 1997 of HIE's Community Land Unit to support community-led purchase or management initiatives was the first tangible sign of this, rapidly followed by the setting up of the Land Reform Policy Group (LRPG) (Section 3.4.2). In 2001 the National Lottery's New Opportunities Fund established the Scottish Land Fund to help rural communities buy, develop and manage land, pledging £10.8 million for the first three years.

It would be a mistake, however, to imagine that local communities throughout rural Scotland are eager to become landowners. Many crofters, in particular, have few quarrels with their landlords. The 1997 Transfer of Crofting Estates Act enables the ownership of the 1,400 state-owned crofting holdings and common grazings to be transferred to crofting communities, but the take-up has been minimal so far (Cameron, 2001). Perceiving the state as a paternalistic and non-intrusive landowner, crofters are content with the *status quo* (MacAskill, 1999). Many have no appetite for the financial burdens and responsibilities of ownership, preferring to stay tenants because of the strong legal rights that this gives them. Community ownership commands widespread support and is in line with the trend towards participatory democracy (Goodwin, 1998a), but it is challenging to operate and certainly no panacea. Hunter (1999: 9) observes:

> As Assynt folk have already discovered, making a go of community ownership is not easy. It is, in fact, extremely difficult.

Crofters in Glendale on Skye who have owned the land since 1956 believe that ownership has damaged the community (MacAskill, 1999). Notwithstanding these caveats, the tide is running strongly in favour of community ownership; it is set to become a feature of rural Scotland and a significant influence on development and land use patterns.

3.4 THE LAND REFORM DEBATE

Land reform was a staple of nineteenth-century radical politics in Scotland, but from the 1920s to the 1990s it virtually vanished from the political map (Cameron, 2001). Now it is back with a vengeance.[2] Land reform is an incendiary topic, as witnessed by bloody revolutions down through history, from France to Russia to Cuba, and by contemporary events in Zimbabwe. Given that, in essence, it is about restructuring established power relations in society it cannot be otherwise. Though no blood has been spilt over Scottish land reform, debates are certainly characterised by high blood pressures and verbal assassinations. Prior to the 1997 General Election many thought that politicians would shy away from grasping the reform nettle because of the costs, the complexity, and the fear of political fallout (Wigan, 1991; Barraclough, 1997). Such convictions have been confounded by subsequent events. Indeed, as the year 2000 approached, land reform became the symbolic new orthodoxy as a new Scotland with a new Parliament faced a new millennium.

3.4.1 The need for reform

Three aspects of land ownership have come in for particular criticism. The first is 'landlordism', the disproportionate influence of large private landowners (notably in the Highlands), together with the abuse of power by a small but high profile minority. As Cramb (1996: 13) puts it, 'the extractive regimes continue – benefiting the owner, failing the environment, doing little for the local community'. This last is a frequent criticism, that landowners' disregard of the aspirations of local people hinders rural development and fosters a sense of powerlessness (Bryden, 1997). The second is feudalism. Scotland escaped the violent social upheavals which erased the feudal system from the rest of the European map. Although most aspects of feudalism ceased centuries ago, Scotland remained the only country in the world in which land ownership at the end of the twentieth century was still legally based on a twelfth century feudal system. Some 80 per cent of Scottish property was not owned outright but on a feudal basis (Nicol, 1997), and the ancient terms 'vassal' and 'superior' were still used in legal documents. Superiors could still demand costly feu duties and impose conditions of ownership.

A third controversial dimension is the unregulated nature of Scottish land sales. In many countries the use and sale of land is tightly controlled, but here there is an unrestricted market in most land. Almost uniquely in Europe, there are no laws controlling the size of land holdings, absenteeism, inheritance, or the extent of overseas ownership. This treatment of land simply as a commodity has been described by a European commentator as 'wild west capitalism' (in Wightman, 1999: 99). Scandinavian systems are often held up as preferable models (e.g. Munro, 1997). However, not all Scottish land ownership is so unconstrained. Wightman (1996a) highlights the anomaly that in crofting areas all aspects of land ownership are specified under the crofting Acts; thus the tenure of 2ha of bog and rock is tightly controlled, while the ownership of 20,000ha is unrestricted.

A variety of issues related to the central matter of land tenure have also exercised commentators, including the inaccessibility of information about owners and properties (Mather, 1995). A range of concerns about land policy (as distinct from land reform *per se*) have also been swept up into the debate, such as public access to private land, and agricultural tenancy arrangements.[3] A case for land reform can be made on economic grounds (MacMillan, 2000), but the passionate nature of the calls for reform springs from perceptions of injustice, both historical and contemporary. Land reform taps into a deep undercurrent of historical grievances stretching back over centuries; the vicious aftermath of the Battle of Culloden in 1746 and the Highland Clearances are two salient aspects of the 'national guilty conscience' which drives land reform (Cameron, 2001: 85). Many people hold little against today's landowners as individuals, believing with Cramb (1996: 190) that 'landowners are guilty of nothing more than human nature'. As a type, however, they are widely reviled, both because of what they represent today and because of what they conjure up historically. 'Popular landlords are as rare as hen's teeth' (Richards, 2000: 11).

In a debate so shot through with the politics of envy and outrage (Wightman,

1996a), language and imagery understandably become emotive (Fig. 3.1). Jim Hunter, a leading reformer, is on record as promising that 'when the sporting estate is dead and buried, I'll lead the dancing on its grave' (*in* Maxwell, 1998), while McIntosh (1997: 91) believes that 'buying back land into community ownership is a bit like securing the stolen family silver'. Wightman (1996a: 216), too, is uncompromising:

> The archaic, secretive, anti-democratic, feudal and exclusive system that constitutes Scottish landownership needs to be finally consigned to the history books.

Equally understandably, those who have much to lose go to opposite extremes in denouncing the reformers as Stalinists, while Wigan (1991: 102) describes the concept of land reform as 'a piece of nostalgic whimsy which takes no account of economics'. The term 'landowner' (as one word) is thus highly charged in Scotland in a way that 'land owner' is not in most other countries (Callander, 1998).

It is often argued by defenders of the *status quo* that the issue of who owns the land, or how much land is owned by an individual, matters less than how the land is managed. Wightman (1996a, 1999) brands this a 'myth', but the Land Reform Policy Group (LRPG, 1998a) leans towards this view, believing that ownership in itself is less important than creating the conditions in which land is managed to a high standard and in ways that benefit locals. A survey of landowners in the Cairngorms came to a similar conclusion (Cairngorms Partnership, 2000). Equally, of course, management styles and objectives depend crucially on who owns the land. The two are inseparably linked. By the late 1990s there was increasing political consensus that both sides of the equation needed attention – that there needed to be greater diversity of ownership, and tighter controls on management.

3.4.2 Towards solutions

Wholesale nationalisation of land is no longer on the agenda as it was when McEwan (1977: 5) called for the 'stranglehold of powerful, selfish, anti-social landlords [to be] completely smashed'. Even since the mid-1990s, when the focus was on a few *causes célèbres* such as Assynt and Eigg, the debate has developed fast, broadening to address the issue of how best to achieve a balance between public and private interests that promotes sustainable development throughout rural Scotland.

On coming to power in 1997, the Labour Party rapidly fulfilled a manifesto commitment by establishing the Land Reform Policy Group. This group swiftly identified the key problems and possible solutions (LRPG, 1998a, 1998b), consulted widely, and then issued its recommendations for action to the Scottish Parliament (LRPG, 1999). These landmark documents moved the debate on from the long-standing media warfare of caricature and polemic into a detailed consideration of the real issues. The LRPG advocated a rolling programme of wide-ranging legislation, but the two key recommendations concerning the way that land is owned and used were increased diversity and increased community involvement. The case in favour of land reform was made on grounds of fairness, and to secure the public good – fairness because opportunities for local enterprise can be stifled under present arrangements, and the public good because of the damage to the natural heritage that can result from poor land management. Proposals to limit corporate, foreign or absentee ownership were not included in the end, but greater accountability in the management of land was advocated. The LRPG (1999: 4) concluded that the overall objective of land reform should be 'to remove the land-based barriers to the sustainable development of rural communities'.

These recommendations formed the basis for a White Paper in 1999 and a draft Bill in 2001 (Scottish Executive, 2001b), due to be enacted in 2002. The proposed legislation consists of three flagship measures: new rights of access (Section 9.4.3), and a right-to-buy for local communities and for crofting communities. Local community bodies in any part of rural Scotland will have a right to buy land before it goes on the open market, as long as they have previously registered an interest and a range of criteria are met.[4] The asking price will be set by a government-appointed valuer, and only if communities fail to raise the necessary sum within six months will the owner be free to sell on the open market. A new power of compulsory purchase is included to deter evasion. The impact of these arrangements will be gradual and long term because the decisions of whether to sell, when, and what area(s) to put on the market remain with the owner. By contrast, the crofting right-to-buy is genuinely radical because the initiative rests with the purchaser: at any time, crofters could potentially take ownership of almost one million hectares of land in the crofting counties.

Predictably, reactions to the current agenda range over a very broad spectrum, from those who feel that they do not go far enough, to those who see the idea of a community right-to-buy and the concept of compulsory purchase as steps towards a Stalinist state. Representative of the former is Wightman (1999: 11, 2001a, 2001b) who argues that the proposals treat symptoms rather than underlying causes, and that they are based on 'flawed analyses, shallow, short-term politics, timidity and poverty of imagination'. He argues that the fundamental problem is the large scale of existing landholdings, and that current reform plans fail to address this. His own far more radical agenda, worked out in considerable detail (Wightman, 1996a, 1999), includes upper size limits on holdings, restrictions on absenteeism and foreign ownership, and the introduction of Land Value Taxation. The reactionary end of the spectrum is given voice by Wigan (1996a, 1998, 1999a) who argues that the reforms will depress land values and lead to a haemorrhage of jobs and investment from rural Scotland, as well as risking damage to the natural heritage by undermining the incentive for long-term, high quality land management. On this basis, he suggests that the land reform movement is motivated more by historical symbolism than by a concern for real livelihoods or environmental stewardship. Such divergent arguments were rehearsed again at the launch of the draft Bill (Ross, 2001).

In the centre ground are many who broadly welcome the land reform agenda but feel that their particular interest has been sidelined. For example, conservation bodies such as WWFS and the NTS want to see more emphasis on the natural heritage and on the quality of natural resource management, not just on the social and equity issues. Certainly a higher priority is being given to sustaining and enlarging rural communities than to environmental protection *per se*. Even the SLF supports the central themes of increased diversity of land ownership and use, and more community involvement, although it takes issue with much of the detail, especially the community right to buy. Others see a need for change but argue that many of the current problems related to the ownership and management of land could be substantially addressed without the need for new legislation or more public money (MacMillan, 2000).

Of concern to all parties have been issues of definition because several of the central aspects are hard to pin down, notably the term 'local community', and how to define good and bad land management (Warren, 1999a). Specifying the membership of 'local communities' is far from easy anywhere, but it is greatly complicated in the Highlands by controversial debates concerning 'incomers' and 'white settlers'. The draft Bill indicates that what is meant by 'community' will depend on the context, but that it will normally consist of all those on the electoral roll; Ministers will have discretion in deciding whether a community body is representative of and supported by local people (Scottish Executive, 2001b). On the issue of land management, politicians have repeatedly stated that good landowners have nothing to fear from land reform, but by what criteria should good and bad management be differentiated? To resolve this issue, codes of good practice have been developed for land ownership and land use, the SLF leading the way in 2000 by launching *A Code of Practice for Responsible Land Management* (SLF, 2000).

There is no doubt that the planned legislation is of considerable symbolic and practical significance. Equally, however, it clearly does not match up to the protestations of radicalism made by the Scottish Executive (Cameron, 2001). Indeed, it is entirely possible that the community right-to-buy will neither empower communities nor effect rapid change in the pattern of landownership (Wightman, 2001a). If so, it will fail to fulfil its *raison d'être*.

3.4.3 Related debates

3.4.3.i The abolition of feudalism

This is arguably the least controversial issue.[5] For over thirty years, there has been broad consensus that the Byzantine, antiquated complexities of the feudal system have no place in twenty-first-century Scotland (Nicol, 1997; Callander, 1998). Though symbolically important, this specific issue is largely legal and technical. Remaining feu duties and the notion of superiors vanished in 2000 under the Abolition of Feudal Tenure Act. However, while welcoming the abolition of feudal tenure, some are concerned that the 'public interest baby' is being thrown out with 'the very dirty feudal bathwater' (Callander, 1998: 194); the fear is that if the concept of conditionality of ownership that is implicit in the feudal system is abolished and replaced with a system of outright ownership, the power of landowners will be enhanced rather than curbed. This concern will be addressed by a Title Conditions Act, and by Codes of Practice for the ownership and management of land, both due by 2002.

3.4.3.ii Community ownership

Relative to the long timescales of land management, community ownership in its current form is a very recent and untried form of tenure which has yet to prove that it will raise standards of stewardship. The harsh reality is that much Scottish land has only modest potential for productive diversification, yet high risk of environmental damage. Sub-division of the big estates into small holdings would not automatically deliver either a prosperous society or a flourishing environment. While increasing diversity of ownership and land use is certainly desirable, it is vital that the management units that are created are viable, and that they are no more (and ideally less) dependent on public support than today's structures (R. Balfour, 1998). Ownership undoubtedly brings opportunities, but also considerable responsibilities, risks and challenges, not to mention financial liability. Given a choice between ownership and a good landlord, many communities would opt for the latter. Surprisingly, perhaps, Wightman (1999: 72) is sharply critical of the fact that 'the fashionable idea of community' has become 'the holy grail of land reform', despite his own advocacy of an expansion of community ownership (Wightman, 1996a).

3.4.3.iii The environmental impact

Integral to much writing on land reform is the view that the environment suffers under the current system and would flourish if the planned reforms were enacted (Hunter, 1995; Cramb, 1996). In fact, the environmental impacts of land reform are likely to be mixed. Reforming the legal framework will probably have less impact on the future wellbeing of the environment than the accompanying 'codes of good practice' and future changes in agricultural, forestry and conservation policies. These, of course, are not the exclusive preserve of the Scottish Parliament but will be driven by European and global forces over which Scottish and UK politicians have little control.

3.4.3.iv The place of private ownership

There is no doubt that the current system has delivered mixed results in terms of environmental management, ranging from enlightened stewardship to the oft-rehearsed examples of environmental abuse. The combination of the abuses and the concentrated pattern of private ownership has fed a tradition of anti-landlordism in Scottish politics (Cameron, 2001). Land reformers used to espouse land nationalisation but no one seriously argues for that any longer; the European Convention of Human Rights would probably prevent expropriation anyway.[6] Most now accept that a majority of landowners are responsible stewards of the countryside and so must be seen as part of the solution, not part of the problem. Moreover, despite the historical rhetoric, it is widely recognised that two injustices do not make a right – that visiting 'the sins of the fathers' on this generation is self-evidently unjust (Hunter, 1995; Dewar, 1998). There is a strong argument for a partnership approach, in which the state, its agencies and the private sector collaborate in providing the financial input necessary to sustain rural Scotland.

At its best, private ownership has many demonstrable strengths, two of which are emphasised by Warren (1999a). The first is long-term stewardship. Land management is a long-term business for which continuity of purpose is vital, especially when it comes to lengthy projects like re-establishing native woodlands. The fact that private landownership entails 'personal responsibility and a duty of care that reaches across the generations' (Scruton, 1999) is an antidote to 'the tragedy of the commons'. Stewardship – an earlier generation's word for sustainability – sums up the sense of responsibility that has long been regarded as part and parcel of owning land in Scotland (Callander, 1998). Significantly, this is a concept which is widely shared by landowners and conservationists alike (Ramsay, 1993; Bryden and Hart, 2000). The second is inward investment. Given the marginal economics of many landowning businesses, owners often subsidise their estates from earnings made elsewhere, an inflow of capital which represents a large financial input to remote rural areas. In such places, 'sporting estates are one of the key building blocks of the local economy' (CWP, 1993: 38). In the Cairngorm area, for example, landowners (including non-private owners) invested some £20 million in their businesses in 1998–9, supporting 500 jobs which are critical to the maintenance of rural community facilities, as well as providing 1,000 houses and offering many free facilities to local communities (Cairngorms Partnership, 2000).[7]

Reasons like these lead commentators such as Callander (1998) to conclude that it is in the public interest to have a vibrant system of private landownership, but that it should be just one element in a diverse pattern of ownership and management, not the dominant partner as at present. A diversity of tenurial options (private, public, community, voluntary) must be advantageous because 'only a pluralistic solution can engage the interest and benefit from the talents of the maximum number of people' (H. Raven, 1999: 150). In fact, once compulsory nationalisation of land is ruled out, diversity of land ownership is the only financially realistic route to take.

3.4.4 Conclusion

Land reform unavoidably addresses power relationships; 'power is still the issue' (Dewar, 1998: 5). The privilege of owning land in Scotland inevitably confers a measure of power and influence on the owner, and history (both ancient and recent) has repeatedly shown that unfettered power is a two-edged gift to entrust to individuals. Its temptations have too often led to abuses. Yet it must be remembered that power – like money or the gift of speech – is ambivalent. It can be used for good or ill. Thus, for example, the very same system of untrammelled landed power which produced Clearances in the Highlands facilitated the eighteenth-century agricultural revolution in the Lowlands (Smout, 1997b; Cameron, 2001). Enlightened land management requires a measure of power teamed with effective decision making if anything is to be achieved. Equally, the corrupting effects of power are not restricted to 'dyed-in-the-tweed' aristocratic landowners; sinners and saints exist in all parts of society, not just in its upper echelons, so codes of good practice must be applied even-handedly to all kinds of land owners and managers. The obstacles which previously kept the genie of land reform in the bottle – lack of parliamentary time at Westminster, and resistance in the House of Lords – no longer stand in the way. The effects of its release on the land and the rural communities of Scotland are likely to take some time to become apparent.

NOTES

1. The historical dimension is helpfully summarised by McCarthy (1998) and colourfully described by McIntosh *et al.* (1994) and Wightman (1999).
2. Much has been said about the land reform debate in the last five years. Commentaries include those by Hunter (1996), Wigan (1996a, 1998, 1999a), Morgan (1996), Bryden (1997), R. Balfour (1998), Dewar (1998), Boyd (1999), H. Raven (1999), Wightman (1999, 2001a, 2001b), Warren (1999a), Bryden and Hart (2000) and Cameron (2001).
3. The involved topic of agricultural tenure arrangements is discussed by Stockdale *et al.* (1996), Stockdale and Jackson (1997), LRPG (1998a) and Wightman (1999). It was left out of the 1999 White Paper on land reform, but a consultation on proposals for legislation to amend the 1991 Agricultural Holdings (Scotland) Act took place during 2000. The new Bill is likely to provide for, *inter alia*, more diverse tenancy options (including limited duration tenancies), new opportunities for diversification, simplified dispute resolution procedures, and a strengthening of the legal position of tenants. In 2000, the NFUS and the SLF agreed on two new types of limited duration tenancies.
4. These criteria are set out in detail in the draft Bill (Scottish Executive, 2001b), as are the full range of proposals and counter proposals for legislation considered during the consultation process.
5. For fuller discussions of the debates discussed in this section, and of related issues, see Warren (1999a) and Wightman (1999).
6. The ECHR also ruled out a variety of other proposals which were suggested for inclusion in land reform legislation (Scottish Executive, 2001b).
7. For a critique of the benevolent, positive interpretation of the role of Highland sporting estates, see Wightman and Higgins (2000).

CHAPTER FOUR

The trees: forest management

4.1 Introduction: time and space in forestry

Scotland is the UK's lead country for forestry, having over half of all the UK's woodland. Trees cover 16.9 per cent of the country. This compares with 11.6 per cent for the UK as a whole and 8.7 per cent for England (FC, 2001). In Scotland's temperate, maritime climate, trees grow fast, faster than almost anywhere else in Europe, an advantage tempered by exposure to Atlantic gales. For historical reasons the distribution of woodland in Scotland (Fig. 4.1) is largely the distribution of commercial plantations. These in turn comprise largely introduced tree species (Table 4.1) because none of Scotland's native species can compete with the growth rates of the conifers which were imported from western North American in the early nineteenth century by pioneers such as David Douglas. These include Douglas fir, Lodgepole pine and Western hemlock, but the tree which out-performs all rivals on these shores is the Sitka spruce from Alaska's Queen Charlotte Islands. When introduced in 1831 by Douglas, he commented with remarkable prescience that since it thrives on poor soils it 'could become a large and useful tree in Britain' (in Edwards, 1999: 26). The Scottish forest industry is dependent to a very great degree on this one species; it now accounts for almost half of all woodland (Smith, 1999).

TABLE 4.1 Percentages of exotic species used in Scottish forestry during various periods of the twentieth century, showing the sustained dominance of Sitka spruce, the variable importance of other conifer species, and the decreasing use of exotic conifers recently. Species are listed in order of importance in afforestation schemes 1981–90. The larch data for 1986–95 refer to all larch species. Data for the first two periods from Malcolm (1991) and for 1986–95 (the most recent data available) from FC (2000a).

Species	*Planting period*		
	1981–1990	1920–1990	1986–1995
Sitka spruce (*Picea sitchensis*)	68%	39%	47%
Lodgepole pine (*Pinus contorta*)	10%	10%	5%
Corsican pine (*Pinus nigra maritima*)	6%	4%	3%
Japanese larch (*Larix kaempferi*)	5%	8%	6%
Douglas fir (*Pseudotsuga manziesii*)	4%	4%	3%
Norway spruce (*Picea abies*)	1%	9%	2%
TOTALS	94%	74%	66%

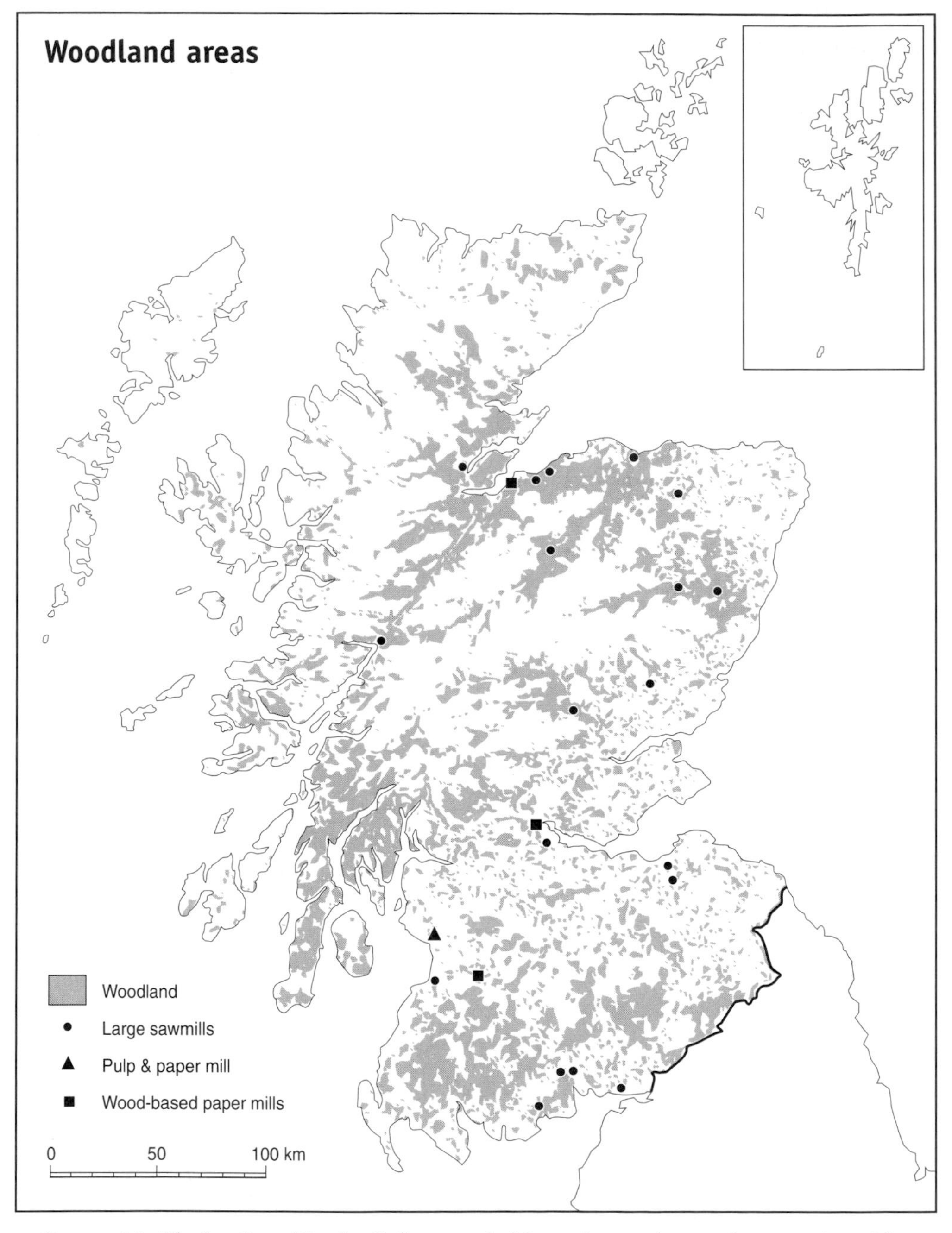

FIGURE 4.1 *The location of Scotland's forests and of the major wood processing operations. After Scottish Executive (2000b), used with the permission of the Forestry Commission.*

In their native lands, all these species form magnificent forests, but the insensitive way that they have been used in commercial plantations in this country has led to their being widely reviled. This was exacerbated by the fact that the inexorably increasing use of exotic conifers marginalised native species. Scots pine (*Pinus sylvestris*), the only commercially useful native conifer, comprised 18 per cent of planting during the mid-twentieth century, but only 3 per cent by the 1980s (Malcolm, 1991). Only two other native species, oak (*Quercus petraea*) and birch (*Betula pubescens*; *B. pendula*), had a place in twentieth-century Scottish forestry, and that was marginal. Technological advances in silviculture during the decades following World War Two transformed the time-honoured rural craft of forestry into an industrial operation, alienating the public (Pringle, 1995). Although forestry in Scotland is now moving in exciting new directions (Section 4.5.1), it has yet to shake off the negative perceptions generated during an earlier era.

Forestry operates over large areas and on long timescales. Like a supertanker, it takes a long time to change direction. Even Sitka spruce rarely matures in less than forty years, and many broadleaved species take over a century. Consequently, although large scale commercial forestry in Scotland dates back about eighty years, many plantations are still in their first rotation. External influences such as economics, markets, politics, and public values change far faster than forests can. Five important consequences flow from this:

1. Forests are inevitably 'out of touch' and behind the times. The fashions and mistakes of previous generations are still writ large in the landscape; 'the sins of the fathers . . . live with us for a very long time' (Lister-Kaye, 1995: 61). Equally, the recent 'forest enlightenment' (Section 4.5.1) will not bear visual fruit for many years to come.
2. Major forest operations occur infrequently, tend to be large scale, and have a big visual impact.
3. Understanding history is more important for forestry than for most other sectors of environmental management because today's woodlands are a product of the past.
4. Policy continuity is vital to give foresters confidence to plan for the long term.
5. More than any other land use, forestry has to incorporate the possibility of long-term climatic and environmental change into contemporary decision making (Moffatt, 1999).

Taken together, the first two of these points have given forestry considerable public relations problems. The length of forestry timescales in relation to human lifespans makes forests feel timeless. Agriculture, by contrast, with short rotations and relatively small areas, can adapt each year. If the number of forest policy changes that have occurred during the life of the early plantations were applied to a single arable crop it would work out at three Acts per week (McCall, 1998), emphasising the tensions inherent in managing a long-lived resource in a rapidly changing policy context. Arable harvests are not halted by public outcries, whereas protesters *do* take to the trees to halt felling programmes. People weep when storms destroy ancient woodlands. Commercial foresters typically regard trees as crops, but to the general public, trees and woodlands have always had a cultural and sacred significance that goes far beyond their basic material usefulness, symbolising longevity, stability and continuity. This is eloquently demonstrated by Rival (1998) and Miles (1999), and it makes the balance between market and non-market values in forestry especially difficult to strike (Section 4.6.2).

4.2 Historical background

Scotland is a deforested country. If human beings had never arrived, it would today be a largely wooded country. The good soils of the lowlands would be covered by oak, ash and elm forests, the poorer upland areas would be dominated by Scots pine, while hazel, rowan, birch and juniper would occupy the more extreme environments. Something approaching this distribution is thought to have existed some 4,000–6,000 years ago (Fig. 4.2). In the subsequent millennia, the co-evolution of forests and people has meant that few if any of today's woodlands are unmarked by some past use (Tipping, 1993). The fascinating history of Scottish woodlands is examined in detail by Smout (1993b, 1997a, 2000a) and by Edwards and Whittington (1997); it essentially consisted of an irregular but inexorable reduction in area, accelerating from about 4,500 years ago onwards (Section 1.2.2.i). Woodlands only survived where and when they were seen as being useful, or in inaccessible locations.

During the eighteenth and nineteenth centuries some extensive afforestation was carried out by the so-called 'planting lairds', notably the Dukes of Atholl. It was they who started the trend for introducing conifers, firstly from Europe (Norway spruce (*Picea abies*), European silver fir (*Abies alba*), European larch (*Larix decidua*)), and then from North America. Net losses continued, however, reducing woodland cover to 4.5 per cent by the early twentieth century (Mather, 1993b). Truly native forest was reduced to small fragments. The neglected condition of Scotland's dwindling woodlands was causing concern by the late nineteenth century, but it took the First World War to translate anxiety into action. Wartime timber shortages led to extensive felling, totalling perhaps 60,000ha in Scotland (Coppock, 1994), and it was this that caused the government to adopt the recommendations of the Acland Report and establish the Forestry Commission (FC) in 1919. Afforestation has remained an important policy objective ever since. As a result, the twentieth century quite probably saw the most rapid expansion in forest cover of any century in post-glacial time.

To a considerable degree, the recent history of British forestry is the story of forest establishment by the FC (Mackay, 1995).[1] The remit given to the FC in 1919 was to plant 720,000ha of new forests in the UK by 1999, two-thirds of it in Scotland. It was presumed that this would be carried out almost entirely by the state. The rationale was primarily strategic, to create a timber reserve for times of war. The approach was to be commercial. The style of forestry which emerged was a direct and logical consequence of this remit:

- Monocultures of exotic conifers are more productive and more easily managed than mixed forests.

- Straight-edged plantations minimise fencing costs (which often make up half the establishment costs).
- Large areas bring economies of scale.
- Hardy pioneer species were a necessity for establishing forest ecosystems on degraded and exposed upland sites.

Consequently, about 99 per cent of FC forests in Scotland are coniferous (D. G. Mackay, 1994), and most are large, unbroken and geometric in shape (Fig. 4.3). Although the strategic objective was dominant, other explicit aims included the provision of employment and forest workers' villages in depressed rural areas. A recreational dimension also emerged early in the FC's life (Section 9.4.1).

Since its inception, the FC has been subject to an unending succession of investigations and redirections, together with threats to its existence (Mackay, 1995). Controversy over the landscape impact of large scale afforestation has also dogged the FC since the 1930s, though this was less of an issue in Scotland than England until more recently. At the outbreak of the Second World War the new FC forests were still immature, so again extensive felling of mature woodlands took place and again Scotland took the brunt. With perhaps a third of the Scottish forest having disappeared during the two wars (Mackay, 1995), the need for further afforestation seemed even more pressing. However, the simultaneous drive for national self-sufficiency in food production (Section 5.1) drove forestry ever further 'up the hill' and into the poorer land of the north and west where species choice and productivity is limited (Mather, 1993b). Nevertheless, successive governments continued to encourage afforestation and increasingly targeted Scotland. Throughout the 1970s and 1980s about 85 per cent of all UK planting took place in Scotland (Mackay, 1995), and from 1988 coniferous planting in the English uplands was prohibited, further focusing commercial forestry in Scotland.

Until the 1950s the private sector showed little interest in new planting, but from then on private forestry has become progressively more important. In 1973, for the first time, more new planting was private than state, and during the 1990s state planting dwindled rapidly (Figs 4.4 and 4.5), hitting zero in 2000 (FC, 2001). The relative importance of the private sector was boosted after 1981 when the government initiated a programme of disposal of FC land. Traditionally, private forestry was the preserve of the large estates, but the rise of the private sector had much to do with the emergence in the 1960s of forest investment companies which acted as agents for private investors who were taking advantage of planting grants and tax concessions. By the mid-1980s 35 per cent of private forestry was in the hands of such companies, with corporate ownership (pension funds and insurance companies) accounting for a further 10 per cent (Mather, 1993b).

The main engine driving private sector planting in this era was a fiscal incentive whereby tax bills could be reduced by offsetting forestry expenditure against capital gains from other business interests. This encouraged many people with little or no involvement in land management to invest in forestry for purely fiscal reasons. The

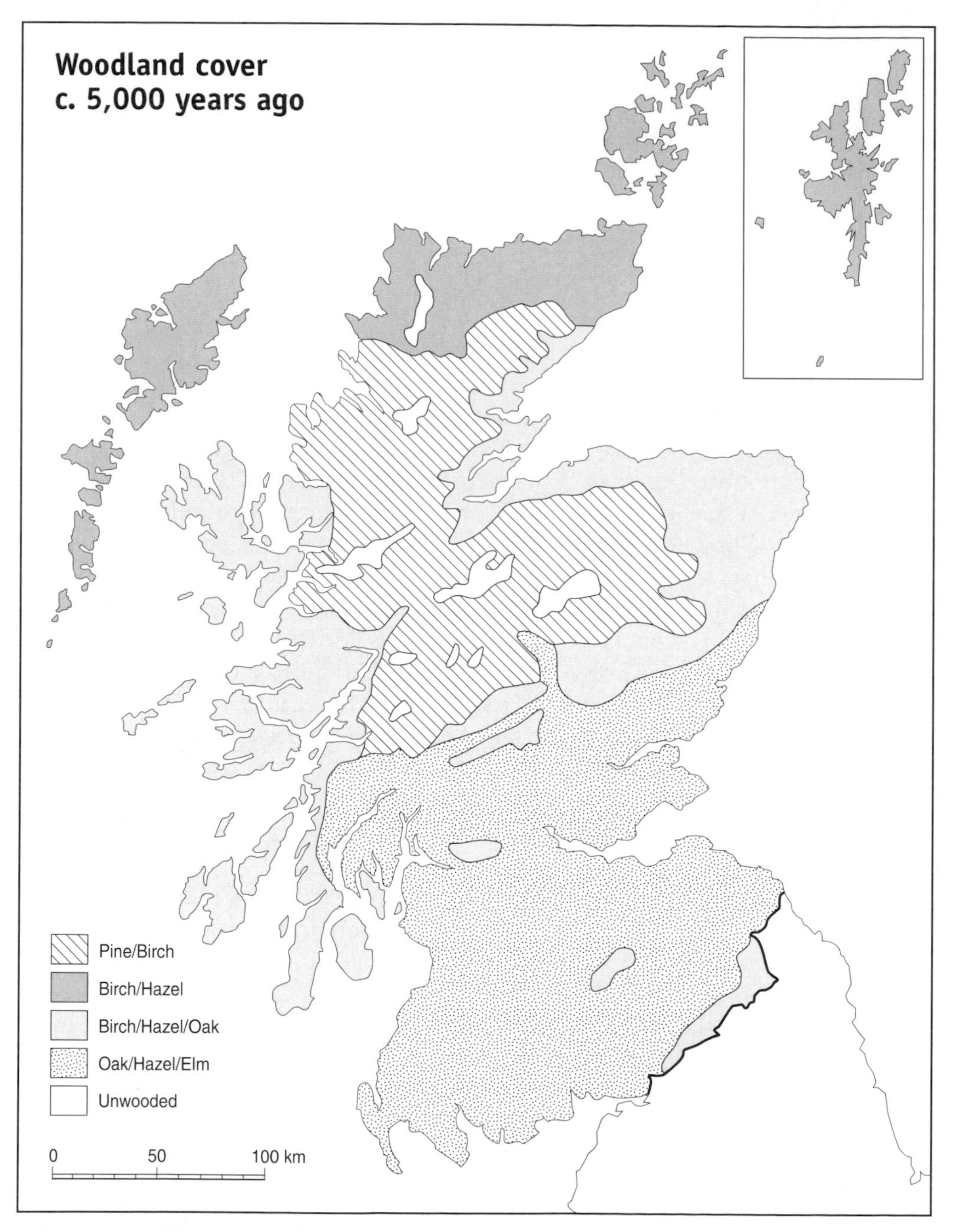

FIGURE 4.2 *Distribution of woodland in Scotland around 5,000 years ago. After Edwards and Whittington (1997).*

result was that a new type of private sector plantation forestry with a particularly hard-edged commercial slant rapidly displaced the smaller scale, more diverse and integrated estate-type afforestation (Mather, 1988; Mackay, 1995).

FIGURE 4.3 *An aerial view of geometric, blanket afforestation near Moffat in the Southern Uplands. Photo: © Patricia & Angus Macdonald/SNH.*

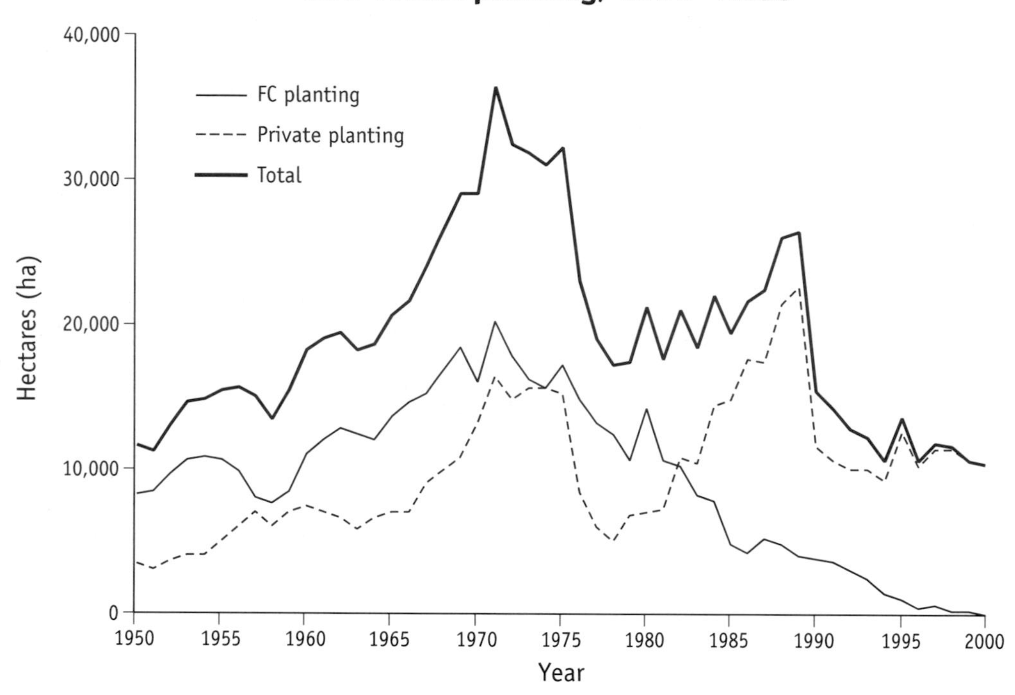

FIGURE 4.4 *New planting in Scotland by the private and state sectors, 1950–2000. Private sector figures include grant-aided and non grant-aided planting. Prior to 1979 the private sector figures include a small percentage of replanting; from 1979 all figures exclude replanting. Data supplied by the Forestry Commission.*

The insensitivity of investment-driven forestry, epitomised by the Flow Country debacle (Warren, 2000a), led to a mounting chorus of disapproval in the 1980s, focusing especially on the impacts of afforestation on birds (Avery and Leslie, 1990). The government responded by largely removing the fiscal incentives in the 1988 Budget, a totally unexpected move which left the Scottish forest sector in shock (c.f. *Scottish Forestry* 43(1)). Much increased grant levels were insufficient compensation, and planting levels crashed by almost 50 per cent (Fig. 4.4). Although forestry had already begun to move away from a 'pure production' mentality, 1988 represents a watershed. The fiscal change, together with the launch of the Farm Woodland Scheme encouraging farm-forestry integration (Section 11.2), symbolise the start of a new era with different priorities (Section 4.5.1).

During the 1990s a recurring theme was the threat of privatisation of the FC estate (c.f. Jeffrey, 1994), which, rather remarkably, resulted in the private sector campaigning to preserve a state enterprise (RSFS, 1993). This was not as strange as it might appear, because state and private forestry have long worked in partnership (Mutch, 1994). The primary argument against privatisation is that a long-term industry needs the long-term security and the (partial) protection from the whims of the market which only state ownership can offer. State forestry can also offer environmental benefits and recreation opportunities which private owners cannot match. It is now generally recognised that market forces are unlikely to deliver forest policies successfully, firstly because many of the benefits are intangible, being enjoyed by a wide public including future generations, and secondly because the land market is so distorted by EU farm subsidies (FC, 1999a).

This last is one example of the way in which supranational influences (amongst which EU membership ranks high) has had an increasing effect on forestry (Aldhous, 1997; Mather, 2001). During the 1990s forestry rose rapidly up the international agenda as a result of global concerns about environmental instability and forecast timber shortages, most notably leading to the Statement on Forest Principles signed at the 1992 'Earth Summit' in Rio. Subsequent pan-European ministerial conferences interpreted these principles for European conditions, giving rise to the 1993 Helsinki Guidelines and the Lisbon Declaration of 1998 which emphasises the social and cultural importance of forests. The government has explicitly adopted these principles of sustainable forestry (Anon, 1994a), and the UK Forestry Standard endeavours to turn them into practical management advice (FC, 1998a; Section 4.5.4). The irony in the application of sustainability principles to forestry is that the concept of sustainability itself arose out of the nineteenth-century silvicultural idea of sustention or maximum sustainable yield. 'Woodland is the concept of sustainable development made flesh' (Magnusson, 1995a: 3).

4.2.1 Twentieth-century forest policy: how successful?

The forest industry is in the unenviable position of having hit almost all of its original targets, only to find that the rules of the game have been re-written. Any concept of negative environmental impact was entirely absent from the Acland Report; the aim was to make 'waste' land (i.e. heath and moor) productive (Mather, 1996a). Judged against its original remit, forest policy has been remarkably successful. The UK planting target was achieved fifteen years early in 1984, and forest cover in Scotland has more than tripled (Fig. 4.5). The growing timber resource created by this dramatic expansion has spawned a world class wood processing industry developed on the strength of current production and on the projected doubling of Scottish timber output by 2015 (Scottish Executive, 2000b).

But Scotland's woodland cover is still only half the EU average of 36 per cent, and is tiny relative to Scandinavia's (e.g. Finland: 65 per cent) (FC, 2001). Moreover, the UK still imports 85 per cent of its timber needs at a cost of £6 billion per year and is the second largest importer of timber products in the world (FC, 1999a). The success of forest policy in fulfilling its wider original goals is also questionable (Mather, 1996b). The socio-economic benefits that were envisaged have not materialised, forest employment has never approached the levels once hoped for, the Forest Workers' Holdings programme has been forgotten, and forestry villages such as Polloch in Lochaber have failed to fulfil their original objectives; they are now occupied by counter-urbanites and second-home owners (Wonders, 1998). Assessed against

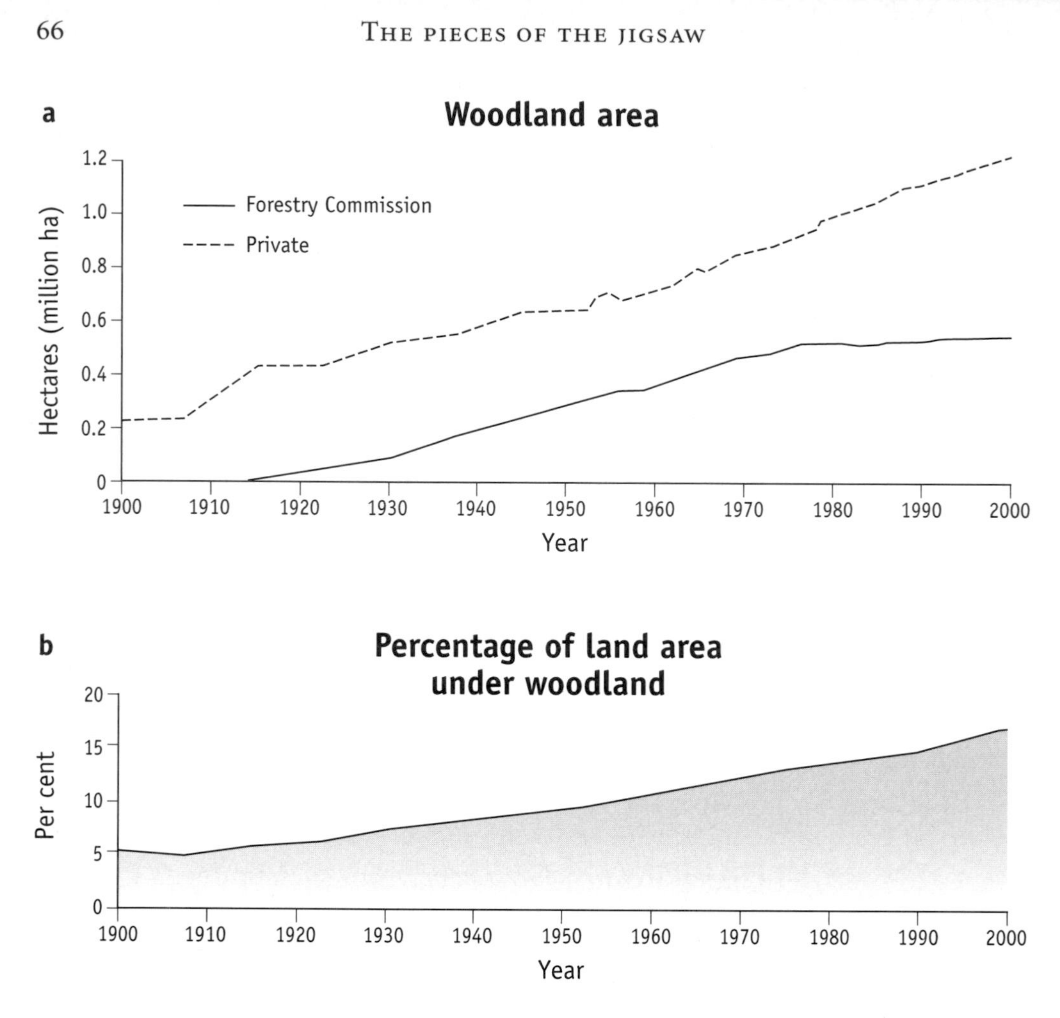

FIGURE 4.5 *The twentieth-century expansion of woodland in Scotland.*
a. The contribution of the state and private sectors to the expanding woodland area, 1900–2000.
b. The rising percentage of land area under woodland, 1900–2000.
After Coppock (1994), updated using Forestry Commission data.

twenty-first-century 'quality criteria', twentieth-century forest policy appears ill-judged and myopic in its one-track focus on quantity.

But it can hardly be fair to judge the ambitious objectives of our forbears by the criteria of a radically changed world and with the benefit of eighty years' hindsight. Success in forest policy is notoriously difficult to assess, not only because the criteria are disputed and constantly evolving, but because it is impossible to maximise the value of all of them simultaneously (Mackay, 1995). There are inevitable trade-offs between, for example, economic efficiency, cultural acceptability, recreation, and wildlife. The perceived blunders of Scottish forestry in terms of low quality and poor location are now being addressed respectively through multi-purpose forestry (Section 4.5.1) and indicative forest strategies (Section 4.5.2).

4.3 THE PROS AND CONS OF FORESTRY IN SCOTLAND

Forestry is increasingly acknowledged as offering a wide range of benefits to people, to the economy, and to the environment (Goodstadt, 1996; Warren, 1998; FC, 1999a).

1. **Forestry for people**. Every year, about two million people who live in Scotland visit woodlands for recreation, making a positive contribution to the health and education of communities (Scottish Executive, 2000b). Forests can help to restore post-industrial landscapes, improve the amenity of urban areas, and reclaim derelict land, all contributing to improved quality of life. In farming areas, woodlands provide shelter from the wind, as well as sporting opportunities. Forests can contribute to the diverse aims of rural development in a wide range of ways (FAPIRA, 1995; Snowdon and Slee, 1998) and offer many non-market benefits to society (Roper and Park, 1999). It is the provision of such benefits that now largely underpins the case for on-going public sector support for forestry.
2. **Forestry for the economy**. Forestry and wood processing contribute some £800 million per year to the Scottish economy, equivalent to 0.5 per cent of GDP, and directly employ some 10,700 people (Scottish Executive, 2000b). Many of these jobs are located in disadvantaged rural areas, and about 85 per cent of them are local (FC, 1999a). The forecast doubling of timber production will boost the economic and employment significance of forestry for Scotland and widen the scope for the niche markets which can be so important locally. Scotland has recently become a net exporter of wood products and is thus helping to reduce the substantial UK and EU timber deficit. Woodlands are also important for tourism, especially in areas such as Galloway, the Trossachs and Speyside where many visitor attractions are forest-related. The burgeoning significance of woodlands as tourist attractions is likely to continue throughout Scotland, particularly as the Argyll and Glen More forest parks become associated with new national parks (Section 8.3.4).
3. **Forestry for the environment**. Trees, woods and forests are an integral part of the character of many of Scotland's much-loved landscapes, as well as providing diverse habitats for plants and animals. The surviving fragments of native woodland are of international conservation significance (Section 10.2.1). Good forest design in upland catchments can help to improve water quality, reduce runoff, and control soil erosion (Section 6.5.2). Worldwide, forests are a renewable but rapidly dwindling resource, so new forests help the government to fulfil its commitments under the Climate Change Convention by sequestering carbon dioxide from the atmosphere. In Scotland, forests absorb about 10 per cent of Scottish fossil fuel emissions, and new forests could contribute 16 per cent towards the country's reduction target under the Kyoto Protocol (Cannell and Milne, 2000). The long-term role of forests as carbon sinks is a complex and controversial issue (F. Pearce, 1999a), but young growing trees do accumulate carbon, and forests store more carbon than other types of land cover. Even the

extensive coniferous plantings have a silver lining in that they ease the pressure on more fragile native forests elsewhere (FoES, 1996; Goodstadt, 1996), and their aesthetics can be improved by the application of natural forestry concepts (Peterken, 1999). Forests can also supply renewable energy (Section 4.5.5).

Given this wide array of positive attributes, why is Scottish forestry saddled with such a negative public image? Partly it is because the forest sector remains a dispersed and uncoordinated industry which struggles to get its arguments across to a largely urban general public. But the main cause of the albatross around forestry's neck was the scale, style and speed of twentieth century afforestation (Warren, 1998). As a result of the FC's zealous pursuit of its remit, 'forestry' (to many people) conjures up images of geometric, monocultural conifer blocks (Fig. 4.3), bulldozed tracks scarring the hillsides, miles of high fencing, dark, impenetrable forests, and the tangled devastation of clearfell sites. The rate of afforestation was remarkable. During the forty years to 1988, the total area of conifer plantations increased by an amazing 696 per cent (Mackey *et al.*, 1998). In parts of Dumfries and Galloway whole landscapes 'disappeared' under blankets of exotic conifers as forestry spread to cover 27 per cent of the region (Smith, 1999; Fig. 4.1). The result of this blizzard of planting is that almost 90 per cent of Scotland's woodlands have been planted or replanted within the last fifty years, and almost half of Scotland's forests are less than thirty years old (Scottish Executive, 2000b). Despite widespread condemnation, many schemes, both state and private, remained rigidly geometric until the late 1980s. The FC acquired a reputation for double-standards, paying lip service to environmental sensitivity from the early 1970s while continuing to fund blanket afforestation (Mackay, 1995). The private sector too has failed to shake off the taint left by the era of investment-driven afforestation. Forestry's former scant regard for conservation, and the negative hydrological impacts which careless forest operations can have (Section 6.5.2), have also left their mark in the public consciousness. Another negative dimension is the growing concern over timber transport. In extensively forested regions such as Argyll, rural roads are already being damaged by timber lorries, roads which will be wholly inadequate to handle the projected increase of production. Increasingly, there are calls for forestry to adopt more sustainable methods of timber transport, such as by water or rail which only currently carry 5 per cent of tonnage.

In summary, it is crucial to differentiate between the various styles of woodland and afforestation. Because there is a world of difference between a commercial Sitka spruce plantation and a native oak woodland, forestry cannot be said to be a 'good thing' or a 'bad thing' in Scotland in a general sense. As Boyd (1993: 161) says, 'forestry in the right hands is gentle, cultured and spiritually uplifting; in the wrong hands it is harsh, uncouth and depressing'. It all depends on getting the right trees in the right places for the right reasons. This sentiment trips easily off the tongue, but designing forest policy to achieve it is no easy task (Section 4.6.2). Nevertheless, as Scotland's forests move into their second and third rotations under management regimes which aim for multiple benefits, they will gradually lose their plantation image and begin to look and function like real forest ecosystems.

4.4 MANAGEMENT FRAMEWORK

The Forestry Commission remains the UK's state forest service. It has always performed a twin role, one arm acquiring, planting and managing forests, and the other administering forest policy and regulating forestry standards (in both private and public sectors). In 1992 these two functions were formally separated with the creation, respectively, of Forest Enterprise (FE) and the Forestry Authority (FA). In 2000, the FA reverted to being called the FC. Although the FC escaped privatisation in the early 1990s, FE was put on a more business-like footing from 1996, and its continued financial survival was made dependent on accelerated sales of land, a policy with implications for forest access (Section 9.4.1). At present, the EU has no forestry competence as such, but it nevertheless has considerable influence on forestry, notably through the CAP (which strongly influences land availability for forestry) and its conservation and freshwater policies. The 1999 Rural Development Regulation recognises and supports forestry as an integral part of rural development, and an EU Forest Strategy is now a likely prospect.

Although Scottish forestry is not subject to full planning controls (Section 2.4.2), it is closely regulated (FC, 1998a). The principal mechanisms for granting forestry approvals are:

- The Woodland Grant Scheme (WGS).
- Felling licences and five-year Forest Plans (for the private sector). Private owners are actively encouraged to develop longer-term forest plans.
- Forest Design Plans (for FE). These take a long-term view of forest cycles and aim to integrate multiple objectives into workable proposals (Hodge *et al.*, 1998). FE currently manages 38 per cent of the forested area in Scotland, and some 70 per cent of this is now covered by approved plans (Scottish Executive, 2000b).
- Consultation procedures which now ensure that all grant applications are published publicly and are scrutinised by local councils, SNH, the Deer Commission for Scotland and other interest groups. Very few woodland schemes are carried out without grant aid, so most are subject to vetting in this way. Additionally, some proposals require an Environmental Assessment. This openness is a far cry from the secrecy which surrounded grant applications until just a decade ago.

A wide range of grants are available through the WGS (Table 4.2). Some 30 per cent of WGS grants are now targeted in some way, either by land type, objective or geographical area, notably through the Challenge Funding scheme (SERAD, 2000a). Notionally, at least, grants are set at levels which will result in the FC's annual planting targets being met (Table 4.3). In broad terms, some £12 million is allocated each year to assist with the costs of creating new woodlands (whether by planting or natural regeneration), and about £5 million towards on-going management costs. That this total represents less than 4 per cent of the funds available to farmers through the CAP partly explains the relative values of agricultural and forest land, and reinforces farmers' lack of enthusiasm for trees (Section 11.2.1).

TABLE 4.2 Woodland Grant Scheme payment structure and rates. Data from FC (2000a).

Type of Grant	*Objectives*	*Rate (£ per ha)*
	New planting grants	
New planting	Conifers Broadleaves in woods <10ha Broadleaves in woods >10ha	700 1,350 1,050
Better land supplement	Conifers or broadleaves on arable or improved grassland	600
Community woodland supplement	Woods close to towns for public recreation	950
Livestock exclusion annual premium	Regeneration of neglected woodland by removing cattle and sheep	80 per annum for 10 years
Locational supplement	New planting in special targeted areas	600
Regeneration	Encouragement of natural regeneration	As for restocking, after establishment
	Restocking grants	
Restocking	Conifers Broadleaves	325 525
	Miscellaneous grants	
Annual Management Grant	Enhancing woodland quality and public access	35
Woodland Improvement Grant	Special projects, e.g. recreation, access, biodiversity	Up to 50% of costs
Farm Woodland Premium Scheme	Compensating farmers for income foregone through woodland establishment	60–300 depending on land type & quality

TABLE 4.3 The Forestry Commission's annual planting targets for Scotland. Data from FC (2000a).

Type of establishment	*Type of woodland*	*Planting target (ha/year)*
Woodland Grant Scheme (new planting & natural regeneration)	Conifers Broadleaves Caledonian pinewood	4,000 4,000 2,500
Restocking by Forest Enterprise	Unspecified	5,000
Restocking by non-state owners	Unspecified	3,500
Farm Woodland Premium Scheme	Unspecified	4,000–5,000
TOTAL		23,000–24,000

Unlike the government's regulatory authorities in agriculture and nature conservation, the FC is still a UK body. Nevertheless, forestry is now a devolved matter. Accordingly, the Scottish Executive (2000b) has developed a Scottish Forestry Strategy. Founded on the principles of sustainable development (Section 13.2.1), this strategy aims to make Scotland 'renowned as a land of fine trees, woods and forests' through the adoption of five 'strategic directions' (Scottish Executive, 2000b: 9):

- maximising the economic value of the growing wood resource
- creating a diverse, high quality forest resource that will always contribute to the country's economic needs
- ensuring that forests make a positive contribution to the environment
- creating opportunities for more people to enjoy trees, woods and forests
- helping communities to benefit from woods and forests.

To turn these aspirations into reality, the strategy identifies a range of priorities for action and adopts a goal of 25 per cent woodland cover by 2050. The development of a forest strategy specific to Scotland is widely welcomed, although the document has not been without its critics (e.g. Henderson, 2000). Arguably, in trying to please everyone it ducks the hard questions and misses the opportunity to give a clear strategic lead.

4.5 RECENT DEVELOPMENTS AND TRENDS

4.5.1 Changing priorities: multi-benefit forestry

One tree-life after the Acland Report, the context in which forestry operates has changed beyond recognition, both nationally and internationally. Much of this paradigm shift has occurred during the last fifteen years. Echoing the trend in Europe and North America, the 'industrial', productivist approach to forestry has now been replaced by a post-industrial ethos (McQuillan, 1993; Mather, 2001). Prior to the mid-1980s the primary aim of afforestation was timber production; other aspects such as landscape value, recreation, and habitat diversity were seen as valuable but subsidiary spin-offs. The Scottish forest industry

> . . . seemed to many outsiders still to be a great juggernaut rolling on oblivious to shrieks of public pain, fuelled by tax breaks and other forms of government assistance, hell-bent on softwood production at any cost, blanketing some of the best conservation areas in Western Europe. (Smout, 1999a: 66)

Since then, there has been a radical shift towards more environmentally friendly practices. Single objective management has been supplanted by multi-purpose, multi-benefit, sustainable forestry, phrases which became the buzz words of the 1990s. Starting in the 1980s, the FC published a series of environmental guidelines,[2] and the private sector organisations began to encourage forest owners to adopt more environmentally sensitive practices. In 1985, an amendment to the Wildlife and Countryside Act required a 'reasonable balance' to be struck between management for timber on the one hand, and conservation and landscape objectives on the other, just as the Agriculture Act imposed a similar balancing act on farming the following year. Timber production remained the key policy priority until 1991, but this is no longer forestry's primary rationale; the provision of environmental services is now paramount. Consequently, forests have increasingly become places of consumption (of amenity, recreation and wildlife observation) rather than of production (Mather, 2001).

More specifically, the new emphases that have come to the fore are these:

1. **Landscape aesthetics and amenity**. The FC felt able to declare in 1992 that 'the days of insensitive monoculture are over' (in Mackay, 1995: 76). As the principles of landscape architecture have been integrated into forest planning (Bell, 1999), there has been a shift away from dense, geometric, even-aged monocultures towards uneven age structures, intimate mixtures of tree species, and open spaces to create mosaics of colour and form. Increasingly, even-aged forests are being restructured through premature fellings to create a multiple age structure. One expression of the trend towards minimising the negative impacts of forestry is the current interest in continuous cover silvicultural systems (Yorke, 1995, 2001; Helliwell, 1999), building on the long-term experiments in irregularly structured forestry at Glentress in the Borders (Whitney-McIver *et al.*, 1992).
2. **Broadleaved and native species**. Exotic conifers dominated almost totally until the 1980s, with broadleaved species being used merely for marginal decoration, if at all. However, the value of broadleaved species was by then widely recognised (Evans, 1984), leading to the introduction in 1985 of grants to encourage broadleaved planting. By the late 1990s, broadleaved species comprised 17 per cent of the woodland area, up from 9 per cent in 1980 (Smith, 1999). Broadleaved species are often championed for their amenity and conservation value but they are also important for their timber value. The frequent popular association of conifers with timber production and broadleaves with environmental benefits is misleadingly simplistic. The pendulum has also swung decisively in favour of native species (Section 10.2; Fig. 4.6). In the 1960s, native hardwoods were routinely felled and replaced (or underplanted) with spruce, but now, in places such as Glen Affric, softwoods are being prematurely felled to make way for Caledonian Pine.[3]
3. **Biodiversity and conservation**. Conservation objectives now carry considerable weight in forest management, a high priority attaching to management for biodiversity (Newton and Humphrey, 1997; Kerr, 1999; Kirby, 1999; Helliwell, 2000). Forest Nature Reserves were inititiated in 1988; there are now six, covering 3,700ha, together with eighteen Caledonian Forest Reserves with a total area of 15,974ha (FC, 2000a). The Biodiversity Research Programme was launched in 1994 (Hodge *et al.*, 1998), and the FC spends around £4 million each year on conservation and recreation in Scotland, including its work on Habitat Action

FIGURE 4.6 *Native oakwood in Taynish National Nature Reserve, Argyll. Photo: © Lorne Gill/SNH.*

Plans (FC, 1999a). Associated with this trend is the introduction of minimum intervention as a management choice, and a move to mimic natural forest ecosystems by, for example, allowing dead and fallen trees to remain (Ratcliffe, 1995a). One approach to the conservation and enhancement of biodiversity is the development of Forest Habitat Networks (Peterken *et al.*, 1995; Hampson and Peterken, 1998). This aims to create the coherence of a large forest area by using patches of interconnected woodland interspersed with open spaces.

4. **Public access and recreation.** The FC remains the largest provider of outdoor recreation in Britain (FC, 1999a), providing facilities in many of its woodlands and managing the five National Forest Parks, but grant-aid has been encouraging private owners too to improve the amenity value of their woodlands and open them to the public (Table 4.2). Under the new access legislation, the public will have a right of access to all woodlands (Section 9.4.1).
5. **Integrating forestry with other land uses.** Having long been squeezed out and up onto marginal land, forestry is now being actively encouraged to come 'down the hill' to help take farming land out of production (Section 11.2.2). There are also bold visions for creating county-scale, open woodlands integrated into the countryside, such as the forests of Mar and Strathspey proposed by the Cairngorms Working Party (CWP, 1993) (Section 11.3).

The adoption of these new emphases has led to dramatic changes in the style and nature of afforestation and forest management. In the 1970s and 1980s new planting averaged around 20,000ha per annum, with exotic conifers comprising some 98 per cent; planting rates in the late 1990s averaged 11,500ha per annum, 70 per cent of it consisting of native Caledonian pinewoods and broadleaves such as oak, ash, birch and willow (FC, 1999a). The 30 per cent of planting that is coniferous and intended for timber production is established within a multi-benefit framework. Forest operations have evolved as well. Establishment by natural regeneration rather than planting is now encouraged, reducing the amount of mechanical ground preparation required. The impacts of large scale clearfelling and mechanical extraction on landscape and water quality have led to a shift (where possible) towards smaller felling coupes which are sensitive to the topography. The use of log chutes and horses for timber extraction is also making a small-scale comeback (Gibson and Warren, 1997; Abbott, 1999).

From a purely commercial stance, almost all these changes are entirely negative, at all stages of the rotation, involving much greater challenges for forest managers (not least in managing deer-related problems (Section 11.1)). It is a bit like asking farmers to plant their fields with mixtures of crops in small, irregular patterns, to leave areas unplanted, and to harvest each crop separately. There is no doubt that the landscape and the environment gain enormously from this new approach, but it makes it far harder to achieve commercial viability. Society reaps the benefits of this commercial compromise, but most or all of the financial cost is borne by the growers (Yull, 1998). Notionally, forests have a very high non-market value (Henderson-Howat, 1993; Roper and Park, 1999), but this cannot usually be realised by the owner. Jean Balfour (1998:66) laments the fact that 'woods which will in future produce timber have become the "poor relation"', and Winter (1996) notes the irony that the forestry lobby may increasingly have to fight to be allowed to cut down the trees that it had to battle so hard to plant.

On the other hand, if public money is used to support forestry, public preferences must be allowed to shape forest policy. Kirby (1999) points out that although management for multiple objectives increases costs, it can also bring in new resources, both of money and of public support. The move away from pure production forestry has seen the fixation with tree quantity replaced with an emphasis on timber quality, a move which will – in the long term – pay dividends. Magnusson (1995a:3) observes that 'we no longer regard it as proper to turn our landscapes into forestry factories, dedicated to material profit . . . for the discardable products of a throw-away society'. But, he goes on, 'that does not mean that there can be no place for Sitka spruce'. The challenge of striking an appropriate balance is discussed in Section 4.6.2.

4.5.2 Indicative Forest Strategies

Prior to the 1990s, afforestation proposals were vetted individually through a set of consultation procedures but there was no strategic direction at regional or national level to guide the location of forest developments. Dissatisfaction with the control procedures grew throughout the 1980s (Bainbridge *et al.*, 1987; Mackay, 1995). As afforestation proceeded apace in the 1970s and 1980s, it became increasingly clear that the case by case approach was entirely inadequate for controlling its *cumulative* impact in a region. This was especially apparent in the Flow Country of northern

Scotland (Warren, 2000a). The Indicative Forest Strategy (IFS) approach was adopted to address these problems. In contrast to the various existing control mechanisms which were all reactive and focused on impacts, IFSs are strategic, proactive and plan-led (Goodstadt, 1996), classifying regions into zones which are 'preferred', 'potential', or 'sensitive' for forestry.

The IFS approach has been 'hailed as a major advance in land use planning' (Mackay, 1994: 129). It is seen as a key element in providing the wider environmental benefits that forestry is now expected to deliver (Goodstadt, 1996) and has played an important part in getting the right trees in the right places (SODD, 1999c). IFSs have not been above criticism, however (Stuart-Murray, 1994). In the Borders, for example, forestry has actually increased in the 'sensitive' areas and decreased in the 'preferred' and 'potential' areas; the latter are too expensive given the low returns offered by forestry (Stuart-Murray *et al.*, 1999). Macmillan (1993) complains that IFSs fail to identify enough suitable land for the forestry expansion which government policy aims for. Critics also ask whether they will be able to cope with the land use consequences of future changes to the CAP (Stuart-Murray, 1999). Like all 'top down' mechanisms, they need to be complemented with the 'bottom up' involvement of local communities.

Because IFSs emerged out of the Flow Country controversy, they were initially a tool for conflict resolution, focusing on the constraints on forestry development (SDD, 1990). The transformations in forest practice during the 1990s have now allowed IFSs to be reshaped into a more positive planning tool, emphasising the benefits that woodlands offer and helping to identify where new woodlands might best be accommodated (SODD, 1999c). Despite this shift, they may still identify areas where no woodland expansion should occur, for example, where considerations of conservation, agriculture, water quality or cultural heritage take precedence.

4.5.3 Community woodlands

The concept of community woodlands emerged in the late 1980s. As originally envisaged, it involved the establishment of a range of multi-purpose woodlands within or near towns, primarily consisting of broadleaves, and including large areas of open land and other land uses. Such woods have a variety of aims but emphasise public access (both for recreation and education), amenity, and the creation of wildlife habitats (Jones, 1999a). In the early years there was much enthusiasm from local authorities, but the lack of tangible progress led to pessimistic prognoses (Crabtree *et al.*, 1994a). Momentum has been gathering recently, however, reflected in a rapid rise in the total area of community woodlands. This has been facilitated by the availability of financial help; FC support under the Community Woodland Supplement (CWS) of the WGS has been running at about £100,000 per annum (FC, 2000a), and funding is also available from SNH's Community Grant Scheme and the Lottery's New Opportunities Fund.

The community woodland vision has broadened and diversified considerably since the early days. At one end of the spectrum are woodlands that are planted and/or managed by an agency or a partnership for the benefit of the community at large. The Central Scotland Forest initiative was the first and largest example. Started in 1989, its aim is to double the forest cover between Glasgow and Edinburgh by planting 17,000ha by 2015 (Edwards and Gemmel, 1999; Evans, 1999a); 4,840ha had been planted by 1999. At the other end of the spectrum are woodlands which are actually managed or owned by local community bodies themselves. Over 100 such groups now exist, together managing almost 23,000ha across Scotland (Gauld, 2001). Examples include the Laggan Forest Partnership, Abriachan Community Forest and the Sunart Oakwood Project (Tylden-Wright, 2000; McGillivary and McIntyre, 2001; McIntyre, 2001). A Community Woodland Association was formed in 2001 to represent the interests of this rapidly expanding movement. A separate category comprises those woodlands (often privately owned) which have received the CWS; these provide access benefits to locals but seldom involve the community in management. The FC has now recognised the importance of local communities within the Scottish Forestry Strategy (Scottish Executive, 2000b) and is actively encouraging local involvement in the management of its forests.

4.5.4 An Accord, a Standard, and Certification

Three developments at UK level during the late 1990s codified the trend towards sustainable, high quality forest management:

1. **The UK Forestry Accord.** This comprises a set of principles for sustainable forest management. It was signed in 1996 after negotiations between forest and environmental interest groups.
2. **The UK Forestry Standard.** This draws together into one document the existing criteria and standards for the sustainable management of UK woodlands (FC, 1998a). In effect, it represents a restatement and a strengthening of existing policies, linking in with international protocols for sustainable forestry.
3. **Certification: the UK Woodland Assurance Scheme.** The UKWAS was agreed in 1999 by woodland owners, processors and environmentalists after tough negotiations. It assures customers that wood products come from certified forests which are sustainably managed according to certain set criteria. Its creation is a world first, since it is the only national standard for sustainable forest management which has the support of all interested parties (Goodall, 2000). Under the certification process, products from such woodlands qualify for a Forest Stewardship Council label, which is the most widely recognised international standard. The aim is to encourage sustainable forest management and give owners access to premium markets, as well as providing consumers with an assurance of responsible management. One of the biggest challenges is that of making the scheme accessible to the owners of small woodlands for whom the costs of participation could prove prohibitive. In the view of the FC's Director-General, the signing of the UKWAS was a supreme test of the forest industry's maturity, requiring recognition of international pressures and of the public's preference for high quality sustainable forest management (Bills, 1999).

4.5.5 Recent initiatives

Amongst the many recent forestry initiatives, three are especially noteworthy:

1. **The Millennium Forest for Scotland Trust.** The bold aims of this scheme, initiated in 1995, were to create 100,000ha of native woodland, and enhance 50,000ha of existing woodland, giving priority to native species, natural regeneration, public access and community involvement. In the event, the Millennium Commission of the Lottery Fund only awarded the MFST £11.3 million of the £50 million applied for, so its ambitions were necessarily scaled back. It has nevertheless supported over seventy woodland projects, including the RSFS's 'Forest for a Thousand Years' at Cashel by Loch Lomond and Trees for Life's restoration work in Glen Moriston and Glen Affric (McGillivary and McIntyre, 2001). By 2001, the woodland projects had established 8,800ha of new woodland (60 per cent by natural regeneration), brought 9,500ha of existing woodland into management, and involved thousands of people throughout Scotland (Hunt, 2000, 2001a). Given the notable successes scored by the MFST, it seems a great shame that it cannot be continued beyond 2005 due to the nature of lottery funding.
2. **Regional partnerships**. Partnership working has been a striking feature of MFST projects, and a variety of regional partnerships with diverse objectives have sprung up all over Scotland in recent years, including Highland Birchwoods, Tayside Native Woodlands, and the Borders Forest Trust. Other examples of creative partnerships include the agreement signed in 1998 between Forest Enterprise and the people of Laggan which facilitates public-private participation in managing Strathmashie Forest (Tylden-Wright, 2000), and the Forests and People in Rural Areas Initiative which focuses on ways of satisfying local needs through forestry-related activities (FAPIRA, 1995).
3. **Forests for energy**. Several pilot schemes exist around the UK in which willow and poplar grown as short rotation coppice, together with forest residues, provide fuel for electricity generation (Coates, 1999). Some of the improved willow varieties can be harvested on a three or four year cycle (Dawson, 1999), and the predicted future surplus of conifer fibre in the UK would be sufficient to support a major development in biomass energy production (Jaakko Pöyry, 1998). This is an emerging technology and unproven economically, but given its green credentials and the government's Non Fossil Fuel Obligation, energy crops could become a significant new dimension of Scottish forestry. The first sign of this was the approval in 1999 of a 12MW biofuel power station to be built near Inverness. This trend could mean that coppicing, an ancient method of woodland management that has dwindled almost to nothing in Scotland, could be given a new lease of life with contemporary objectives (if, that is, coppice woodlands can be protected from deer damage (Putman and Moore, 1998; Section 11.1.1)).

As always, the bottom line for these and other initiatives is funding. Until the mid-1990s the FC was the only significant source of money to support woodland projects, but a raft of other options have become available. These include EU LIFE funds, EU structural funds, the Heritage Lottery Fund and the Landfill Tax. UK-wide, funding from such sources may now exceed that from traditional forestry grants (Kirby, 1999).

4.6 Current issues and debates

4.6.1 Recycling: how green?

Following the Earth Summit in 1992, and the 1994 EU Directive on waste management which set a 50 per cent recycling target for paper waste by 2001, the pulp and paper industry in the UK has invested heavily in improving environmental safety and in paper recycling (FC, 1998a). Landfill and collection taxes have also promoted the use of recycled fibre. In the UK, more wood products are now 'produced' from recycling than from trees, supplying 18 per cent of domestic demand against timber's 15 per cent (Tickell, 1996). Recycled fibre and processing residues are an under-exploited resource. Their use is likely to increase by 2–3 million m^3 over the next twenty years, with Scotland well placed to benefit (Jaakko Pöyry, 1998; Selmes, 1999). Given the controversial nature of new forestry planting, would it perhaps be more sensible to subsidise recycling than new plantations? This question is not easily answered because it depends on the spatial and temporal range of the factors that are incorporated. For example, Life-cycle Analysis (LcA), an approach which assesses environmental impact from 'cradle to grave',[4] suggests that high levels of recycling could actually increase levels of carbon dioxide and sulphur dioxide in the atmosphere (Collins, 1996). Increased recycling reduces the demand for new wood, compromising the profitability of forestry and leading to falling standards of forest management, hardly a desirable outcome (Pearce, 1997).

Leach *et al.* (1997) also use LcA to criticise the accepted view that recycling is the most beneficial option environmentally. They argue that some types of waste paper can be re-used in a more environmentally friendly and economic fashion in energy production (incineration) rather than in pulp, given that the latter requires de-inking and subsequent effluent disposal. That this iconoclastic conclusion has, in turn, been trenchantly criticised by FoE (MacGuire and Childs, 1998) emphasises how complex and emotive this issue is. What does seem likely is that, as a result of LcA, the value of wood relative to steel, concrete and plastics will increase, as it has already in parts of Europe (HGTAC, 1998). At present, European and UK policies are firmly in favour of recycling, but in the case of paper at least (a product derived from a renewable resource), it seems that this policy bias may need critical re-examination (Collins, 1996; Pearce, 1997). Certainly the balance that is adopted between recycling of wood products and the production of new wood has far-reaching economic and environmental knock-ons.

4.6.2 Forest policy choices: how much forestry, where, and what kind?

These three questions dominate today's debates about forestry, and have done for many years (Cunningham, 1991). They need to be answered if Scottish forest policy is to have a clear and coherent direction during its second century, something which many felt was lacking at the end of its first (e.g. Lister-Kaye, 1998). But answers are not easily found, especially without clarifying what the ultimate objective of Scottish forest policy should be. Is the currently wooded area of 17 per cent enough? Should it be doubled, or is the Scottish Forestry Strategy's target of 25 per cent adequate? Why, where and how should aspirations for increased woodland cover be realised? Since 1991, forest policy in the UK and Scotland has aimed for 'the sustainable management of existing woods and forests, and a steady expansion of tree cover to increase the many diverse benefits that forests provide' (Anon., 1994a: 7; Scottish Office, 1995). It thus endeavours to combine socio-environmental aims with further afforestation for timber production. Steering a course between commercial 'tree farming' and an exclusively green focus seems eminently sensible, and the Scottish Forestry Strategy endeavours to do just that. But are economic and environmental sustainability compatible? Can it be 'both and', or is it, in reality, 'either or'?

On the one hand, political, social and environmental pressures are requiring forestry to conform to the new emphases (Section 4.5.1), most of which compromise commercial viability. Pure production forestry that makes no concessions to broader objectives is no longer acceptable to the public. On the other hand, it would be folly for forest policy to lose sight of market realities. For one thing, commercial activities underpin the non-market benefits of forestry (Watt, 1999). Gale (1998: 24) stresses the point that 'without a commercially viable wood-using industry, none of the multi-benefits of forestry as a long-term land use can be supported nor sustained'. As the saying goes, 'the forest that pays is the forest that stays'. Secondly, if the forest industry is to survive in an extremely competitive international market it cannot afford to sacrifice economics on the multi-benefit altar. This applies especially to the forest products sector which has invested heavily in Scotland and employs almost 3,000 people, more than the FC's workforce of 2,000 (FC, 2001). The collapse in log prices by almost 50 per cent since the mid-1990s to their lowest recorded level makes a productive balance between market and non-market objectives even more elusive. Within the forestry world, the most widely shared view in both the public and private sectors is that the best balance should comprise production forestry with a strong emphasis on other benefits (Bills, 1999; Bloomfield, 1999), a combination dubbed 'consensus forestry' by the Duke of Buccleuch (1996). But this apparently sensible compromise still leaves key questions unanswered.

4.6.2.i How much forestry?

There is 'a broad consensus that the 'right' balance between forestry and other land uses in Scotland would include more trees' (Henderson-Howat, 1996: 67). In a recent survey, 77 per cent of the population said that they wanted more woodland near them, with large minorities wanting Scotland's woodland area to be doubled or increased by half, and only 3 per cent wanting less (FC, 1999c). The RSFS (1993) and FoES (1996) advocate woodland cover of 25 per cent and 30–50 per cent respectively; FoES calculate that 28 per cent cover would make Scotland self-sufficient in wood products. A dissident note is sounded by Smout (1999a) who doubts whether the aim of doubling forest cover is either feasible or rational.

The reality is that market-orientated planting rates have collapsed in recent years because grant levels are inadequate to stimulate enough private sector planting, and state planting has been wound down (Fig. 4.4). The timber processing industry, foreseeing a deficit opening up after 2025, is agitating for a reversal of this trend. To avoid future shortfalls in timber supply, some 17,000ha needs to be planted annually over the next twenty-five years, 12,000ha of which needs to be production-orientated (J. Balfour, 1998). This stands in stark contrast to annual rates of new conifer planting of around 5,500ha in the late 1990s (Scottish Executive, 2000b). Accordingly, the SLF (1999) argues that planting rates of market-orientated woodlands by the private sector should be increased to at least 10,000ha per annum.

For anything approaching such rates to be realised, there would need to be a combination of substantial increases in forestry grants and profound reform of the CAP. Neither seem politically likely in the near future. Moreover, the competitive advantage that Scotland has enjoyed in having some of the fastest tree growth rates in Europe has been steadily diluted by the globalisation of the forest industry (Mather, 1994). Conifers may reach a harvestable size nearly three times faster in Scotland than in Scandinavia, but in countries such as Chile, Brazil and New Zealand rotations are even shorter and operational costs are lower, so Scottish growers struggle to compete (Watt, 1999). There is also the public's rejection of 'industrial' forestry to contend with. There are thus strong economic and political forces which make any return to large scale commercial afforestation unlikely in the foreseeable future.

Nevertheless, certain factors point towards a continuation of the twentieth-century trend of forest expansion, both in Scotland and across Europe. Chief amongst these are the release of land from agriculture as a result of CAP reform, and the fact that the EU is a net importer of forest products. Except in the uplands, expansion of forest cover has not been a feasible policy priority during the era of massive CAP subsidies because the resulting distortion of land values has made farming a more lucrative option (Goodstadt, 1996). Any real reform of the CAP which resulted in a more level playing field would inevitably benefit forestry. Smout (1999a), however, points out that if present agricultural trends continue, woodland of a rather different kind could spread rapidly by natural regeneration on abandoned farmland; a drive through the eastern USA shows how extensively this can happen. One way or another, the woodland area is set to continue expanding but not in the massive leaps seen in the second half of the last century.

4.6.2.ii Where should new forests be located?

Substantial expansion of the forested area would bring Scotland much more in line with European norms, and would restore woodland to a deforested land, but it is hard to see where it would all go. Zonation is one obvious solution, concentrating commercial forestry in certain extensive core areas. Such 'forest regions' were recommended by the FC's 1943 report *Post War Forest Policy* which suggested the Moray Firth and the Borders as candidate areas (Mackay, 1995). Wightman (1992), too, recommends a tripartite zonation: natural (unmanaged) forests; extensive forests with the stress on natural regeneration and community ownership; and intensive, commercial forests. Today, large scale zonation for commercial afforestation would probably suffer from NIMBY-syndrome opposition and, depending on the areas selected, from environmental protest. A more acceptable vision is smaller scale integration of forestry within the countryside, an approach which would tie in with the government's desire to see forestry playing a 'prominent but sensitive role' within a mosaic of rural land uses (SODD, 1999c: 6). However, this would not satisfy the market needs of the forest products sector in the long term. It might not be best ecologically either. Arguing that a few large forests would be superior to a scattering of woodland patches, Peterken (1996) advocates expansion of existing forest districts in preference to the creation of new ones.

One much-touted avenue for future expansion is the use of agricultural land, either as farm woodlands or through agroforestry (Section 11.2). The SLF (1999), for example, recommends that the Farm Woodland Premium Scheme should become the main element for promoting private forest expansion, with a shift of focus away from better quality farmland towards Less Favoured Areas. If 'agriforestation' were actually to happen on a significant scale, much of the conflict between forestry, on the one hand, and upland conservationists and archaeologists on the other might disappear (Smout, 1999a). Further potential for expansion involves woods in and around towns. HGTAC (1998) recommends 100,000ha of new urban woodland in the UK over the next twenty years, and the recognition of the importance of such woodlands in Scotland has led to numerous initiatives (c.f. Wilson, 1999).

4.6.2.iii What kind of forests for the future?

What types and structures of forests should multi-purpose forest management be aiming for? Given the strong preferences amongst the Scottish public for mixed and broadleaved woodland (including Scots pine as an 'honorary broadleaf'), for ancient and native woodlands, and for non-market objectives, public money is likely to become progressively more targeted towards these aims (FC, 1999c; Kirby, 1999). It can be argued (as above) that the pendulum has already swung too far from market to non-market objectives. If so, further targeting of the latter could fatally undermine the forest industry. From this, and the foregoing, it seems that the long-term fears of the forest products sector are far from groundless. It looks as if the space for purely commercial forests will remain rather restricted.

The woodland vision which probably has the strongest public support is that of restoring and extending native woodlands (Section 10.2.1). Alongside this widespread enthusiasm, there has been increasing interest in recreating the many natural forest environments which are now 'extinct' or rare in Scotland. These include:

- Treeline forests or montane scrub (Scott, 2000). The present treeline in much of Scotland is well below the biological limit. The dwarf pine/juniper woodland at almost 650m on Creag Fhiaclach in the Cairngorms is one of the few treeline woods which remain (Hale *et al.*, 1998; Fig. 4.7). Grant aid is now available to encourage extension of woodland cover 'up the hill' in this novel sense (Dunsmore, 1998), and some high altitude colonisation is occurring (Ramsay, 1996; Bayfield *et al.*, 1998).
- Riparian and floodplain woodlands, and other wetland forest environments such as muskeg (Gilvear *et al.*, 1995; Peterken, 1996; SNW, 2000). Extending and diversifying riparian woodlands could enhance the conservation and amenity value of many river corridors and confer a wide range of other benefits (Brogan and Soulsby, 1996). Since 1998 the EU-funded Wet Woods LIFE Project has been restoring bog woodlands and alluvial forests.
- Old growth climax forests (Lister-Kaye, 1995).
- Pasture woodlands or woodland pastures (depending on the balance between trees and open space) (Begg and Watson, 2000; Quelch, 2000, 2001).
- True wildwood (Featherstone, 1997; Ashmole, 2000).

These kinds of woodland, together with more familiar types, could be linked together to form forest networks, an approach which would reverse the historic trend of increasing fragmentation and isolation of native woodlands (Hampson and Peterken, 1998). Exciting though such visions are, getting from here to there necessarily involves

FIGURE 4.7 *The treeline woodland at Creag Fhiaclach in the Cairngorms. Photo: © Lorne Gill/ SNH.*

working with what exists, and the present forest resource is dominantly young, even-aged, coniferous and exotic. The scope for substantial short-term structural change may therefore be limited, but if specific, medium term objectives can be agreed upon, forest managers can work towards them. Over the next twenty-five years half Scotland's forests are scheduled for harvesting and replanting, offering an opportunity for extensive transformation (Scottish Executive, 2000b). Even today, with the widespread use of native species, there is a compartmentalised approach to management which lacks any vision of how native and non-native woodlands can be integrated within the landscape (Mason *et al.*, 1999). It is important to get away from this black-and-white legacy, and instead aim for an integrated continuum of woodland types. All kinds of woodlands can produce timber, and all can deliver multiple benefits.

Forest planning, management and research are all now focusing on how to maximise the non-market benefits of forests without compromising wood supply requirements. As Hodge *et al.* (1998) conclude, clear priority setting is vital if the competing demands are to be resolved and if the maximum environmental benefits are to be gained per unit sacrifice in wood production. If twentieth-century forest policy aimed for (and achieved) quantity, quality looks set to be the watchword in the twenty-first.

NOTES

1. The origins and development of the FC are discussed by Mutch (1994), Mackay (1995) and Pringle (1995). Aldhous (1997) sets out the achievements of the first seventy years of British forestry. Useful reviews focusing on the Scottish dimension include Anderson (1967), Malcolm (1991), Mather (1993b) and D. G. Mackay (1994, 1995). This section draws extensively on these commentaries.
2. The FC's 'Guidelines' series covers forests and water; forests and soils; landscape design for forests, community woodlands and lowland landscapes; forest nature conservation; forest recreation; and forests and archaeology.
3. Native semi-natural woodlands, and the issues surrounding them, are discussed in detail in Section 10.2.
4. For an introductory discussion of Life-cycle Analysis, see Mitchell (1997) pp.123ff.

CHAPTER FIVE

The fields: agriculture and crofting

Agriculture cannot fail to have an impact on the environment because in fulfilling its fundamental objective – the production of food and other commodities – it must use and modify the natural world. But the idea that production is farmers' primary *raison d'être* may not survive very far into the twenty-first century because the future shape and purpose of Scottish agriculture is hanging in the balance. Farming is fast losing the pre-eminent position in the countryside and in rural policy that it has held for so long, but it is far from clear how it will evolve. The way in which the current debates are resolved will have a profound effect on Scotland's natural heritage (Thomson, 1993a), not to mention on farmers and rural communities, simply because over three-quarters of the land area is at present farmed in some way.

Scotland is not well endowed with good agricultural land, 69 per cent of the agricultural area consisting of rough grazing and just 11 per cent of arable (SNH, 1995a). Very little of the country is classed as top quality agricultural land, and those small areas are concentrated in the east (Fig. 5.1). Elsewhere, agriculture is strongly constrained by climatic and topographic factors (Robinson, 1994a), the land largely consisting of inhospitable or impossible farming country where the growing season is short and the soils are poor, leached by high rainfall. The regional distribution of activities is largely dictated by this natural endowment, the productive arable areas confined to the dry eastern lowlands, dairying focused in the south-west where high rainfall promotes good grass growth, and sheep farming common in the north and west on the less productive grasslands (Figs 5.2 & 5.3). Overall, livestock systems dominate, especially in the uplands, both in area and output terms. They produce 60 per cent of total output compared to 29 per cent from crops (SEERAD, 2001a).

Notwithstanding these challenging natural circumstances, Scottish agriculture utilises no less than 6.1 million ha, produces output worth £2 billion each year (1.3 per cent of GDP), and directly employs 69,525 people (including working occupiers, working spouses, and full- and part-time employees) (SEERAD, 2001b; UoA/MLURI, 2001). Although this is barely 3 per cent of the workforce, in rural Scotland as a whole this figure rises to 8 per cent and much more in some farming areas (32 per cent in upland Grampian, for example (Shucksmith, 1992)). Farming remains the single biggest reason why people can continue to live in Scotland's more remote communities (SERAD, 2000b).[1]

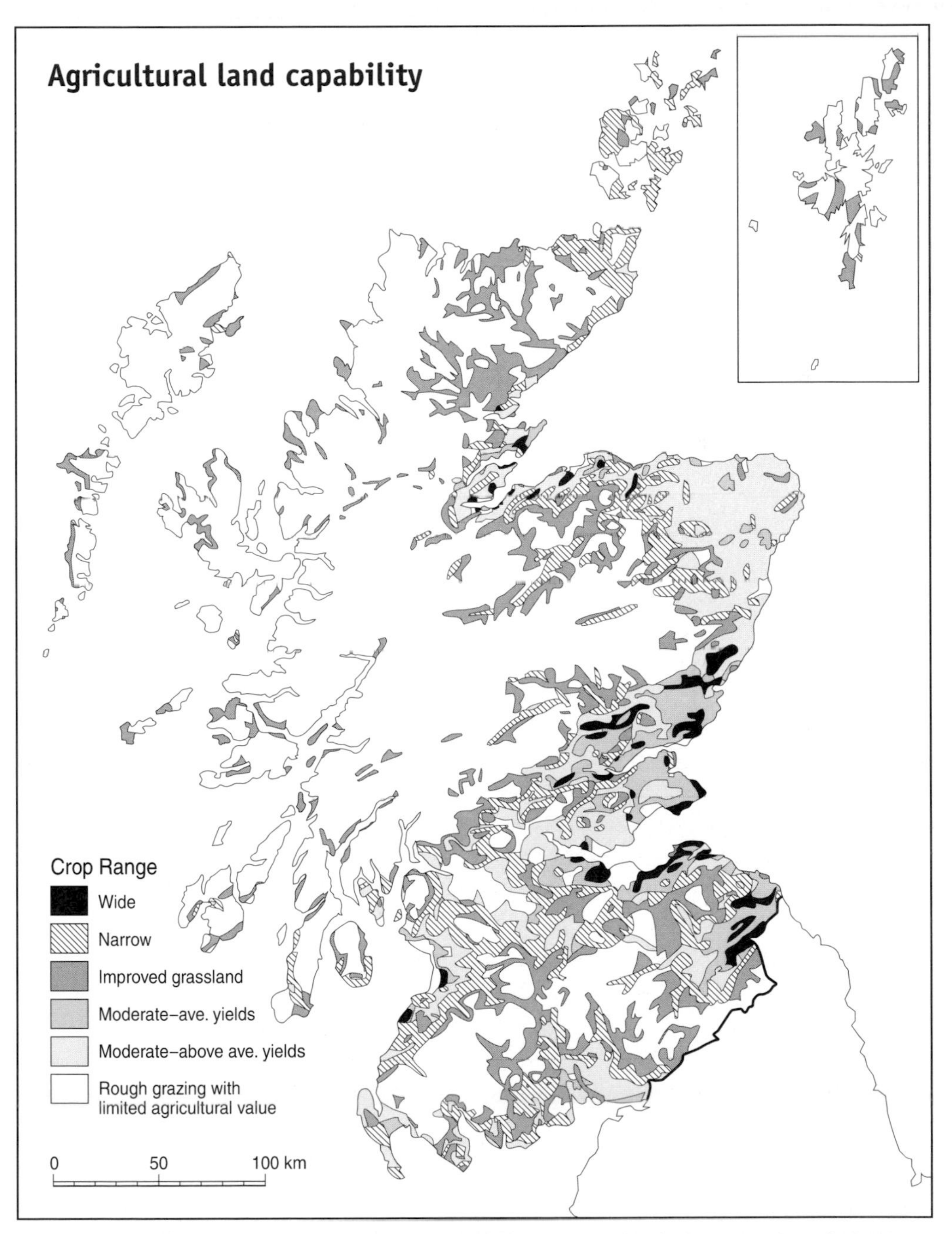

FIGURE 5.1 *The capability of farmland in Scotland. Reproduced with the permission of MLURI.*

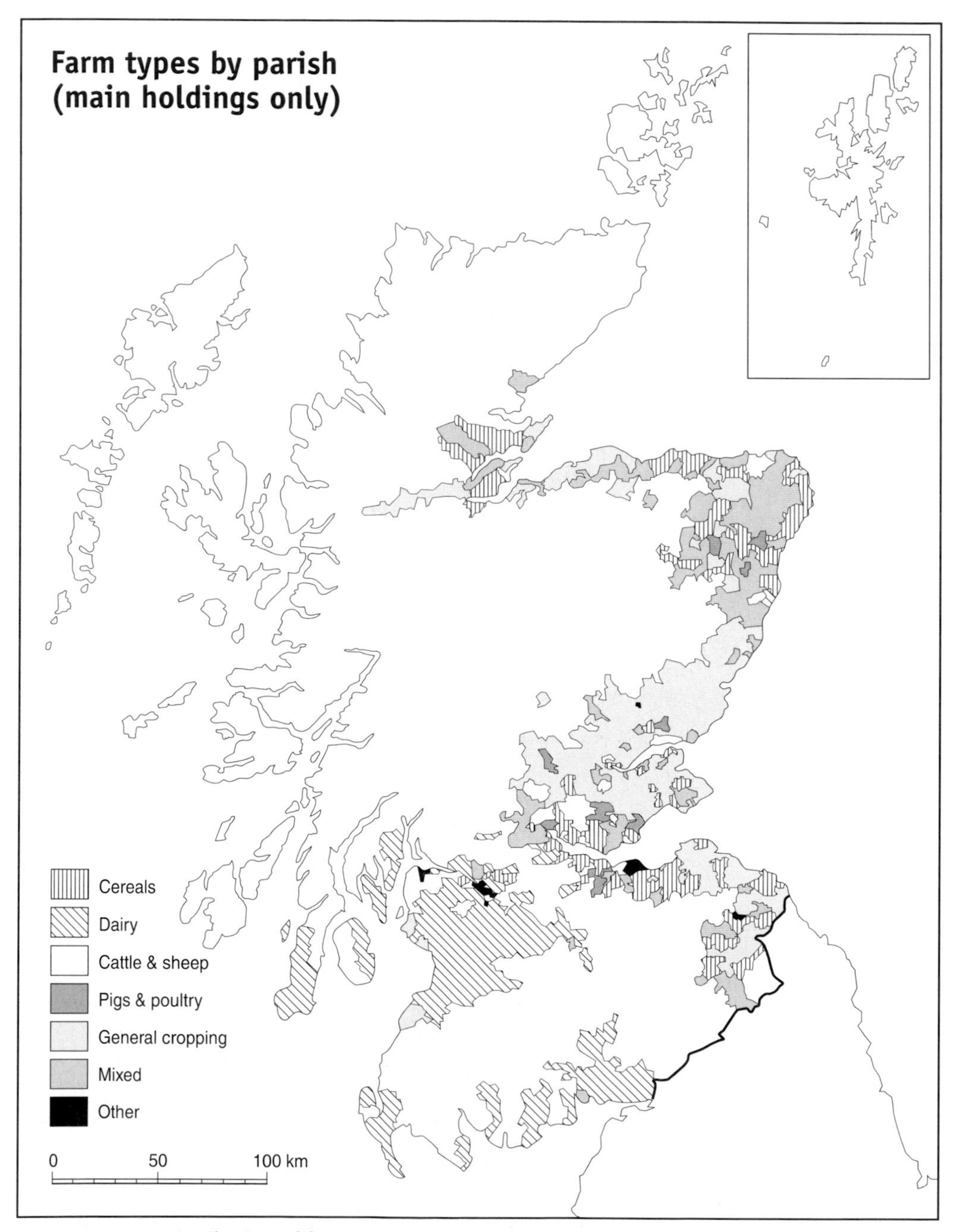

FIGURE 5.2 *Distribution of farm types. After SERAD (2000c), reproduced with the permission of SEERAD.*

5.1 HISTORICAL BACKGROUND

Despite its largely unpromising nature, the land has reluctantly supported livestock and yielded crops throughout the millennia of human occupation. For much of the 6,000 years since agriculture was introduced to Scotland (Ritchie and Ritchie, 1997) it consisted of local subsistence farming based on the runrig (ridge and furrow) system. From the eighteenth-century 'agricultural revolution' onwards, this was progressively replaced with more commercial operations, especially in the east. There was extensive enclosure of fields, and the size of land holdings grew. The evolution of Scottish farming from that time onwards, and the forces driving it, have been widely discussed.[2]

Throughout the twentieth century until the late 1970s, British agricultural policy was singularly focused: politicians who had endured rationing during two world wars aimed for self-sufficiency in food production (Waters, 1994). This single-minded drive to 'produce for the nation' led to increasingly intensive farming practices, mechanisation, huge increases in productivity, and the 'improvement' or reclamation of large areas of marginal land, all generously supported from the public purse. As a result, Scottish agriculture and farmland has changed dramatically during the last fifty years (Egdell, 1999), a period described by Allcock and Buchanan (1994: 365) as 'the agricultural equivalent of the original industrial revolution'. Judged by its stated objectives the expansionist policy was hugely successful, especially in terms of improved productivity and crop yields. Scottish agriculture is now amongst the most efficient in Europe and has the highest average farm size, almost ten times higher than the EU average (McCarthy, 1998; A. Raven, 1999). Managing to increase agricultural production while forestry was simultaneously trebling in area was a remarkable achievement (Mather, 1996a). Socially, however, the policy resulted in a steady (and on-going) fall in agricultural employment at around 2 per cent per annum from 1945 (Inskipp, 1997), while the unforeseen environmental implications turned out to be largely negative (Section 5.2.2). Recently, a dualistic farm structure has been emerging (UoA/MLURI, 2001), the large commercial farms standing in marked contrast to smaller farms associated with pluriactivity (Section 5.3.3.i).

The UK's accession to the then EEC in 1972 was a watershed for Scottish agriculture. Membership brought farmers under the aegis of the Common Agricultural Policy (CAP) and had an instant effect on Britain's international trade obligations; a preference for the Commonwealth was replaced by the obligations associated with membership. Since then, Scottish agriculture has been dominated by the CAP, becoming largely dependent on EU payments. In fact, Scottish agriculture has only remained economically viable through the combination of direct subsidies to farmers and market price support. Many sectors, notably hill sheep farming, are now critically dependent on subsidies. Some 84 per cent of agricultural land in Scotland qualifies for extra support, being classified by the EU as a Less Favoured Area (LFA). The agricultural economy has thus become distorted and entirely artificial, the plaything of politicians.

FIGURE 5.3 *Four aspects of Scottish agriculture:*
a. *Harvest time in the lowlands: rolling arable land near Cupar, Fife. Photo: © the author.* (above)
b. *Hill sheep farming, Glen Garry. Photo: © Lorne Gill/SNH.* (below)
c. *Dairy farming in Dumfries. Photo: © Alan Aitchison/SNH.* (opposite top)
d. *A Highland cow and calf on Rum. Photo: © the author.* (opposite below)

5.2 RECENT DEVELOPMENTS

5.2.1 Reform of the CAP

5.2.1.i The MacSharry reforms, 1992

By the late 1980s the CAP was spiralling out of control, creating huge food surpluses and dominating the EEC budget. Various measures taken during the 1980s had done little to improve matters making further reform unavoidable (Winter, 1996). Radical measures proposed by the Agriculture Commissioner Ray MacSharry were eventually pushed through in a much watered-down version in 1992. Despite public declarations at the time that environmental protection was to be made an integral part of the CAP, the reforms were not motivated primarily by environmental concerns (Egdell and Badger, 1996; Lowe and Baldock, 2000). They were driven by the need to control overproduction, especially in arable, beef and sheep, and by the stark inequities of CAP payments; 80 per cent of support was going to 20 per cent of farmers, mostly the larger arable farmers in the lowlands rather than the needier upland livestock farmers (Thomson, 1993b). The looming GATT talks, with their pressure for global free trade, were a further incentive for reform.

Three aspects of the MacSharry reforms stand out. Firstly, market intervention was scaled down and partly replaced with direct payments to farmers. This shift from product price support to area or headage (per animal) payments, combined with quotas, was intended to reduce the incentive for intensive farming. Secondly, a compulsory set-aside scheme and restrictions on stock densities were introduced. These required most arable producers to set aside 15 per cent of their cropped land as a condition for receiving payments, while livestock farmers had to comply with (rather weak) maximum stocking densities. These were 'finger in the dyke' measures to control production, not positive steps promoting either agricultural or natural heritage objectives (SNH, 1994a). Significantly, however, they introduced the concept of cross compliance, the 'strings attached' idea that those benefitting from the CAP should undertake positive environmental action in return (Waters, 1994). Thirdly, the Agri-Environment Programme (AEP) was introduced (Section 5.2.2). This brought together the disparate, voluntary agri-environmental measures into a more coherent, EU-wide, mandatory programme. It provided incentives for habitat creation, set-aside management, reduced stock densities, organic farming, and reduced use of agrochemicals. Despite its name, however, the AEP was never a purely environmental policy but was strongly motivated by socio-political considerations like the support of farmers' incomes and the control of over-production (Potter, 1998a).

The MacSharry reforms constituted a watershed, putting the CAP on the fraught path towards trade liberalisation. Ironically, however, despite their budget-cutting objectives, their net effect was to increase the cost of the CAP (Curry and Owen, 1996). Moreover, the reforms left the basic support mechanisms in place, including headage payments. There was thus no incentive for livestock farmers to reduce animal numbers, despite problems from overgrazing: more sheep equalled more money. This explains why sheep numbers in Scotland remained obstinately above nine million throughout the 1990s, even increasing somewhat after a low in 1996. Cattle numbers, too, remained steady just above two million. Since 1996 subsidies have exceeded farm income for all the main farm types (Fig. 5.4). This is a manifestly topsy-turvy situation. It will also be unsustainable in the medium term as the EU expands eastwards to incorporate nations whose agricultural sectors are far more needy than Scotland's. The MacSharry reforms, then, were clearly inadequate and did less for the environment than appearances suggested (Winter, 1996, 2000). At best, they were 'a faltering step in the right direction' (Whitby, 1996a: 237).

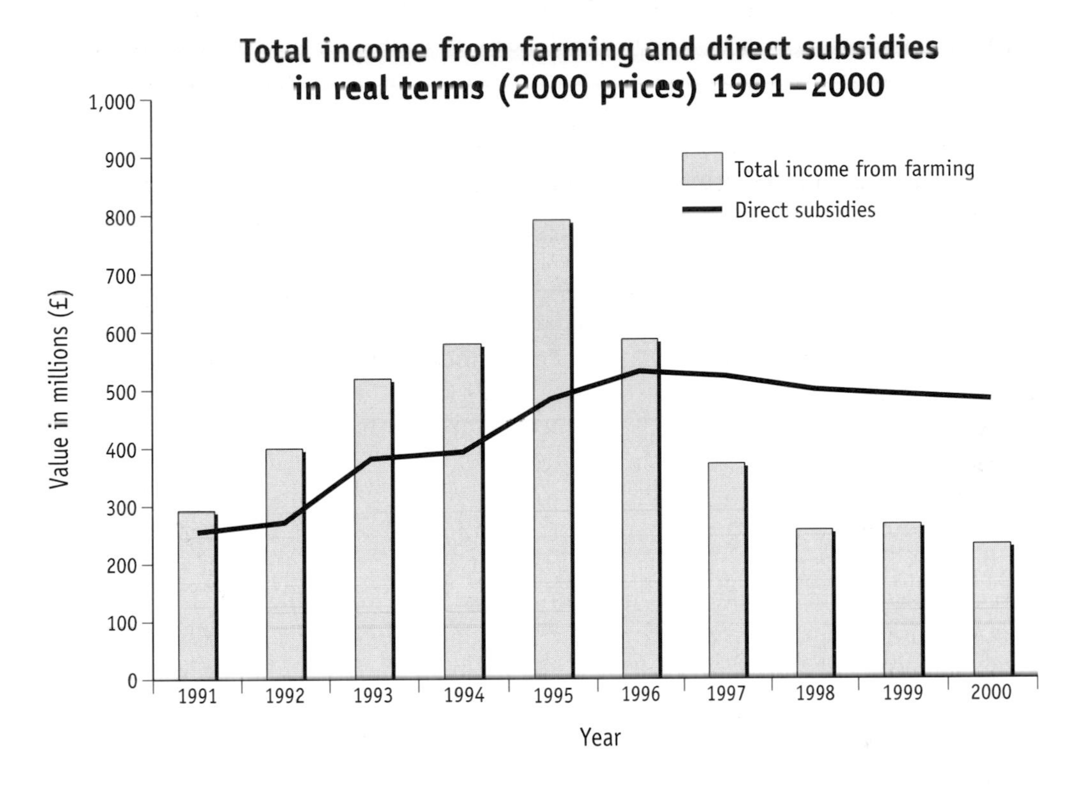

FIGURE 5.4 *Net farm incomes (the Total Income From Farming (TIFF) per farm unit in real terms) and direct subsidies per farm, 1991–2000. After SERAD (2000b), updated with data provided by SEERAD.*

5.2.1.ii Agenda 2000

The Agenda 2000 package constituted a major attempt at further overhaul of the CAP in the face of continuing budgetary pressures and a further round of negotiations of the World Trade Organisation (WTO) (GATT's successor). Bold intentions were again shipwrecked on political rocks. The package finally agreed in 1999 was a pale shadow of the 1997 proposals, much less radical than the UK and France (amongst others) had argued for. For example, while direct payments within LFAs changed from a headage to an area basis to tackle over-grazing, other livestock subsidies are still paid per animal. Reform of the dairy regime, whose intensive production contributes to nitrate problems and farm waste pollution, was postponed (Lowe and Brouwer, 2000). The package continues the direction of the 1992 reforms, reducing support prices but compensating farmers by increasing direct subsidies (Scottish Office, 1999a).

A significant innovation is the Rural Development Regulation (RDR) which integrates agri-environment and rural development programmes under one regulation and shifts support away from agricultural production (Rutherford and Hart, 2000). It aspires to be a 'second pillar' of the CAP (the first being commodity management).

One mechanism for achieving the RDR's wider remit is so-called 'modulation' which allows an increasing proportion of direct farm subsidies to be redirected towards non-agricultural activities, such as forestry, marketing and diversification. In Scotland, this proportion is set to reach 4.5 per cent, or £36m, by 2006. The RDR offers new ways of boosting funding of the AEP and of paying farmers for the non-market benefits that agriculture brings to society. However, even by 2006 the RDR will account for less than 11 per cent of the CAP budget (Lowe and Brouwer, 2000), so unless or until it is adequately resourced, it will remain a decidedly anorexic second pillar by comparison with the ample proportions of the first.

Although several environmental objectives are incorporated within Agenda 2000,[3] the main aim is to rein in CAP spending before the new eastern European countries join the EU. However, instead of reducing CAP expenditure, the measures merely stabilise it, and they fail to decouple payments from production. Consequently, they have been widely criticised as another wasted opportunity (Lowe and Brouwer, 2000). Given that they are based on similar instruments to the 1992 reforms and that these had only limited environmental benefits, it is unlikely that Agenda 2000 will deliver significantly more positive results (Winter and Gaskell, 1998). Radical, rather than incremental, reform of the CAP remains urgent and inevitable (Section 5.3.3.ii) because 'current agricultural policy is unsustainable politically, economically and environmentally' (A. Raven, 1999: 134). Unfortunately, the EU's Nice Summit in 2000 saw the prospect of such reform receding still further.

5.2.2 The origins and development of agri-environmental policies

Ironically, although Britain is hardly renowned as a leader in Europe (Section 2.2.2), it was Britain that pioneered the reform of European agricultural policy (Baldock and Lowe, 1996). By the beginning of the 1990s, the UK had arguably the most comprehensive system of agri-environmental incentives of any EU member state, a remarkable state of affairs given that ten years earlier British agricultural policy was still resolutely expansionist. Farming has always fulfilled three functions: an economic function (food production), a social function (rural employment; recreation), and an environmental function. Until the early 1980s, UK agricultural policy heavily emphasised the first of these; the social benefits were recognised and valued as a spin-off, while the environmental dimension was either ignored or assumed to be benign. Around that time, however, four factors combined to initiate a rapid shift towards a more environmentally sensitive agriculture.[4]

1. **Food surpluses and the CAP budget crisis**. These concentrated minds and accelerated the pace of reform. The budgetary implications of a high price, protectionist regime, maintained for thirty years in the face of mounting criticism, could no longer be ignored (Potter, 1998a). A greening of farm policy would have come eventually, but the budget crisis was the spark which ignited the reform process by making policy-makers much more receptive to the arguments of the agri-environmental reformers. By the late 1980s, 70 per cent

of the entire EU budget was being swallowed by the CAP, and the regular dumping of surpluses onto the international market depressed world prices and antagonised trading partners (Potter, 1998a). The post-war fixation with self-sufficiency was finally supplanted by the realisation that reducing the productivity of agriculture would not constitute a threat to food supplies (N. Russell, 1994; Waters, 1994).

2. **Agri-environmental damage**. The increasing recognition of the environmental impacts of high-input/high-output 'industrial' agriculture driven by the CAP led to a mounting public outcry, informed and stimulated in no small measure by Shoard's (1980) powerful polemic. In Scotland the key concerns have been water pollution from agrochemicals (Section 6.5.3), drainage of wetlands, the replacement of small, mixed farms with larger, specialised enterprises, loss and fragmentation of habitats due to intensification (e.g. heather moorland from overgrazing), and the on-going losses of valued elements of the countryside such as hedges, traditional field walls, and hay meadows. Reviews of the direct environmental impacts of agriculture across the UK (Winter, 1996; Skinner *et al.*, 1997; Krebs *et al.*, 1999) show that pesticides, nitrates and farm livestock wastes all constitute significant environmental hazards, especially with regard to surface water and groundwater quality, and to biodiversity. Indeed, within Scotland, agriculture is now second only to sewage as a source of water pollution, with 300–400 agricultural pollution incidents every year; it is predicted to be No.1 in ten years (Egdell, 1999; SEPA, 1999). The indirect or external environmental costs of farming are also very large (F. Pearce, 1999b). There is abundant evidence of the profound impacts that agricultural practices have had and continue to have on Scotland's natural heritage (SNH, 1995a; Egdell, 1999). Thus, for example, the decline of cornfield and hay meadow species, the dwindling numbers of farmland birds such as grey partridge and skylark, and the reduction of the corncrake (*Crex crex*) population are all a direct result of changing farming practices. Agriculture is also responsible for much of the damage to Sites of Special Scientific Interest. A major legislative response to these rising concerns came in the 1981 Wildlife and Countryside Act, but during these last years before the supertanker of agricultural expansionism began to change direction, policy makers were attempting to obviate the need for a change of policy by accommodating the environmental demands. Thus farmers were paid compensation for profits foregone when they agreed not to carry out damaging operations, but all the agricultural support mechanisms remained intact.
3. **Contradictory policy**. The consequence of these compensation arrangements was that one part of government (agriculture departments) was offering farmers financial inducement to do things which other parts of government (environmental/conservation departments) had to pay them not to do. Such an absurdly illogical situation, in which compartmentalised public bodies were so obviously working at cross-purposes (Whitby, 1996b; Potter, 1998a), was unsustainable. The case for fundamental reform of the agricultural support edifice itself was becoming irresistible.

4. **Farming as environmental stewardship.** On a more positive note, the final factor prompting reform was the realisation of farmers' potential to be transformed into the role of environmental custodians. The existing system, in which farmers had to threaten to do something damaging before they could claim compensation payments, was negative both practically and psychologically (N. Russell, 1994). How much better, it was widely argued, if public money was spent on achieving positive environmental outcomes – so-called 'green payments'. Despite the damage done by agriculture, Scotland still has a higher proportion of extensive farming systems than many European countries. Much of its farmland is of high conservation value, supporting many species of national and international importance such as the corncrake, golden plover (*Pluvialis apricaria*), and several birds of prey (Badger, 1996; Egdell, 1999). Spending priorities, it was recognised, should reflect this non-commercial but highly valued dimension.

By the mid-1980s both national and European policy-makers saw the inevitability of fundamental change. The need for reform was especially acute with respect to hill farming where support over the previous three decades had encouraged a steady intensification of land use. Current policies were not only causing environmental damage (through overstocking especially), but were failing to support those most disadvantaged farmers who were the intended recipients. In the UK the beginning of the resulting transition towards a more environmentally-sensitive agriculture was signalled by the 1986 Agriculture Act which for the first time promoted a balance between the various functions of farming. This act is widely seen as symbolising the end of the post-war productivist era. The years since then have seen experimentation with a wide range of ideas aimed at the greening of agriculture, notably Environmentally Sensitive Areas, the Scottish Countryside Premium Scheme and the Organic Aid Scheme.

5.2.2.i Environmentally Sensitive Areas

The first and undoubtedly the most significant of these experiments was the creation of Environmentally Sensitive Areas (ESAs) (Whitby, 1994; Table 5.1). These were pioneered by the UK within the EU from 1987. The ten Scottish ESAs cover 19 per cent of the agricultural area (Fig. 5.5) and consume some 65 per cent of the agri-environment budget. They vary considerably in character and in objectives, but they essentially represent the marriage of a farm survival policy with positive conservation (Morris and Robinson, 1996). In effect, farmers within ESAs who choose to participate are paid to be custodians of the landscape – to 'produce countryside' – and to farm in traditional ways which are welcomed by the public and seen as more environmentally friendly (Skerratt and Dent, 1996; Gourlay and Slee, 1998). Basic payments are offered on a per ha/per annum basis for avoiding environmental or archaeological damage, and higher rates are available in support of active measures which enhance the conservation interest of the farm (Robinson, 1994b).

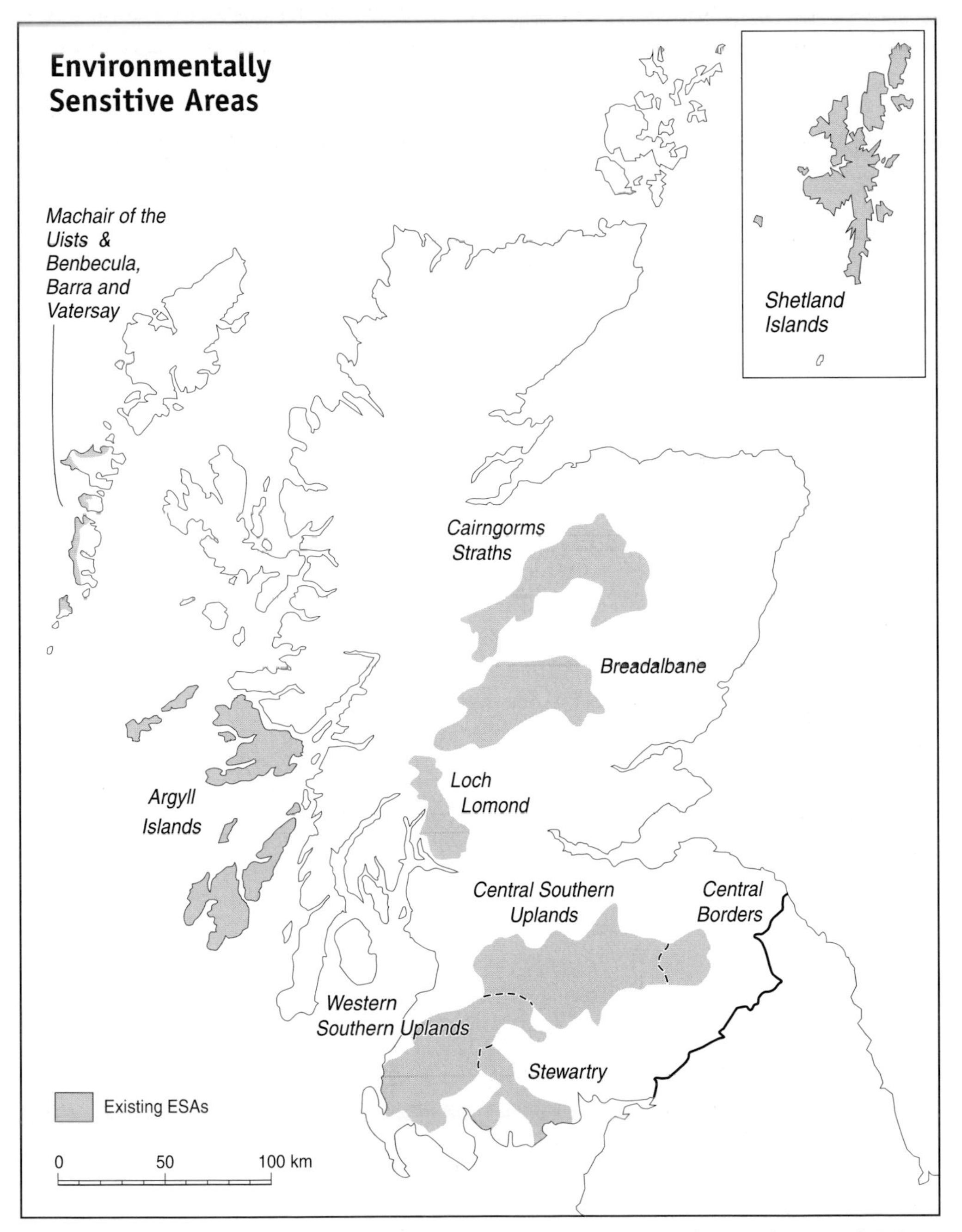

FIGURE 5.5 *The location of Environmentally Sensitive Areas in Scotland. From the start of 2001, ESAs became part of the Rural Stewardship Scheme.*

TABLE 5.1 Participation in Scottish agri-environmental schemes as at 31st May, 2001. From the start of 2001, ESAs and the SCPS were combined to form the Rural Stewardship Scheme. Data provided by SEERAD.

	Start date	*Number of participants*	*Area covered by agreements (ha)*	*Funding in 2000/1 (£ million)*
Environmentally Sensitive Areas	1987	2,815	780,956	9.7
Organic Aid Scheme	1994	490	237,281	3.4
Scottish Countryside Premium Scheme	1997	1,478	Not available	5.3

How successful has this flagship scheme been? Egdell and Badger (1996) argue that, at least until the mid-1990s, the environmental benefits of agri-environmental policy proved less evident than its ability to offer farmers an alternative source of income, sometimes by doing next to nothing (Colman, 1994). A study of the ecological and conservation impacts of a Welsh ESA found that the scheme had been successful in preserving habitat quantity (i.e. in halting habitat reduction at the landscape level), but had done little to improve quality (Wilson, 1997). Moreover, farmers' attitudes to environmentally-sensitive farming remained largely unchanged. Steward-minded farmers constitute a small minority of those participating in ESAs, and there is little evidence that involvement causes farmers to become more positively disposed towards conservation (Potter, 1998a).

The great difficulty in appraising the ESA scheme is the near-impossibility of establishing what farmers would have done in the absence of the scheme (Whitby, 1996b). Yet the 'audit culture' demands evidence of value for money from the scheme. Sceptics argue that a clearer demonstration of public benefits is needed before more public money is committed to ESAs, and that better targeting is required (Gourlay and Slee, 1998). Enthusiasts counter with the claim that until such schemes are more generously funded they will not deliver the more tangible results that are being demanded of them. It is perhaps too soon to say that ESAs have succeeded in bringing about environmental improvements that would not have taken place anyway (Potter, 1998a). Despite their finished appearance and relative longevity, they are still essentially experiments that have yet to prove their worth beyond reasonable doubt. A more fundamental critique of ESAs is one which can be levelled at all area-based designations, namely that even if all goes well within the special areas, things may go from bad to worse in the wider countryside (Morris and Potter, 1995). Site-based policies imply that 'the rest' has only limited value, when in fact the health of special areas depends on their environs (Section 8.2.2.iii). Biodiversity rarely survives on islands within a hostile sea. Incentives for environmentally friendly farming are needed everywhere, not just within ESAs, and this was the rationale behind the creation of the Scottish Countryside Premium Scheme (SCPS) in 1997.

5.2.2.ii Other agri-environmental initiatives

During the four years of its operation, the SCPS provided grants for farmers and crofters outwith ESAs for positive management of habitats and biodiversity. Local priorities were agreed for each region. The aim was, at the very least, to maintain the natural heritage, and ideally to enhance it by making Scottish agriculture more environmentally sensitive. The various previous schemes which the SCPS superseded had not proved popular with farmers (Egdell, 1999), whereas the new scheme was consistently oversubscribed, demonstrating both its popularity and its limited budget (Table 5.1).

Under Agenda 2000, ESAs and the SCPS were merged in 2001 to form the Rural Stewardship Scheme. As with all schemes under the AEP umbrella, this is co-financed with European and UK money. Funding has been rising rapidly since 1995, but in comparison to the on-going price and production supports flooding out to farmers through the CAP, AEP payments represent a tiny trickle, 'an under-funded afterthought' (A. Raven, 1999: 135; Table 5.2). The 4 per cent of the budget allocated to the AEP compares unfavourably with figures of 20 per cent or more in other EU countries. Moreover, until 2001 Scotland was the only part of the UK where there was a ceiling on what a farmer could receive from agri-environment payments. These restrictions inevitably limited the attractiveness and take-up of AEP schemes, even where farmers were receptive to their objectives.

Table 5.2 Spending (in £ million) on agri-environmental schemes and on agricultural output grants and subsidies since 1992. Data provided by SEERAD.

Year	*Agri-environmental schemes**	*Output grants and subsidies*	*Agri-environment payments as a % of grants/subsidies*
1992	1.2	215.4	0.56%
1993	1.0	292.5	0.34%
1994	1.0	306.1	0.33%
1995	2.0	394.0	0.50%
1996	3.4	451.2	0.75%
1997	5.7	462.2	1.23%
1998	11.2	451.9	2.48%
1999	14.5	439.4	3.30%
2000	17.2	443.5	3.88%

*Includes Environmentally Sensitive Areas payments, the Scottish Countryside Premium Scheme, the Organic Aid Scheme, and elements of the Habitats and Heather Moorland schemes which closed to new applicants at the end of 1996.

In addition to the AEP, there are a variety of other green initiatives such as TIBRE (Targeted Inputs for a Better Rural Environment). This is a working partnership between SNH, academics, industry, and farmers that has been developing since the mid-1990s. It promotes new 'precision farming' technologies and techniques that meet farmers' needs without damaging the environment. These include field-scale soil mapping to enable fertiliser to be applied only where needed, precision sprayers, and gel-based chemicals which minimise leaching into groundwater. Ideally, these allow farmers to use less to achieve more. The operation of all the various agri-environmental schemes has been facilitated to a considerable degree by the Farming and Wildlife Advisory Group (FWAG) which provides practical conservation advice to farmers.

It is clear, then, that agricultural policy has evolved from a focus on 'production for the nation' to incorporate a broader ethos encompassing environmental values. As in forestry, a single-purpose drive for commercial production has been replaced with a multi-functional policy. This is emphasised in the development strategy for rural Scotland (Scottish Office, 1998d) which stresses that the three components of sustainable development – economic, social and environmental – should be at the heart of agricultural policy-making. However, the rapid policy changes of the last fifteen years will take time to filter through; the long-held attitudes and beliefs of farmers don't change overnight (Shucksmith, 1992). Because the gap between the policy rhetoric and the persisting reality on the ground is considerable (Winter, 1996), many argue that the long-term objective of agri-environmental policy should be a transformation of farming culture towards a more sympathetic attitude to the environment, and that environmental considerations should become a core objective of mainstream support (A. Raven, 1999). If this is not achieved then agri-environmental measures will continue to be seen as 'temporary bribes, shallow in operation and transitory in their effect' (Morris and Potter, 1995: 52).

Looking ahead, it is conceivable that agri-environmental measures could one day swallow up agricultural policy altogether because green payments of some sort are likely to become one of the few politically sustainable forms of state support for agriculture in years to come (Potter, 1996, 1998a). The RSPB has pointed out that greater funding of the AEP makes logical sense; in delivering support for farm incomes as well as conservation benefits, they give more to the public for their money than mainstream CAP payments which only achieve the former (Badger, 1996). Although agri-environmental measures are still to a great extent the Cinderella of agricultural policy, they are here to stay and to expand.

5.2.3 The crisis in Scottish agriculture

Scotland's farmers, especially hill farmers, have been facing severe problems since the 1980s (Wigan, 1991), but in the last five years Scottish agriculture has been hit by a succession of hammer blows which have driven it into an ever-deepening crisis, probably the worst since the 1930s. After a sharp peak in the mid-1990s, farm incomes have tumbled (Fig. 5.4), falling so far that over a third of all farms made net losses in 1998/99 (SERAD, 2000c). The total income from farming fell by a further 12 per cent (£31 million) in 2000 (SEERAD, 2001b). No sector has escaped, but the beef industry has been hit particularly hard by the BSE ('mad cow disease') outbreak, the resulting closure of the beef export market from 1996 to 1999, and the beef-on-the-bone ban. Some farmers replaced cattle with sheep, a move which has damaging knock-ons for vegetation because sheep graze more intensively. The collapse of sheep and pig prices has led to business closures and to animals being destroyed on farms because it is simply not worth transporting them to market. With incomes in freefall and subsidies changing little, subsidies represented an ever greater percentage of net farm income (Fig. 5.4; Table 5.3). Then, on top of all this, came the outbreak of foot-and-mouth disease in 2001. This had devastating impacts throughout rural Scotland, hitting not only farmers but tourism and the entire rural economy. It necessitated a total export ban, leaving livestock farmers staring into the abyss.[5]

Three general factors have exacerbated the crisis. One is the relative strength of sterling, which puts Scottish farmers at an export disadvantage in Europe. Another is low world prices due to oversupply in commodity markets. A third is the highly capitalised state of most farm businesses. Levels of farm debt of the order of £50,000–100,000 are not uncommon, and although these levels are low compared to the capital assets of most farms, they are depressingly high compared to current farm income levels (Egdell, 1999). Bank borrowings stand at a record £1.2 billion, and the ratio of bank interest to total income from farming is 2:1 (SEERAD, 2001a). Some regions are particularly hard hit by the crisis, notably the Borders, parts of which are facing economic ruin (Ramsay, 1999). Agriculture provides 20 per cent of the region's GDP, so as farm businesses collapse, the whole rural economy is dragged down, raising the spectre of farm abandonment and rural depopulation. For Scotland as a whole, incomes dropped 80 per cent in the four years to 1999, and surveys indicate that 14 per cent of farm businesses will fail over the next five years (Maxwell, 1999). Traditional hill farming, in particular, is likely to continue suffering, with average net incomes for LFA livestock farms dropping to a mere £2700 in 1999/2000 with only small increases forecast (Table 5.3). It is no surprise, therefore, that suicide rates amongst farmers are tragically high. Politicians have provided a series of emergency support packages, and the Scottish Executive (2001c) has endeavoured to chart a path towards future prosperity in its 'forward strategy for Scottish agriculture', but light at the end of the tunnel is dim and distant.

5.3 Current issues and debates

5.3.1 The future of crofting

The history of crofting has been compellingly told by Hunter (1991, 2000) and McIntosh *et al.* (1994). They describe the anguish of the Clearances and how the crofters' lack of rights drove them to take the militant action which led to the Crofters' Holdings Act of 1886 (Cameron, 1986). They also chart the subsequent progressive increase in crofters' legal rights to the point where, today, a croft is often described as 'a piece of ground entirely surounded by regulations'. The historical and cultural importance of crofting cannot be doubted (Parman, 1990). But what place should this historic arrangement have in Scotland's rural future?

Although crofting incorporates elements of the old Highland system of farming, in its current form it only dates back to the eighteenth century (Whyte and Whyte, 1991). It consists of small scale, usually part time, traditional farming comprising a small (typically 1–10ha) arable area for fodder crops, and extensive common grazings

TABLE 5.3 Net Farm Incomes (NFI) and direct subsidies by farm type, 1998/9–2000/1 (forecast). The NFI figures are net of subsidy; without subsidies the average farm income for 2000/1 would be –£22,663. From SEERAD (2001a).

Farm type	*1998–9*			*1999–2000*			*2000–1*		
	Net Farm Income	*Direct subsidies*	*Subsidies as a % of NFI*	*Net Farm Income*	*Direct subsidies*	*Subsidies as a % of NFI*	*Net Farm Income*	*Direct subsidies*	*Subsidies as a % of NFI*
	£/farm	£/farm	%	£/farm	£/farm	%	£/farm	£/farm	%
Cereals	4,300	25,000	590	10,600	27,500	260	5,000	27,700	560
General crops	16,900	29,100	170	3,600	30,200	830	9,200	30,200	330
Dairy	5,900	8,400	140	1,900	8,000	430	1,000	8,800	850
LFA sheep	2,500	27,200	1,100	**–200**	24,500	n/a	300	23,800	n/a
LFA beef	5,100	28,700	570	4,100	27,000	660	4,700	27,700	590
LFA mixed	5,500	37,100	680	2,700	33,200	1,240	3,700	33,200	910
Mixed	**–2,100**	32,800	n/a	3,600	32,000	890	2,200	33,200	1,490
All	5,200	27,100	520	3,900	26,300	670	3,800	26,600	700

FIGURE 5.6 *Crofting in the Outer Isles.*
a. Crofts on the Isle of Lewis. Photo: © the author.
b. Aerial view of crofts on the machair of North Uist. Photo: © Patricia & Angus Macdonald/ SNH.

mostly for sheep (Fig. 5.6). It also contributes to the economy through fishing, fish-farming and tourism. But crofting is far more than just a type of land use; it is 'a unique social system in which small scale agriculture plays a unifying role' (Rennie, 1997: 123). It forms 'the mortar that binds communities together and keeps people on the land' (MacLeod *in* Inskipp, 1997: 29). It is in the 'crofting counties' of the Highlands and Islands (Fig. 5.7) that Gaelic thrives as an everyday language, and it is largely due to crofting that the distinctive Nordic culture of the Orkney and Shetland islands persists today (Crofters Commission, 1998). Parman's (1990) evocative description of a crofting township on the west coast of Lewis shows the extent to which crofting is a way of life.

There are some 33,000 crofters and about 17,700 registered crofts, 83 per cent of which are tenanted rather than owner-occupied (Crofters Commission, 1998). Crofts comprise about 10 per cent of Scottish agricultural land, and the common grazings covering some 5,000km^2 (12 per cent) of the Highlands and Islands (Rennie, 1997; Crofters Commission, 1999). Crofting areas have been the least affected by mechanisation and intensification, allowing traditional practices to continue. This has preserved several environments and species now recognised as having great conservation and cultural value, notably the machair. These fertile, low-lying, dune meadows, found mainly in the Outer Hebrides (Fig. 5.6b), are a priority habitat under the Habitats Directive and support some of the highest densities of breeding waders anywhere in the world (Crawford, 1997; SNH, 1998a). The Western Isles machairs constitute an ESA (Fig. 5.5), and it is no coincidence that the last UK stronghold of the corncrake is in these crofting areas where hay meadows still survive (Green and Riley, 1999).

As recently as the 1970s there was general pessimism about the future of crofting, with many believing that the 1976 Crofting Reform Act would bring about new Clearances through widespread 'decrofting' (Parman, 1990), but optimism has steadily grown in recent years. Crofting is more in tune with current rural trends than is high-intensity agriculture, and seems to be well placed to benefit from CAP reform (Crofters Commission, 1999). It is community orientated, much of its produce is organic, and it offers a sustainable, low-intensity base for remote settlements while maintaining aesthetically pleasing, diverse, and highly valued cultural landscapes. On this basis it is held up as 'an excellent model of sustainable development' (Crofters Commission, 1998: 3). 'Far from being a mildly embarrassing relic from the distant past, crofting points the way to the diversified rural economy which is being sought on all sides' (Rennie in Hunter, 1991: 21). Having long been marginalised and scorned, crofters are now widely viewed as cultural standard bearers occupying the moral high ground in rural affairs (Shucksmith *et al.*, 1996). Both Hunter (1991) and McIntosh *et al.* (1994) believe that crofting may fare better in the twenty-first century than it did in the twentieth. Certainly the demand for croft land is now high, with a long waiting list. Accordingly the Crofters Commission, having found that over 20 per cent of crofters are absentees, is making energetic efforts to tackle the chronic problems of underuse and absenteeism in order to make more land available, and to safeguard the future by stemming the long-standing outflow of young people

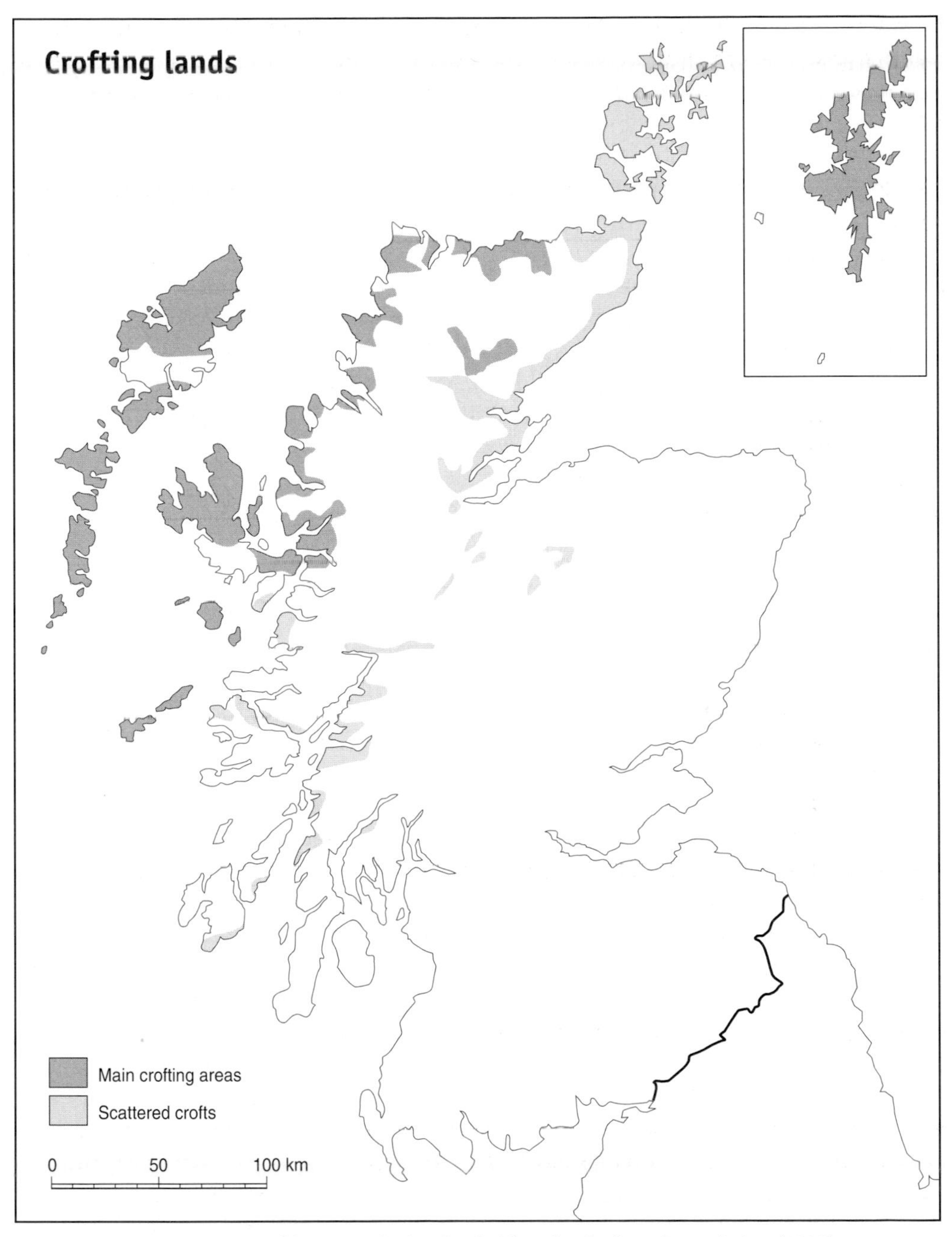

FIGURE 5.7 *Crofting areas in Scotland. After the Crofters Commission (1998).*

(Crofters Commission, 1998, 1999). Eight crofts were created at Balmacara in Wester Ross in 2000, the first new croftland for twenty years.

The resurgent confidence within crofting was boosted by the formation of the Scottish Crofters' Union in 1985 and is best symbolised by the recent trend towards crofting buyouts – crofters forming trusts to buy their land and bring it under community control. The trail was blazed by the Assynt crofters in 1993 (MacAskill, 1999), and several other communities have followed suit (Section 3.3.3). The 1992 Crofter Forestry Act, which allowed diversification into woodland, has also provided a welcome new facet to crofting (Section 11.2.2), and land reform legislation is now set to give crofting communities a right to buy their tenanted land and common grazings (Section 3.4.2).

It would be misleading to infer from all this that all is rosy and optimistic in the world of crofting (Shucksmith *et al.*, 1996). The Crofting Commission's extensive examination of the role of crofting in rural communities revealed much concern about the future (Crofters Commission, 1998). Further, it identified a range of challenges relating to employment opportunities, housing, community empowerment, land use, regulation, and the preservation of the crofting culture. There is also general agreement on the need for simplification and modernisation of crofting legislation, scheduled for 2002. Nevertheless, a crofting revival is on-going despite the general crisis afflicting Scottish agriculture (Section 5.2.3). In the light of this, and given the increasing stress on the social and environmental functions of agriculture, there have been calls for an extension of the crofting concept across large parts of Scotland (e.g. Kenrick, 1993; Wightman, 1996a). It will be intriguing to see whether this historic way of living with the land does indeed turn out to offer rural solutions for the twenty-first century.

5.3.2 Organic v. *GM crops*

This debate, which only came into focus in the late 1990s, goes far beyond the practicalities of Scottish farming to raise ethical issues with global dimensions (Mayer, 2000). Organic farming abhors the apparatus of the 'industrial' agricultural establishment with its intensive mechanisation and dependence on agrochemicals, adopting 'natural' approaches wherever possible. In the current social climate, this enables it to claim the moral high ground (albeit on sometimes questionable premises (Young, 2000)), to charge premium prices, and to capture a rapidly increasing market share. The enhancement of biodiversity on organic farms further improves its profile (Azeez, 2000). Mirroring exponential growth across Europe, the organic sector has grown fast, especially since the launch of the UK Organic Aid Scheme in 1994 which offers farmers and crofters transitional payments during the five year conversion period to organic production. Government support for organic farming recognises its contribution to the policy objectives of reduction of surpluses, environmental improvement and rural development (Lampkin *et al.*, 2000). UK sales are rising at 40 per cent per annum and are expected to reach £1 billion by 2002 (Young, 2000). Within Scotland, 237,281ha are now entered into the scheme (Table 5.1), but the rocketing demand for organic produce far outstrips homegrown supply, and most is still imported. Continuing rapid expansion seems inevitable.

Genetically Modified Organisms (GMOs) represent the other end of the spectrum, the ultimate techno-fix. Precise genetic modification of organisms to achieve desired outcomes (e.g. resistance to herbicides, pests and drought; enhanced flavour) became technically possible in the early 1970s, and has been subject to precautionary regulation under the Environmental Protection Act since 1990, but widespread public concern (largely to do with GM foods) only arose in the late 1990s. The highly charged debates surrounding GMOs (Mayer, 2000; Porritt, 2000) are too complex and wide-ranging to be discussed here. However, in a detailed examination of the issues from a Scottish perspective, Bruce and Bruce (1998) argue that the biotechnology 'baby' should not be thrown out with the 'bathwater' of its environmental risks. They advocate careful, small-scale testing to enable the real risks and potential to be evaluated, and recommend concentrating on applications which confer strong human or ecological benefits rather than trivial consumer preferences. Greenpeace and FoE, on the other hand, want a five-year moratorium on all GMO farm trials (Coghlan, 1999).

In a world in which many nations struggle to feed their burgeoning populations, the ethical debate over GMOs is far from clear cut. The unknown risks of irreversible genetic contamination of the wider environment ('superweeds' and damage to biodiversity) have to be balanced against the unknown potential of GMOs to help feed the world's neediest people, tackle pollution, or even replace non-renewable energy sources. Within Scotland, the tide is currently flowing strongly in favour of organic produce and powerfully against GM crops. In fact, since 1998 public concern over the latter has led to sharply rising demand for the former. Given the pace of change, however, it would be unsafe to project these trends very far into the future.

5.3.3 Agricultural futures

5.3.3.i Farmers as environmental managers?

Now that food production is no longer an overriding priority, and agriculture (even with massive subsidies) is barely economically viable across much of Scotland, what services should the countryside deliver? Should it be managed to please wildlife, local communities, visitors or taxpayers, or simply to please the eye? Society faces a profound rethinking of what it wants from the countryside, and is prepared to pay for (Potter, 1998b, 2001). Environmentally sensitive agriculture is highly valued by the general public (Hanley *et al.*, 1998), but although the social and environmental benefits are very real they are hard to quantify, so most farmers receive little tangible reward. The public expects farmers to look after the countryside, and are used to reaping the benefits free, yet farmers cannot be expected to be 'custodians of the countryside' unless they can make a living and find satisfaction in that role (Smout, 1993c).

There seem to be three possible directions in which farming could evolve. One option is simply to use taxpayers' money to pay farmers to maintain the *status quo*, but this would turn the countryside (and farmers) into a museum piece, an artificial landscape frozen in time. At the opposite end of the spectrum is the 'cold turkey' option: remove all subsidies and let agriculture find its real level in the global marketplace. For the moment at least, this is politically unacceptable because of the presumed socio-economic havoc that it would wreak across great swathes of rural Scotland; perhaps as much as 50 per cent of the land could move into non-agricultural uses and land values would tumble (Potter, 1998b). The environmental impacts would also be profound, albeit mixed, producing a very different countryside from today's managed rural space (Potter, 1998b; Warren, 2000b). It is interesting to note, however, that similar forecasts of disaster were not fulfilled when New Zealand removed all its state subsidies at a stroke in 1985. Very few farmers went bankrupt, agricultural pressure on the environment was reduced, and 'for conservation it was the best thing since sliced bread' (Stephenson, 1997: 24; Legg, 2000). Voluntary conservation work by farmers actually increased after grants and subsidies were removed, suggesting that grants may actually work against conservation.

There may come a day when all subsidies are removed, but in the meantime the third and most likely direction is an extension of the ESA approach: offering direct incentives from the public purse for green agriculture and sensitive environmental management. This goes beyond the simple maintenance of the *status quo* by encouraging farmers to become positive environmental managers, creating habitats and caring for wildlife. It is a well established and strengthening trend; the Scottish Executive (2001a) expects the number of participants in agri-environmental schemes to increase by at least 15 per cent annually. As a means of accelerating the adoption of green practices, some advocate 'strong' cross compliance whereby participation in an agri-environmental scheme (and perhaps the drawing up of 'whole farm' conservation plans) would be a condition of receiving other CAP aid (Egdell, 1999). Some commentators go one step further and advocate 'green recoupling', the idea that support should be detached from agricultural production altogether and coupled instead with conservation measures (Potter, 1998a). The latest idea, borrowed from France, is Land Management Contracts. This is a new, integrated scheme of 'whole farm' support which will be designed to reward the full mix of economic, social and environmental benefits provided by farmers and crofters (Scottish Executive, 2001c).

Paying farmers to produce environmental 'goods' that taxpayers want, rather than surplus food, seems self-evidently better use of public money. However, it ignores farmers' psychology. Like anyone else, they need a sense of purpose and motivation. Most farmers love the land and enjoy caring for it, but want to make their primary living from their produce (Shucksmith, 1992). They are uncomfortable with the new 'multifunctional' roles being expected of them (UoA/MLURI, 2001). Living life on a subsidy cheque is uncomfortably akin to a life on the dole. Will such an existence attract a new generation of farmer-conservationists onto the land when today's ageing farmers retire? At present, farmers' perceptions of nature, the land and conservation can diverge strikingly from those of conservationists (Green, 1993; McHenry, 1996, 1998; Beedell and Rehman, 1999).

Given the long-term pressure on farm incomes, and that these can only intensify, farmers have had to make an increasing proportion of their living from non-agricultural income (Mowle, 1997). The writing has been on the wall in this regard for at least fifteen years (Selman, 1988a). On the farm, pluriactivity can involve charging for countryside services related to the land, such as farm tours and off-road vehicle courses, or converting land or property for other uses (golf courses, bed-and-breakfasts). Increasingly farm incomes are also supplemented from off-farm part-time employment.

In the last decade there has been a sharp increase in pluriactivity. Non-farm income in the late 1990s averaged £7,500 per household, double the figure of the early 1990s (UoA/MLURI, 2001). However, most Scottish farming families are still largely dependent on agriculture for their income (SEERAD, 2001b), and such families have been hit much harder by the current crisis than pluriactive households (UoA/MLURI, 2001). For struggling upland farmers, farm forestry and agroforestry represent possible new options (Section 11.2), as does wind farm development, but diversification is simply not an option for many remote farms, nor for many of the tenant farmers who work over a third of agricultural holdings (Wightman, 1996a). Nevertheless, the Scottish Executive (2001c) is making £70 million available over the next five years to help farmers restructure and diversify their businesses, and an innovative, proactive approach to diversification is being taken in the Borders by the Foundation for Rural Sustainability (Armstrong, 1999).

The days when farmers could concentrate solely on farming their land are already long gone. Where they once used to base their decision making simply on prices at the local market, they now need to incorporate European and global economic factors. 'Diversify or die' may become the harsh motto of the immediate future. Certainly government thinking has been refocused from an exclusive emphasis on agriculture to the wider context of rural policy. But whatever Scottish politicians and environmental groups advocate, the dominant influence on the future of Scotland's farmland – and hence on the future of much of the natural heritage – will be the way in which the CAP evolves.

5.3.3.ii The rural environment under a changing CAP

The monolithic CAP has created a high-cost agriculture which prevents European farmers from competing on world markets without subsidy. It still consumes around half the EU's budget, despite the fact that farming accounts for only 5 per cent of EU employment and less than 2 per cent of GDP. It has neither halted the decline in rural employment nor safeguarded the environment, and is not sustainable even on its own terms (Allanson and Whitby, 1996). Morris and Robinson (1996: 67) comment that 'as tools for aiding disadvantaged areas, agricultural policies are fairly blunt instruments'. For all these reasons, its demise in its present form is long overdue. Future support must be designed in a way which rewards the mix of benefits required by the public and which does not mask market signals (Scottish Executive, 2001c).

But how can rural economies be sustained while CAP subsidies are slashed? It is the intractability of this dilemma that has kept European policy makers prevaricating and compromising for so long. Despite all the agri-environmental initiatives there is a long way to go in the greening of European agriculture. The CAP is still geared towards production, and politicians have yet to grasp the nettle of an inherent contradiction: environmental measures have been bolted onto agricultural policies which have created the very environmental problems that the measures are designed to counter.

Following Agenda 2000, environmental goals should become more important objectives for the CAP (though still running a distant second to the support of farm incomes), and the AEP is to be given a more prominent, better resourced role. The considerable overlap between LFAs and areas of high conservation value has now been recognised, and, in line with earlier proposals by SNH (1994a), it has been suggested that LFA payments could be transformed to promote low-input farming systems. All these developments are welcome from an environmental viewpoint, but recent history suggests that not all the good intentions will survive the political mangle. Within Scotland, Egdell (1999) suggests that the crucial agri-environmental policy questions concern future funding levels, the extent to which farmers will participate in the new Rural Stewardship Scheme, whether it can be linked with Local Biodiversity Action Plans (Section 8.2.3), and whether it can incorporate such diverse initiatives as TIBRE (Section 5.2.2.ii), goose management (Section 7.7.3) and nitrate management (Section 6.5.3).

The reform of the CAP is not just an EU debate. Because agriculture is now included within the WTO process, agricultural and agri-environmental policy making rests firmly within the international debate about the liberalisation of agricultural trade. This process provides a strong and perhaps irresistible impetus for further reform (and perhaps eventual abolition) of the CAP as agricultural policies converge under internationally agreed rules. Sooner or later, European farmers thus face a decisive reduction in production-orientated support (Potter, 2001).[6] This could have dire consequences for the occupancy and management of land in Scotland. Tempering this pessimism to some degree is the concept of the 'double dividend'. On the one hand, a reduction in production subsidies should reduce the incentives for intensive farming, leading to extensification with its associated environmental benefits. On the other, the green recoupling of support should release some production-related subsidy money to boost funding for environmental programmes, benefiting the environment and cushioning farmers from the full effects of reduced support. Although neither side of the double dividend equation is as straightforward or as certain as each appears (Potter, 1998a), the concept is compelling, not least for policy makers trying to reconcile the international pressures for reduced support with the social (and hence political) imperatives of safeguarding rural economies. One of the strongest arguments in favour of green recoupling is that without some substantial compensating measures, the net effect of liberalisation in economically marginal areas could be a new era of clearances which would remove much of the human capital that is necessary for the effective conservation of biodiversity and amenity in the countryside.

In concluding his detailed review of agri-environmental policy, Potter (1998a: 162) highlights the uncertainties which cloud our view of the future:

> Important questions remain . . . about taxpayers' willingness to fund government programmes which may not always deliver an immediate or measurable environmental benefit, and of the willingness and ability of farmers themselves to take on the role of environmental stewards that is being prepared for them. Particular uncertainty surrounds . . . the extent to which policy makers will be

able to extract a conservation dividend from the withdrawal of conventional farm support.

At present, environmental arguments remain peripheral to EU policy making (Winter, 2000). The much heralded post-productivist era is having a difficult birth.

NOTES

1. A wealth of descriptive and factual information about the present farming scene is provided by Robinson (1994a), McCarthy (1998), SERAD (2000b) and SEERAD (2001a, 2001b). The contribution of farming to society, the economy and the environment is explored in detail by UoA/MLURI (2001).
2. For discussions of the evolution of farming in recent centuries, see, for example, Franklin (1952), Clark (1983), Whittington (1983), Fenton (1997), McCarthy (1998) and Devine (1999).
3. Details of the environmental provisions under Agenda 2000 are given by Egdell (1999) and the Scottish Office (1999a), and are discussed by Lowe and Brouwer (2000). LFA policy, in particular, has been significantly reorientated towards environmental objectives.
4. The story of the development of European agri-environmental policy is told in detail by Potter (1996, 1998a) and by Baldock and Lowe (1996). The volume edited by Brouwer and Lowe (2000) provides detailed examinations of the impacts of the CAP on the environments of EU member states, showing that not all have been negative.
5. The issues raised by the foot-and-mouth outbreak are complex and wide-ranging. At the time of writing, with new cases still being confirmed, it would be premature to examine the implications of the epidemic for rural Scotland. It raises questions about intensive farming (with its high use of chemicals and long distance transport of animals), the power of large retailers, and the relative merits of international *versus* local food supply chains. It also reinforces the importance of treating farming and crofting as integral parts of rural development, not as separate enterprises (Scottish Executive, 2001c).
6. One suggested means of facilitating the transition to a subsidy-free agriculture is the notion of the 'transferable bond'. See Potter (2001).

CHAPTER SIX

The waters: freshwater resource management

6.1 Introduction: the nature of the resource

Scotland is richly endowed with freshwater resources (Figs 6.1 & 6.2).[1]

> The lochs and rivers of Scotland are famed all over the world in poetry and in song as a national resource of breathtaking beauty and inestimable value . . . No image of Scotland can be properly conjured up without including her lochs, her lochans, her rivers and her burns. (Maitland, 1997a: 157)

To enumerate the richness of that endowment is to encounter statistics of abundance (Maitland, *et al.*, 1994; McCarthy, 1998). There are some 31,000 lochs and almost 7,000 river systems comprising perhaps 109,000 streams. Water covers 2 per cent of the country, 5 per cent of the Highlands and Islands, and 10 per cent of the Western Isles. In a UK context, Scotland takes most of the water resource 'prizes'. Of the freshwater bodies, it has the largest (Loch Lomond: 71km^2), the longest (Loch Awe: 41km), and the deepest (Loch Morar: 310m). Scotland has 91 per cent of the volume of standing freshwater in Britain, the volume of Loch Ness alone (7,452 million m^3) exceeding the combined volume of all the lakes and reservoirs in England and Wales. The 193km-long River Tay, with a catchment of 5,000km^2, has a greater average flow than any other UK river (194 cumecs – cubic metres per second), and during damaging floods in January 1993 its flow peaked at 2,269 cumecs. The River Spey, set to become a Special Area of Conservation, has the largest intact area of floodplain wetland in Britain – the Insh Marshes (Fig. 6.2b) – and probably the largest floodplain forest (Leys, 2001).

Scotland's extremely varied geology, topography and climate produces strong regional contrasts in hydrology. For example, the deep lochs of the north and west tend to be oligotrophic and acidic whereas the sedimentary rocks of the eastern lowlands produce shallower, nutrient-rich lochs. Rivers in the wet west are typically short, steep and very responsive to precipitation, while those flowing east have larger catchments and more moderate fluctuations in flow. Seasonal contrasts can also be very large. Summer flows in a typical upland river may be less than 15 per cent of the mean flow, while flood discharges may be 300 times greater than dry weather flows (Gilvear, 1994). Variability is the essence of the resource.

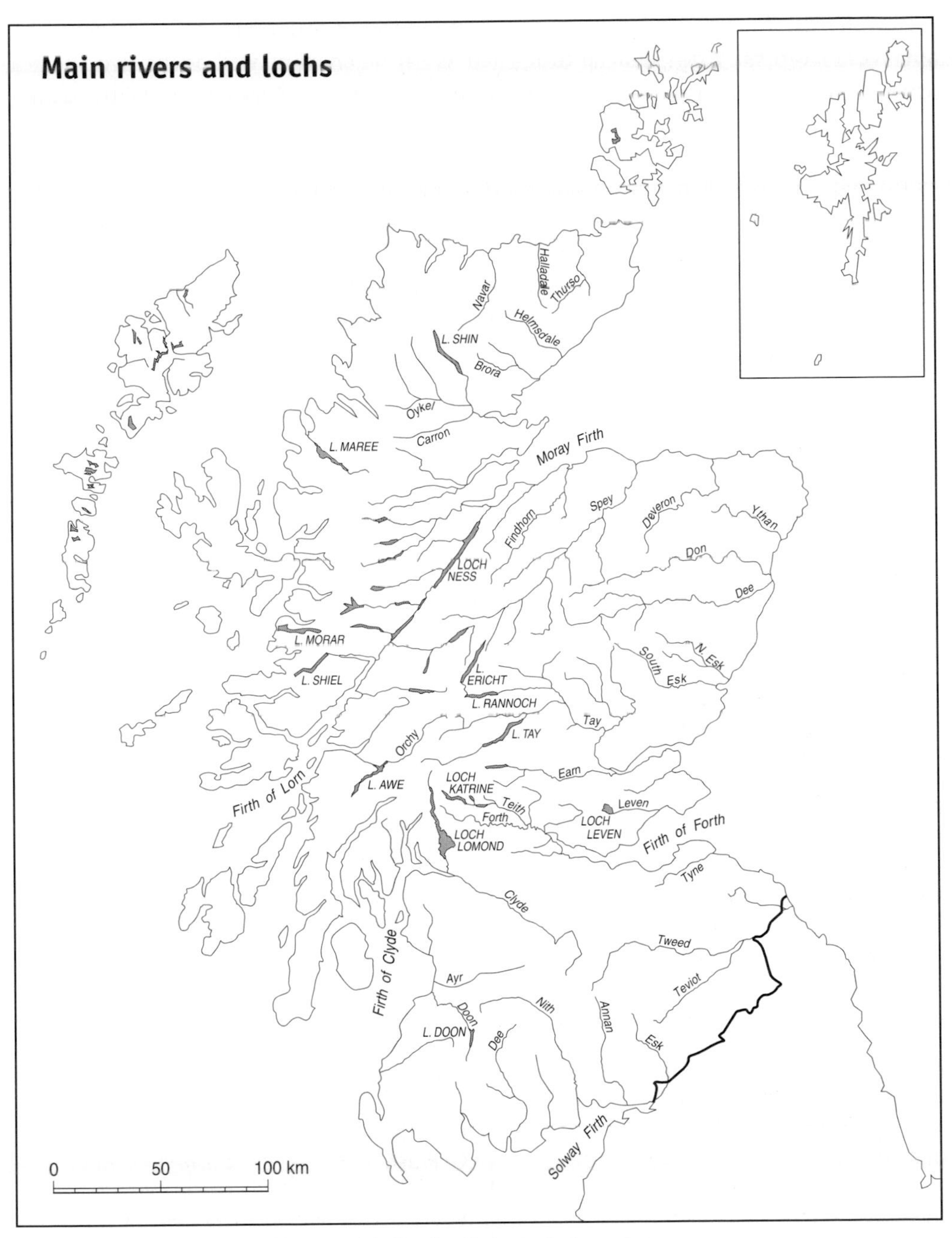

FIGURE 6.1 *Scotland's main lochs and rivers.*

This same diversity produces a wide range of freshwater habitats – rushing mountain torrents, placid lochs, extensive wetlands and valuable riparian (water's edge) habitats – variety which allows a wide diversity of faunal and floral species to thrive in, on or around fresh waters (Maitland *et al.*, 1994). Of the fifty-six species of freshwater fish found in Britain, forty-one are found in Scotland (Table 6.1), including survivors from early post-glacial time such as Arctic charr as well as the two species most closely associated with Scotland, salmon and trout (Maitland, 1997b). Game fishing has made rivers such as the Tay (Fig. 6.2c), Tweed, Spey and Dee famous the world over and is important for the economy of certain rural areas. Rod angling is estimated to be worth over £400 million per annum to the Scottish economy (SOAEFD, 1997), and salmon fishing alone supports some 3,360 full-time jobs (Maitland, 1997a). The number of fish species has been slowly increasing (Table 6.1), both as a result of natural northwards migration and through artificial introductions from England and overseas; no less than nineteen species are found in Loch Lomond alone (Maitland *et al.*, 2000).

TABLE 6.1 Fish species occurring in Scottish freshwaters. Species in bold type have protected status under either UK or EU law. Vendace were extinct but have now been reintroduced. Sources: Maitland (1994); SNH (1995a).

Original colonisers		*Later arrivals*	
Sturgeon	*Acipenser sturio*	By 1790	
Sea lamprey	*Petromyzon marinus*	Pike	*Esox lucius*
River lamprey	*Lampetra fluviatilis*	Roach	*Rutilus rutilus*
Brook lamprey	*Lampetra planeri*	Stone loach	*Noemacheilus barbatulus*
Allis shad	*Alosa alosa*	Perch	*Perca fluviatilis*
Twaite shad	*Alosa fallax*	Minnow	*Phoxinus phoxinus*
Atlantic salmon	*Salmo salar*	By 1880	
Brown/Sea trout	*Salmo trutta*	Brook charr	*Salvelinus fontinalis*
Arctic charr	*Salvelinus alpinus*	**Grayling**	*Thymallus thymallus*
Powan	*Caregonus lavaretus*	Tench	*Tinca tinca*
Vendace	*Caregonus albula*	Common bream	*Abramis brama*
Smelt/Sparling	*Osmerus eperlanus*	Chub	*Leuciscuc cephalus*
Eel	*Anguilla anguilla*	Crucian carp	*Carassius carassius*
Three spined stickleback	*Gasterosteus aculeatus*	By 1970	
Nine spined stickleback	*Pungitius pungitius*	Rainbow trout	*Oncorhynchus mykiss*
Sea bass	*Dicentrarchus labrax*	Pink salmon	*Oncorhynchus gorbuscha*
Common goby	*Pomatoschistus microps*	Common carp	*Cyprinus carpio*
Thick lipped mullet	*Chelon labrosus*	Goldfish	*Carrasius auratus*
Thin lipped mullet	*Liza ramada*	Gudgeon	*Gorbo gobio*
Golden mullet	*Liza aurata*	Rudd	*Sarkinius erythorophathamus*
Flounder	*Platichthys flesus*	Orfe	*Leuciscus idus*
		Dace	*Leuciscus leuciscus*
		Bullhead	*Lottus gobio*
		By 1990	
		Ruffe	*Gymocephalus cernua*

FIGURE 6.2 *Illustrations of Scotland's rich diversity of freshwater resources.*
a. *Loch Maree in the north-west Highlands. Photo: © the author.* (above)
b. *The Insh Marshes on the River Spey south of Aviemore. Photo: © SNH.* (below)
c. *The River Tay near Dunkeld. Photo: © the author.* (overleaf p. 114)
d. *Pool systems and blanket bog in the Sutherland Flow Country. Photo: © SNH.* (overleaf p. 115)

Scotland is one of the few remaining European strongholds for otters (*Lutra lutra*) and has perhaps 60 per cent of the EU's freshwater pearl mussels (*Margaritifera margaritifera*) (Green and Green, 1994; Scottish Executive, 2001a). Many lochs provide safe roosting places for wintering wildfowl, including several species of ducks and geese (Section 7.7.3), while rare birds such as red- and black-throated divers (*Gavia stellata*; *G. arctica*) breed in the intricate pool systems of wetland areas in the north and west (Fig. 6.2d). These wetlands and associated expanses of blanket peat bog (notably the Flow Country (Section 6.5.5.i)) are some of Scotland's most precious conservation resources. Although probably all the lochs and rivers have been affected by human activity, many can be regarded as semi-natural, being relatively undisturbed and unpolluted. In sum, then, Scotland has an abundance of diverse, high quality freshwater resources with which to meet its domestic, industrial, recreational and environmental requirements.

6.2 Historical background

Scotland's fresh waters have always been important for human settlement, culture and defence, as well as being a source of food and power. In the days before roads they also acted as key transport routes. The construction of major canals in the eighteenth and nineteenth centuries, such as the Forth & Clyde and the Caledonian, heightened the importance of water transport, and to this day the major river valleys remain the primary routeways around the country. Water is needed to satisfy a great diversity of human needs, primary amongst these being domestic supply, sewage treatment, agriculture, aquaculture, industry, hydro-electricity, inland navigation, recreation,

conservation and landscape value. Over the centuries, satisfying these needs has involved extensive modification of rivers, lochs and wetlands.

> The most striking thing that we have done to water . . . is to reorder it. Once it has fallen from the sky, it lies in quite different places and behaves in quite different ways from what it once did. (Smout, 2000a: 90)

In particular, this modification has resulted from regulation and from the use of freshwaters for waste disposal.

6.2.1 Regulation: water supply, flood management and HEP

6.2.1.i Water supply

The regulation of rivers and lochs to provide water for domestic, industrial and agricultural purposes has a long history (Gilvear, 1994; Macdonald, 1994). Rainfall is generally abundant, but the areas with the greatest amounts of water (the west and the uplands) are not those with the greatest need (the central belt and the lowlands). Transporting water from source to need has necessitated considerable engineering effort. As the size of settlements rapidly increased, especially Edinburgh and Glasgow in the nineteenth century, so water supply became an important concern. Reservoirs began to be built from the end of the eighteenth century, and large lochs began to be used from the mid-nineteenth century. The opening of the Loch Katrine system in 1859 was the trigger for numerous smaller schemes all over Scotland (Maitland and Hamilton, 1994), and capacity has steadily increased ever since (Fig. 6.3). This has involved both the construction of new systems and the augmentation of nineteenth-century schemes. A recurring theme has been the negotiation of compensation agreements as downstream users deprived of water by these schemes demand the provision of compensatory flows and freshets (artificial 'mini-floods' to stimulate fish movement in angling rivers). Such agreements sometimes necessitated the construction of further reservoirs, such as Loch Venachar downstream of Loch Katrine.

Of the ten largest water supply sources, six have been constructed in the last forty years, including a scheme which has, since 1971, effectively turned Loch Lomond into a regulated reservoir (Jolley, 2000). A feature of the recent extension of capacity has been increasing exploitation of springs and boreholes, notably for the burgeoning bottled water market. Until the 1970s, public water supplies came almost exclusively from surface sources but more recently the groundwater resources in Fife, Dumfries and Moray have been exploited, and there are now more than 30,000 private groundwater supplies. Despite recent growth, only 3.5 per cent of public water supplies originate from groundwater (SEPA, 1999). Controversially (and unusually), water abstractions from groundwater and surface sources remain unregulated.

6.2.1.ii Flood hazard management

Over the centuries numerous methods have been employed for controlling floods, including river straightening and the construction of embankments (levées) and

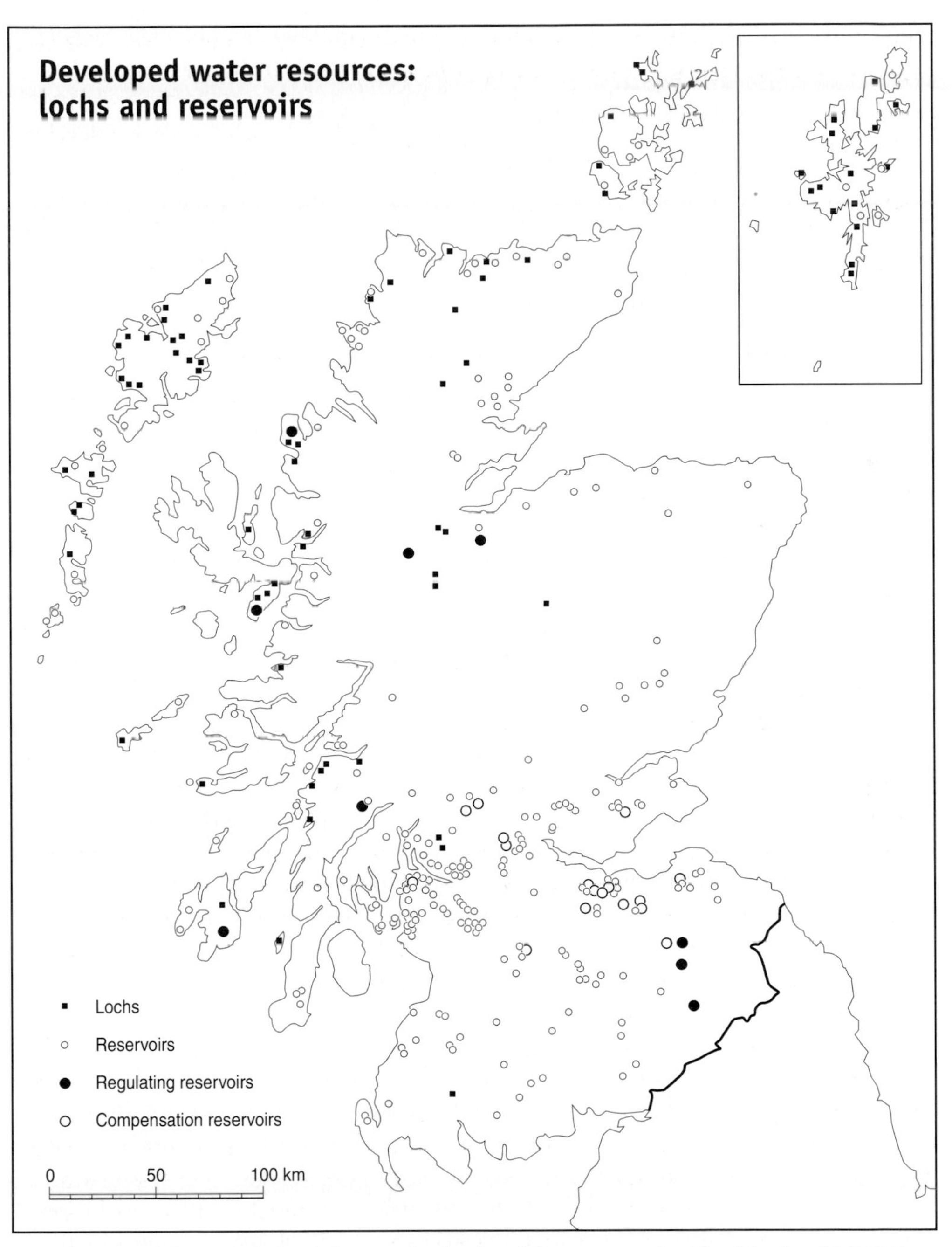

FIGURE 6.3 *The main lochs and reservoirs developed for water supply. After Macdonald (1994).*

sometimes reservoirs. During the eighteenth and nineteenth centuries, however, modification of rivers and wetlands occurred on an unprecedented scale as a consequence of the increasing demand for land for agriculture and industry, and little thought was given to the hydrological consequences. Floodplains were reclaimed, marshes drained, riparian woodlands and reed beds cut down, and in a few cases (e.g. Loch Leven, Fife) loch levels were lowered. One result of such changes was the loss of many fish stocks, especially of migratory and estuarine species (Gilvear *et al.*, 1995). Another was considerable exacerbation of the flood hazard as the intensive development of floodplains increased the physical and economic damage caused by floods. Piecemeal protection works were progressively carried out along many of the more flood-prone reaches, typically involving channelisation and levées. However, both these measures tended to worsen flooding downstream, and flood levées proved to offer false security, being repeatedly breached during large floods (Section 6.5.1.ii).

6.2.1.iii Hydro-electric power

As a mountainous country with abundant rainfall, Scotland has great potential for hydro-electric power (HEP), potential which has been extensively tapped. In the Highlands some 20 per cent of river catchments have been modified for hydropower generation (Fig. 6.4) satisfying around a quarter of Scotland's electricity needs (Butt and Twidell, 1997). The heyday of HEP development was 1945–65 when twenty-eight schemes were constructed, some involving remarkable dam, aquaduct and tunnel systems which allowed the combined potential of several catchments to be tapped (Johnson, 1994). The large storage reservoirs have proved helpful in managing flooding, often enabling peak flows to be significantly reduced, as well as supplementing low flows. After a lull in construction arising from economic factors and the increasing strength of the conservation interest, hydropower is experiencing a revival as a consequence of the emphasis on renewable energy and the Scottish Renewables Obligation. The number of private suppliers selling electricity via the national grid is increasing (Butt and Twidell, 1997), and the first major hydro scheme for thirty years, a 3MW plant near Ullapool, received planning consent in 2000. Many of the new developments are run-of-river schemes which require no dam. Theoretically, 1,000 MW of potential hydropower remain to be developed, but this will never be realised because of the impacts it would have on the landscape and on nature conservation. Nevertheless, the number of small scale, community-orientated schemes may well multiply in the future.

The net effect of meeting the needs of water supply, flood management and HEP is that many rivers and lochs have been extensively modified (Gilvear *et al.*, 1995, 2002). Natural river flows have been controlled through impoundment, inter-basin transfer, channelisation, and abstraction from springs, rivers and groundwater. The levels of many lochs have been raised, water levels are frequently changed at unnaturally fast rates, and river flows below HEP stations are subject to large, rapid and frequent changes. All these modifications have impacts on the flow regimes and quality of rivers and lochs, in turn altering habitats and affecting the species which depend on them (Gilvear, 1994). Obvious examples include the obstruction of fish migration by dams, and reduced river flows below HEP reservoirs. These problems are commonly mitigated, respectively, by fish passes and by compensation flows and freshets, but fluctuating water levels in reservoirs devastate the littoral flora and fauna. The outcome of two centuries of river and catchment management has been to change rivers 'from

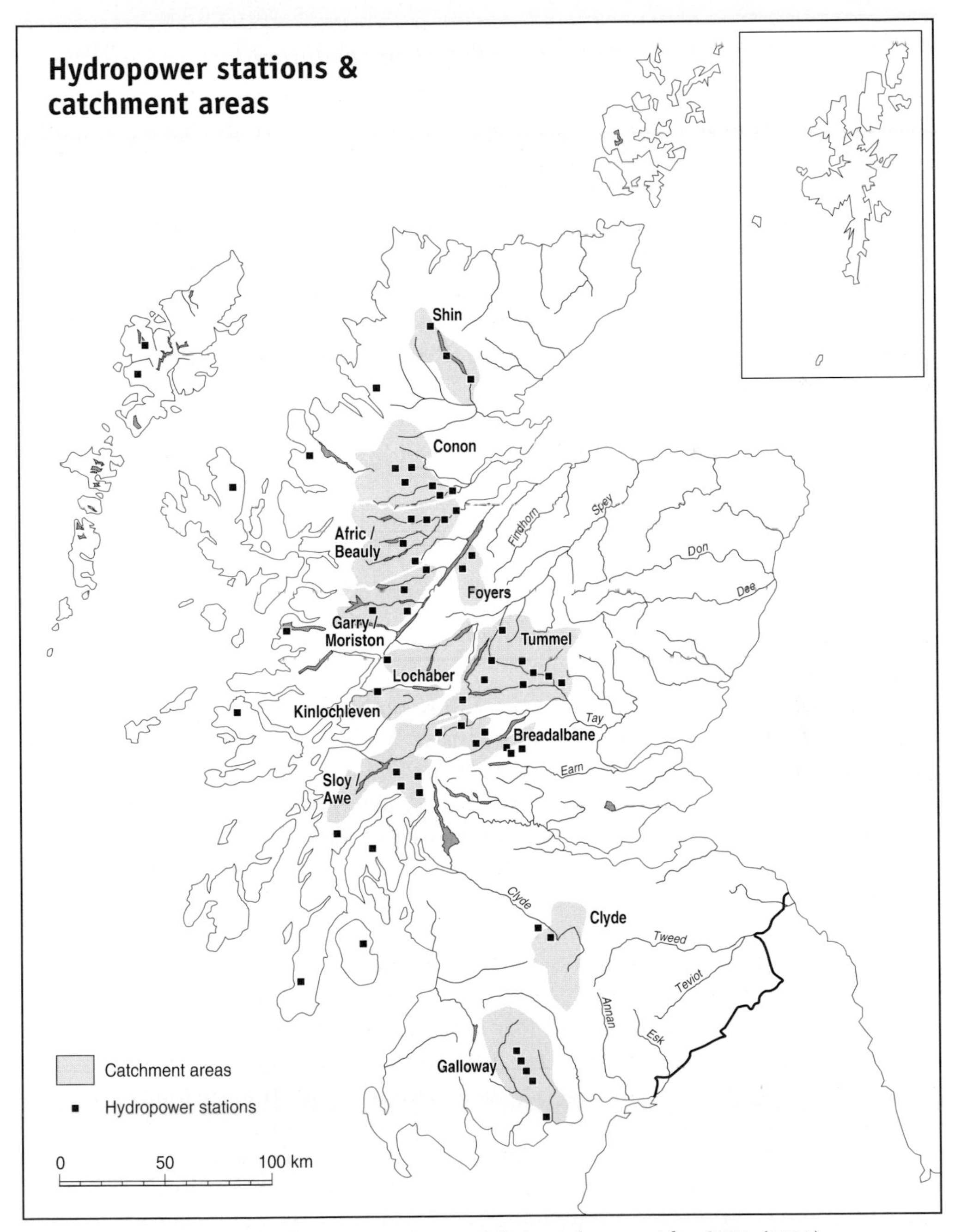

FIGURE 6.4 *Hydropower stations and their catchments. After SEPA (1999).*

sustainable ecological systems to highly controlled environments with few natural processes and habitats' (Gilvear *et al.*, 1995: 18). Wild rivers have been strait-jacketed.

6.2.2 Waste disposal: tackling pollution

The industrial revolution brought severe pollution problems to lowland Scotland as effluents from industry and the growing urban centres were discharged raw into rivers, lochs and estuaries. By the mid-nineteenth century many rivers in populous catchments had effectively become open sewers, devoid of all aquatic life; the Clyde was one of the worst affected (Hammerton, 1994). Great strides were taken during the later twentieth century in reducing point source pollution, especially following the creation in the 1950s of the River Purification Boards (RPBs). During the forty years of their existence, the RPBs made steady progress in cleaning up Scotland's rivers (D. W. Mackay, 1994), and during the 1990s several important steps were taken towards a more coordinated approach to tackling pollution. Notable amongst these were the introduction of Integrated Pollution Control under the 1990 Environmental Protection Act and the creation in 1996 of the Scottish Environment Protection Agency (SEPA). Control of point source pollution resulted in rapid improvements in water quality (Hammerton, 1994; SEPA, 1999) (Section 6.5.1.iii), with striking results such as the return of Atlantic salmon to the River Clyde and of salmon and sparling to the River Forth since the mid-1980s. The need to meet ever more stringent water quality requirements has necessitated the construction of more and progressively more efficient waste treatment plants, both by local authorities and by big industrial concerns such as BP.

Simultaneously, however, the effects of diffuse pollution from agriculture, forestry and atmospheric acidification grew in intensity, leading to increasing problems of eutrophication and acidification (Sections 6.5.2; 6.5.3). The spectacular growth of aquaculture in the last two decades has also caused water quality problems.[2] Fish farming (primarily of salmon) has been a remarkable success story in technical, economic and social terms. It is now an industry worth around £260 million per annum employing 6,500 people at almost 500 licensed sites; its annual output of 127,000 tons dwarfs the 200 tons of wild salmon taken by rod and net (Carrell, 2000, 2001). However, it has been increasingly dogged by controversies over its environmental and aesthetic impacts (Hambrey, 1997). Fish farms can introduce disease to wild fish, and the regular mass escapes of farmed fish (440,000 fish during 2000 alone)[3] pose a threat to the genetic integrity of native fish. In particular, aquaculture causes substantial nutrient pollution (MacGarvin, 2000), and solids from waste food and faeces cause deoxygenation and siltation, harming shellfish cultivation. Mounting concern has now led SNH to call for an independent inquiry into the industry (Carrell, 2001). In combination, the effects of industry, farming, forestry and aquaculture continue to downgrade the quality of freshwater in parts of Scotland. One consequence is that the fish communities of many large rivers are still substantially impoverished (Maitland, 1994). Despite considerable progress towards reversing the damaging effects of the centuries during which freshwaters were used indiscriminately for waste disposal, water quality problems remain.

6.3 MANAGEMENT FRAMEWORK

All aspects of the Scottish water industry remain in the public sector (in contrast to the privatised utilities in England and Wales), but many organisations and bodies are involved (Table 6.2). The array of stakeholders with an interest in water resource management is perhaps greater than for any other sector. This gives policy makers a dilemma. Management can either be carried out by an all-embracing statutory agency or by a series of specifically targeted organisations. The former approach, adopted south of the Border, can offer integrated management but runs the risk of being compromised by unwieldy bureaucracy. The latter, which allows for focused, locally orientated management but lacks a wider holistic perspective, has characterised the Scottish approach.

TABLE 6.2 A hierarchy of principal players in water resource management.

European Union	European Commission European Parliament
National and local government	UK Government Scottish Parliament Scottish Executive Rural Affairs Dept. (SEERAD) Local authorities
Government agencies and public bodies	Scottish Environment Protection Agency (SEPA) Scottish Natural Heritage (SNH) Water Authorities (East, West and North of Scotland Water) District Salmon Fisheries Boards Forestry Commission Government research institutes, e.g. CEH, MLURI
Non-Governmental Organisations	Conservation groups, e.g. WWFS, SWT, NTS Landowning interests, e.g. SLF, NFUS
Companies and private interests	Industrial interests, e.g. CBI; Alcan Business interests, e.g. whisky distillers; brewers Electricity companies, e.g. Scottish Hydro-Electric Local communities Riparian owners/occupiers, e.g. farmers Individuals

Although the arrangements for managing Scotland's freshwaters differ significantly from those in England and Wales, there has been a measure of convergence in recent years (Wright, 1995; Cook, 1998). South of the Border a considerable degree of integrated management has long been the rule, but until the mid-1990s water resource management in Scotland was piecemeal, a situation characterised as 'event-led chaos' (SWCL, 1994: 9). The RPBs were responsible for water quality, District Salmon Fishery Boards for salmonids, landowners for other fish, the Scottish Office for drainage, and local authorities for water supply and flood control in urban areas (Campbell *et al.*, 1994). This situation improved somewhat in 1996 with the creation of SEPA from the merger of the RPBs and Her Majesty's Industrial Pollution Inspectorate. This brought together control of water, air and land pollution under a single management structure, and created a proactive environmental body in place of its reactive, regulative forerunners (Lloyd, 1999).

However, despite the many calls for SEPA to be given comprehensive responsibility for water resource management (SWCL, 1994; Howell, 1994), its remit was limited primarily to pollution monitoring and control; it was not given a duty to implement integrated, sustainable water management (although it has adopted the principles of sustainable development (SEPA, 1998)). Its relative lack of local democratic accountability has also been criticised (Lloyd, 1999). Responsibility for water supply and sewerage was given to three new Water Authorities – East, West and North of Scotland Water – which came into being alongside SEPA in 1996 and which are not part of local government. Responsibility for the conservation of aquatic landscapes, habitats and biota is shared between SEPA, the Water Authorities and SNH, while flood defence and land drainage remain primarily the responsibility of landowners and local authorities. Despite the major restructuring and rationalisation in 1996, water resource management therefore remains highly fragmented (Werritty, 1997).

6.4 Recent developments and trends

6.4.1 Conflict with the EU

The UK's membership of the EU has had a profound impact on the management of water resources. The traditional British approach to environmental regulation contrasts markedly with the European style (Section 2.2.2), and there are few arenas in which this has been more apparent than in the management of water resources. In Ward's (1998: 244) view, 'Britain's relationship with European water quality policy has been one of the most turbulent and controversial aspects of the Europeanisation of environmental policy'. The UK has twice been found guilty of being in breach of EU water quality laws, and almost every EU Directive has been fought tooth and nail by the UK. The recent Water Framework Directive was no exception.

The UK has frequently argued that emission standards should be set by national governments according to local environmental conditions, and has objected to the EU's penchant for uniform emission and water quality standards. In practice, however, the EU's 'one size fits all' approach has repeatedly prevailed over the UK's flexible style. Similarly, the monitoring required under EU water Directives has forced the UK to shift its focus from controlling point sources of pollution to addressing less visible, chronic forms of diffuse pollution, especially those caused by farming (Section 6.5.3) (Ward, 1998). If controversy and resistance were the hallmarks of the first two decades of the UK's membership of the EU, the 1990s saw an increasing convergence of principles and approaches between the UK and the rest of Europe in terms of water policy. This reduction of acrimony has been aided by increasing amounts of prior consultation by the Commission, and now by the rationalisation of the plethora of Directives under the Framework Directive (Section 6.4.4).

6.4.2 Integrated Catchment Management

Of all types of natural ecosystem, fresh waters are perhaps most in need of integrated management because they are subject to a wider range of use than almost any other, making conflicts between competing uses inevitable (Boon, 1994). For the same reason, however, truly integrated management is perhaps hardest to achieve, requiring consideration of so many disparate issues. Integrated catchment management (ICM) is nevertheless the best long-term strategy because consideration can be given to all forms of water and land use within a river basin (Maitland and Morgan, 1997; Cresser and Pugh, 1999). There are many merits in managing water resources at the catchment scale rather than at the level of individual rivers, lakes and wetlands (Werritty, 1995, 1997; Newson, 1999; Tané, 1999). Firstly it recognises the indivisible nature of hydrological systems. Second, it highlights the ways in which the physical environment determines both flow regime and water quality. Thirdly, it constitutes a decisive step away from sectoral, site-based perspectives, and towards a holistic conception of drainage basins as ecosystems. Fourthly, it allows land and water to be managed together. Lastly, it is probably the only effective approach to tackling the problem of diffuse pollution.

The aim of ICM is 'to conserve, enhance and . . . restore the total river environment through effective land and resource planning across the whole catchment area' (Gardiner and Cole, 1992 *in* Werritty, 1997: 492). The importance of managing drainage basins as single, integrated entities has been internationally accepted for many years (Jamieson and Sheldon, 1994; Werritty, 1995). The example of the Loch Katrine catchment, managed in an integrated fashion since the late nineteenth century, demonstrates the effectiveness of this approach (Campbell and Maitland, 1999). Although none of the statutory agencies have had a duty to promote ICM, holistic water resource planning has nevertheless been adopted in places by both the private and public sectors, as well as through public/private partnerships. SEPA is involved in twenty such co-operative management schemes (SEPA, 1999), successful examples of which include the River Valleys Project (Edwards-Jones, 1997a, 1997b) and the ICM plan for Loch Leven (Section 6.5.3). Most of the numerous private sector initiatives have focused on enhancing salmonid fisheries (Werritty, 1995).

But despite several on-going success stories, and notwithstanding the widespread support for the concept of ICM (SWCL, 1994; FoES, 1996), it has had a hesitant, sporadic start in Scotland. Compared to the challenge of integrated management in the world's large, multi-nation catchments (Newson, 1999), ICM in Scotland should be relatively easy, but major barriers exist. These include the lack of expertise in multi-sectoral coordination, the difficulty in persuading stakeholders to compromise for the general good, and (as is so often the case) the ingrained habits of sectoral thinking and practices (Section 2.4.3). Until these can be overcome, 'repeated chanting of the mantra of ICM will not yield success in terms of the sustainable development of freshwater resources' (Werritty, 1997: 504). This gulf between aspirations and reality is now set to be bridged because of the requirement (under the EU Framework Directive (Section 6.4.4)) for an ICM system to be in place by 2003.

In parallel with this move towards a holistic management perspective has been the opening up of the policy making process to a much wider range of actors. This has been driven by a combination of on-going Europeanisation and the current emphasis on participatory democracy (Ward, 1998). Prior to the 1980s, water resource planning was the preserve of a closed policy community of technical specialists and regulatory officials, whereas today participants include government officials (at EU, national and local level), representatives of industry associations, environmental groups, consumer groups and local communities. Much greater public and political debate now fashions water policy, and, as in so many sectors, the partnership approach has come to the fore. The ongoing work of voluntary stakeholder partnerships such as the Tweed Foundation should progressively shrink the gaps between sectoral perspectives.

6.4.3 Natural management

Recent years have seen a growing preference for natural approaches to managing freshwater resources in place of the 'hard engineering' solutions that have often been favoured in the past. Historically the overriding objective has been to improve the hydraulic efficiency of channels using techniques such as river resectioning (deepening, widening and straightening channels), re-profiling, and the reinforcement and raising of river banks. But in removing sediment, vegetation and meanders, and in engineering artificial channels and banks, aesthetics and ecology are sacrificed on the altar of efficiency, sometimes turning rivers into little more than drains (Cook, 1998). Moreover, the traditional 'hard engineering' responses to river problems are expensive both to create and to undo, and tend to accelerate flow velocity and erosion. Far from solving problems they simply pass them on downstream. Increasingly, therefore, they have been criticised as being economically, ecologically and aesthetically unsustainable (Leighton, 1999; Gilvear *et al.*, 2002). In their place, 'soft engineering' approaches are advocated whereby natural processes and features are harnessed to provide solutions. For example, the idea of 'letting floodplains perform their natural function' is gaining currency, and (where it does not compromise other land uses) rivers are being allowed – and sometimes encouraged – to rediscover their natural courses.

Examples of this move towards natural, sustainable solutions include the following:

- WWFS's *Wild Rivers* project (Gilvear *et al.*, 1995; Leighton, 1999). This river restoration initiative, part of their Europe-wide *Living Rivers* project (WWF, 1999), adopts a holistic perspective towards river management, and aims for a reduction in land use intensity in catchments, especially on floodplains. The objective is to reverse the long-standing trend towards ever greater control of rivers – to remove the strait-jacket.
- Two recent SEPA initiatives: the Habitat Enhancement Initiative (HEI) and Sustainable Urban Drainage Systems (SUDS).[4] The first of these, launched in 1998, aims for measurable improvements in the quality of habitat management, focusing on the conservation of biodiversity in aquatic and riparian habitats. The HEI is closely integrated with Local Biodiversity Action Plans (Section 8.2.3) and also with SUDS. The latter represent sustainable alternatives to conventional

urban drainage systems. They aim to reduce the risk of pollution and flooding, and to improve amenity and biodiversity in urban areas. Like the soft engineering approach, it does this by constructing or utilising 'natural' structures such as retention ponds, detention basins and constructed (storm water) wetlands, all integrated into a network of habitats and wildlife corridors.

- Scottish Native Woods' initiative to restore riparian woodland (SNW, 2000). Floodplain forests are now some of Scotland's rarest habitats but are rarely the focus of conservation efforts (Leighton, 1999).
- Various 'green' approaches to the treatment of minewater effluent and sewage effluent such as the successful use of reed beds at Valleyfield in Fife.

6.4.4 EU Water Framework Directive

Agreed in 2000 after three years of negotiation, this is one of the most important and wide-ranging policy intitiatives ever undertaken by the EU. It came into force at the end of 2000 and will be transposed into Scottish law by 2003. In replacing many of the current Directives, it is set to become the main framework for the sustainable management of all surface and groundwaters for decades to come (Rogers, 2000). It shifts the focus of water management from chemical quality targets to ecological effects, and its ambitious aims go beyond the purely scientific to incorporate social dimensions (Pollard, 2000). A prime aim is to manage human pressures on the hydrology of rivers and lochs so effectively that their ecological status is as near-natural as possible. This will require a comprehensive system of controls on both point and diffuse sources of pollution, and (for the first time in Scotland) on water abstractions and impoundments (Marsden, 2000). It is likely that some or all of the considerable costs of achieving the exacting new standards will be passed on to consumers.

Significantly, the Directive will require the development of ICM through River Basin District Plans. These will involve wide-ranging programmes of monitoring, consultation and measures to attain specified objectives. They are to be drawn up by 2010 and thereafter will be reviewed every six years. In Scotland, there will probably be three districts with sub-basin plans at the scale of large individual catchments. The Framework Directive provides a unique opportunity for 'joined-up' environmental policy making, but problematic issues of definition, authority and resourcing will make its implementation extremely challenging, both technically and politically (Marsden, 2000; Rogers, 2000).

6.5 CURRENT ISSUES AND DEBATES

6.5.1 Water quantity and water quality

6.5.1.i Water supply

At a national level, water supply is unlikely to be a problem in the medium term future. Only 1 per cent of available runoff is used to satisfy public demand (FoES, 1996). Exploitable water resources are equivalent to 16,000m^3 per annum per person, almost eight times the figure for the UK as a whole (SEPA, 1999). Total demand of

2,206 Ml d^{-1} (million litres per day) compares with available supplies from all sources of 3,560 Ml d^{-1}, giving a demand/yield ratio of 0.62 (Cook, 1998). Combined with an average annual growth in demand of just 1.7 Ml d^{-1}, this suggests little need for action to augment the quantity of existing resources, although there is always room for improvement in quality and in efficiency (Wright, 1995). However, there are regions where complacency is inappropriate, notably around Glasgow and Edinburgh where systems are already operating at capacity and are in need of substantial upgrading to meet EU standards.

6.5.1.ii Flood management

The situation concerning the management of flood risk is much less rosy, with land use change and changing climatic patterns greatly worsening an already far from satisfactory situation. Afforestation, changing agricultural practices and urbanisation all exacerbate the flood risk. The last two decades have also been the wettest on record in western and northern Scotland (Werritty, 2002). After the relatively 'flood-poor' 1960s and 1970s, the period since 1988 has seen eight of Scotland's sixteen largest gauged rivers experience their maximum recorded flows, with a succession of extreme floods on the Tay, in Moray and in central west Scotland causing cumulative financial damages in excess of £150 million (Black and Burns, 2002; Table 6.3). Furthermore, climate change scenarios suggest increasing flood risk in future, especially in winter (Werritty, 2002; Wright, 2000).

TABLE 6.3 The maximum recorded floods for the largest rivers in Scotland prior to 1988 and greater subsequent maxima. The table shows data for gauging stations at the outfall of all catchments with areas greater than 500km², listed in decreasing size. 1 cumec = 1 m³/second. After Black and Burns (2002).

River	*Year record began*	*Maximum flood to 1988*		*New maximum since 1988*	
		Discharge (cumecs)	*Date*	*Discharge (cumecs)*	*Date*
Tay	1948	1,890	2–1950	2,268	1–1993
Tweed	1962	1,556	1–1982		
Spey	1952	1,594	8–1970		
Ness	1930/73	594	?–1937	669	2–1989
Clyde	1958	670	9–1985	830	12–1994
Dee	1934	1,134	1–1937		
Don	1969	286	8–1970	313	9–1995
Deveron	1960	521	5–1968		
Annan	1967	473	10–1977		
Nith	1957	986	1–1962		
Leven	1963	151	1–1974	197	3–1990
Findhorn	1958	2,410	8–1970		
Earn	1951	305	2–1948	358	1–1993
Helmsdale	1975	273	12–1985	287	10–1993
Teith	1963	247	1–1975	374	1–1992
Whiteadder	1969	280	11–1984		

Such events and evidence concentrates minds. By the mid-1990s there was general acknowledgement that the flood problem needed urgent attention, and the result has been rapid progress in flood hazard management (Riddell, 2000). In particular, NPPG7 – *Flooding and Planning* (SOED, 1995) – and the 1997 Flood Prevention and Land Drainage Act represent major steps forward. Planning authorities now have to take flood risk into account in the planning process, and can refuse planning permission on that basis. But despite this progress, responsibility for flood control remains piecemeal, and flood management is frequently reactive. Local authorities often do not have adequate knowledge, expertise or (until very recently) sufficient public funds to carry out their duties under the 1997 Act. Furthermore, although SEPA has a remit to investigate flood risk and operates twenty-eight flood warning schemes, its effectiveness in this regard is limited by inadequate resources (Riddell, 2000). Although local authorities have certain flood prevention duties in urban areas, the prime responsibility for reducing flood risk still ultimately rests with property owners.

6.5.1.iii Water quality

Maintaining and improving water quality is of great importance.[5] Drinkable water is obviously a basic need for all animal species (humans included!), but high quality water is important in many other ways. Its importance for industrial operations such as paper-making is obvious, but, less tangibly, Scotland's freshwaters have an international image of purity which underpins the success of many sports and industries, most notably game fishing, whisky production and bottled water (CWP, 1993). Furthermore, this perception of Scotland as a land of unspoilt lochs and rivers teeming with wildlife is a priceless asset for the tourist industry.

Strictly speaking, water quality is determined by natural factors (climate, vegetation, lithology) combined with land use and water use (farming, industry, domestic), and has traditionally been defined using physical, chemical and biological criteria. More recently, however, 'quality' has come to embrace a much broader agenda because a lot of water use is now non-consumptive (Boon and Howell, 1997). These wider criteria include, *inter alia*, wildlife, recreation, amenity and landscape. Increasingly these are being incorporated in official water quality assessments through, for example, the use of River Habitat Assessment methodology (SEPA, 1999). If judged against such criteria, Edwards-Jones (1997a) argues that water quality has been declining as a result of the progressive modification and regulation of lochs and rivers.

However, if defined in chemical and biological terms the quality of Scottish freshwaters has improved dramatically in recent decades. Excellent or good water quality now characterises 91 per cent of the rivers and 95 per cent of the lochs monitored by SEPA, even though problems remain in some of the most intensively populated and farmed regions (SEPA, 1999). These problems derive from a number of well known sources of pollution (Table 6.4). Of these, the two most serious are diffuse agricultural pollution and sewage effluent which, between them, affect 60 per cent of polluted rivers and 53 per cent of polluted lochs. The numerous impacts of agricultural practice on water quality are discussed below. Most sewage pollution results from water authority discharges. Since the water authorities raise their revenues *via* water charges, most of the estimated outlay of £5 billion which is required to improve and/or replace antiquated sewage treatment plants and water infrastructure will have to be borne by their domestic and business consumers (FoES, 1996).

TABLE 6.4 The nine most important causes of freshwater pollution in Scotland in 1996, listed in order of importance for river pollution. These sources account for some 99% of polluted waters. The percentage figures sum to more than 100 because some water bodies are classified as polluted for more than one reason. After SEPA (1999).

Type of pollution	*Source of pollution*	*Polluted water bodies affected*	
		Rivers	*Lochs*
1 Sewage effluent	Sewage treatment works; sewage overflows	34%	19%
2 Agriculture: diffuse sources	Runoff of pesticides, fertilisers, herbicides, organic wastes, soil, and nutrients	26%	34%
3 Acidification	Deposition of acidic compounds from combustion processes	12%	53%
4 Urban drainage	Drainage from roads, roofs & paved areas contaminated by metals and oils	11%	0.1%
5 Mine drainage	Discharge from abandoned mines – highly acidic water containing dissolved metals	9%	0%
6 Agriculture: point sources	Leaks from silage, slurry & fuel oil stores; spillage of organophosphate sheep dips	6%	0%
7 Industrial effluent	Waste water from industry	2%	0%
8 Forestry	Runoff of fertilisers, soil, nutrients, heavy metals & atmospheric pollutants	0.4%	9%
9 Fish farming	Waste food, faeces, nutrients & chemicals	0.3%	16%

6.5.2 Forestry and water

All stages of forestry, from ground preparation and planting through to felling and restocking, have impacts on fresh waters and their biota (Maitland *et al.*, 1990; FC, 2000b). In general terms, planting trees within a catchment will reduce both the quantity and (at times) the quality of river discharge. The primary effects are hydrological and chemical, changes which have implications for aquatic fauna and flora (Best, 1994; Johnson, 1995; Nisbet, 2001).[6]

6.5.2.i Hydrological effects

Afforestation often results in increased water losses through interception, evaporation and transpiration, and reduced runoff during droughts, but also higher and more rapidly attained flood peaks resulting from efficient forest drainage networks. It is estimated that each 10 per cent of a catchment that is planted will result in a 2 per cent reduction in annual runoff (Best, 1994), so the amount of water available for power generation or water supply can be significantly reduced (Johnson, 1994). Forest schemes close to areas of blanket peat and wetlands can lower the water table and damage the habitat (Shotbolt *et al.*, 1998). Ground preparation, forest road construction and accelerated runoff all mobilise sediments and can trigger erosion, resulting in increased sedimentation in streams and lakes. As forests mature the problems of erosion and excessive runoff diminish, but summer water temperatures may then be lowered by tree shading. Felling operations again increase nutrient loading and sediment yields.

6.5.2.ii Chemical effects

Forestry is often associated with increasing nutrient loads in runoff as a result of leaching of exposed soils and of applied fertilisers (most commonly rock phosphate, but also nitrogen and potassium). However, one of the greatest concerns has been the acidification of fresh waters following afforestation, and the release of harmful levels of dissolved aluminium (Best, 1994; Morrison, 1994). Given the dominantly acid geology and limited soil buffering capacity across much of Scotland, many fresh waters have a naturally low pH and/or are susceptible to acidification (Fig. 6.5). Afforestation can all too easily be the trigger which pushes the pH below a critical threshold. This is because tree canopies enhance the dry deposition of contaminants from the atmosphere and scavenge pollutants from low cloud and mist (especially at forest edges), so runoff from coniferous forests can be significantly acidified. Acidity increases as forests mature. Work in Galloway shows that afforestation can result in acidification not just at the catchment level but regionally (Puhr *et al.*, 2000).

6.5.2.iii Impacts on biota

All these hydrological and chemical changes have an impact upon freshwater plants and animals, often making conditions more extreme (Best, 1994; Morrison, 1994; Maitland and Morgan, 1997). Increased turbidity and siltation affects aquatic plants, invertebrates and fish. Aluminium becomes soluble in acid waters and is toxic to fish even at concentrations below 0.1 mg/l. Elevated nutrient levels lead to increases in primary productivity and changing species composition, as well as excessive algae which can be of concern in water supply reservoirs. Invertebrate diversity tends to be lower in afforested catchments, and fish catches tend to decline with increasing acidity. In the worst cases, acidification can eliminate all invertebrates and most plants. Although the impacts on fish populations have been much less marked than in Scandinavia, a high proportion of waters have become acidified to some extent, significantly affecting many nursery streams for salmonids. It is possible to counteract

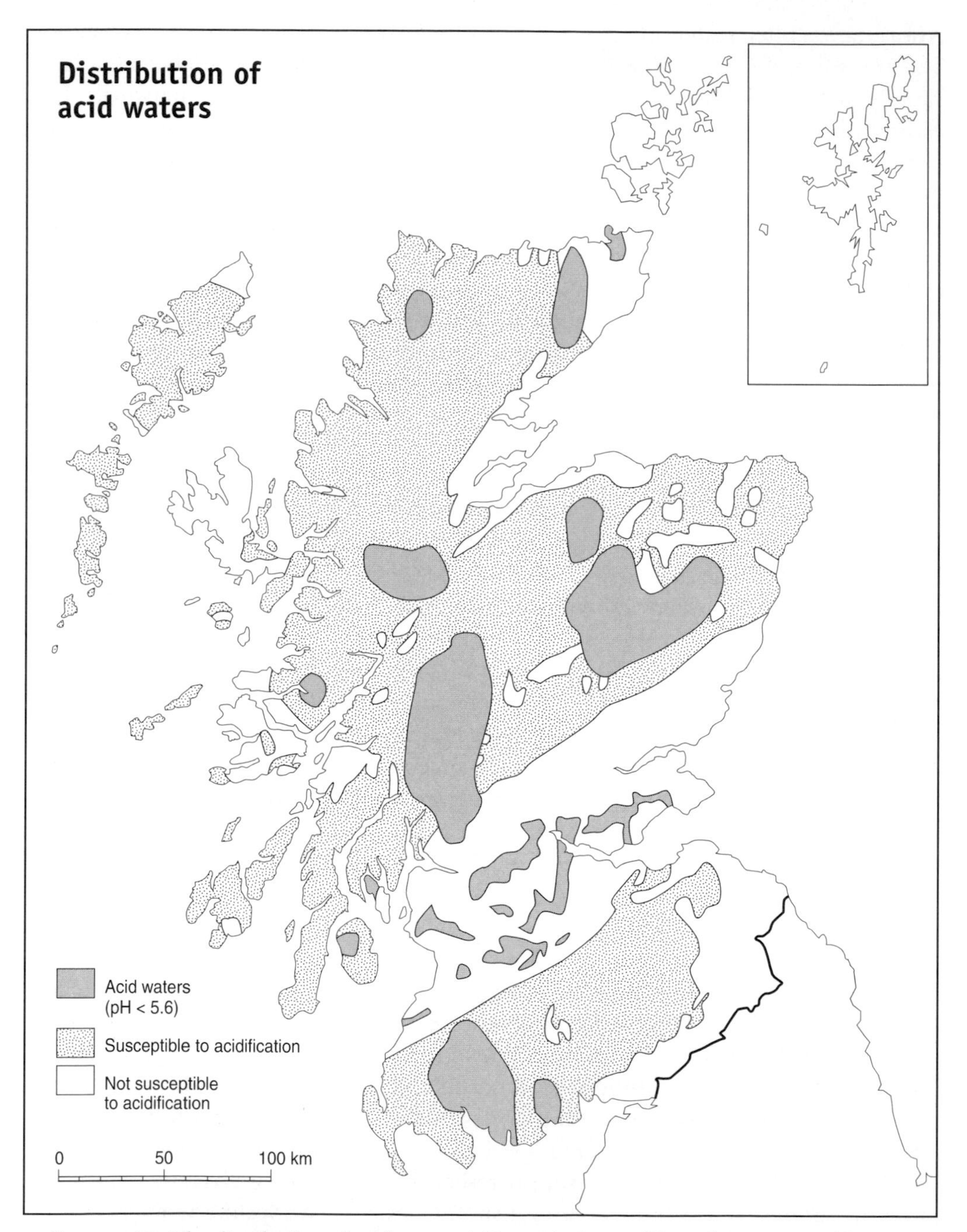

FIGURE 6.5 *The distribution of acid waters (pH <5.6). Susceptible and non-susceptible areas are classified on the basis of solid geology and of regions with extensive areas of deep peat with acid runoff. After Morrison (1994).*

the effects by applying calcium carbonate to the catchment, as at Loch Dee and Loch Fleet in Galloway in the 1980s, but such restoration projects are expensive (£1.5 million for Loch Fleet) and the effects often temporary (Maitland and Morgan, 1997). They may nevertheless have a role to play in saving important local stocks of fish such as the Arctic charr in Loch Doon.

The increasing weight of evidence that coniferous plantations can have this wide range of effects led to the formulation of the Forestry Commission's *Forests and Water Guidelines* (FC, 2000b). Since first published in 1988 they have had a significant effect on the design of forest planting schemes and have helped to minimise the impacts of afforestation on water quality. *Inter alia*, the guidelines recommend a 'minimal cultivation' approach, and buffer strips of unplanted and undrained ground between the forest edge and watercourses; these trap silt and reduce nutrient inputs to streams. Scrutiny of forestry grant applications now takes full account of freshwater impacts, especially to ensure that the critical load of pollutants into soils and waters is not exceeded; this is particularly important in many parts of western and northern Scotland where buffering capacities are low. Available evidence, including a study of the sensitive upland catchment of the River Halladale in Sutherland, suggests that the guidelines are effective in reducing diffuse pollution from forestry (Nisbet, 1999, 2001).

Most research and hence most of our knowledge of the hydrological and chemical impacts of afforestation has addressed coniferous plantations. Only recently have the impacts of broadleaved woodlands on catchments begun to be addressed, accompanying the growing focus on broadleaved species during the last fifteen years (Institute of Hydrology, 1998). In contrast to the acidifying potential of coniferous woodland, new broadleaf woodland may ameliorate the pollution of surface and groundwaters from agricultural, domestic and industrial sources. On the other hand, if large areas of broadleaf woodland are planted in the drier lowland areas of eastern Scotland, the cumulative impact on water quantity could become significant. Short-rotation coppice, in particular, has very high transpiration rates (Institute of Hydrology, 1998).

6.5.3 Agriculture and water

Farming has many effects on water resources, both as a substantial consumer of water and through its impacts on water quality (Winter, 1996; Thompson, 1999). Scottish agriculture annually consumes some 50 million m^3 of water through such uses as irrigation, animal consumption, cleaning and crop spraying; the total area irrigated doubled between 1982 and 1996 (Egdell, 1999). The increasingly intensive nature of farming has impacted river catchments in several important ways, notably through land drainage, abstraction, soil erosion and pollution (Gilvear *et al.*, 1995). Land drainage, which used to be grant-aided, can increase flood peaks downstream and reduce the time lag between rainfall and peak runoff. It can also lead to lower flows in dry weather which can increase the need for irrigation. In turn, this can lead to over-abstraction from watercourses and groundwater, leading to lower flows and a reduction in the capacity of rivers to dilute pollutants. Rapid runoff tends to remove nutrients and agrochemicals from the fields, as well as soil, with adverse downstream effects. Soil erosion problems have been exacerbated by increased cultivation of winter cereal crops and by widespread overgrazing by sheep which has, amongst other impacts, severely reduced riparian vegetation cover. Point source pollution of surface and groundwaters is a serious issue (Table 6.4), especially resulting from spillage of sheep dip and of farm wastes such as silage liquor and slurry which have a high biochemical oxygen demand. These can cause rapid deoxygenation of watercourses and the death of invertebrates and fish (Allcock and Buchanan, 1994).

Arguably the most pervasive impact of farming on fresh waters is diffuse pollution (Table 6.4) (Skinner *et al.*, 1997). This is set to be the most important cause of river pollution by 2010 (SEPA, 1999). Concentrations of nitrogren and phosphorus are often high in agricultural catchments (Edwards *et al.*, 2000), both in surface and groundwaters, and this not only puts aquatic life but human health at risk. In particular, it can lead to eutrophication (excess nutrient enrichment). In turn, this causes blooms of blue-green algae, deoxygenation, and a host of damaging effects on aquatic biota including fish extinctions (Bailey-Watts, 1994). Almost eighty lochs and several of the smaller east coast rivers are prone to eutrophication (SEPA, 1999), but two especially high profile examples in recent years have been Loch Leven in Fife (caused by phosphates) and the River Ythan north of Aberdeen (due to nitrates) (Fig. 6.1).

1. **Loch Leven**. Loch Leven is a world-famous brown trout fishery and is also a NNR on account of its importance for breeding and migratory wildfowl, but it is prone to algal blooms in warm summers (Crofts, 1995). Runoff from the surrounding farmland and from sewage works is rich in phosphates which then build up in the shallow loch. The only solution is to reduce these inputs. The practical difficulty of achieving this emphasises the dictum that 'prevention is better than cure' (Bailey-Watts, 1994). Since 1995 the efforts of the Loch Leven Catchment Management Project have led to falling phosphorus concentrations but biological water quality is still not good (SEPA, 1999).
2. **The River Ythan**. The Ythan catchment is 95 per cent agricultural. In 1993 the entire 698km^2 catchment was proposed for designation as a Nitrate Vulnerable Zone (NVZ) under the EU's 1991 Nitrates Directive. This was in response to high nitrate levels in the river and the increasing occurrence of eutrophic conditions in the estuary which is an internationally recognised bird sanctuary and part of a SPA (Inskipp, 1997). These conditions were widely blamed on agricultural practices on the basis that 20 per cent of the fertiliser applied to the land ends up in the estuary and that nitrate concentrations have been steadily rising since 1958 (Balls *et al.*, 1995; Gilvear *et al.*, 1995). Establishing a link between high levels of nitrate and increased growth of algae proved problematic, however, and that, combined with opposition from the farming lobby, delayed NVZ designation until 2000. The Scottish Executive then consulted in 2001 on a proposed Action Programme. A second, much smaller NVZ exists around Balmalcolm, Fife, to protect a groundwater source from the impacts of intensive horticulture. In both cases, applications of agrochemicals to the land are closely controlled, and grants are available to offset farmers' expenditure on

water quality measures. The Scottish Executive is trying to resist EU pressure to create further NVZs in several other east coast catchments in which nitrate levels are high, such as the River Eden in Fife.[7]

The combination of the use of the NVZ designation and the 1997 *Code of Good Practice for the Prevention of Environmental Pollution from Agricultural Activity* is likely to reduce the incidence and severity of both point source and diffuse pollution from farming in the future. Further reductions might be obtained by applying the buffer zone approach adopted in forestry (Section 6.5.2). Several other current trends are proving to have positive knock-ons for water resource management and for soil and water conservation. Among these are the spread of farm woodlands (Section 11.2.2), organic farming (Section 5.3.2), community forestry (Section 4.5.3) and sensitive land management in Environmentally Sensitive Areas (e.g. low stocking densities; zero fertiliser regimes (Section 5.2.2.i)). Nevertheless, SEPA (1999) believes that major improvements in water quality will not come unless or until the CAP is reformed and the incentives for intensive agriculture are reduced.

6.5.4 Recreation and water

The perception of Scotland's lochs and rivers as pristine and beautiful has been attracting tourists for almost two centuries, but recent decades have seen an explosion of interest in recreation on, in and around water bodies. Rivers and lochs are now some of the most frequently visited destinations for recreation, and the popularity of water sports continues to grow. The range of pursuits is considerable (Table 6.5). Many of them need shore-based facilities such as jetties, boathouses or hides, and most require access (either pedestrian or vehicular) to the water's edge. All this puts inevitable pressures on the landscape and the environment (Sidaway, 1994; Walker, 1994; Bannan *et al.*, 2000; Dickinson, 2000a, 2000b). Such pressures include disturbance of wildlife, bank erosion, increased turbidity, pollution, damage to vegetation, and

TABLE 6.5 The main types of water-related recreation.

Water-based		*Waterside*	
Swimming & diving	Sailing	Angling	Walking
Rowing	Windsurfing	Wildfowling	Jogging
Canoeing	Power-boating	Bird watching	Camping
White-water rafting	Water skiing	Picknicking	Bicycling
Gorge walking	Jetskiing	Driving	Pony trekking

litter (with damaging impacts on wildlife from, for example, discarded fishing lines and poisoning by lead fishing weights). A less tangible but often important consequence of recreation is the diminished sense of naturalness, whether from visual or audio intrusion, which devalues the recreational experience. Although such impacts can be significant locally, especially in the most popular areas, they are not substantial when compared with those from other land and water uses (such as those discussed above). In terms of environmental impact, Dickinson (2000a: 47) therefore suggests that 'recreation is not the villain of the piece, except in a very limited number of locations'.

Conflicts that arise in connection with recreational uses of water are often more to do with friction between different forms of recreation than with environmental damage *per se* (Gittins, 1999). Some recreational pursuits are clearly mutually exclusive (bird watching and jetskiing, for example), and increasing visitor numbers can spoil the recreational experience. Competing demands continue to cause numerous conflicts, friction between anglers and canoeists being an oft-cited example, and the new legal right of responsible access to inland waters (Section 3.4.2) will probably result in increased visitor pressure on the resource. The site at which conflicts have been most intense is Loch Lomond (Fig. 6.6). The competing demands on and around the loch were judged complex enough by 1991 to warrant a government-appointed working party which produced a management strategy (LLTWP, 1993). Since that time the pressures have continued to grow. Peak numbers of recreational craft on the loch have nearly doubled since 1989, and visitors now make perhaps two million visits each year (Bissett *et al.*, 2000; Dickinson, 2000b). Impacts are exacerbated by the very concentrated nature of visitor use, both in time (summer weekends) and space (popular loch shore locations). Resolving the long-standing land use and recreational conflicts will be one of the most pressing tasks for the new Loch Lomond and the Trossachs National Park Authority (Section 8.3.4).

6.5.5 Conservation and water

6.5.5.i Freshwater ecosystems

Freshwater ecosystems and their biodiversity are every bit as deserving of conservation as their terrestrial counterparts, but the approaches used to conserve the latter may be inappropriate or even counterproductive for the former (Moss, 2000). This is because the medium (water) rather than the organisms determines the structure of freshwater ecosystems, and because it is even harder to draw boundaries around areas of high quality. No individual species can be conserved without giving thought to the protection of the entire system.[8] At present, however, much effort in conserving freshwater biodiversity is species-orientated or site-based (SSSIs, Ramsar sites, SACs), with insufficient reference to the sources of water on which the sites depend, and are thus vulnerable to changes elsewhere in the catchment (Moss, 2000). Although this criticism is relevant to conservation in standing waters, the SSSI mechanism has proved to be particularly inappropriate and ineffective for running waters (Leighton, 1999). SSSIs do at least have a 'spotlight' effect, drawing attention to the value of river environments (Boon, 1994), but only 1 per cent of SSSIs are focused on rivers. All of this emphasises the importance of adopting a 'whole catchment' perspective (Section 6.4.2).

Threats to the conservation interest are numerous (Maitland and Morgan, 1997), including many discussed in earlier sections – damaging land use practices (agriculture, forestry and aquaculture), eutrophication, acidification, recreation and pollution

FIGURE 6.6 *Intensive recreation at Milarrochy Bay, Loch Lomond. Photo: © SNH.*

(both chemical and thermal). In addition, the impacts on aquatic habitats of road construction, gravel extraction and the regulation of rivers and lochs can be considerable, damaging or destroying the natural character of rivers. Of these diverse threats, Gilvear *et al.* (2002) identify land use change and river regulation as the two types of human activity which cause most ecological damage, and this partly explains the recent search for more natural answers to fluvial problems (Section 6.4.3). Indeed, there is a growing realisation that conservation can no longer be an optional add-on but should underpin the management of all freshwater resources (Leighton, 1999). The traditional, demand-led approach is becoming discredited.

An increasingly significant subset of water resource conservation is the management of wetlands and blanket peat bogs. These are an important part of Scotland's natural and cultural heritage (SNH, 1995b; Bragg, 2002) yet their true value has only recently been appreciated. Prior to this realisation they were subject to centuries of exploitation and damage caused by peat extraction, drainage for farming, and blanket afforestation. The greatest expanses of wetlands lie in the Flow Country of Caithness and Sutherland, the largest and most unspoilt area of blanket bog in the northern hemisphere (Fig. 6.2d). A haven of biodiversity, the region is widely regarded as a globally important ecosystem (Lindsay *et al.*, 1988; Parkyn *et al.*, 1997). Ironically, it was the spectre of its loss to afforestation which brought its conservation value to international attention in the 1980s (Stroud *et al.*, 1987) and which subsequently led to the designation of 136,000ha of peatland SSSIs (Warren, 2000a). As mire systems and peatlands have undergone a perceptual transformation from 'useless bogs' to 'precious wetlands', so research has revealed their physical and biological complexity (Cox *et al.*, 1995; Lindsay, 1995) and enthusiasm has grown for their protection and restoration (Parkyn *et al.*, 1997; Stoneman and Brooks, 1997).

6.5.5.ii Managing and conserving fish populations

Despite the diversity of fish species in Scottish freshwaters (Table 6.1), angling and fishery management are focused to a remarkable degree on game fishing for salmonid fish, primarily salmon and trout (Campbell *et al.*, 1994).[9] With some 400 salmon rivers, Scotland has one of the largest and most diverse populations of Atlantic salmon in Europe and hosts an internationally important salmon fishery. However, in common with the entire North Atlantic region, salmon and sea trout catches have been declining for thirty years (Fig. 6.7). For salmon, this has largely consisted of steep declines in the net fisheries catch (partly through decreased fishing effort); the number of salmon caught by rod-and-line has increased slightly and now constitutes almost half of the total catch (SOAEFD, 1997). Both the numbers and age of salmon caught in Scotland have fluctuated significantly over the last two centuries, partly as a result of changing sea conditions (Summers, 1993; Tapper, 1999). Such natural oscillations provide context for the recent declines but are no cause for complacency because salmon catches in the late 1990s were the lowest on record, and stocks are now thought to be below their safe biological limits (SERAD, 2000d). If present trends continue, extinction looms.

Stocks had been partially depleted by overfishing as long ago as the early nineteenth century, but salmon have now entirely disappeared from many rivers (Hambrey, 1997). Because of their great reproductive power, overfishing by itself would probably not severely reduce stocks, but when combined with the wide range of other pressures on fish populations (Table 6.6) the effects have been catastrophic. The primary factors affecting the abundance of salmon and sea trout are identified by the Scottish Salmon Strategy Task Force (SOAEFD, 1997) and Gilvear *et al.* (2002) as follows:

- Habitat, river geomorphology and water quality, and the effects on these of pollution, abstraction, regulation and land use. A total river length of 36,658km is now designated for salmonids under the EU Fresh Water Fish Directive which should guarantee minimum water quality standards.
- Predation, notably by grey seals, otters, mink and cormorants. The rapidly expanding population of grey seals eats three times as many salmon as are caught each year.
- Stocking, introductions and transfers.
- Fish farming (Section 6.2.2).
- Disease and parasitism.
- Threats during the marine stage of their lifecycle.

Sea trout also declined dramatically during the 1990s (Fig. 6.7), both in numbers and weights, particularly in the north west, a collapse which has been linked to the rise of fish farming (Edwards, 1998). Voluntary catch-and-release schemes are operating in some places, especially for spring salmon which have declined most markedly. However, more radical, far-reaching changes in fisheries and catchment management are needed to improve the fortunes of salmonids and, indeed, of all Scotland's freshwater fish fauna. This is because fisheries management currently involves many practices which are harmful not only to the aquatic environment but also to the fisheries themselves (Maitland, 1992).

TABLE 6.6 Main pressures facing fish in freshwater habitats. After SNH (1995a).

Pressure	*Effect*
Industrial & domestic effluents	Pollution; poisoning; blocked migration routes
Acid deposition	Acidification; release of toxic metals
Land use (farming & forestry)	Nutrient enrichment; acidification; sedimentation
Industrial development	Sedimentation; habitat & species loss; metal pollution
Warm water discharge	Deoxygenation; temperature gradients; intake on cooler screens
Channel obstructions & river regulation (dams, barrage)	Blocked migration routes; sedimentation of spawning beds; unnatural hydrology
Infilling, drainage & canalisation	Loss of habitat, shelter & food supply
Water abstraction	Loss of habitat & spawning grounds; transfer of species
Fluctuating water levels (reservoirs)	Loss of habitat, spawning grounds & food supply
Fish farming	Nutrient enrichment; introductions; disease; genetic changes
Angling & fishery management	Pesticides; introductions; litter
Commercial fishing	Overfishing; genetic changes
Introduction of new species	Elimination of native species; diseases; parasites

'The general picture around the world is that the conservation of most aquatic organisms has been sadly neglected' (Maitland and Morgan, 1997: 180). Scotland is no exception. There is no comprehensive monitoring of key fish stocks, little research on most species, and there are no effective controls on the movement of fish (other than salmonids) or on the introduction of exotic species (Maitland, 1997b). Neither is there a national organisation with authority to protect fish and fisheries as there is in England and Wales. The closest equivalent is the District Salmon Fishery Board system, but this is only concerned with salmonids, and only sixty-two of the 100 districts have a Board (although, of the major rivers, only the Clyde lacks a Board). Two of the most pressing issues affecting the viability of fish populations are introductions of non-native fish and movements of fish between catchments. Introductions, which may be intentional for sporting purposes or unintentional *via* live bait, can have serious effects such as the introduction of diseases and parasites, alteration of predator-prey

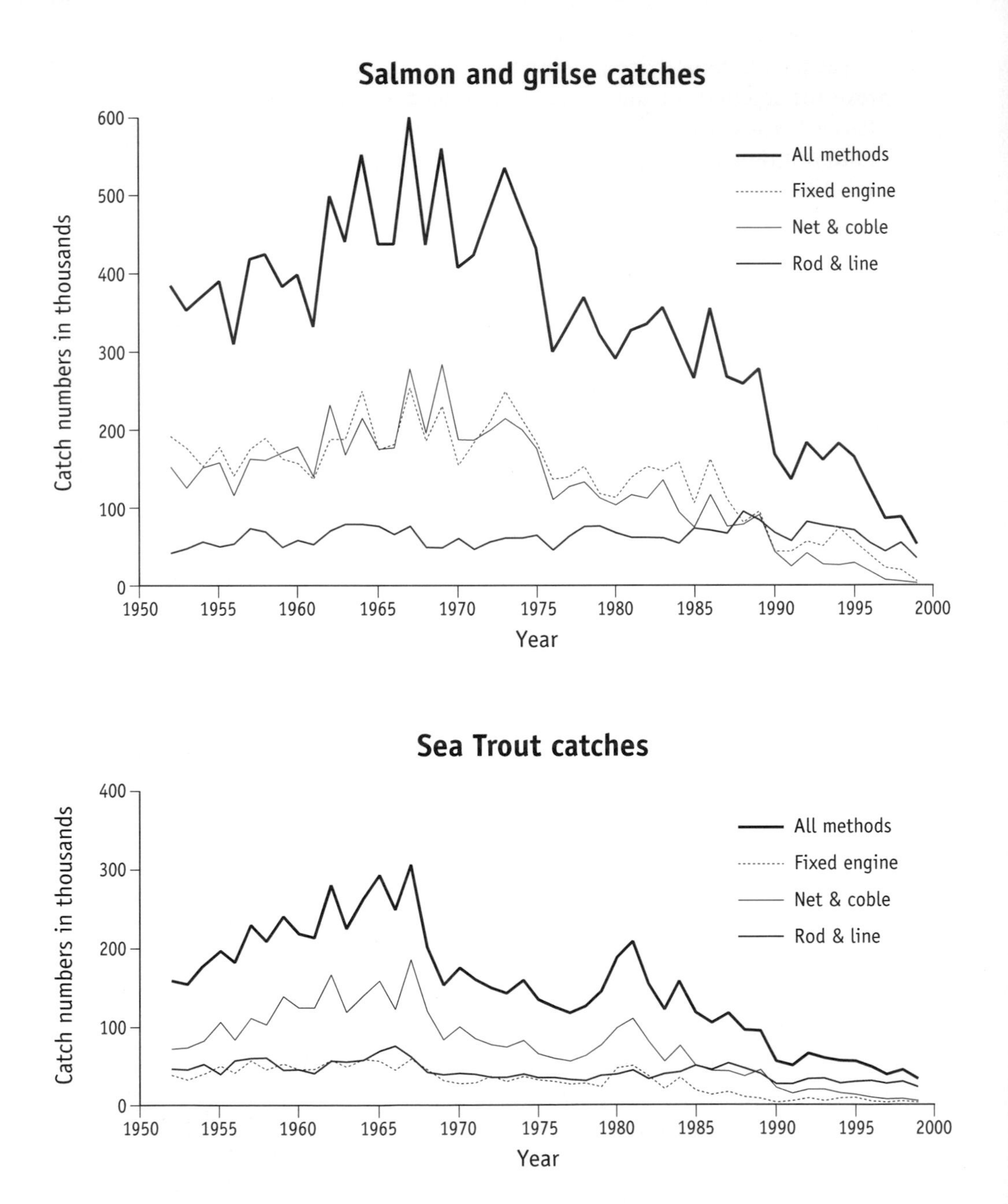

FIGURE 6.7 *Total numbers of salmon and grilse and of sea trout caught and retained, 1952–99. After SERAD (2000e).*

relationships, and changes to the genetic make-up of native species. Moreover, once exotic populations are established they are almost impossible to eradicate. For example, coarse fishermen's use of live bait has resulted in four exotic species becoming established in Loch Lomond since 1980, and one of these, ruffe, is now so abundant that it is threatening the unique race of river lamprey and one of Scotland's two remaining populations of powan (Maitland *et al.*, 2000). Growing fears about declining populations of freshwater fish led to the launch in 1999 of *Angling for Change*, an unprecedented joint campaign by angling and conservation groups (Cairns, 1999b).

The situation has been improving in recent years. A range of mechanisms have been put in place both to protect the species themselves and also to protect water bodies and habitats (SERAD, 2000d). Further recommendations for improving the management structures have been proposed both by the Scottish Salmon Strategy Task Force (SOAEFD, 1997) and by Maitland (1997b). If adopted, and if effectively resourced and empowered, such structures might prevent such threats to biodiversity as the progressive loss of Arctic charr caused by the introduction of non-native stocks by charr farmers. They might also allow salmon and sea trout populations to recover. Some improvements will inevitably flow from the EU Water Framework Directive (Section 6.4.4) since it will require monitoring of native fish stocks, but the Scottish Executive is pressing for a more holistic approach, encompassing not only all fish species but the whole freshwater environment (SERAD, 2000d). The 2001 Salmon Conservation Act broadens the range of measures available.

The primary long-term means by which successful conservation of any aquatic species is achieved is the protection of habitats, but (as for conservation more generally) there are significant political and economic obstacles. Adequate experience and research-based information now exists to guide conservation objectives and practice, but the leap from theory to practical application is always a challenging and expensive one. Nevertheless, in the light of recent progress and current trends, it is tempting to join Gilvear *et al.* (2002) in believing that the future for the natural heritage status of Scotland's freshwaters is rosy.

NOTES

1. The most comprehensive discussion of the nature, significance and management of Scotland's freshwater resources remains the volume edited by Maitland *et al.* (1994). Boon and Howell (1997) provide in-depth discussions of water quality issues.
2. Although there are over 100 freshwater fish farms, aquaculture raises issues which are more properly addressed in the context of the marine environment; they therefore lie beyond the scope of this book. The development of the industry and the associated impacts and controversies are described by, *inter alia*, Williamson and Beveridge (1994), Allcock and Buchanan (1994), Hambrey (1997), SOAEFD (1997) and MacGarvin (2000). The unfolding fish farming sagas (especially following the outbreak of Infectious Salmon Anaemia in 1998) have been followed in detail in *SCENES*.
3. Reported in *SCENES* 158: 5, 2001.
4. Full details of HEI and SUDS are available in a series of guidance and promotional leaflets produced by SEPA which can be viewed on SEPA's website.
5. SEPA (1999) provides an in-depth discussion of contemporary water quality issues in Scotland, as well as setting out ambitious targets for improvement.
6. The single greatest research effort to investigate the effects of forestry on hydrology and ecology in Scotland was the paired catchment study at Balquhidder (Calder, 1993; JoH, 1993; Johnson, 1995).

7. See *SCENES* 143: 4, 1999 and 147: 2, 2000.
8. The challenges of managing freshwater habitats for conservation are discussed comprehensively by Maitland and Morgan (1997).
9. The long history of fish populations and angling in Scotland is charted by Campbell *et al.* (1994). A wide-ranging review by SERAD (2000d) includes discussion of Scotland's fish resource and the threats to the sustainability of fisheries, and sets out the existing conservation mechanisms and management structures. The Report of the Scottish Salmon Strategy Task Force (SOAEFD, 1997) discusses the biology, management and conservation of salmon, and proposes new management arrangements.

CHAPTER SEVEN

The animals: wildlife management

7.1 INTRODUCTION: THE IMPORTANCE OF GAME MANAGEMENT

It would be foolhardy to attempt in a single chapter a comprehensive discussion of the many diverse wildlife issues in Scotland. Fascinating tales and important management issues attach to almost every animal and bird species (Harris *et al.*, 1995; Kitchener, 1997; Minns, 1997; Yalden, 1999). Three of these (the controversies surrounding red squirrels, mink and geese) are briefly discussed at the end of the chapter, but the main focus here is the management of red deer and red grouse. The justification for this narrow focus is that these two species dominate the wildlife management scene in three ways: in the area of the country that they inhabit, in the impacts that they have on the uplands, and in the amount of research and literature that their management has generated. They are also the *sine qua non* of sporting estates which, alongside forestry and sheep farming, constitute one of the three principal land uses in the uplands. Such estates are crucial to the economics of large parts of rural Scotland in both direct and indirect ways. Direct expenditure by the providers of grouse shooting and deer stalking in Scotland totals some £25 million per annum, supporting 2,200 full-time equivalent (fte) jobs in remote rural areas where little alternative employment exists (SLF, 1995; CRC, 1997). Taking multiplier effects into account, grouse shooting alone supports £17 million worth of GDP in Scotland (FAI, 2001). Game sports also attract tourists. Visitors to Scotland come primarily to see the landscape, but second on their priority list is seeing wildlife in general and game species in particular – deer, salmon and grouse, in that order (McCall, 1998).

Game management is thus an important subset of wildlife management and remains a significant activity today throughout rural Scotland. Game species are usually considered to be 'wild mammals, birds or fish now or formerly hunted, shot or caught for sport' (Thompson *et al.*, 1997: 199). They can be regarded as a sustainable resource. Historically, game sport has sometimes led to extinctions but it has also had unintended conservation benefits. For example, several native species have been preserved because of the cultural value attached to them for hunting. Red deer are a prime example, as is the capercaillie which was reintroduced to Scotland from Sweden in 1837 for hunting but is now a prized conservation species (Kitchener, 1998). Game value was also a factor in the preservation of the red fox (*Vulpes vulpes*) and the otter (*Lutra lutra*), despite their being regarded as pests (Yalden, 1999).

Game management usually involves some or all of the following:

- habitat manipulation and/or creation (e.g. muirburning for grouse; planting woodland for pheasant shooting; creation of duck flighting ponds)
- pest control (e.g. shooting of foxes and crows which prey on game birds)
- the release of animals (e.g. rearing and release of pheasants)
- supplementary feeding (e.g. winter feeding of red deer)
- disease and parasite control (e.g. reducing parasitic worm burdens in gamebirds)
- control of poaching
- reduced used of agrochemicals on adjacent/surrounding farmland.

Given this range of activities, and the fact that gamekeepers manage some 8.2 million - ha or 34 per cent of the UK (BASC, 2000), there is no doubt that game management is a significant influence on biodiversity, habitats and landscapes.

The effects, however, are spatially variable and difficult to measure (Thompson *et al.*, 1997; BASC, 2000). On the plus side, management for game species such as pheasant, partridge, grouse and duck can have benefits for conservation, as well as for the landscape and the local economy (CRC, 1997; McCall, 1998; Hankey, 1999; Phillips, 2001). Thus, for example, the preservation of heather moorlands has largely been due to their value for sport; the least losses of heather moorland have been in grouse shooting areas (Tapper, 1999). At a more local level, the conservation benefits of game management for a wide range of floral and faunal species (including rare wild flowers, songbirds, butterflies and beetles) have been quantified by the Game Conservancy Trust (GCT) (Tapper, 1999). Farmers with game interests have an incentive to plant and maintain more woodlands and hedgerows, and to leave wider headlands around fields, than those without, creating habitats which benefit many species in addition to the game (Yalden, 1999; Swift, 2001). Finally, land management for game helps to maintain an irregular, mixed land use pattern that is much favoured by the public.

This positive view of game management is not uncontested, however. There are those who deprecate any form of sport which involves killing, whatever the environmental fringe benefits may be. Moreover, the selective manipulation of habitats and populations to promote certain species at the expense of others operates to the detriment of naturalness in the environment. The over-large, female-biased red deer population is a prime example (Section 7.5.1). Most controversially, the (over-) zealous control of predators to promote game survival sometimes runs counter to current conservation interests and the law, notably in the case of the persecution of birds of prey in the single-minded pursuit of large populations of red grouse for sport (Section 7.6.2).

Unlike forestry and agriculture which are heavily subsidised, game sports receive almost no public money. On the contrary, until the abolition of sporting rates in 1995 they were taxed, a situation which in some regards discouraged environmentally sound management (McGilvray and Perman, 1992). Contemporary debates about the place of game sport (including the possible reintroduction of sporting rates) incorporate a heady mix of themes, ranging from conservation issues to the maintenance of the social fabric of rural areas, spiced up further with injections of politics, social comment and passionately held single-issue viewpoints. In these debates the game sport traditions accumulated over two centuries are sometimes regarded as the cause and sometimes as the solutions to contemporary problems in the uplands.

7.2 DEER IN SCOTLAND

There are four main deer species in Scotland (Table 7.1):

TABLE 7.1 Estimated populations of deer species in Scotland.

Species	*Estimated population*	*Source*
Red deer (*Cervus elaphus*)	350,000	DCS, 2000a
Roe deer (*Capreolus capreolus*)	>200,000	DCS, 2000a
Sika deer (*Cervus nippon*)	>20,000	McLean, 2001a
Fallow deer (*Dama dama*)	8,000	DCS, 2000a
Reindeer (*Rangifer tarandus*)	80	Harris *et al.*, 1995
Muntjac (*Muntiacus muntjak*)	?	DCS, 2000a

1. **Red deer** (Fig. 7.1a) have been present in Scotland throughout post-glacial time. They are Scotland's largest land mammal and are symbolic of the Highlands, most famously in Landseer's painting, *The Monarch of the Glen*. Scotland's wild red deer constitute the largest population in Europe and range over some 3 million ha or 40 per cent of the country (SNH, 1994b; Fig. 7.2). Naturally a forest animal, they seek woodland for shelter and nutrients but have adapted to the open hill. As a consequence, they are smaller, lighter and less fecund than European woodland red deer. There are large spatial contrasts in population density. Many parts of the open range carry about 15 deer/km^2 (Staines, 1999a) but densities reach 30–50 per km^2 in places (Trenkel *et al.*, 1998).
2. **Roe deer** (Fig. 7.1b). These 'slender, graceful, impudent . . . and infinitely beguiling animals' (Prior, 1995: 7) are also native to Scotland. Primarily woodland animals, they are active colonisers and are present throughout Scotland, from sea level to 700m, except on some islands (DCS, 2000a; Fig. 7.2). Widespread afforestation in the twentieth century greatly assisted their spread (Ratcliffe and Mayle, 1992), and none of their main predators (wolf, lynx, wolverine, bear) remain in Scotland to check their expansion. Because of their small size and secretive nature, they are generally regarded as being uncountable, so estimates of population size vary between 200,000 and 400,000. The recorded annual cull is now around 40,000 (DCS, 2000a). Although they do significant damage both to agricultural crops and to forestry plantations (Section 11.1.1), no fencing keeps them out for long.
3. **Sika deer** were introduced to deer parks in the nineteenth century. Their offspring escaped or were released, and their descendents are now breeding in the wild in large parts of the uplands (Putman, 2000; Fig. 7.2). They are also hybridising with red deer (DCS, 1998a), a fraught situation which is discussed in Section 10.3.

FIGURE 7.1. *Scotland's two main deer species.*
a. A red deer stag. Photo: © Laurie Campbell. (opposite)
b. A roe buck. Photo: © Neil McIntyre. (above)

4. **Fallow deer** have been established in Britain for about 900 years (Harris *et al.*, 1995) and have a limited, dispersed presence in Scotland (Fig. 7.2). The small population is thought to be on the increase in woodland areas.

The two other deer species are both of minor significance. Reindeer are not present in the wild. Originally native to Scotland but dying out naturally around 9,500 years ago, they were reintroduced to the Cairngorms from Sweden in 1952 (Harris *et al.*, 1995). Muntjac were reported in the Borders in 1994 but it is still not clear whether this, the UK's smallest deer, is established in the wild in Scotland (DCS, 2000a).

7.2.1 The history of red deer management

Red deer have been a favourite quarry of hunters since Mesolithic times or earlier because they provided not only food, but clothing (from skins) and implements (from bone and antler). By the sixteenth century they were protected by law, largely due to the requirements of royal hunting, and but for this they would almost undoubtedly have followed Scotland's other large mammals into extinction (Yalden, 1999). From the seventeenth to the mid-nineteenth centuries red deer were present but largely incidental; their numbers declined, and commercial sheep farming came to predominate in rural Scotland (Smith, 1993). But with the advent of refrigeration and

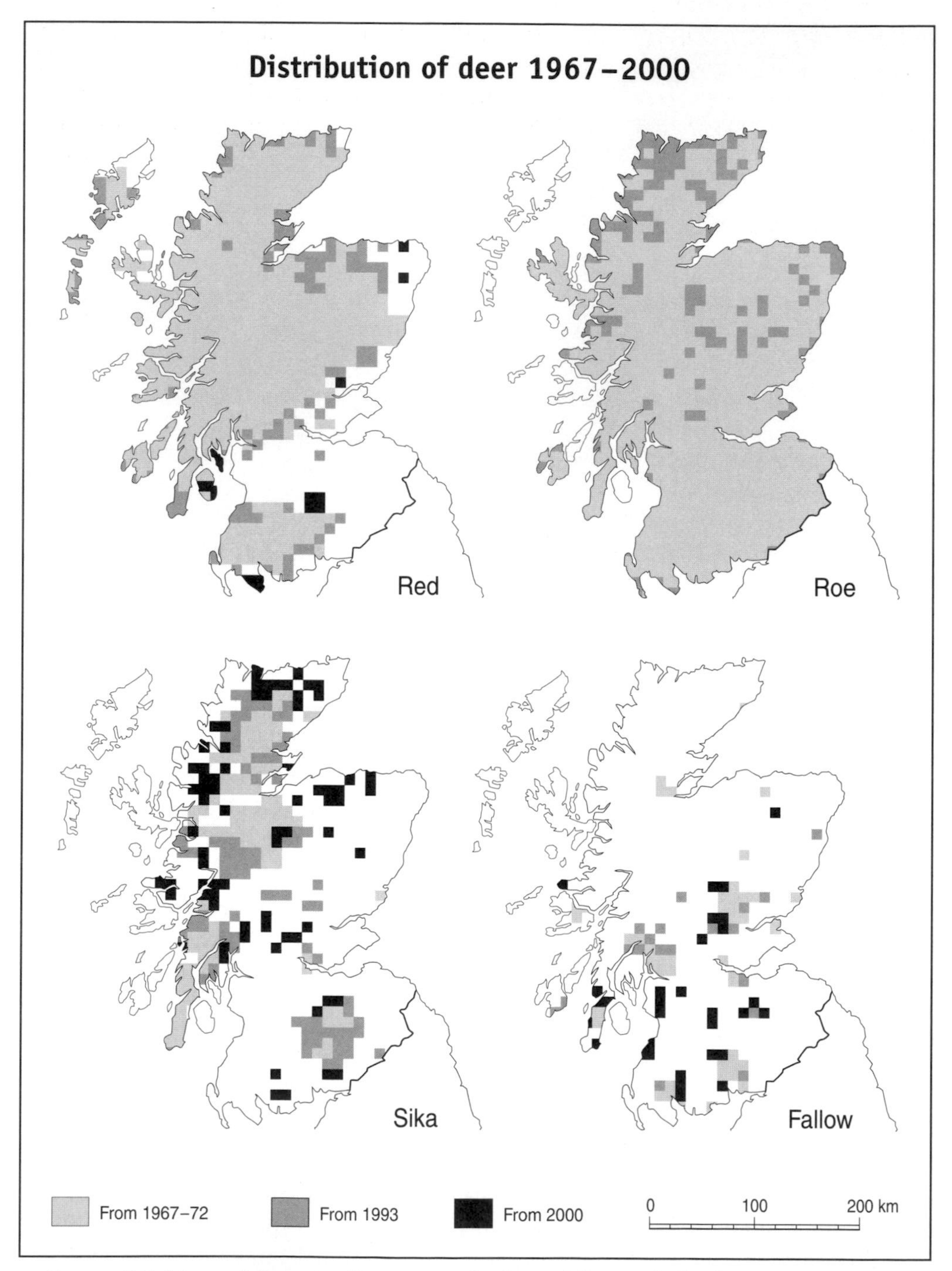

FIGURE 7.2 *Maps of the expanding ranges of red, roe, fallow and sika deer since 1967. Data supplied by the British Deer Society and used with permission.*

imports of antipodean lamb and mutton in the 1870s, prices collapsed, bringing a sudden end to the sheep farming boom. This, combined with the rising popularity of stalking, led to the rapid conversion of large tracts of marginal sheep land into deer forests. 'An unstoppable red deer fashion had been born and received royal approval' (Lister-Kaye, 1994: 16), a process dubbed the 'Balmoralisation' of Scotland (Toogood, 1995). By 1912 deer forests covered 1.5 million ha or 20 per cent of Scotland, having doubled in area in just thirty years to become the principal land use of the Highlands and Islands (Callander and MacKenzie, 1991). (The use of the phrase 'deer forest' to describe areas which are mostly treeless is a survival from medieval days when 'forest' was an all-encompassing term for any land harbouring wild game.) Although the deer forest area shrank somewhat during the twentieth century, to around 15 per cent, red deer management is still the dominant game-orientated land use in the Scottish uplands (Thompson *et al.*, 1997).[1]

Conflict and controversy have surrounded red deer for over 200 years, the primary frictions being with farmers and crofters who objected to the damage that marauding deer can do to crops. Between 1872 and 1954 there were no less than seven government inquiries into red deer conflict (SNH, 1994b), but only in 1948 were farmers given the legal right to kill marauders on enclosed land, and not until 1959 was the Red Deer Commission (RDC) established. It was given responsibility for the conservation and control of wild red deer, especially in relation to agriculture and forestry. Legal close seasons were introduced at the same time, but no thought was given to the impact of red deer on the natural heritage, despite expressions of concern in the earlier twentieth century (SNH, 1994b). In subsequent decades deer-related controversy widened beyond agriculture to encompass forestry and conservation. It became increasingly clear that a holistic, integrated approach to red deer management was needed, incorporating other deer species, other land uses and the natural heritage interest (Callander and MacKenzie, 1991; CWP, 1993) in place of the narrow focus on sporting value and venison production. This realisation led to the 1996 Deer (Scotland) Act, legislation which renamed and redefined the RDC. As the Deer Commission for Scotland (DCS) it has responsibility for furthering the sustainable management of all species of wild deer in Scotland. Its powers include the instigation of deer control schemes when deemed necessary for the protection of the natural heritage or for the benefit of forestry and farming.

Since the last wolf was killed some time in the seventeenth century, deer have had no predator except human beings. Left unmanaged, deer populations would increase until limited by environmental and population factors, a stage at which both their habitat and their health would be suffering badly. At their current high levels they are already being negatively affected by density-dependent factors (SNH, 1994b). Human management of deer (and of the natural heritage more generally) is thus an ecological necessity because of the extent to which humans have altered the environment (Budiansky, 1995; Sections 1.3 & 13.3.1.ii). If we shirk this inherited responsibility the results are unlikely to be to society's liking.

Deer and their habitats may be managed for at least six reasons (Ratcliffe and Mayle, 1992):

- to protect timber and agricultural crops, and sites of conservation importance
- for sport and venison production
- to preserve plant assemblages and their dependent animals
- to conserve deer populations
- to regulate population size and prevent starvation
- to enhance the aesthetic appeal of the environment and to provide recreational opportunities.

The aims of deer management vary widely, depending on the overarching objectives of the land manager, but they will almost always include three facets, albeit with different emphases (Prior, 1995): maintaining a healthy deer herd in balance with their habitat, controlling damage, and seeking to offset the costs of damage and management against income from a sustainable cull.

7.3 Grouse in Scotland

7.3.1 The grouse family and other game birds

Like 'the monarch of the glen', 'the famous grouse' has become quintessentially Scottish in its associations, symbolic of a particular historical-romantic vision of Scotland as a sportsman's playground. Because of its sporting and economic significance throughout the uplands of eastern and southern Scotland and northern England, the red grouse (*Lagopus lagopus scoticus*) (Fig. 7.3) has been the subject of intensive

FIGURE 7.3 *The four main grouse species in Scotland.*
a. Red grouse. Photo: © Laurie Campbell. (opposite)
b. A male black grouse. Photo: © David Whitaker. (above)
c. A male capercaillie displaying. Photo: © Neil McIntyre. (overleaf p. 150)
d. Ptarmigan in winter plumage. Photo: © Neil McIntyre. (overleaf p. 151)

research since late Victorian times, resulting in almost unparalleled knowledge of its population biology (Hudson, 1992). Grouse numbers fluctuated greatly throughout the twentieth century, but the trend was inexorably downward, stabilising at a lower level after 1987 (Smith *et al.*, 2000). Superimposed on this long-term trend in some regions are dramatic cyclic fluctuations every 4–8 years primarily caused by heavy infestations of parasitic threadworms. The number of grouse shot for sport declined by 50 per cent during the century, with especially steep declines in the 1930s and 1970s (Thirgood *et al.*, 2000a; Fig. 7.4). This has halted shooting on 30 per cent of grouse moors since 1950 (UKRWG, 2000). The explanation for this decline appears to be multi-factorial (Hudson, 1992; Phillips, 2001); weighting the various factors is proving to be highly controversial (Section 7.6.2).

There is no doubt that the red grouse is the single most important game bird in the uplands, but it is only one of the four members of the grouse family in Scotland (Fig. 7.3), and grouse are only one of a number of game birds (Table 7.2).[2] In the lowlands, pheasant (an introduced species) and duck are the most significant. Pheasants are the

most widespread gamebird in the world. The 12 million shot for sport in Britain each year comprise 83 per cent of all game shot (Tapper, 1999), although their relative significance is lower in Scotland.[3] Because all these species are classified as game birds, *Homo sapiens* is their most significant predator, but also preying on them are foxes, crows and several species of raptor (birds of prey) (Table 7.3). Being ground-nesting species, all grouse are vulnerable to predation and disturbance, especially when incubating. For red grouse, the most significant raptors are hen harriers, peregrine falcons and golden eagles. Populations of many raptors have recovered from the low point caused by widespread use of organo-chlorine pesticides in the 1950s and 1960s, and they have now reached levels at which they can produce local conflicts with game management (Section 7.6.2).

TABLE 7.2 Primary game bird species in Scotland: population sizes and trends. Populations are given as numbers of individuals (not as breeding pairs) and do not include over-wintering individuals. Starred figures are UK population totals; for these species the size of the Scottish subset is unknown. Sources: SNH, 1995a; Tapper, 1999; Petty, 2000; Smith, 2000.

Species	*Population*	
	Size	*Trend*
THE GROUSE FAMILY		
Red grouse (*Lagopus lagopus scoticus*)	350,000	Strongly down
Ptarmigan (*Lagopus mutus*)	20,000	Down
Black grouse (*Tetrao tetrix*)	9,500	Down
Capercaillie (*Tetrao urogallus*)	1,073	Strongly down
PHEASANT (*Phasianus colchicus*)	c. 1 million	Stable
DUCK		
Mallard (*Anas platyrhynchos*)	*250,000	Stable
Teal (*Anas crecca*)	2,900	Up
Wigeon (*Anas penelope*)	750	Stable
WOODCOCK (*Scolopax rusticola*)	*30,000	Down
SNIPE (*Gallinago gallinago*)	*110,000	Stable
JAY (*Garrulus glandarius*)	*320,000	?
GREY PARTRIDGE (*Perdix perdix*)	15–30,000	Stable

TABLE 7.3 Breeding raptors in Scotland: population sizes and trends. Populations are given as numbers of individuals (not as breeding pairs). Sources: SNH, 1995a; Tapper, 1999; Smith, 2000; UKRWG, 2000.

Species	*Population*	
	Size	*Trend*
Peregrine falcon (*Falco peregrinus*)	1,320	Strongly up
Merlin (*Falco columbarius*)	720	Down
Kestrel (*Falco tinnunculus*)	25–35,000	Down
Hen harrier (*Circus cyaneus*)	1,200	Stable
Marsh harrier (*Circus aeruginosus*)	10	Strongly up
Goshawk (*Accipiter gentilis*)	100	Strongly up
Sparrowhawk (*Accipiter nisus*)	14,000	Stable
Golden eagle (*Aquila chrysaetus*)	850	Stable
Sea eagle (*Haliaeetus albicilla*)	20	Stable
Osprey (*Pandion haliaetus*)	170	Strongly up
Red kite (*Milvus milvus*)	10	Strongly up
Buzzard (*Buteo buteo*)	12–15,000	Strongly up
Honey buzzard (*Pernis apivorus*)	6	Stable

The populations of the other members of the grouse family are small and all are declining (Table 7.2). **Capercaillie**, the world's largest grouse, is now facing extinction in Scotland for the second time. From a population of 20,000 in the 1970s, its numbers declined to 2,200 in 1994 and to just 1,073 in 1999 as a pincer movement of factors conspired to reduce breeding success and frustrate energetic conservation efforts (Moss and Picozzi, 1994; Kortland, 2000; Petty, 2000). These factors include the fragmented nature of the mature Scots pine forests in eastern Scotland which are its preferred habitat (Summers *et al.*, 1995), mortality through collisions with forest fences (Section 11.1.4), increased predation, runs of wet summers, and (ironically) disturbance by birdwatchers during the lek (courtship displays). Similar factors, especially habitat fragmentation, explain the long-term and on-going decline in populations of **black grouse**, not only in Scotland but throughout the UK and Europe (Cayford, 1993; Tapper, 1999). Because of their threatened status, there has been a voluntary moratorium on sport shooting of capercaillie and black grouse for the last decade. The declines continue, however, so the Scottish Executive has decided to introduce a statutory ban on shooting capercaillie. **Ptarmigan** are birds of the high tops, frequently by-passed unseen because of their unsurpassed camouflage. Though a quarry species, they are rarely shot, and then only by the keen and fit.

7.3.2 Heather moorland management: history and value

Moorlands dominated by heather (*Calluna vulgaris*) are one of Scotland's most important biological resources, yet they are not natural but semi-natural habitats – cultural landscapes – largely created by human agency. Human management of the uplands over five millennia or more, using fire and grazing, progressively transformed formerly wooded areas into a mosaic of grassland, dwarf-shrub heaths, scrub and woodland (Stevenson and Birks, 1995). From Victorian times large swathes of these areas came to be devoted to a single sporting pursuit, grouse shooting, creating today's treeless heather moorland. The same factors that led to the establishment of deer forests from the 1870s (Section 7.2.1) meant that the sporting value of red grouse came to rival grazing, so grouse progressively replaced sheep. From being incidental inhabitants of the uplands, red deer and red grouse became vital elements of the economy of the proliferating sporting estates. Accordingly, armies of gamekeepers wiped out vast numbers of what were then called 'vermin', a sweeping description which included many of today's prized bird species such as eagles, kites, hawks and ospreys. The records collated by Smout (1993c: 31) of the sheer scale of these 'Victorian holocausts' beggar belief.

Despite the long-term decline of grouse populations, there are still almost 500 grouse moors operating in Scotland, averaging 4,300ha in size (McGilvray and Perman, 1992). About 250,000 red grouse are shot annually, although there is much variation in the annual bag, both in time (Fig. 7.4) and space. After a sharp decline in the early 1990s, average bags rose substantially in the late 1990s to reach 815 birds shot per estate, accompanied by a dramatic increase in participants (FAI, 2001). On most estates grouse shooting is one part of a multiple land use system including sheep farming, fishing, forestry and deer stalking. Running a grouse moor brings pleasure and prestige to owners (and to their family and friends) and enhances the capital value of the estate (Section 3.2.1), but it is usually a loss-making enterprise. Grouse moors operate at an average annual loss of almost £17,000, and in a majority of cases

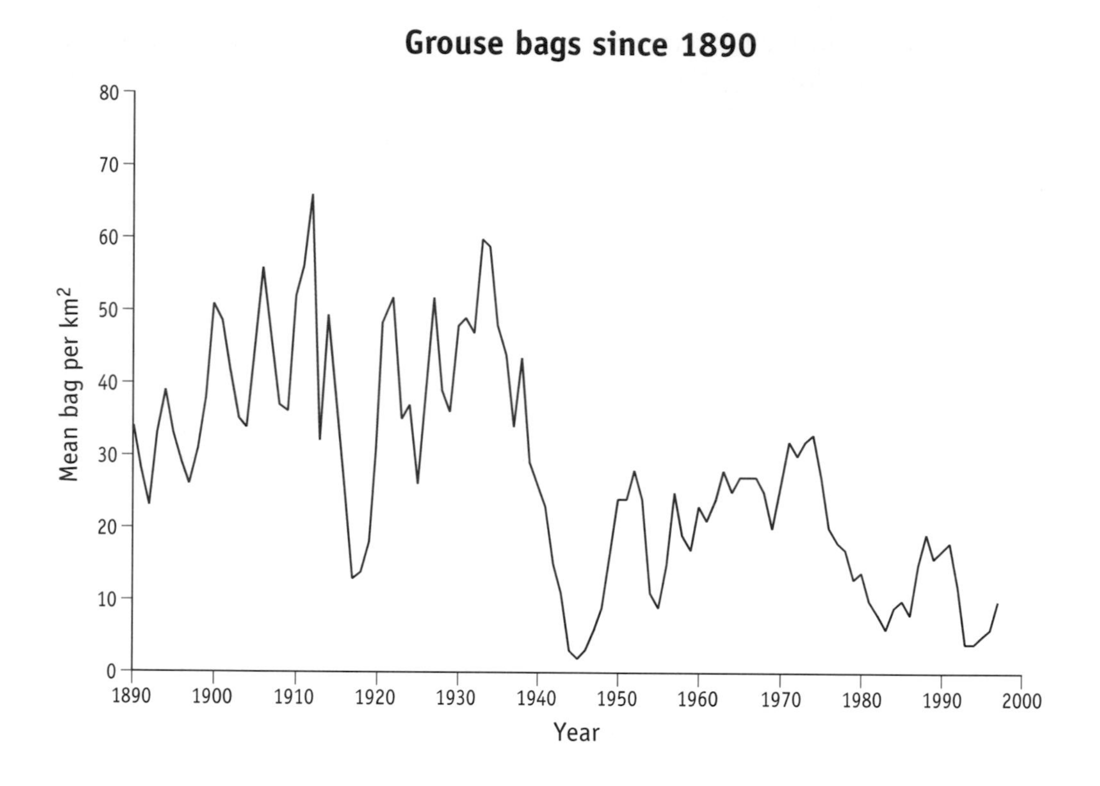

FIGURE 7.4 *Red grouse numbers shot on managed estates in Scotland, 1890–1997. The long-term decline is probably greater than suggested by these figures because results were collected only from estates which continued grouse shooting; many former grouse moors are now used for sheep grazing or forestry. Data provided by the Game Conservancy Trust.*

grouse shooting contributes less than 25% of an estate's income (FAI, 2001). Nevertheless, this contribution is crucial in maintaining the viability of many estates, and across Scotland grouse shooting supports some 940 fte jobs (FAI, 2001), a total which is small in absolute terms but which represents a significant contribution to rural employment.

The fact that red grouse is the only species primarily dependent on heather attaches a direct cash value to heather moorland (Phillips and Watson, 1995); much of today's rolling moorland owes its creation and preservation to a management regime designed to enhance grouse populations.[4] Heather dominates some 15 per cent of Scotland, and it is present on some 39 per cent of the land (3.08 million ha) (Thompson *et al.*, 1997; Mackey *et al.*, 1998). Estimates vary, but about 50–60 per cent of this (about 1.75 million ha) consists of grouse moor, most of which is concentrated in the drier east of the country (Fig. 7.5). Commenting on the importance of the eastern moorlands, the Cairngorms Working Party (CWP, 1993: 24) describes them as 'one of the glories of the Highland scene':

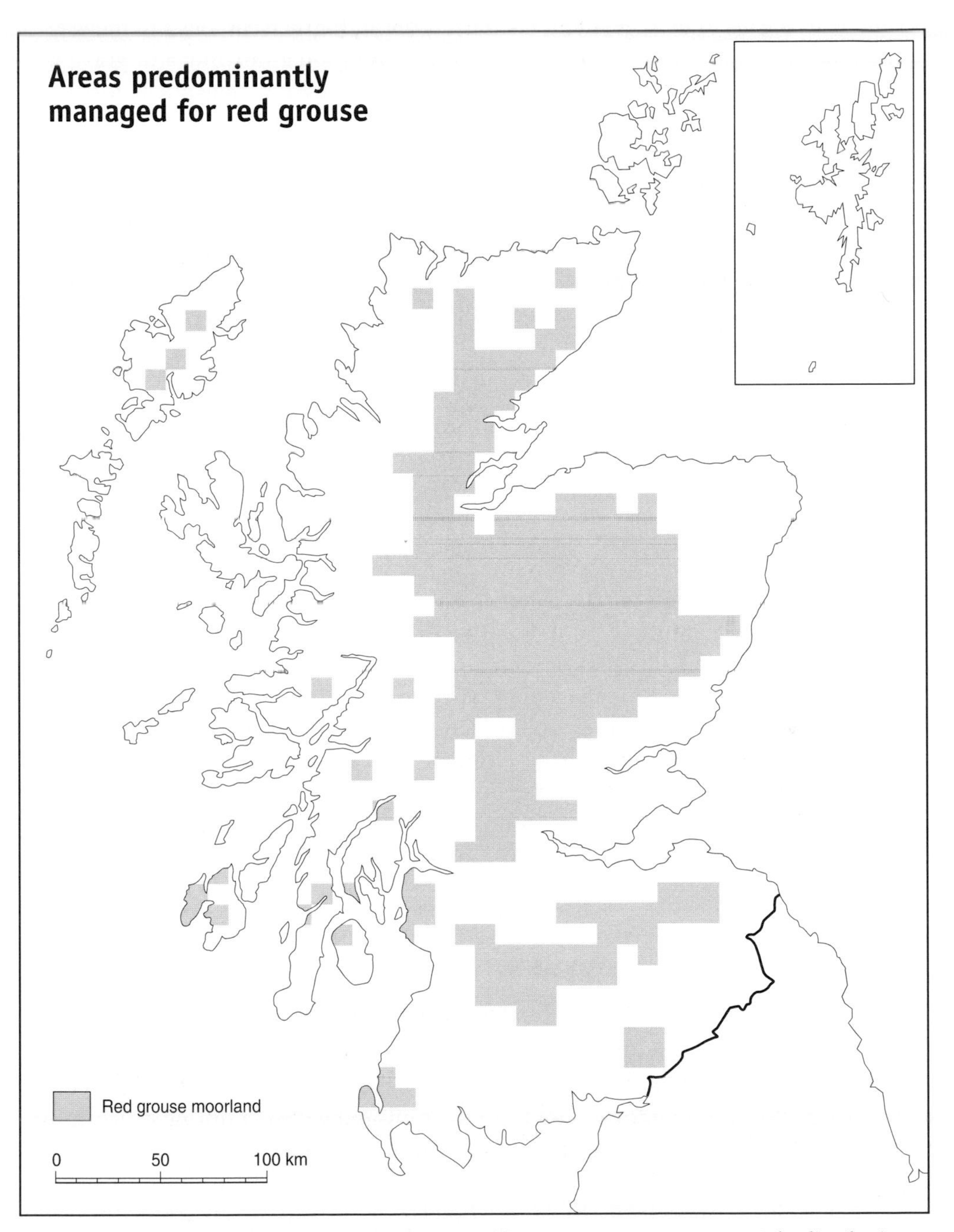

FIGURE 7.5 *Areas of upland which are dominated by grouse moor management. The distribution is based on 10km squares with more than 50 per cent cover of managed grouse moors. After Thompson et al. (1997).*

> There is no landscape nor ecosystem quite like it anywhere else in the world, not only with regard to its vegetation, habitats, birds and insects, but also in the way these heather moorlands are managed through rotational burning and grazing . . . We recognise the pivotal importance of grouse moor management for the economy, ecology and social fabric of the Scottish uplands.

These heather moorlands are of great conservation significance internationally (Thompson *et al.*, 1995b; Smith *et al.*, 2000). Their bird life is seen as especially valuable (Ratcliffe, 1990; Brown and Bainbridge, 1995), with thirty-four species breeding on them and large numbers of golden plover (*Pluvialis apricaria*), curlew (*Numenius arquata*) and several raptor species using them for breeding and feeding. Mountain hares (*Lepus timidus*), many songbirds and a wide range of invertebrates also thrive in these habitats (Usher and Thompson, 1993). In addition to their conservation significance, they are important in terms of their landscape, archaeological, aesthetic and tourism value.

Concerning the impacts of grouse moor management on biodiversity, the effects are mixed (Thompson *et al.*, 1997). On the one hand, it was only their value for grouse shooting which prevented large areas of moorland being commercially afforested in the 1970s and 1980s, a land use transition which would have been damaging for many upland breeding birds and for the wider conservation interest. Grouse shooting thus helps to maintain a multiple land use system in the uplands (Phillips, 2001). On the other, many estates illegally persecute raptors, greatly reducing their numbers. Nevertheless, grouse moor management is regarded as more benign in its impact on biodiversity than some other upland land uses. Because of these conservation spin-offs, and because managed grouse moors used to represent a bulwark against encroaching afforestation, some common goals have been shared amongst grouse interests, conservationists and recreationalists. However, this alliance has been an uneasy one (Brown and Bainbridge, 1995), and it is now under severe pressure (Section 7.6.2).

7.4 PRESENT MANAGEMENT: FRAMEWORK AND PRACTICE

7.4.1 Deer management

Overall statutory responsibility for deer management rests with the DCS which works closely with deer management groups (Section 7.5.2) and individual estates, providing assistance, guidance, and information from their on-going deer counting programme. Practical responsibility for open range deer lies almost entirely with stalkers employed by private estates, while the increasing population of woodland deer are managed, according to forest ownership, by Forest Enterprise, private forest companies, and private estates. The primary management activity is the annual cull, stags being taken in the autumn and hinds through the winter. Open range stalking as practised in Scotland is relatively unusual in a European context and is highly valued as a result (Bullock, 2001). It is a demanding and skilful undertaking. Considerable patience and fieldcraft skills are required to approach unseen, unheard and unsmelled to within 100m of alert, fleet-footed creatures, and the selection process depends on accumulated deer knowledge (Whitehead, 1996). Normally, the aim is to take those animals that wolves would take: the sick, the weak and the old. Once shot and gralloched (gutted), the stalker is responsible for getting the carcass back to a deer larder (no mean challenge in the rougher, wilder parts of the Highlands) and for preparing it for a venison dealer according to increasingly exacting EU standards. In addition to the cull itself, stalkers also undertake winter feeding, population counts, muirburn and predator control, together with a host of general estate duties such as the maintenance of paths and tracks, and the management of public access.

7.4.2 Grouse and moorland management

In contrast to deer management, there is no state oversight of moorland management for grouse. Within the limits imposed by conservation designations, the private estates are largely autonomous. There are two types of grouse shooting. For driven grouse, beaters drive grouse over a static line of guns (shooters). This is a sporting tradition unique to the UK, attracting many from overseas. For a viable driven shoot the minimum grouse density is 60 birds/km^2 (Hudson, 1995). The alternative is walking up, when the guns and their dogs walk across the moor and shoot at the grouse as they are flushed. Walking up involves lower management costs but brings in far less revenue. A successful day's driven shooting can generate £10,000 for the owner whereas a typical walked-up day will bring in a tenth of that (Thirgood *et al.*, 2000b).

Grouse populations are controlled by a combination of habitat characteristics affecting food and shelter, and by levels of predation, cycles of disease, and weather patterns. Management by gamekeepers therefore focuses on the habitat, the predators and the grouse themselves. The main habitat management tool is rotational muirburning which aims to create a patchwork of young nutritious growth intermixed with older growth for nesting cover (Fig. 7.6); this also benefits sheep and deer, so grazing levels need to be controlled to prevent a transition to acidic grassland which is of much lower conservation value. Muirburning is a skilled and weather-dependent job which, a century ago, used to involve many people and produce numerous small burned areas. In these days of reduced keepering (Section 7.5.3) the traditional ideal of a 10–15 year rotation is rarely attained, and the result – less frequent burning of larger areas – is less beneficial for grouse. The second management focus is control of predators and disease. Legally, predator control is restricted to foxes, mustelids (except otters), feral cats, most corvids and some gulls. Illegally, it all too often includes protected birds of prey (Green and Etheridge, 1999).

7.5 RECENT DEVELOPMENTS AND TRENDS

7.5.1 Increasing numbers and range of red deer

During the last 100 years deer populations have been increasing throughout the northern hemisphere (Clutton-Brock and Albon, 1992). In Scotland, red deer numbers

FIGURE 7.6 *Aerial view of the patterns created by muirburning near Wanlockhead in the Southern Uplands. Photo: © Patricia & Angus Macdonald/SNH.*

have risen steadily since 1950, and sharply from 1970 (Fig. 7.7). Deer numbers are notoriously difficult to estimate, all too often being based on 'hot air and guesswork' (McLean, 2000), but by the early 1990s the Scottish red deer herd was thought to have reached 350,000 (Harris *et al.*, 1995). All the evidence points towards continuing growth since then. Although a few deer management groups are managing to reduce their population, elsewhere both numbers and densities continue to rise, in places dramatically (Table 7.4). The precise overall figures may be debatable but there is no doubt that the population is growing rapidly (McLean, 2001b). Most of this population increase has been amongst hinds while stag numbers have remained more constant. The result is that the hind:stag ratio is commonly 2:1 and has reached 5:1 in some areas (Bullock, 1999).

TABLE 7.4 Changes in red deer population size in open range areas which were counted by the DCS in 1999/2000. Source: DCS, 2000b.

Counting block	*Area (ha)*	*Date of last DCS count & total counted*		*Number counted in 1999/2000*	*Density in 2000 (deer/km²)*	*Change between DCS counts*
Ben Nevis	42,300	1990	2,090	3,953	9.3	+89%
Blackmount	61,000	1990	5,635	8,544	14.0	+52%
Cabrach/ Glenbuchat	44,900	1992	2,276	1,563	3.5	–31%
E. Grampians	34,000	1994	7,915	8,652	25.0	+9%
Glenartney	12,500	1993	4,668	3,901	31.2	–16%
S. Perthshire	75,600	1986	3,546	5,359	7.1	+51%
S. Uist & Benbecula	14,000	1983	38	398	2.8	+947%
N. Uist	18,000	1996	670	869	4.8	+30%
W. Loch Shiel	17,200	1994	2,418	2,477	14.4	+2%
Harris/Lewis	65,000	1993	2,690	4,248	6.5	+58%
West Ross	14,300	1999	3,132	3,138	22.0	0%
Dunrobin/ Morvich	10,900	1999	1,227	975	9.0	–21%
Rannoch	20,000	1999	3,048	2,210	11.0	–27%
Rum	10,050	1999	1,432	1,341	13.3	–6%
Balquhidder	40,000	1997	2,436	2,398	6.1	0%
Trossachs	21,300	1998	1,092	2,212	10.4	+103%

The expansion has involved increases in both density and range. Population density is locally increased by new afforestation schemes from which deer are fenced out; critically, these often exclude deer from their wintering ground at lower altitudes, exacerbating the long-standing problem of winter concentrations. In the Cairngorms, winter densities can reach 150 deer/km² (Youngson and Stewart, 1996). Conversely, range expansion has been facilitated by increasing colonisation of older forestry plantations as forest fences become permeable or collapse. There are now probably no deer in Scotland without access to woodlands (McLean, 2001b). The rising woodland population is even harder to estimate than that on the open range, and must now be

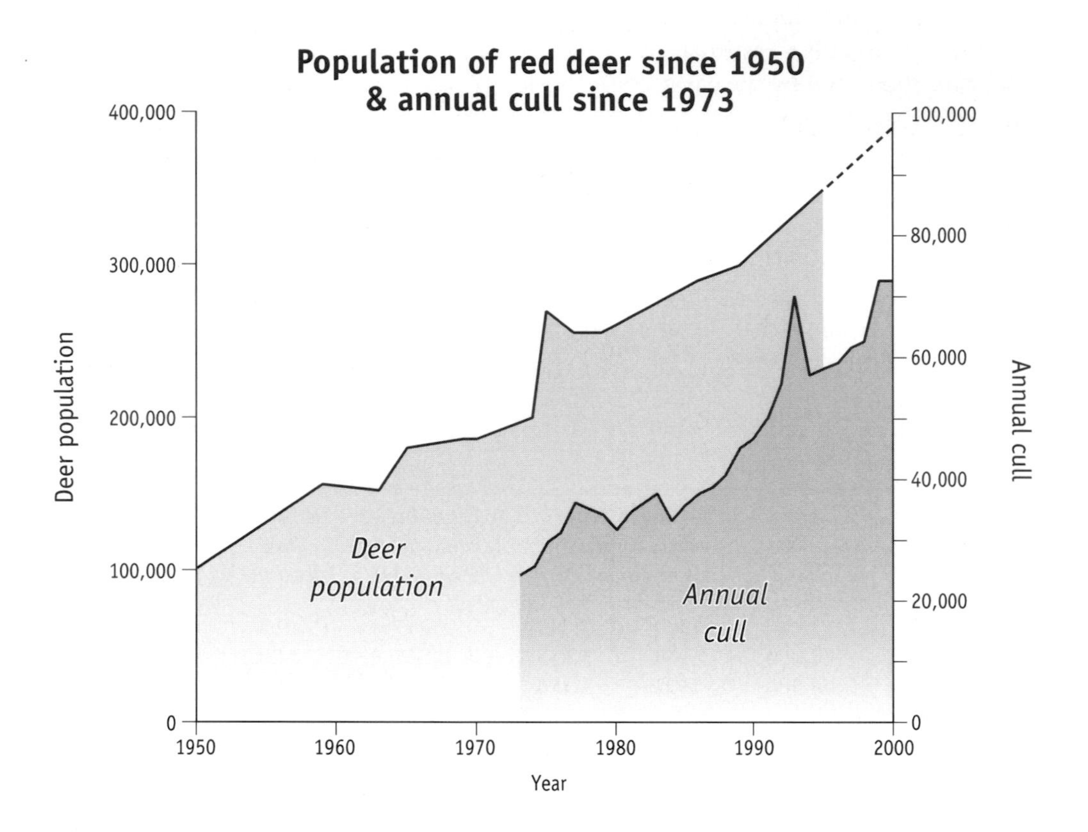

FIGURE 7.7 *Scottish red deer population estimates since 1950 and annual culls since 1973. Whitehead (1996) argues that the population estimates for the 1950s and 1960s were too low. Sources: SNH, 1994b; RDC/DCS Annual Reports.*

well in excess of Ratcliffe's (1987a) estimate of 50,000; certainly the annual woodland cull more than doubled during the 1990s, reaching 14,389 in 1999/2000 (DCS, 2000b). The relatively greater fecundity of woodland deer will only exacerbate the population problem (Whitehead, 1996). Managing deer in woodlands raises a range of distinct issues which are discussed in Section 11.1.

The causes of population growth have long been a matter of debate. The main factors are thought to include the following (Clutton-Brock and Albon, 1989; McLean, 2001b):

- An inadequate cull. To achieve a stable population the cull must at least match the recruitment rate (the rate of population increase by birth and in-migration). Annual culls have been rising since the early 1970s (Fig. 7.7) but despite unprecedentedly large culls during the 1990s in response to increasing pressure from the DCS, the population continues its apparently inexorable growth. Hind numbers in particular remain obstinately high, partly because heavy hind

culling leads to higher calving rates. Underculling of hinds is regarded by the DCS (1999) as the prime reason for increasing deer stocks.
- Declining hill sheep numbers, freeing resources to sustain more deer.
- Favourable weather patterns. The last two decades have seen sequences of dry summers and mild winters which promote deer health, increase calving percentages and minimise winter mortality.
- Habitat change. A more questionable issue is whether the deer themselves increase the carrying capacity of their habitat. The suggestion is that their browsing, dunging and trampling leads to the replacement of heather with palatable grasses.
- Colonisation of plantation forests resulting in reduced mortality and increased rates of reproduction.

There is less uncertainty about the impacts of population growth (Staines *et al.*, 1995). These include the prevention of natural regeneration of native woodlands (Section 11.1.4), the loss of dwarf shrub and tall herb communities, and damage to rare plant species, especially alpines on high plateaux such as the Cairngorms. Dense concentrations of deer (especially when combined with high sheep numbers) are thought to cause soil erosion and localised declines in the area of heather moorland through their grazing and trampling (Milne *et al.*, 1998; Staines, 1999b). Furthermore, the growing numbers and range of red deer lead to increasing competition (and friction) with other land uses, especially farming, forestry, grouse moors, conservation and recreation. (Curiously, although agricultural damage from marauding deer has long been at the heart of the red deer problem, there has still been no research on the direct effect of deer damage to crops (Scott and Palmer, 2000).) Deer are also a frequent cause of road traffic accidents in rural areas (Staines, 2000).

However, two observations about deer impacts need to be made. Firstly, the pattern of population growth and its associated impacts are spatially variable, not always mirroring the national trend. Deer-related problems tend to be most severe in the east where numbers have increased at dramatically above-average rates in places. In the East Grampians, for example, the population increased by two and half times in the period 1966–86, and in the Cairngorms/West Grampians it tripled between 1962 and 1983, although both populations have since stabilised (DCS, 1999). Local problems are related to local densities and need local solutions. Secondly, the habitat impacts of sheep are similar to those of deer. Sheep share much of the deer range and outnumber red deer by 8:1 in the uplands (Milne *et al.*, 1998), so the impacts of the two species are hard to disentangle (Staines *et al.*, 1995), especially when the effects of other herbivores such as rabbits, mountain hares and roe deer are added to the equation.

The crux of the red deer problem, in Staines' (1999b) view, is the use of the same piece of land for conflicting purposes. Opinions about the ideal red deer population size differ amongst farmers, foresters, hillwalkers, conservationists and sporting estates, but there has been general agreement for some time that the current population is too high to be compatible with *any* of the various management goals (Abernethy, 1994a).

The pressing issue of how to solve this 'red deer crisis' (Wigan, 1993) is discussed in Section 7.6.1.

7.5.2 Progressive integration: Deer Management Groups

From the 1960s, groups of estates began to form themselves into deer management groups (DMGs) with the aim of collaborating, sharing data, and setting cull targets for their combined area rather than simply for estates in isolation. The rationale for this joint approach is unarguable. Except for the very largest estates, deer roam over a far greater area than single deer forests, so without collaboration the policies of one estate can be nullified by those of its neighbours. The pooling of resources also facilitates counting programmes and the on-going battle against poaching. In addition to those directly involved in deer management, DMGs have increasingly included representation from woodland, conservation and recreation interests. This is symptomatic of the broadening arena in which deer management now operates; having been a private, single-objective sporting pursuit for 150 years, it is now a matter of increasing public concern and is adopting multiple objectives (DCS, 1999). DMGs thus serve a broader function than initially envisaged, providing a much-needed local forum in which different and sometimes competing objectives can be discussed and reconciled. This is replicated at national level by the Deer Management Round Table, a twice-yearly meeting of over thirty organisations which share an interest in deer (DCS, 1999).

Estates' naturally conservative 'island mentality' meant that DMGs were slow to catch on initially, but they have spread rapidly since the early 1980s with strong encouragement from the DCS. Typically they involve private owners, the FC, the DCS, and sometimes SNH. There are now forty-five, covering almost the entire red deer range (DCS, 2000b). SNH (1994b) sees DMGs as a key to providing integrated management of deer and their habitat, while Bullock (1999: 248) believes that they constitute 'a significant step forward in that they represent an acknowledgement by owners of their dependence upon a common resource'. Their weakness is that they rely on peer pressure and persuasion; thick-skinned owners and recalcitrant stalkers can plough their own furrow with virtual impunity. Moreover, DMG meetings tend to steer around 'hot' topics like cull levels, preferring gentlemanly discussions about easier subjects like venison marketing and carcass tagging. The selection of a culling strategy which satisfies the multiple objectives represented within a DMG and which takes into account the specific population dynamics of the area's red deer is a complex challenge which is rarely addressed (Trenkel, 2001).

A logical extension of the DMG rationale to the national scale led to the formation of the Association of Deer Management Groups (ADMG) in 1992. This body aims to be a voice for deer management and deer managers at European, national and regional level on all matters relating to wild deer, and in this it has had notable success. In particular, ADMG has been a major player in the access debate, being a prime mover behind the Access Concordat (Section 9.3.1.ii), and it has worked hard to promote a coordinated approach in the notoriously fragmented field of venison marketing. These efforts finally bore fruit in 2000 with the establishment of a Scottish Game Dealers and Processors Association.

A significant current development is the formulation of Deer Management Plans. Such plans were piloted in seven areas in the late 1990s, and the DCS is now encouraging all DMGs to formalise their management aims and objectives in this way. These plans focus not only on deer numbers and sporting objectives, but also (in response to the 1996 Act) on the quality of the habitat. Using rapid habitat survey techniques, estates are beginning to gather (for the first time and at their own cost) the basic habitat data without which informed decisions about carrying capacity cannot be made. These and other data form crucial inputs for HillDeer, a decision support software package developed in the late 1990s by the DCS and ADMG. This tool allows managers to explore the long-term consequences of different management strategies for deer populations and their habitats (Buckland *et al.*, 1998; Gordon and Hope, 1998). The hope is that the combination of management plans, habitat surveys and HillDeer will allow strategies for sustainable deer management to be developed at DMG level throughout the deer range. Taken together, these developments represent a decisive professionalisation of a field hitherto guided almost entirely by tradition.

7.5.3 Decreasing management of moorland

Falling numbers of gamekeepers and consequent reduction in the intensity of moorland management has been a long-term if irregular trend. Overall, the number of keepers in upland counties fell by 85 per cent between 1901 and 1981 (Hudson, 1992). During the two world wars most keepers went to fight (c.f. Fig. 7.4) and many never returned; moorland management and grouse numbers did recover, but not to former levels. This, combined with economic forces, led to a general decline in the fortunes of grouse moors. During the 1970s and early 1980s, when grouse populations were in steep decline (Fig. 7.4) and afforestation was fiscally prudent (Section 4.2), many grouse moors were sold for forestry. Afforestation not only causes direct losses of moorland area but has negative edge effects on adjacent moorlands. Forests close to grouse moors constrain muirburning (for fear of starting a forest fire) and harbour foxes and crows which prey on grouse. Bird strikes on forest fences also cause heavy mortality amongst grouse (Section 11.1.4). Afforestation can thus have a 'domino effect' as reduced grouse productivity compromises the economic viability of nearby moors and triggers further land sales for forestry (UKRWG, 2000). Moorland owners and managers have become trapped in a vicious circle: most grouse shooting enterprises are loss-making so on those estates where profitability is important keepers are laid off or employed part time. As the intensity of management declines so predator control slackens off and muirburning becomes less frequent resulting in rank heather growth. The habitat deteriorates, and grouse populations fall below the viability threshold for driven grouse, further undermining the financial situation. Habitat deterioration has been accelerated by rising sheep numbers and falling shepherd numbers, exacerbating the overgrazing problem. Given these pressures it is hardly surprising that traditional, intensive grouse moor management is now a rarity. Phillips and Watson (1995) estimate that only about five moors practise muirburning at a level approaching that of the early twentieth century.

7.6 Current issues and debates

7.6.1 Controlling red deer populations

Of the many issues related to red deer management, three stand out. One is the interaction between deer and forestry, and particularly the impact of deer on native woodland regeneration. Another is the friction between stalking and mountain recreation. These issues are discussed in Sections 11.1 and 9.2.1 respectively. The third is the need to control and then reverse the population explosion. Essentially this comes down to finding ways to increase the annual cull, especially of hinds. There has been talk of reintroducing the deer's natural predator, the wolf, as a supplementary means of population control (Spinney, 1995), but this is most probably unworkable (Section 10.4.1).

Since the early 1990s both the DCS and SNH have been calling for a reduction in the overall red deer population by 100,000 animals as a first and urgent step towards integrated management (SNH, 1994b). Such a reduction is no small challenge, not least because of density dependent aspects of the population biology of deer. As population is reduced, reproduction rates increase and mortality declines, both of which work to increase the rate of population growth (Ratcliffe, 1998). Most DMGs do not need persuading that the deer population must be reduced (not only for ecological but political reasons) but the pressure to increase culls is coming at a difficult time as deer managers find themselves caught between economic realities and conservation demands. Most estates are loss-making, commercial sporting lets and venison prices are notoriously fickle, and there has been increasing disturbance of stalks by walkers in the popular mountain areas (Section 9.2.1). Estates are often reluctant to reduce deer numbers because of the link with capital values (Section 3.2.1), and because initiating a policy of heavy culling is an expensive, long-term decision with no guarantee of personal benefit (Bullock, 1999).

Moreover, amongst deer forest owners there remains a residual reluctance to heavy hind culling deriving from a clutch of widely held beliefs which date back to Victorian times. Reynolds (1995: 9) observes that 'tradition, not biology, has dominated the management of this precious wildlife resource'. One such belief is that stags are attracted by large numbers of hinds, and thus that fewer hinds will result in fewer stags, lower revenues, reduced capital values and enforced reductions in employment. Another is the supposed 'vacuum effect' whereby heavy culling on one estate leads to mass immigration of deer from neighbouring ground and consequent inter-estate friction. The biological facts do not support such shibboleths (Staines *et al.*, 1995; Toogood, 1995). On the contrary, at lower densities growth rates are faster, body weights and antler weights are higher, and fecundity and survival both improve. Reduced quantity therefore leads to improved quality without compromising either venison production or the stalking value of the deer population (Reynolds, 1995; Staines, 1999b). Research on Rum has even shown that increasing the hind cull from 6 per cent to 16–18 per cent could *increase* the sustainable cull of mature stags by some 30 per cent (Clutton-Brock and Albon, 1992). Increasing population densities on Rum led to a decreasing proportion of males born each year and to a reduction in the survival of stags, while reductions of hind numbers led to increases in stag numbers (Clutton-Brock and Thomson, 1998; Kruuk *et al.*, 1999). It seems, then, that large, female-biased deer populations are bad not only for the environment but also for the sporting estates, and, conversely, that a much reduced deer stock would benefit the natural heritage and maintain the value of stalking (Staines *et al.*, 1995). What evidence there is also suggests that the 'vacuum effect' is a fiction because deer (especially hinds) are strongly hefted to a piece of ground (Staines, 1999b).

Cherished traditions die hard, however. Until the recent research is more widely publicised and assimilated, these beliefs will continue to create a gulf between stated intentions and effective action, delaying the universal adoption of heavy culling. Insufficient culls are caused primarily by reluctance, not inability (McLean, 2001b). Nevertheless, given the economic and psychological barriers, the success in achieving high culls in the late 1990s (Fig. 7.7) is rather remarkable. It demonstrates either a costly altruism by private landowners, or a fear of legislation (better to jump before being pushed), or most probably a bit of both. Traditional barriers *are* coming down as well. For example, stalking is no longer exclusively focused on stags; a majority of large estates now offer hind lettings as well (Trenkel *et al.*, 1998). But even with the unprecedented recent efforts, only twice during the 1990s did the hind cull exceed estimated recruitment (DCS, 1999).

Consequently, from 1998 the DCS started setting advisory target culls for DMGs. Initially, these are being calculated on the basis of sporting requirements, but as habitat information becomes available they will increasingly incorporate carrying capacity. To help meet these targets, the DCS can negotiate voluntary Control Agreements with DMGs which set population and cull targets, and can allow an extension of culling into the statutory close season. There are now twelve of these in place covering a total of 356,830ha (DCS, 2000b). Under the 1996 Act, the DCS have the power to carry out emergency culling themselves, but believe that this should be used only as a last resort. Bringing deer populations back into balance with their environment and with other land uses is seen by some as a test case for the controversial voluntary principle (Section 13.3.5). Others interpret the continuing population increase (despite the constant cajoling of landowners by the DCS for over thirty years) as evidence that the voluntary principle has already been tried and found wanting (MacMillan, 2000), and urge the DCS to utilise the more draconian powers that it now has.

7.6.2 Red grouse population dynamics and the future of heather moorland

7.6.2.i Causes of grouse population decline

The question 'why have grouse populations declined?' is deceptively simple. Teasing out answers from the entangled web of possibilities has proved to be a tough challenge, made all the harder by the strength of the various vested interests involved. This acronym-ridden range of interests includes the landowners themselves (represented by the SLF), SNH, the GCT, the RSPB and a range of other conservation charities. Further complexity is added by the spatially and temporally variable nature of the factors involved.

Possible explanations for declining grouse numbers include the following (Hudson, 1992, 1995; Phillips and Watson, 1995; UKRWG, 2000):

- Habitat loss: the area of heather moorland has dwindled by 23 per cent since the 1940s (Mackey *et al.*, 1998) through the combined effects of upland afforestation and the replacement of heather with grassy sheep walk. In addition to reducing the amount of heather moorland directly, afforestation has negative edge effects for grouse moors (Section 7.5.3).
- Habitat deterioration: poor management, combined with overgrazing by sheep and (to a lesser extent) deer, leads to a decline in heather quality and ultimately heather loss. Heather only persists with stocking densities up to 1.5 ewes/ha (Thompson *et al.*, 1995b). A long-term decline in soil fertility has also been postulated.
- Disease and parasites: notable amongst these are tick-borne louping ill and strongylosis, the first of which kills 80 per cent of infected birds on the small number of moors affected (Hudson, 1992).
- Reduced keepering: the reduction in the numbers of gamekeepers has inevitably resulted in reduced quantity and quality of management as lone individuals struggle to manage extensive swathes of upland. Critical elements are the reduced amount and quality of muirburning (contributing to habitat deterioration), and inadequate predator control. Greater keeper density in England is thought to be a partial explanation of the higher, better sustained grouse bags there (Smith *et al.*, 2000).
- Bad weather: runs of bad years can compound the effects of all the above.
- Overharvesting and disturbance.

There is broad consensus now that the *long-term* trend of population decline is primarily explained by the sustained losses of heather moorland area (Thirgood *et al.*, 2000b). Thus population declines have been greatest in the wetter west where heather losses have been most extensive (Hudson, 1992). Equally, there can be little doubt that efforts to improve the habitat would help to maximise grouse numbers on the grouse moors which remain. As Thompson *et al.* (1997) observe, there is no possibility of a return to 'Edwardian' grouse bags on degraded, fragmented, smaller moors. The barrier to habitat improvement, of course, is financial; maintaining a high quality grouse moor is a labour-intensive task. But when it comes to the question of what controls *short-term* fluctuations in grouse numbers, finding any consensus has proved far more difficult.

7.6.2.ii The 'predation trap' and the Joint Raptor Study

On the basis of extensive research, Hudson (1992) proposed a 'predation trap' theory. At high densities, grouse mortality is strongly associated with parasites, but at low densities an inverse density dependent predation rate appears to keep grouse populations low. The theory proposes that, because generalist predators take a relatively constant number of grouse, when grouse population densities are low the proportion of the population killed by predators is high, preventing a recovery of numbers. Grouse populations thus become 'trapped' at small sizes. Predator-prey relationships are usually density dependent, so the discovery of an apparently inverse density dependence was intriguing, as well as running counter to earlier studies.

It was also politically charged. The findings gave the SLF and grouse moor owners a scientific basis for their campaign for a legal raptor cull. SNH and the RSPB argued that raptors are threatened (not least as a result of endemic illegal persecution on grouse moors (Green and Etheridge, 1999)), and dismissed the predator trap as a myth (Edwards, 1996). To examine the relationship between raptor predation and grouse populations, the Joint Raptor Study (JRS) (1992–7)[5] was set up by an unlikely consortium of conservation, shooting and landowner interests. It was carried out on six grouse moors in south-west Scotland, centred at Langholm (a moor which holds the Scottish record of 2,523 grouse shot in a single day on August 30th, 1911). Throughout the JRS, foxes and crows were rigorously controlled but raptors were allowed to breed unchecked. The result was an increase in the number of breeding hen harrier females from two to twenty and a doubling in peregrine numbers.

The JRS exonerated raptors from any blame for the *long-term* decline of grouse populations, stressing that grouse numbers were declining during many decades in which raptors were rare or absent (Redpath and Thirgood, 1997; Thirgood *et al.*, 2000a). Instead it pointed to the 48 per cent reduction of heather-dominant vegetation since 1950 (due to sheep grazing) and to the increasing fragmentation of heather areas as the probable explanation. In the *short term*, however, the JRS provided convincing empirical support for the predator trap concept, showing that predation by raptors can reduce grouse populations by more than 50 per cent within a single breeding season and hold them below levels at which grouse shooting is economically viable (Thirgood *et al.*, 2000c). Grouse population cycles at Langholm had previously been in step with those on nearby estates with low raptor densities. A predicted peak in grouse numbers duly occurred on the neighbouring estates in 1997 but not at Langholm (Fig. 7.8), leading to an enforced cessation of driven shooting and net losses to the estate of around £100,000 per annum. All five keepers have since been stood down (Hart-Davis, 2000). The JRS data further indicated that removal of raptors would increase autumn grouse densities by 3.9 times within two years (Thirgood *et al.*, 2000c). It seems, then, that there would be no predator trap without habitat change, but equally that the trap would not be lethal without raptors.

The RSPB and SNH, understandably wishing to exonerate raptors, emphasise the long-term results. Equally understandably, moor owners and keepers who see their livelihoods threatened stress the short-term impacts of raptors. Potts (2000: 9) even accuses the RSPB of 'manipulating the truth' and argues that 'it is illogical to manage everything in the environment except raptors'. Long-term and short-term causes are to some extent related, in that increasing fragmentation of heather cover increases the densities of the small mammals and passerines (songbirds) on which hen harriers prey, thus enabling harrier numbers to increase. In fact, habitat may affect the interaction between grouse and their predators in a variety of direct and indirect ways (Smith *et al.*, 2000). The JRS provided some of the scientific answers but it was not tasked to address the fraught political and management questions. This poisoned chalice was given to the government's UK Raptor Working Group (UKRWG).

7.6.2.iii Raptors versus *red grouse: management solutions*

The elusive objective is to find means whereby raptor populations can increase nationally while their densities on grouse moors are kept low enough to permit grouse shooting to continue. In the light of the JRS results, grouse shooting interests see a reduction of raptor numbers (whether by culling or relocation) as the only short-term solution, and point out that keepers at present have no incentive to stay within the law. Since gamekeepers' remuneration is performance-related, in effect raptors prey on their wage packets; the fear of 'another Langholm', with redundancy following, is enough to drive many to break the law. Unsurprisingly, however, conservationists view the culling of protected species as utterly unacceptable (RSPB *et al.*, 1998). They argue that legal culling of raptors is not justified when raptor populations are still under pressure from egg collectors and illegal control. While such persecution is rightly deplored, the irony is that allowing raptors to breed freely might not be in the best long-term interests of raptor conservation. If birds of prey were to make driven grouse shooting unviable, the consequent reduction of keepering would allow fox and crow numbers to increase and habitat quality to decline, both of which could reduce raptor populations (UKRWG, 2000). This scenario seems to be unfolding already at Langholm, with hen harrier numbers having halved since the end of the JRS (Hart-Davis, 2000; Potts, 2000). It therefore seems to be in everyone's interests to maintain grouse stocks at densities which permit driven shooting. If legal culling of raptors to achieve this end is not politically acceptable, then other management options will have to be explored.

A variety of possibilities have been under discussion. These include rearing and release of grouse, relocation of raptors (either by live capture or through 'nest management'), raptor 'quotas', intraguild predation (letting raptors control raptors) and diversionary feeding (Tapper, 1999; Thirgood *et al.*, 2000b; UKRWG, 2000). Tapper (1999: 7) argues that the legal protection given to raptors under the 1981 Wildlife and Countryside Act was a necessary first step, but that 'in the future we will need management as well as species protection if we are to retain a rich and diverse predator fauna as well as prolific game stocks'. Such management may require a change in perception whereby raptors come to be seen as a renewable resource (Thirgood *et al.*, 2000b). It remains to be seen whether any of the suggested compromises can be effected (legally and financially) and effective in practice.

In the long term, the best solution would be improved management of the habitat (Smith *et al.*, 2000; UKRWG, 2000). Even this, however, may not offer a complete answer (Thirgood *et al.*, 2000b), and few moorland owners are likely to set out in pursuit of such a distant and costly goal in the present climate of confrontation and uncertainty. Clearly what is needed is a combination of immediate amelioration measures and a start on long-term habitat restoration. Only such a two-pronged approach will ensure a sustainable future for grouse shooting, rural communities and upland conservation (RSPB *et al.*, 1998). The solutions that are adopted to resolve

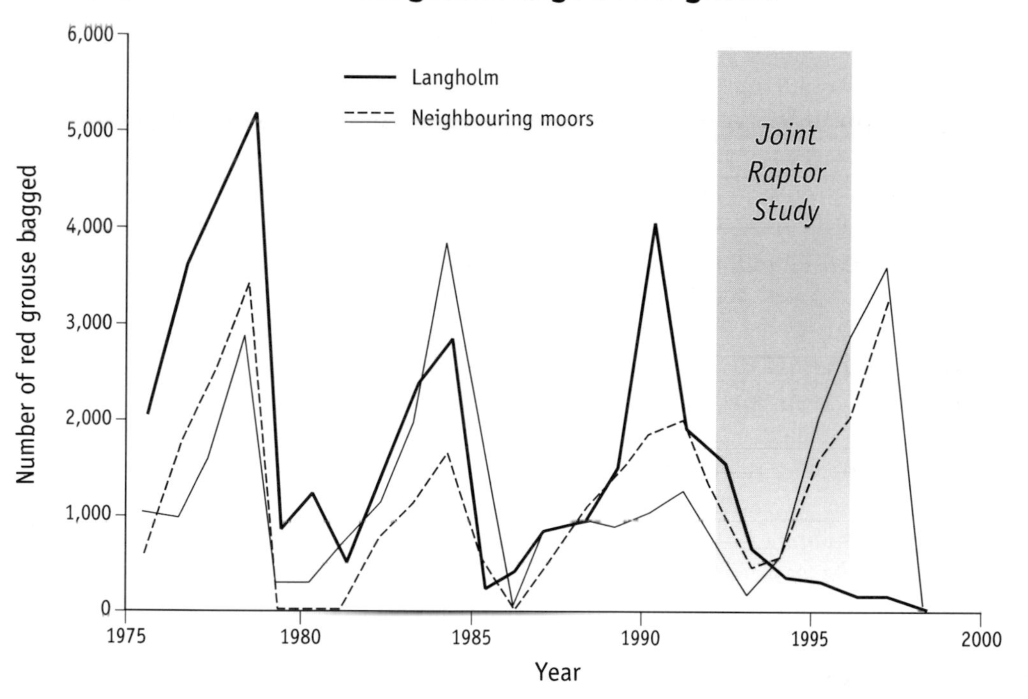

FIGURE 7.8 *Red grouse bags at Langholm since the mid-1970s compared with two neighbouring moors. All three showed similar cyclic variation until the period of the Joint Raptor Study when Langholm bags did not increase to the expected cyclic high. After UKRWG (2000).*

these vexed issues will have a profound effect on the future of large parts of the uplands, affecting resident and visiting populations of both wild animals and human beings.

7.6.2.iv What future for heather moorland?

Heather moorlands are highly valued cultural landscapes. How should they be managed for the future? Should they be conserved and extended, or should the long-term trends of diminishing area and replacement with other land uses be allowed to continue? Of the spectrum of options available these are the end points. Both have their advocates. The first option naturally has the support of grouse moor managers but also of many conservationists because, with the glaring exception of illegal persecution of raptors, grouse management seems to be beneficial for conservation (Thirgood *et al.*, 2000b). It is ironic that the product of what was for the Victorians a practical game management regime is now advocated by some as the ideal natural heritage prescription for large parts of the uplands (Phillips and Watson, 1995; Magnusson, 1995b). The patch structure which was designed to maximise grouse numbers for sport turns out to be excellent for maintaining biodiversity and a distinctive landscape character (Usher and Thompson, 1993; Thompson *et al.*, 1995b), although the relative merits for conservation of various patch structures and of the means for creating them (muirburning *versus* cutting or layering) remain open questions (Macdonald *et al.*, 1995).

A negative argument which complements these positives is that the abandonment of grouse moor management would have damaging effects. There would certainly be economic damage, but the ecological consequences are not easy to predict (UKRWG, 2000). Twenty-five years ago the cessation of grouse shooting usually led to afforestation, but now the land use consequences are less obvious. Conservationists prefer heather moorland to plantation forests, but if the alternative to heather moorland is naturally regenerated scrub and native woodland then the pros and cons become more finely balanced (Smith *et al.*, 2000). It is certainly true, though, that without management most of the better moorland would peter away as heather naturally gave way to other dominant species such as grasses, trees or bracken (Gimmingham, 1995). It also remains the case that management for grouse shooting provides the greatest sustainable, unsubsidised income from moorland areas (Thompson *et al.*, 1995b); all other alternatives require significant injections of public funds. Using such arguments, Phillips (2001) makes a radical 'pro-heather' recommendation, namely that in the drier, eastern parts of Scotland some of the afforested moorlands should be restored to heather moorland when the current forestry rotation ends.

The second option – allowing grouse moors to dwindle – is advocated primarily by the native woodland movement (as well as by anti-blood sports, animal rights and 'back to nature' groups). Much of today's moorland could (and once did) support a forest cover, and would revert to forms of woodland if grazing and burning were halted. Extending the area of native woodland is an objective that has widespread support (Section 10.2), and the uplands are one of the obvious candidate areas for large scale expansion. Allowing natural regeneration of woodland to take its course would unquestionably be the most natural option, and the re-establishment of woodland and scrub as part of a mosaic of upland habitats would enhance biodiversity and provide shelter for animals (Hester and Miller, 1995). However, if this were to happen on a large scale, locals and tourists alike would doubtless object to the loss of familiar and much-loved landscapes, and there could be job losses with knock-ons for the rural economy. Moreover, it is hard to envisage very many private estates warming to this hands-off prescription unless handsome incentives were attached.

Obvious middle ground exists between these two extremes, involving zonation and diversification. Multi-purpose management must presumably be the way to reconcile the various desirable but competing objectives (Thompson *et al.*, 1997). Gimmingham (1995) argues that there is room in the uplands for extensive areas of moorland, for more native woodlands, and for the creation of patchy heath areas which combine heather with scrub and trees. Increasing the area of heather moorland mosaics (as distinct from a heather monoculture) is a widely shared objective. Consequently, high quality management of the habitat and of grazing herbivores should be encouraged on the remaining active grouse moors, producing an expanding mosaic of habitats which will sustain grouse populations for shooting but also enhance the conservation fringe benefits. Elsewhere, a diversity of other land uses (agriculture, forestry, recreation, conservation) could flourish. As ever, the practical problem is finding an acceptable balance, and specifying the local particulars of such a broad-brush, unexceptionable vision. What is going to happen where? Who is going to organise and carry it out? Who chooses? And who is going to pay?

At present government policy seems to be encouraging all three of the above options. In the middle ground, diversified, multi-purpose management is the watchword. On one flank, generous public support is available for native woodland establishment (Section 10.2). On the other end, the high conservation value of heather moorlands (especially given their decline elsewhere in Europe) has been recognised with a UK Biodiversity Habitat Action Plan (HAP) (Section 8.2.3). This aims to extend the total area of upland heathland by 5 per cent, with an emphasis on reducing fragmentation and creating or maintaining areas greater than 10km^2 (UKRWG, 2000). Such targets are contingent upon major reform of the Common Agricultural Policy (Section 5.2.1) (particularly to reduce sheep densities) and the provision of substantial public funding. Annual implementation costs are likely to approach £10 million in the early years and to double later on. Linked to the HAP for Upland Heathland, the UKRWG (2000) recommends a national campaign for heather moorland restoration.

Given that most moorland areas are in private ownership, the issue of public support is a crucial one. At present there are no incentives or policies specifically designed to encourage high quality heather moorland management. Such incentives (or, in the short term, a refocusing of existing schemes) are likely to be required if HAP targets are to be met, and they are advocated by the UKRWG (2000). This is not a new idea. McGilvray and Perman (1992), Hudson (1995) and RSPB *et al.* (1998) have all suggested that providing support of some kind for grouse moor managers would be a cost-effective approach to conserving heather moorland. However, it is the first time that a government study has proposed a major incentive scheme for moorland management *per se*.

Debates surrounding grouse have a long and acrimonious history, with deeply entrenched views and suspicions in many quarters. Peace and consensus have yet to break out; there is even a lack of consensus about the degree of consensus! However, given the amount of serious attention being given to the issues, and the many recent signs of collaboration and partnership,[6] there are grounds for optimism about the medium-term future both for grouse moors and heather moorlands.

7.7 OTHER HIGH-PROFILE WILDLIFE ISSUES

7.7.1 Preserving the red squirrel

Red squirrels (*Sciurus vulgaris*) (Fig. 7.9) arc native to all parts of Scotland north of the central belt but were driven to almost complete extinction in the late eighteenth century before recovering as a result of reintroductions from England and Scandinavia (Kitchener, 1998).[7] They are now under threat again as the range of the grey squirrel (*S. carolinensis*) steadily increases (Fig. 7.10). Grey squirrels, natives of

FIGURE 7.9 *A red squirrel in a Scots pine. Photo:* © *Niall Benvie.*

eastern North America, were first introduced beside Loch Long in 1892 and have spread ever since (Kitchener, 1998). They typically exist at much higher population densities than reds and can cause serious economic damage in forests through bark stripping. The 121,000-strong population of red squirrels (75 per cent of the British population) is now dwarfed by 200,000 grey squirrels (Harris *et al.*, 1995), threatening their long-term survival in Scotland and making them one of Britain's most endangered mammals. Concerns about displacement of red by grey squirrels were raised in the 1930s, but concerted conservation efforts only began in the late twentieth century. Despite a Species Action Plan for red squirrels the decline in their range has yet to be halted.

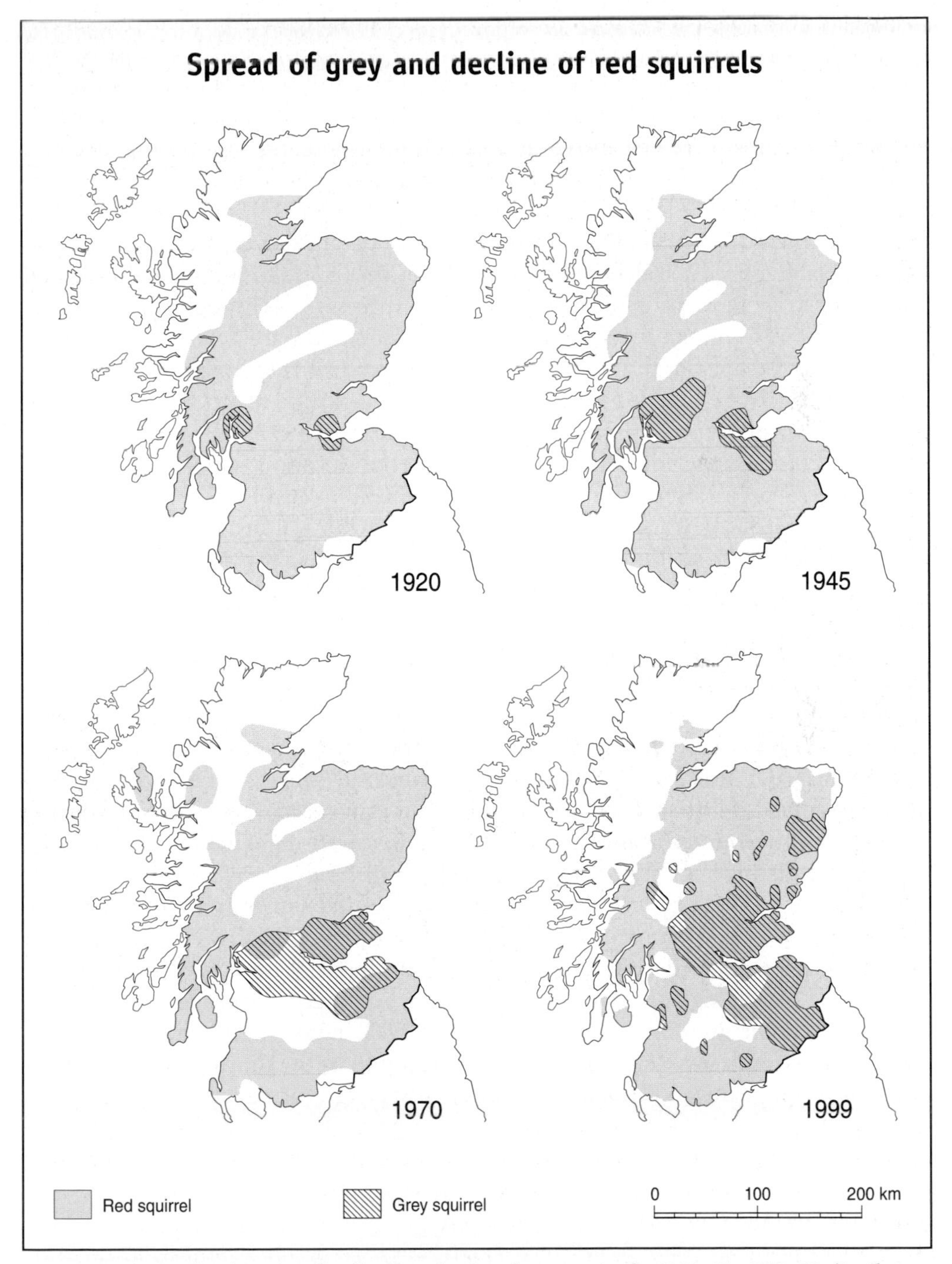

FIGURE 7.10 *The changing distribution of red and grey squirrels, 1920–99. 1920–70 distributions after Yalden (1999); 1999 distribution after Pepper et al. (2001).*

The complex question of why greys usually outperform reds and yet sometimes coexist for decades is reviewed by Skelcher (1997) and Yalden (1999). The two species rarely fight. Hypotheses to explain the decline of red squirrels include disease and environmental change, and competition with grey squirrels involving aggression, mating behaviour, passive avoidance, and competition for resources. Grey squirrels outcompete red squirrels in mixed and deciduous woodland by utilising available food supplies more effectively, and through their ability to detoxify foods such as acorns which poison reds. It seems likely that the larger greys displace reds passively. Outwith Scotland a contributory factor may be the parapox virus which is benign to greys but fatal for reds. Skelcher (1997) concludes that competition between red and grey seems to act more strongly through breeding success than through adult survival.

Eradication of grey squirrels is not a feasible option. They are here to stay, whether we like them or not (and of course many people *do* like them). Conservation and research projects are therefore exploring how best to tip the competitive balance in favour of red squirrels (JNCC, 1996a; Gurnell and Lurz, 1997; Pepper *et al.*, 2001). Possible techniques include habitat management, translocation, supplementary feeding using 'red-only' hoppers, or (more controversially) control of grey populations. This last may be achieved either by using bait laced with warfarin poison (currently illegal in Scotland) or through immunosterilisation whereby treated bait causes infertility, a more publicly acceptable but unproven approach (Bonner, 1996). Shooting and trapping are only short-term, local solutions. Unfortunately, many of the strategies for promoting red squirrels are not feasible in the short term, and they are compatible neither with best forestry practice, nor with each other nor with other wildlife conservation objectives (Bryce and Balharry, 1995; JNCC, 1996a). For example, their ideal habitat is extensive areas of coniferous forest (especially Scots pine and Norway spruce) (Summers *et al.*, 1995), so the current emphasis on broadleaved and mixed woodlands (Section 4.5.1) works against red squirrel conservation by favouring greys. A helpful counterbalance to this, though, is the strong encouragement being given to the planting of native Scots pine (Section 10.2.1).

Harris *et al.* (1995) believe that large numbers of red squirrels are only likely to survive in a few large (bigger than 2000ha) coniferous woodlands. Recent evidence, however, suggests that replacement of reds by greys may not be as swift or inevitable in Scotland as in England and Wales, and that prolonged coexistence of the two species may sometimes be possible (Bryce, 1997). Certainly the impact of greys on reds has been much less drastic in Scotland than south of the Border; in some forests red squirrel populations remain stable after thirty years of coexistence with greys (Pepper *et al.*, 2001). There are nevertheless no grounds for complacent optimism about the future of red squirrels. Since a grey squirrel population is now expanding in northern Italy, it is not just British red squirrel populations that are potentially threatened but the entire Eurasian population.

7.7.2 Limiting the spread of mink

The American mink (*Mustela vison*) was imported to Scotland for fur farming in 1938, but escapees thrived in the wild so that by the 1970s mink were established throughout much of the country (Yalden, 1999). By the mid-1990s the population was estimated to have reached 52,250 (Harris *et al.*, 1995), and it is continuing to increase both in range and size. This success can probably be put down to the absence of natural predators, to the absence of a native mink species (unlike in most of Europe), and to the low numbers of likely competitors such as otters and polecats (Green and Green, 1994); as populations of these species recover, so the spread of mink may eventually be slowed. In the meantime, however, they continue to visit devastation upon Scotland's native wildlife.

Mink are semi-aquatic and feed on chicks, eggs, small mammals and fish (including the economically valuable salmon (SOAEFD, 1997; Section 6.5.5.ii)). They sometimes kill far more than they can eat and can prevent whole bird colonies from breeding. Predation by mink since the mid 1980s has drastically reduced populations of gulls and common terns (*Sterna hirundo*) in some west coast areas (Edwards, 1997). Such rampant and damaging success constitutes a prime example of the impact that exotic species can have (Section 10.1). Perhaps most serious ecologically, mink have found their way to the Outer Hebrides, a stronghold for ground-nesting birds due to the absence of native ground predators. The subsequent disappearance of the moorhen (*Gallinula chloropus*) on Lewis and Harris is attributed to mink (Tapper, 1999), and they are preying on a wide range of waders, ducks and seabirds, including Arctic terns (*Sterna paradisaea*). In many crofting areas free-range poultry has been wiped out. In the view of David Bellamy, they are causing 'nothing short of carnage on the killing fields of the Western Isles'.[8] In response, the Mink Eradication Scheme Hebrides (MESH) was set up in 1999 with the aim of removing the estimated 12,000 mink by trapping, the projected cost being £10 million or more.

Even more pressing, however, is the need to prevent mink spreading through the rest of the Outer Hebrides. To general consternation their presence was confirmed on North Uist in 1999, prompting a flurry of emergency action and the establishment of the Uist Mink Group to eradicate them. The fervent but vain hope had been that they would not cross the Sound of Harris. On the Uists mink threaten the endangered corncrake, a conservation priority, as well as vulnerable populations of many other bird species such as oystercatcher (*Haematopus ostralegus*), ringed plover (*Charadrius dubius*), dunlin (*Calidris alpina*) and redshank (*Tringa totanus*). At particular risk are five SPAs, as well as fish farms, domestic poultry and tourism. Consequently, SNH is coordinating a five-year eradication plan likely to cost £1.65 million.[9] Given that the Uists offer the elusive mink an ideal habitat, and given that they have already reached Benbecula, containment rather than eradication may be all that can be realistically hoped for (Cairns, 1999a). This will require sustained vigilance in the long term and (the only small silver lining) perhaps a dozen extra jobs.

7.7.3 Managing migratory geese

Wild geese are an important part of Scotland's natural heritage, but there has been a long history of conflict between protected geese and agricultural needs.[10] Six species of geese breed or over-winter in Scotland (Table 7.5), but at the centre of the controversy are white-fronted and barnacle geese, both internationally protected species which breed in the arctic but spend their winters in middle latitudes. Legal protection was granted in the days when their numbers were low, but in recent decades goose populations have staged dramatic recoveries, the result of the improving quality of grasslands and cereals, greater protection of roosts, and more restrictions on shooting (Egdell, 1999). The number wintering on the island of Islay has increased fivefold since the mid-1970s (Wigan, 1999b). In all, some 80,000 of these geese now over-winter on Islay, on other west coast islands, and along the sheltered Solway Firth (NGF, 2000), attracted by farmland which has been improved with public subsidies. The recovery is seen as a conservation triumph, but as the already large populations continue to grow, damage to agricultural crops inexorably increases. On Islay alone the estimated cost of goose damage in 1999/2000 was £560,000, averaging £11,500 per affected farm (MacMillan *et al.*, 2001b), so farmers and crofters understandably regard geese as serious pests. On the other hand, the huge winter congregations of geese are a magnet for large numbers of birdwatchers and wildfowlers whose presence is estimated to bring in £5.4 million to local economies and to support the equivalent of 100 full-time jobs (NGF, 2000). Farmers are rarely able to gain much from this human influx, however; they bear the burden without reaping the benefits, and so are paid compensation through goose management schemes, compensation which costs SNH almost £750,000 per annum.[11] This is a classic case of funds from one government pot being used to counteract the effects of subsidy from another (Crofts, 1995).

TABLE 7.5 Population sizes and trends of geese which over-winter in Scotland. The starred figures are three year rolling averages for 1996/7–1998/9. The bracketed rates of population change are predictions based on Population Viability Analyses. Sources: SOAEFD, 1996; NGF, 2000.

Species		*Global population*	*Scottish 'population'*	*Population trend*	*Quarry listed*
White-fronted (Greenland)	*Anser albifrons flavirostris*	30,000	20,000 (Islay: 12,267)*	Increasing (7.2% p.a.)	No
Barnacle (Greenland)	*Branta leucopsis*	40–45,000	32–37,000 (Islay: 32,134)*	Increasing (3.6% p.a.)	No
Barnacle (Svalbard)	*Branta leucopsis*	24,000	24,000	Increasing	No
Pink-footed (Greenland & Iceland)	*Anser brachyrhynchus*	235,000	223,450*	Increasing	Yes
Greylag (Iceland)	*Anser anser*	80,600	80,600*	Increasing (1% p.a.)	Yes
Greylag (native/naturalised)	*Anser anser*	10,000	10,000		
Bean (Taiga)	*Anser fabalis fabalis*	80,000	150	?	No
Canada	*Branta canadensis*	n/a	1,000	Increasing	Yes

The government's policy objective has been to meet the UK's international nature conservation obligations while minimising the economic loss to farmers and maximising the value of public money spent on goose management (SOAEFD, 1996). A National Goose Forum (NGF) was set up in 1997 with the aim of building a consensus amongst the various interested parties. Farmers believe that, despite compensation, they are being disadvantaged commercially (not to mention severely inconvenienced), and this they resent (Mitchell, 1999). Bird conservationists meanwhile refuse to countenance suggestions that goose numbers should be controlled, arguing that goose populations are still at risk internationally. Crofts (1995), however, controversially suggests that there is a need to challenge the 'protect birds at all costs brigade', recommending that population viability and species diversity rather than sheer numbers should be the key objective. More confrontationally still, Wigan (1999b: 70) advocates the culling of supernumerary geese, bluntly saying 'eat them, don't protect them'.

The latest evidence suggests that most goose populations are not now at risk (NGF, 2000), so some difficult questions have had to be faced. Should these species remain protected and inviolate? Or should the amount of licensed shooting be increased, and goose meat be sold on the open market for the first time in thirty years? If the latter, what are the optimum populations that should be aimed for? If the former, should levels of compensation be 'index linked' to increasing goose numbers? Much to the discomfort of the RSPB, the NGF's report recommends that licensed shooting of several of the main goose species should now be authorised (or increased) in order to prevent further serious damage to agriculture (NGF, 2000). The report also recommends that payments for goose management schemes should be made on an area rather than a headage basis (as at present), but it leaves open the question of the sale of goose carcasses. It warns that the new policy framework is likely to require extra funds.

This on-going saga raises a wider issue, namely the novel but increasingly pressing question of how to respond when protection succeeds too well. Wigan (1999b: 70) regards the goose controversy as a prime example of 'the curse of conservation success' – the problem whereby once threatened populations grow to sizes at which they conflict with other land uses. Three other current examples of such controversies involving protected species are these:

- The continuing population explosion of grey seals (*Halichoerus grypus*) to unsustainable levels (Lambert, 2001a). The population almost tripled between 1984 and 1999, reaching 109,100 (Scottish Executive, 2001d).
- The growth of raptor populations on grouse moors (Section 7.6.2).
- The expanding range and numbers of reintroduced sea eagles (*Haliaeetus albicilla*). As they spread from Rum to colonise Skye and Mull, they are causing increasing livestock losses on farms, notably by taking lambs.

Fishing interests and moorland owners have long campaigned for a legal cull of seals and raptors respectively. Conservation failure was a familiar motif of the twentieth century, but conservation success is set to become a serious environmental management

highest use of nature is no use at all. Equally, whereas debates about the 'deterioration' of the uplands used to refer to physical productivity (especially for hill sheep farming), they now focus on the natural heritage interest (Mather, 1992). The diametric character of this attitudinal transformation becomes clear in current nature conservation thinking. Lister-Kaye (2001), for example, advocates an inalienable conservation ethic which recognises the authority of nature and wildness; once we used to subjugate nature for our purposes but now we should subordinate our needs to nature's bidding. Such shifts in attitude can occur not just over centuries but decades; many green views which were regarded as the preserve of 'cranks' only twenty years ago have become the new orthodoxy (Pepper, 1996).

Smout (1993c) identifies and charts the evolution of six attitudes to the natural world. Through history, nature has been seen as:

- a resource for natural produce
- a resource for game sport
- a resource for industry
- a space for recreational sport
- a spiritual/romantic resource
- the home of flora and fauna.

The first two of these have an ancient pedigree and were the only attitudes known prior to the eighteenth century. The third came into its own from the time of the industrial revolution. The idea of the environment as recreational space grew gradually from the late nineteenth century and then took off in the later twentieth century. The romantic, Wordsworthian view of nature has grown steadily since the nineteenth century. Finally, the scientific, ecological perspective has experienced dramatic growth since the Second World War. Conservation, in particular, has come to lean heavily on science, although the irony here is that many conservationists are driven primarily by aesthetic, emotional and spiritual motivations – love of nature and dreams of wilderness. This revolution in society's view of the natural world has led to radical transformations in the values attached to it. For example, many birds of prey which were once persecuted as vermin are now seen as prized and protected species. 'Useless bogs' are now precious wetlands (Smout, 1997d). More generally, these changes have produced what Thomas (1983: 301, 303) describes as a central 'human dilemma' of our times:

> . . . how to reconcile the physical requirements of civilization with the new feelings and values which that same civilization has generated . . . [This conflict] is one of the contradictions upon which modern society may be said to rest.

8.1.2 The rise of the conservation movement

Because conservation has become increasingly internationalised, the picture within Scotland and the UK cannot be understood in isolation from its historic and contemporary context internationally, both within Europe (Section 2.2) and globally.

Conservation has experienced a dramatic journey in the last half century, transforming from a fringe concern into a mainstream, global phenomenon as environmental concern and 'green consciousness' have moved centre stage (Pepper, 1996; Connelly and Smith, 1999). One practical outworking of this has been the swelling edifice of agreements, laws and treaties at European and global level by which the UK is now bound (Table 8.1). Another has been the dramatic growth of conservation organisations (Fig. 8.1). This rising tide of environmental concern has been associated with two long-term trends in developed societies, namely increasing affluence and urbanisation. Only the rich can afford to leave land uncultivated. Historically, this operated at the individual level, with well-off landowners displaying their affluence by turning fields into landscaped parks and gardens. Subsequently it came to apply to nations, as developed countries led the way in setting aside increasing proportions of land for non-commercial use. Progressive urbanisation, and accompanying degradation of land by industry, has also promoted conservation. Once people are isolated from the countryside, the 'otherness' of unspoilt nature becomes special and appealing in a way it never can for a predominantly rural population. The following quotations reflect on these trends:

> Nature parks and conservation areas . . . are fantasies which enshrine the values by which society as a whole cannot afford to live. (Thomas, 1983: 301)

> Wild things . . . had little human value until mechanisation assured us of a good breakfast. (Leopold, 1949: vii)

TABLE 8.1 The primary international legal and policy instruments for nature conservation which apply to the UK. Based on Hossell *et al.* (2000).

Instrument	*Date*	*Common name*
Global Treaties		
Convention on Wetlands of International Importance especially as Waterfowl Habitat	1971	Ramsar Convention
Convention on the Conservation of Migratory Species of Wild Animals	1979	Bonn Convention
Convention on Biological Diversity	1992	Biodiversity Convention
European agreements, policy and law		
Council of Europe 'Convention on the Conservation of European Wildlife and Natural Habitats'	1979	Bern Convention
Bonn Convention Agreements (various)	Various	–
Council Directive 79/409/EEC on the conservation of wild birds	1979	Birds Directive
Council Directive 92/43/EEC on the conservation of natural habitats and of wild fauna and flora	1992	Habitats Directive

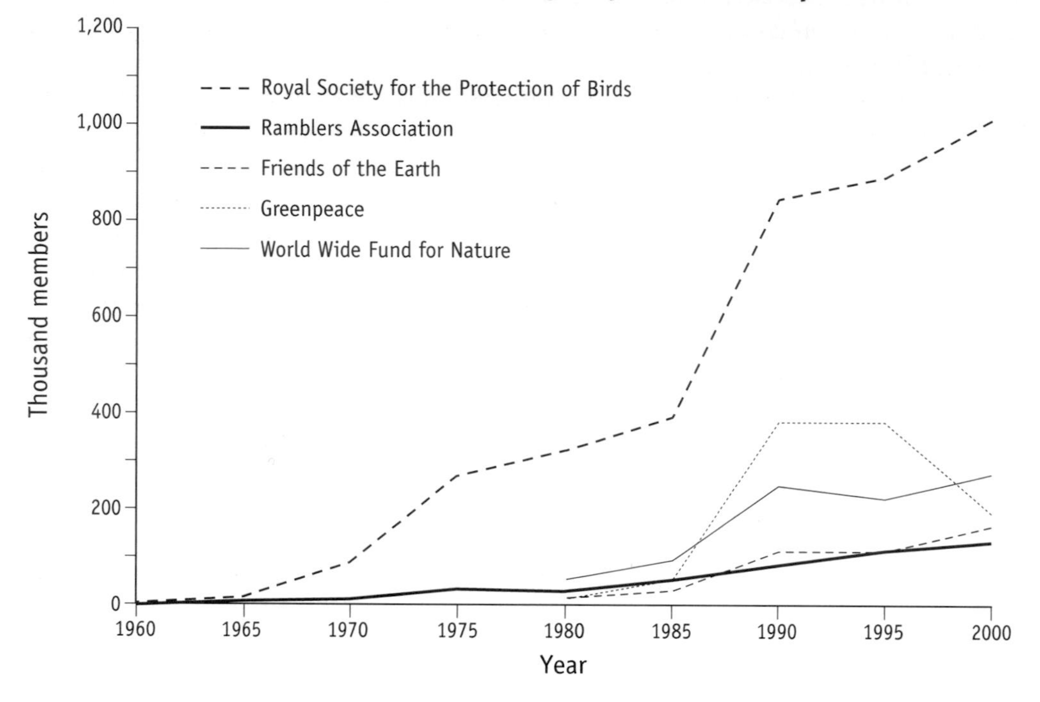

FIGURE 8.1 *The growth of some leading environmental organisations in the UK since 1960 as a proxy for the growing significance attached to environmental issues in UK society. See also Table 8.2. Data are at five year intervals, and are sourced as follows: 1960–80 from Winter (1996); 1980–95 from Connelly and Smith (1999); 2000/2001 from correspondence with the organisations.*

In the last three decades, four events stand out as milestones along the road of the global rise of conservation. The first, in 1972, was the UN Conference on the Human Environment in Stockholm. Although it focused on the urban scene and was 'widely pilloried as a dialogue of the deaf' (Jeffries, 1999: 1), its significance lay in the fact that it was the first time that a global conference had addressed environmental issues directly. The second was the launch in 1980 of the World Conservation Strategy which called for a new ethic embracing plants and animals as well as people. In 1987 the World Commission on Environment and Development (WCED, or 'the Brundtland Commission') published its landmark report *Our Common Future* which promoted the then little-known concept of sustainable development (Section 13.2.1). Perhaps the most significant event came in 1992: the UN Conference on Environment and Development (UNCED) in Rio de Janeiro, universally known as 'the Earth Summit' or simply 'Rio'. With as many national agendas as nations attending, Rio risked schism over the same issues that had divided the international community in Stockholm twenty years earlier. It was therefore remarkable that UNCED successfully (if painfully) gave birth to six major agreements (Koch and Grubb, 1993; O'Riordan, 2000b):

conventions on climate change, biodiversity and diversification; the Rio Declaration, consisting of twenty-seven principles of sustainable development; Agenda 21, an action plan to implement these principles globally during the twenty-first century; and the Statement of Forest Principles.

It is interesting to note the telling progression in the titles of these milestone events. The pendulum has swung inexorably from an anthropocentric towards an ecocentric emphasis, from the 'human environment' to the 'Earth Summit'. This swing has been paralleled by the rise of postmodernism and alternative philosophies which have progressively challenged the modernist hegemony of positivist, scientific, rationalistic viewpoints (Section 13.1). In practical terms, the UK's response to Rio was to publish, in 1994, four White Papers dealing with sustainable development, biodiversity, climate change and sustainable forestry.[2] The need to integrate environmental considerations into many other policy areas is now widely recognised, even if not implemented with great effectiveness.

8.1.3 Conservation in the UK and Scotland

8.1.3.i Background

The UK has one of the oldest and strongest nature conservation movements in the world (Evans, 1997; Dixon, 1998).[3] It comprises three broad and overlapping arenas, namely the statutory agencies, science, and the voluntary sector. The balance between these is constantly evolving. Initially, the state took the lead in conservation matters, but increasingly the agenda has been set by a number of well resourced, confident and rapidly growing voluntary organisations (Fig. 8.1). The RSPB alone now has a larger UK membership than all the political parties put together. The importance of the voluntary sector is now considerable, in terms of its expanding landholdings (Section 3.3.2), its research and demonstration role, and its political influence.

Conservation has long operated with twin foci: a focus on sites and a focus on species – the protection of space and the protection of species. There is also a division, almost unique in Europe, between *nature* conservation and *landscape* conservation (Reynolds, 1998). The latter has been weaker than the former, in part because the protection of species and habitats is perceived as resting on objective, scientific criteria, whereas the aesthetics of landscape are inherently more subjective (Smout, 2000a). Until 1991, nature conservation was managed on a UK-wide basis, whereas landscape conservation in Scotland was the responsibility of the Countryside Commission for Scotland (CCS). This separation accounts for the evolution of a distinctively Scottish approach to landscape protection, both at a national and regional level (Moir, 1997).

For much of the twentieth century there was comparatively little emphasis on conservation in Scotland because of the perceived abundance of resources (Selman, 1988b). Economic and social recovery were seen as the pressing needs in the countryside; nature could look after itself. In Scotland generally and in the Highlands especially, concern for conservation has always lagged behind the rest of the UK (Smout, 1993c), a fact which is illustrated by the proportionate memberships of conservation organisations (Table 8.2). Although the rapid growth of the voluntary sector south of the Border has been paralleled in Scotland, it has been at a lower level. It has also long been the case that support for scenic and historic conservation runs much deeper in Scotland than support for nature conservation alone (Smout, 1993c). Nevertheless, the importance of conservation has steadily grown in recent decades, so that it is no longer a peripheral curiosity but a leading player and significant land use in its own right (Davidson, 1994). The total Scottish membership of the environmental groups in Table 8.2 is 377,000, a significant constituency when compared, for example, with membership of the National Farmers Union of Scotland which by 2001 had fallen to 10,700.

TABLE 8.2 Membership of leading environmental organisations in Scotland and (where applicable) in England and the UK in 2000/2001, and the membership totals as proportions of the population. Population totals in 1999 were: Scotland – 5.1 million; England – 49.8 million; UK – 59.5 million. Membership data were obtained directly from the organisations; all are for 2001 except for the RSPB and the RA which are for 2000.

	Scotland		*England*		*UK*	
	Members	*% of population*	*Members*	*% of population*	*Members*	*% of population*
National Trust for Scotland	239,000	4.7%	–	–	–	–
Royal Society for the Protection of Birds	70,300	1.4%	886,900	1.8%	1,011,400	1.7%
World Wide Fund for Nature	28,000	0.5%	–	–	272,000	0.5%
Scottish Wildlife Trust	17,000	0.3%	–	–	–	–
John Muir Trust	7,600	0.1%	–	–	–	–
Ramblers Association	6,100	0.1%	117,300	0.2%	130,000	0.2%
Friends of the Earth	5,000	0.1%	–	–	163,000	0.3%
Woodland Trust	4,000	0.1%	68,000	0.1%	75,000	0.1%

8.1.3.ii Legislative and administrative evolution

Site-specific nature conservation in the UK began in 1948 with the establishment of the Nature Conservancy (later the Nature Conservancy Council (NCC)), the world's first statutory nature conservation body. From the start it was charged with undertaking both research and conservation, but it evolved from a science-based agency into a campaigning organisation (Mackay, 1995). A complex, multi-layered framework of statutory measures for safeguarding the natural heritage has progressively developed, using both conservation and planning legislation (Table 8.3). The 1949 National Parks and Access to the Countryside Act established National Parks in England and Wales (but not in Scotland (Section 8.3.4)), and introduced Sites of Special Scientific Interest (SSSIs) and National Nature Reserves (NNRs). In so doing, it placed the 'key area' concept at the heart of UK conservation. SSSIs have proved to be an enduring bedrock for nature conservation, not only being significant in their own right (covering a greater area than any other designation) but also forming the foundation for several other designations, including NNRs and Natura 2000 sites (Section 8.2.1). They therefore play a crucial role in the protection of species and habitats, and of geological and geomorphological features.[4] The 1981 Wildlife and Countryside Act was of huge significance for conservation throughout the UK in numerous ways, not least in greatly strengthening SSSIs (Lowe *et al.*, 1986). For the first time, public funds were available to pay landowners for *not* doing things, compensating them for profit foregone as a result of leaving habitats undisturbed. Equally controversial was the shifting of the burden of proof; previously, conservationists had had to demonstrate that environmental damage was being done, but subsequently the onus has rested on landowners and the government to show that damage is not occurring. Farmers are now guilty until proven innocent.

TABLE 8.3 The main elements of the statutory framework for protecting the natural heritage. Based on SODD (1999a).

Act of Parliament	*Key elements*
National Parks and Access to the Countryside Act 1949	Introduced SSSIs and NNRs. Gave powers to local authorities to create Local Nature Reserves.
The Countryside (Scotland) Act 1967	Strengthened 1949 Act. Imposed on public bodies a conservation duty towards the natural heritage.
The Wildlife and Countryside Act 1981	Strengthened SSSIs. Additional safeguards for certain types of area. Protection for many wild animals and plants.
The Natural Heritage (Scotland) Act 1991	Established SNH. Charged it with protecting, enhancing and facilitating enjoyment of Scotland's natural heritage.
The Town and Country Planning (Scotland) Act 1997	Consolidated the development control framework. Requires development plans to include an environmental dimension.

In 1991 the NCC was broken up and its responsibilities devolved to three new agencies – English Nature, the Countryside Council for Wales, and Scottish Natural Heritage (SNH). The Joint Nature Conservation Committee (JNCC) was set up to coordinate these new agencies and to maintain a UK-wide perspective in nature conservation. Under the Natural Heritage (Scotland) Act 1991, SNH was given a testing remit:

TABLE 8.4 The main mechanisms for the conservation of landscapes and species, and their number and total areas as at 31st March, 2000. Data from SNH (2000a). For the 'x + x pending' figures, the first figure refers to new sites and the second to extensions to existing sites. For further information concerning the functions, practical management and legislative basis of these and other designations, see SNH (1995a) and Scottish Office (1998c). Full lists of sites are given in SNH (2000a). Not included here are the two proposed national parks (Section 8.3.4).

Designation	*Number of Sites*	*Area (ha)*	*Summary description*
Nature conservation			
Sites of Special Scientific Interest (SSSIs)	1,458	990,809	Sites notified because they are special for their plants, animals, habitat, geology, geomorphology, or combinations. Cover 12.6% of Scotland. Mostly on private land.
National Nature Reserves (NNRs)	71	114,271	Statutory nature reserves of national importance. Cover 1.4% of Scotland. All NNRs are also notified as SSSIs. One-third of total NNR area is owned by the state.
Ramsar Sites	48 (2 + 3 pending)	240,487 (14,643)	Wetlands (marsh, fen, peatland or water) of conservation importance, especially for waterfowl.
Special Protection Areas (SPAs)	113 (12 + 4 pending)	391,234 (25,766)	Classified under the EU Birds Directive for protecting the habitat of rare, threatened or migratory species.
Special Areas of Conservation (SACs)	134 (3 pending)	717,862 (50,419)	Designated under the EU Habitats Directive for protecting biodiversity by conserving flora and fauna.
Local Nature Reserves	30	9,309	Sites of special local natural interest; valuable for environmental education & recreation.
Biosphere Reserves	9	28,768	Sites for studying human influence on the natural environment. Most are also NNRs.
Biogenetic Reserves	2	2,388	Reserves to conserve representative examples of European flora and fauna and natural areas, primarily for biological research.
World Heritage Sites	1 (2 proposed)	853	Natural and cultural sites of global significance. (Existing natural site: St Kilda; proposed natural sites: the Flow Country; the Cairngorm Mountains)
Landscape conservation and recreation			
National Scenic Areas (NSAs)	40	1,001,800	Nationally important areas of outstanding natural beauty. Cover 12.7% of Scotland.
Regional Parks	4	86,160	Extensive areas of countryside providing facilities for informal recreation.
Country Parks	36	6,481	Small areas of countryside near towns managed for public enjoyment & recreation.

> To secure the conservation and enhancement of, and to foster understanding and facilitate the enjoyment of, the natural heritage of Scotland . . . [The natural heritage includes] the flora and fauna of Scotland, its geological and physiographic features, its natural beauty and amenity.

The Act specifically made sustainability SNH's guiding principle. This challenging combination of aims, incorporating not just nature conservation but landscapes and people, reflect the fact that SNH was formed in the context of new international thinking on sustainable development in the run-up to the Earth Summit (SNH, 1993). The creation of a Scottish conservation agency had several potential advantages (Crofts, 2000): integrating nature conservation with landscape conservation and access, bringing decision making about Scotland to Scotland, and improving understanding and management of the natural heritage. Concerning the last of these, although the NCC had been successful from a conservation viewpoint, it had been dogged by accusations of insensitivity and 'scientific colonialism' (Mather, 1993c; Mackay, 1995), a problem inherent in much of conservation (Section 8.2.2). This was one reason why SNH's remit includes a specific charge to incorporate the needs of people within its decision making.

SNH was formed through the merger of the CCS with the Scottish part of the NCC, thereby officially integrating landscape and nature conservation and ending the 'great divide' which many had seen as artificial and counter-productive (Bishop, 1997). Although the two cultures are taking time to mesh, considerable effort since the merger has been invested in landscape conservation. This was primarily because much less was known about the landscape resource than about nature conservation, but also in recognition of the fact that society is typically more concerned about 'natural landscapes' (however artificial they actually are (Section 1.2.2)) than about the 'bugs and beasties' which scientists often focus on. As Smout (2000a: 27) puts it, 'more of the public care about tranquility and fresh air than about obscure species in the mud'. The immediate problem, of course, given that 'beauty is in the eye of the beholder', was finding dispassionate, repeatable and defensible ways of assessing the value of landscapes. This has been addressed through the development of Landscape Character Assessment (LCA) techniques (Hughes and Buchan, 1999; Usher, 1999a). Though never entirely devoid of subjectivity, LCA has proved to be a useful tool for managing change, and, for example, in defining Natural Heritage Zones (Thin, 1999; Section 8.2.2.iii).

When SNH was created, some saw a chance for a new start and a clean slate. Many, however, opposed the break-up of the NCC, fearing a sell-out to developers' and landowners' interests, expecting cuts in budgets and influence, and worrying that it would weaken the international standing of UK conservation (Marren, 1993; Bishop, 1997; Evans, 1997). Some interpret subsequent events as proof that such fears were justified (Ratcliffe, 1998). Another concern, given the relative sizes and very different cultures of the CCS and the NCC, was that the merger would result in the voice of landscape being drowned by the voice of nature conservation. Raemaekers and Boyack (1999) suggest that this concern persists, citing the approval of the Cairngorm funicular (Section 12.1) as an example of the relative weakness of scenic protection, and advocating the creation of a new scenic designation equivalent in strength to Natura 2000 sites (Section 8.2.1). One incontrovertibly positive outcome of the formation of SNH, however, has been a substantial increase in the resources (both financial and human) devoted to conservation (Bishop, 1997).

Given the nature of its task, a measure of unpopularity and controversy is bound to surround any statutory nature conservation body, and SNH is no exception. It is regularly accused by farmers and crofters of bureaucratic, high-handed interference, and by conservationists of ineffectiveness. Some, however, give it credit for developing a much more holistic view of environmental problems and solutions than the NCC and CCS could have done separately (Bishop, 1997), and many believe that it is discharging its complex and often thankless responsibilities as effectively as can be expected within the limits of its annual budget. After all, it will always be the case that 'conservation is a compromise between that which is a biological necessity and that which is politically expedient' (Balharry, 1990: 12).

Although it is still too soon to see the implications clearly, the devolution settlement of 1999 is already beginning to have signficant ramifications for conservation in Scotland. At a practical level, responsibility for the environment and rural affairs now rests with Scottish ministers, ensuring a much more detailed and focused consideration of specific Scottish issues than was ever possible from Westminster. It also raises some new and vexing questions. For example, when protecting sites of national importance, should 'national' mean Scotland or the UK? If the former, some fear that sites or species that are significant for the UK and internationally, but which are not especially rare within Scotland, may not get the attention that they arguably deserve. But if devolution means anything, it must allow for divergent paths.

8.2 RECENT DEVELOPMENTS AND TRENDS

8.2.1 The proliferation of designations

Almost a fifth of Scotland is now covered by some form of conservation designation (Table 8.4; Fig. 8.2 a–f). Some of these sites are state-owned, but most are in private hands and are overseen by SNH through management agreements. All designations, in themselves, have a sound rationale, but there is much overlap and confusion, with many new areas and categories having been added recently in an *ad hoc* fashion (Bishop *et al.*, 1997). In the areas of highest nature conservation value, such as Ben Lawers or the Cairngorms (Fig. 12.2), a multitude of designations can apply on the same site. St Kilda, for example, is a NNR, a SSSI, a NSA, a Ramsar site and is Britain's only 'natural' World Heritage Site, while much of Caithness and Sutherland has quadruple protection: SSSIs, SACs, SPAs and Ramsar sites (Fig. 8.2). Scotland simply has too many designations, so a degree of simplification has become desirable (Scottish Office, 1996).

The work of managing existing sites and designating new ones is on-going. Much of the recent effort (and controversy) in this regard has related to the latest pair of designations, SPAs and SACs. Together, these constitute the Natura 2000 series. Both

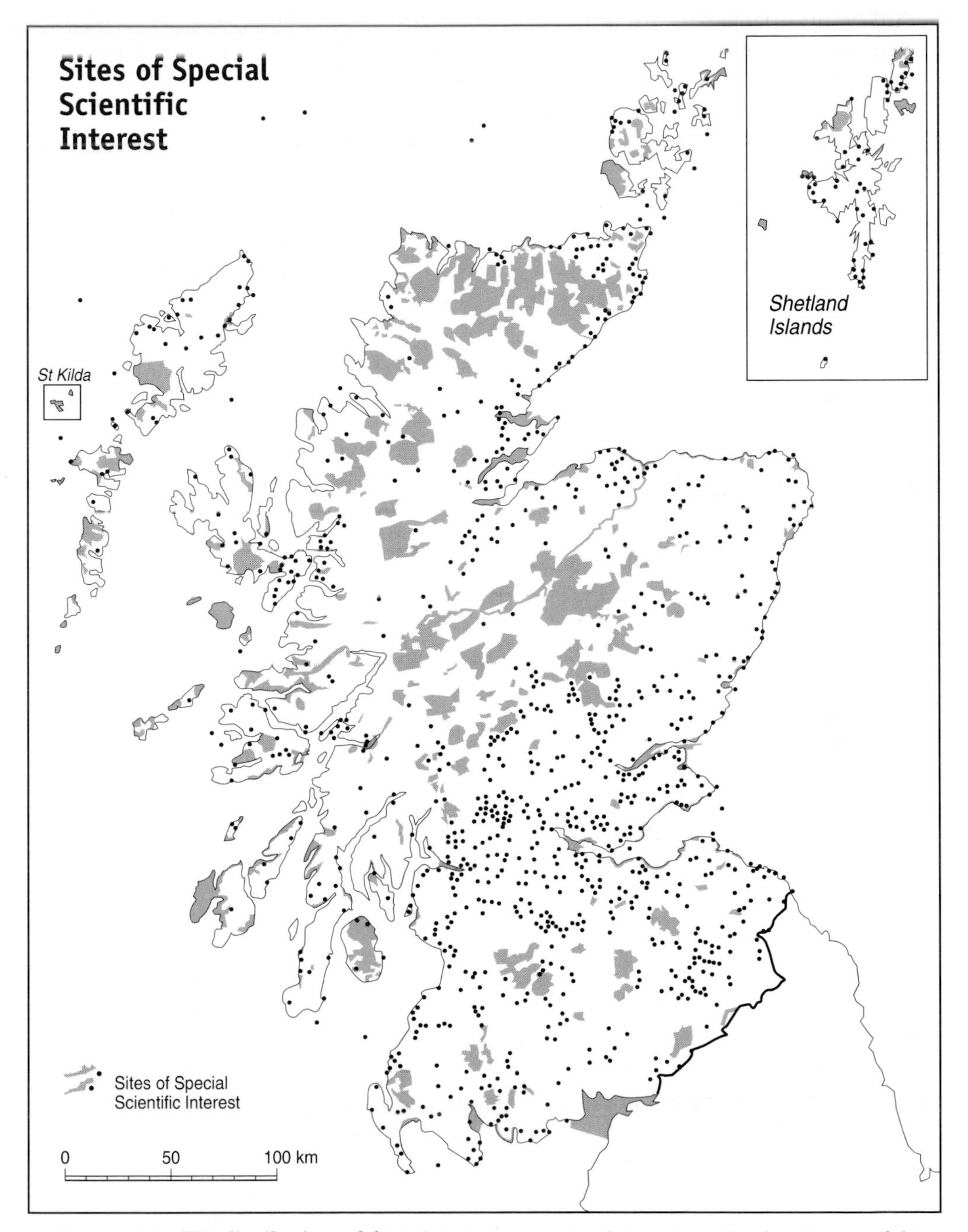

FIGURE 8.2a *The distributions of the primary conservation designations. For descriptions of the designations see Table 8.4. Maps based on SNH (2000a), and used with permission. Data for NSAs used with the permission of the Scottish Executive.* (FIGURES 8.2b–f on following pages)

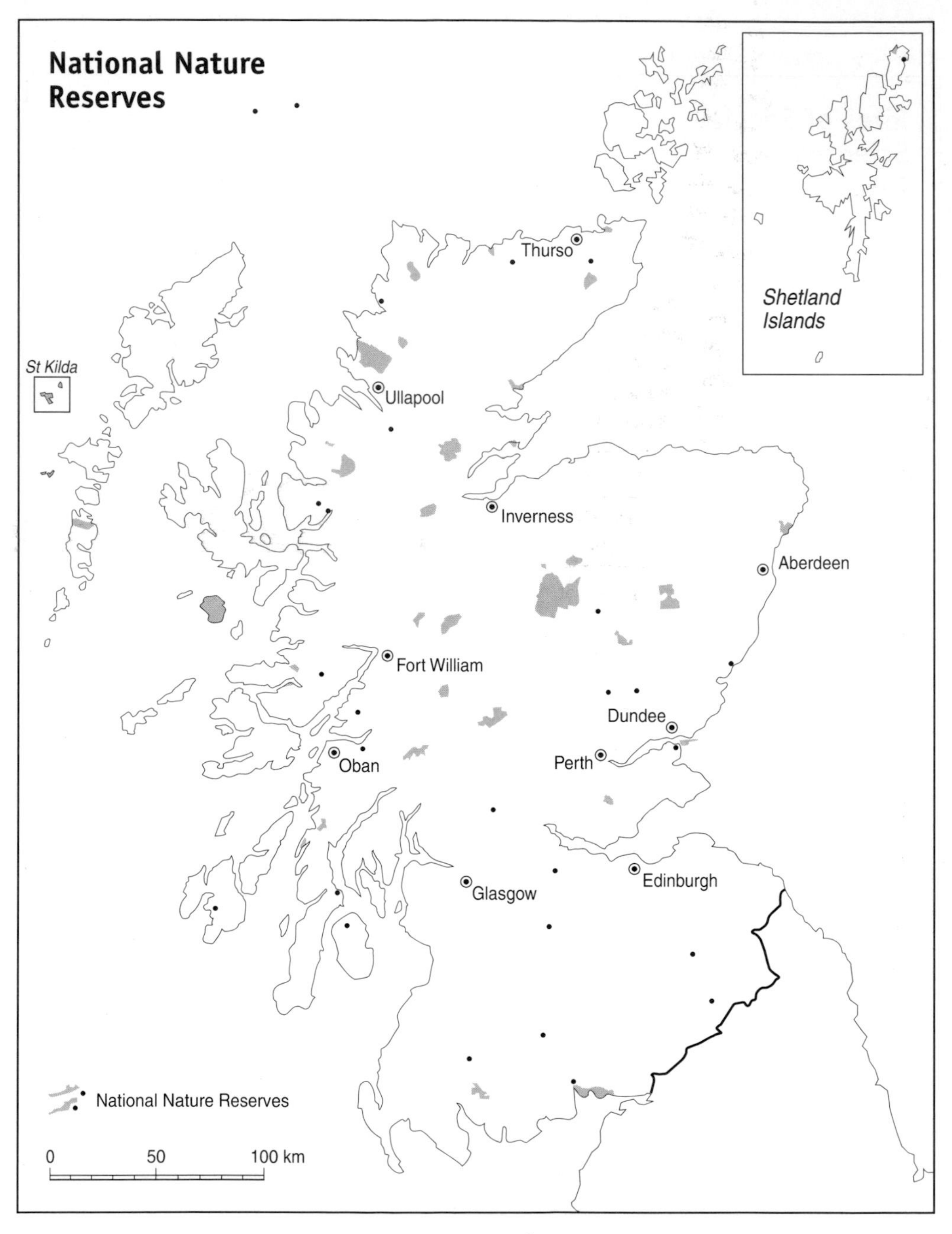

FIGURE 8.2b

are EU-wide designations but are usually underpinned by SSSIs in the UK. They are being given high priority as the EU's major policy response to the 1992 UN Convention on Biological Diversity, the aim being to secure a network of important natural heritage sites across Europe (Ledoux *et al.*, 2000). Scotland's contribution to this network

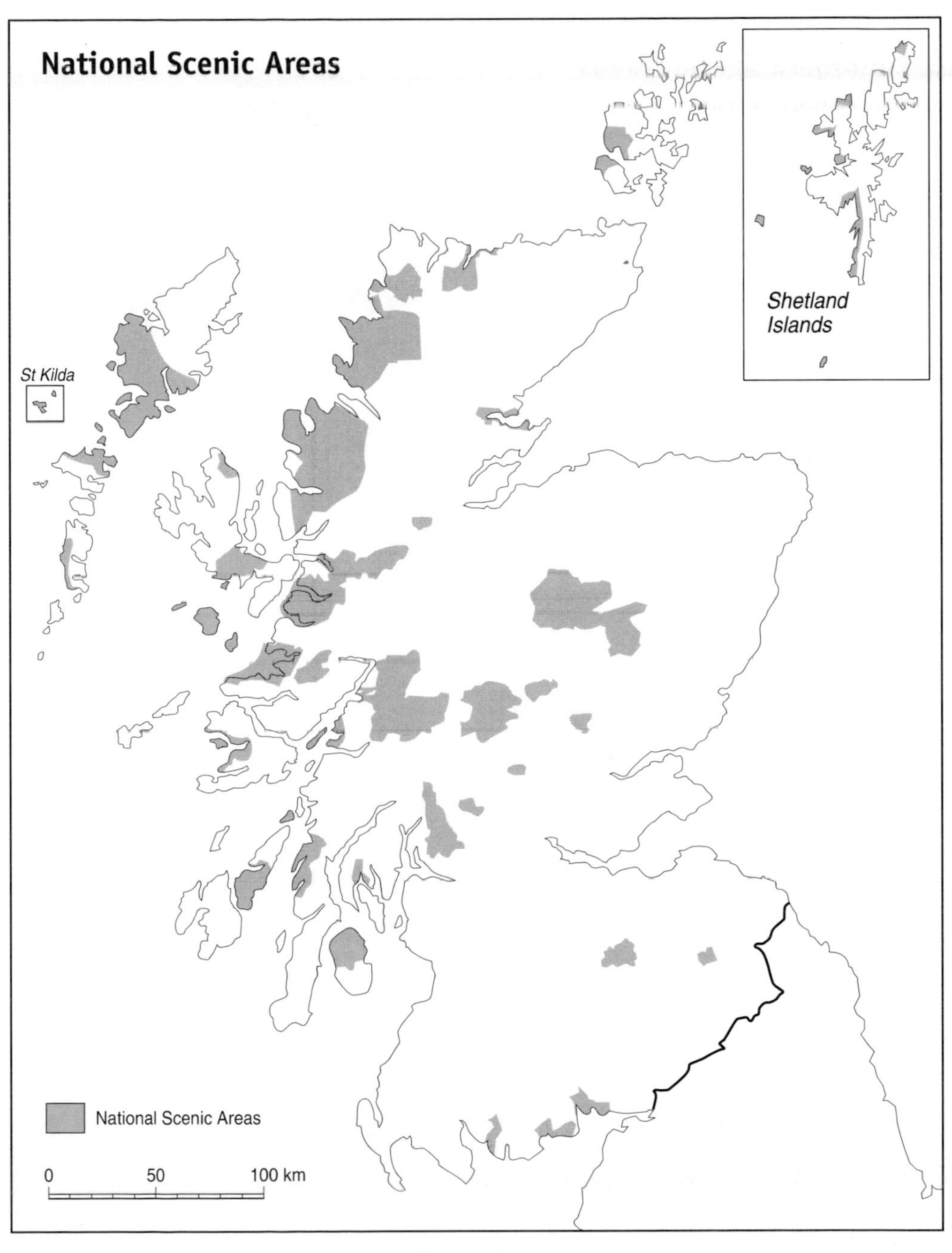

FIGURE 8.2c

includes such distinctive habitats as Caledonian pinewoods and active raised bogs, and species such as barnacle geese (*Branta leucopsis*) and corncrakes (*Crex crex*).

Most designations operate through development control rather than through positive management, and this frequently leads to restrictions on the nature and timing

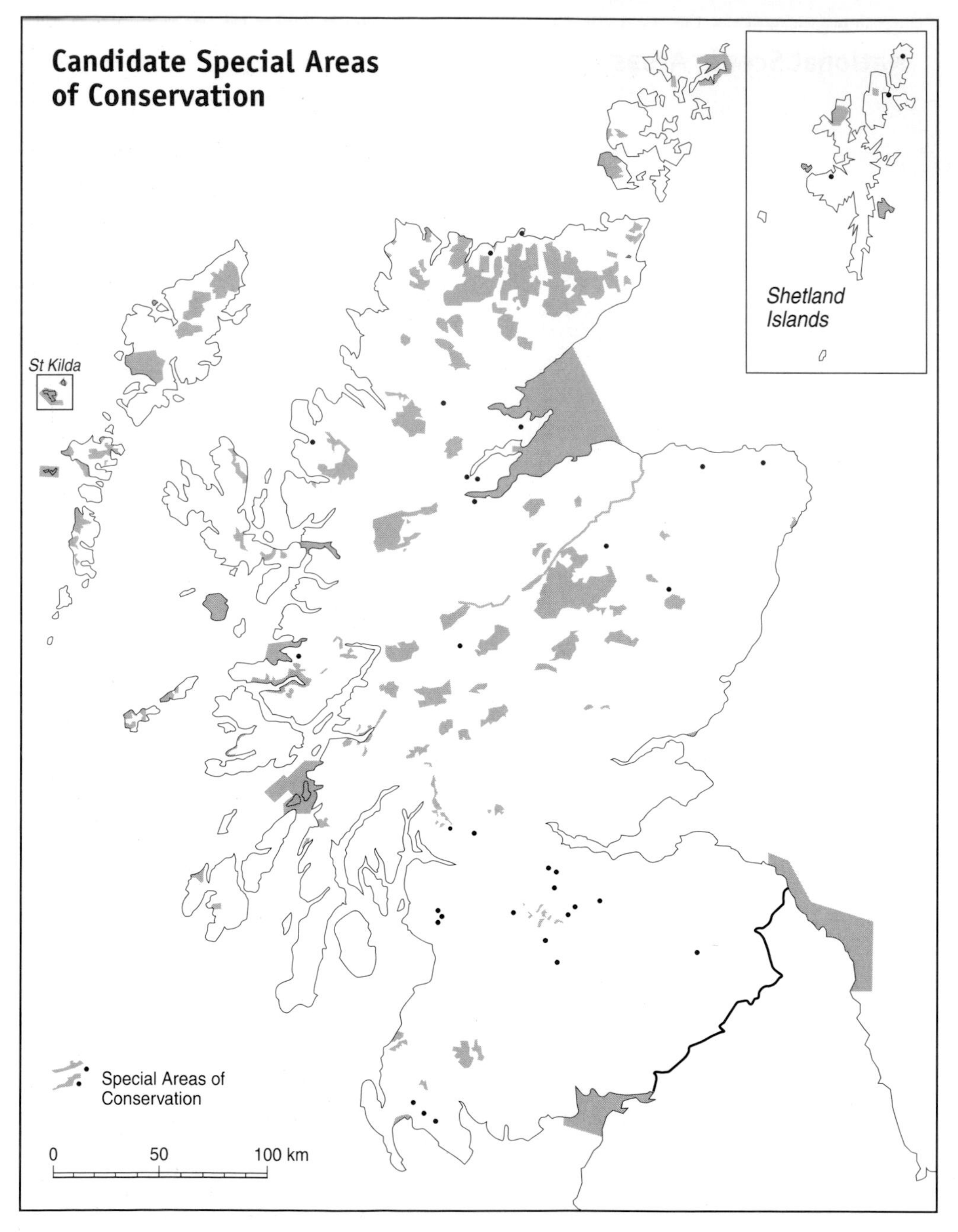

FIGURE 8.2d

of land management operations. Farmers and land managers with designated sites on their land are caught up in time-consuming bureaucracy and considerable frustration (sometimes dubbed 'death by acronym'). Consequently, although many have some sympathy for conservation, a majority regard the avalanche of designations as unwelcome

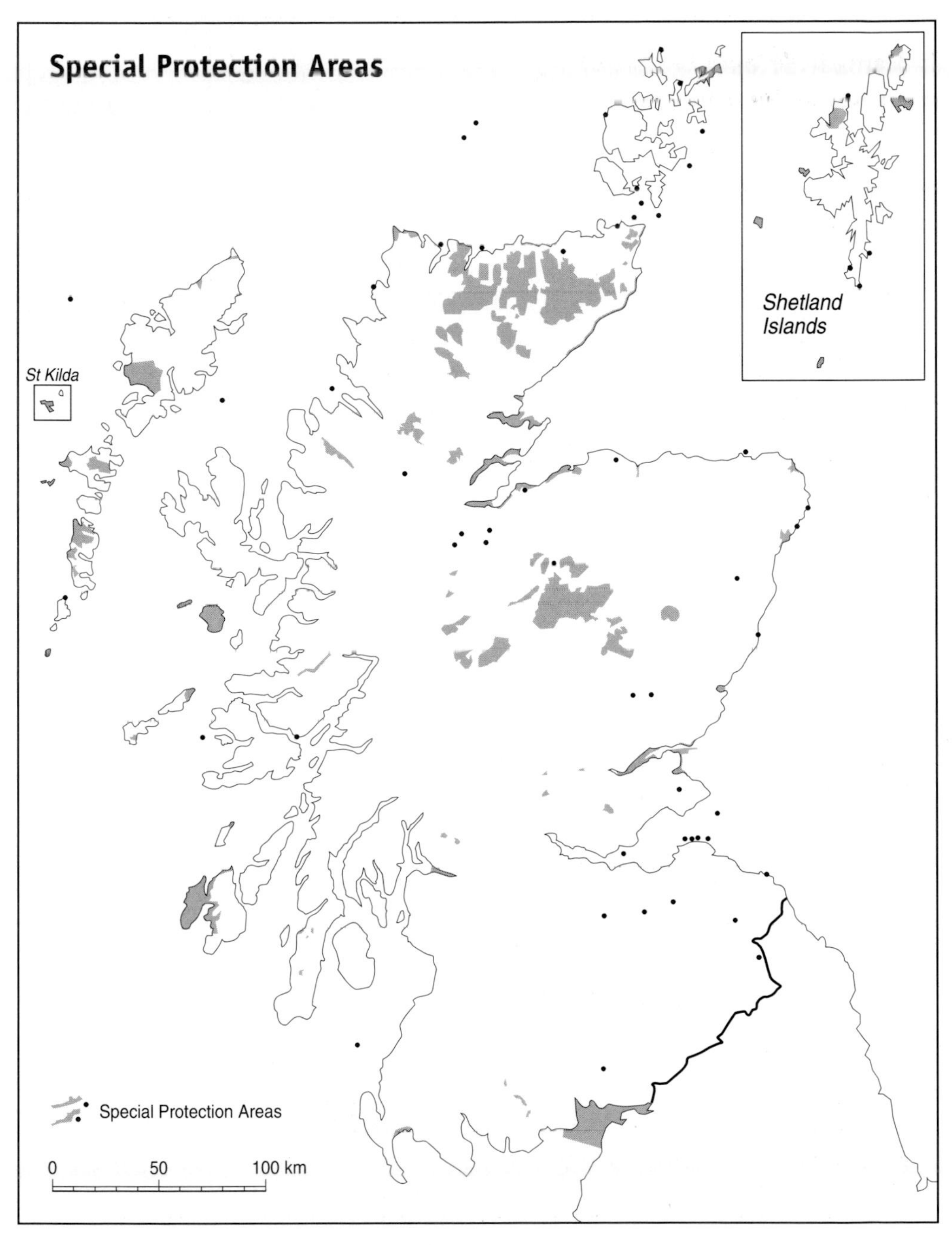

FIGURE 8.2e

interference and meddling (Mitchell, 1999). SSSI designation has even been described as 'land nationalisation by the back door' (Linklater, 2000). Meanwhile, from conservationists' viewpoint, the system is simply not working, its weakness and ineffectiveness shown by its failure to halt the continuing degradation of semi-natural habitats.

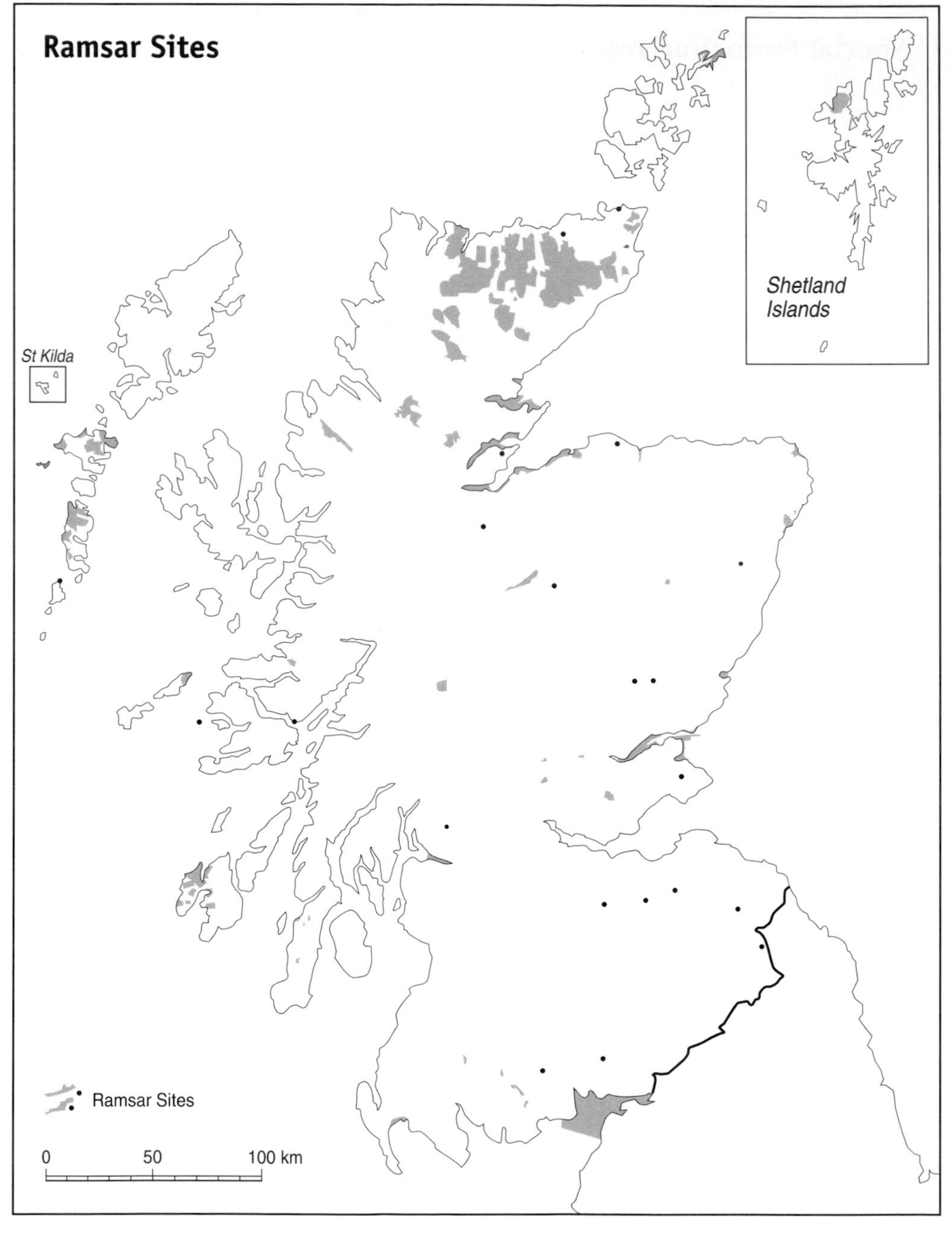

FIGURE 8.2f

The system of protected areas that should conserve and enhance Scotland's natural heritage remains incomplete and deeply flawed. . . . The apparently complex designation system is an expensive failure in practice. (SWCL, 1997: 1, 4)

In some places, overlapping designations are mutually enhancing, but this is mainly a fortuitous accident, and where their objectives are in conflict there is no simple remedy (Bishop *et al.*, 1997). Given such widespread dissatisfaction, it is not surprising that moves are afoot to overhaul the present system (Section 8.3.3).

8.2.2 Multi-purpose conservation

As in agriculture and forestry, and indeed throughout environmental management, single-issue policies are being replaced by a more inclusive, integrated approach with multiple objectives. In conservation, this has involved at least three linked developments: a broadening of the criteria used to select sites, the inclusion of local people in conservation management, and the progressive replacement of the key area concept with a focus on the so-called wider countryside.

8.2.2.i Broadening criteria

Although conservation had its roots in a love of nature, statutory conservation policy has always rested on precise, objective, scientific criteria. Arising as it did during the modernist era, and competing with the 'white heat of the technological revolution', it could do no other. During the 1980s and 1990s, however, society's faith in science as an omnicompetent guide faltered as the old, confident certainties were undermined by postmodernism and by the environmental damage caused by technological 'solutions'. The use of exclusively scientific criteria led to conservationists being accused of elitism and of ignoring the values of local people who felt alienated and disempowered by technical discussion (Harrison, 1993; Harrison and Burgess, 2000). The 'expertocracy' did not manage to convey why diversity and rarity actually *matter*, and the system manifestly failed to win the hearts and minds of those living and working in the countryside. In response, conservation thinking and practice has progressively begun to incorporate more cultural criteria, better reflecting what matters to the general public (Goldsmith, 1991). Adams (2001), however, argues that there is still a long way to go if conservation is to lose its technical, exclusive image and gain mass appeal.

In Scotland, local hostility to institutionalised conservation and to the NCC, its chief representative, had been building for years, part of the long-running 'insider *v.* outsider' conflict (Section 8.3.2), but in the late 1980s it erupted into the open dramatically during the Flow Country controversy. This bitter showdown between conservation and commercial forestry was one of the key factors that led to the dismemberment of the NCC and its replacement in Scotland with SNH (Mather, 1996b; Warren, 2000a). On Orkney the NCC's opposition to the conversion of moorland to improved pasture also stirred local resentment. Its campaigning approach led to accusations of 'scientific colonialism', and it became increasingly clear that its narrow focus was actually proving counter-productive, leading to an unsustainable form of conservation (Mather, 1996b).

The new thinking that this realisation has engendered is that:

> . . . conservation should cease to be a self-contained crusade, wilfully pushing all

> other considerations aside . . . and should take its place alongside other interests, seeking to understand their point of view. (Mackay, 1995: 119)

Nature conservation and landscape conservation, both traditionally single-purpose pursuits, have broadened in scope and vision to incorporate multiple objectives and benefits (Bishop *et al.*, 1995). Thus, for example, SNH has (controversially) adopted community development as one of its management objectives for the Rum NNR (Samuel, 2001). Science will always be important for conservation, but conservation is no longer driven by science alone (Adams, 2000). There has been a dawning realisation that although scientific and utilitarian arguments for conservation will convince some, appealing to people's hearts and spirits is far more effective (Holdgate, 2000b). The adoption of broader, more 'user friendly' criteria is allowing the impacts of 'scientific colonialism' to be ameliorated through dialogue, local participation and accountability.

8.2.2.ii Participatory conservation

It is now universally acknowledged that local support is a *sine qua non* for conservation, and that the needs and aspirations of people must be incorporated within the conservation agenda (Crofts, 2000).

> If local people do not support protected areas then protected areas cannot last. . . . Protected areas cannot last for very long with communities that are hostile to them. (Ramphal, 1992 in Hunter, 1995: 161)

The 1992 Caracas Declaration therefore urged governments to involve local populations as partners in the planning, establishment and management of protected areas (Dower *et al.*, 1998). As society has progressively lost faith in national state institutions and has increasingly questioned the role of the expert, so the new ethos of participatory conservation has come to the fore (especially since the Earth Summit), shifting the balance of power from technical and political elites towards 'ordinary' people (Goodwin, 1998a; Warburton, 1998). The growth of the Local Agenda 21 movement is a prime example of this (Selman, 2000). Belatedly catching up with the far-sighted vision of John Muir, Aldo Leopold and Fraser Darling, the term 'holistic' has now been stretched beyond ecosystems to incorporate human communities. Thus 'natural heritage conservation today is not just about the saving of natural communities and species, but . . . about ensuring the place of people within the landscape' (Johnston, 2000: 11). Instead of an exclusive (and excluding) focus on non-human nature, conservation now aims to strike a balance between the needs of people and the needs of nature, promoting integration in place of polarisation. This has been termed 'sustainable conservation', and it reflects three key aspects of the sustainable development agenda, namely the empowerment of local people, self-reliance, and social justice (Mitchell, 1997). The hope is that conservation will start 'growing from below . . . rather than dripping from above as local versions of a master-narrative' (Adams, 2000: 3).

Two obvious examples of this trend stand out. The first is the partnership approach whereby all stakeholders are involved in a consensual decision making process (Gemmel, 1996). The highest profile experiment is the Cairngorms Partnership which has built a broad consensus for its visionary management strategy (Cairngorms Partnership, 1997) and has attracted plaudits from conservation organisations (Austin, 1997; SWCL, 1997). Partnerships are also integral to the implementation of Biodiversity Action Plans (Section 8.2.3) and are in vogue at European level too (Dixon, 1998). The second example is the increasing amount and depth of public consultation that government agencies engage in. This involves not just the production of accessible, colourful consultation documents in plain English, but also local public meetings and roadshows. This, for example, has been a facet of the recent National Park consultations (Section 8.3.4).

There is no doubt that participatory and partnership approaches are helping to build bridges over the gaping chasms of yesteryear, fostering understanding and cooperation, and facilitating community involvement in the natural heritage (SNH, 1999b), but they are not uncontroversial or unproblematic (Section 14.1.2). In particular, some conservationists fear that because local people are ill-equipped to recognise the national or international significance of local sites, participation could put important sites at risk (Goodwin, 1998b). Local participation invariably makes decision making a longer and more complicated process. This investment of time is usually worthwhile, however, because it minimises conflict (Mitchell, 1997), especially if there is participation at the agenda-setting stage, and not just in the implementation of centrally-set objectives (Goodwin, 1998a). Gemmell (1996: 219) comments that although it is hard to make partnerships work because they require much time, patience and skill, they are 'currently essential and the only viable path to take'. Equally, however, participation cannot solve everything, and there will always be a need for expert advice (Warburton, 1998).

8.2.2.iii Beyond the key area: conservation of the wider countryside

Until recently, conservation focused on 'holding the line' – protecting the best habitats, the rarest species and the finest landscapes – but the key area approach is now recognised as being severely limited (Bishop *et al.*, 1995; Yalden, 1999). In the mid-twentieth century it seemed appropriate because agriculture and forestry were perceived as environmentally benign; conservation's task was simply to protect the 'crown jewels'. The CCS and the NCC therefore became largely preoccupied with the classification and designation of particular sites, while the rising threats to the majority of the countryside went unchecked (Mackay, 1995). Protected areas became increasingly isolated. Belatedly, the problem has been recognised, with fragmentation and isolation coming to be regarded as the biggest threat to conservation (Hampson and Peterken, 1998). The emergence of biodiversity as a dominant concept in conservation (Section 8.2.3), with its stress on the interconnectedness and value of all elements of the natural world, is driving a further nail into the coffin of the key area concept and is broadening people's perceptions of conservation.

Consequently, the protectionist mentality of so-called 'fortress conservation' is being abandoned.

> It is now widely accepted that biodiversity management cannot be achieved by simply establishing reserves as islands of protection in a sea of unregulated agriculture, forestry, fisheries and urban development. (Mackey *et al.*, 1998: 246)

In other words, the hitherto watertight compartments separating working landscapes from conservation landscapes must become permeable, and conservation must escape from its 'prescriptive, site-based timewarp' (Adams, 2000: 1). Because nature reserves perpetuate a damaging view of nature as somehow separate from our ordinary lives (Cooper, 2000), the importance of conserving the wider countryside has come to the fore, together with the need to do this by adopting integrated management strategies which bring together all interested parties (Selman, 1996; Scottish Office, 1996; Crofts, 2000).

As the inadequacies and ineffectiveness of an exclusive reliance on protected areas have been acknowledged, so SNH has increasingly sought to promote conservation as the matrix of life and work throughout the countryside.

> The conservation message will only have succeeded when the natural heritage is respected and cherished *everywhere*... There is a need for especial care *everywhere* and in *all* we do... Conservation must become a principle of *all* social and economic activity. (SNH, 1995a: 157,158; emphasis added)

Thinking within Scottish planning circles has also begun to pay attention to the wider countryside. For example, NPPG 14 (SODD, 1999a) states that fragmentation or isolation of habitats should be avoided, and that links which have been broken should be restored. It also makes it clear that the natural heritage should be considered in planning decisions whether or not a designated area is likely to be affected.

At the same time there is renewed interest internationally in understanding whole ecosystems, not just isolated fragments of the environment (Budiansky, 1995), and to address them on an appropriate geographic scale within natural boundaries. This has been dubbed the bioregional approach. It seeks to reduce the isolation of protected areas by recognising the part that each component of a region plays in the functioning of the whole (Crofts, 2000). A major practical problem that impedes movement in these directions in Scotland is that most existing legislation relates to sites. Nevertheless, the bioregional concept is being explored through the delimitation of Natural Heritage Zones (NHZs), a landscape scale approach developed during the late 1990s by SNH as a step towards long-term, integrated environmental management. NHZs build on the previously defined biogeographical zones (Usher and Balharry, 1996), and are intended to provide a new framework for SNH's operations. In particular, they will facilitate the evaluation of environmental changes driven by either natural or human processes. The twenty-one zones (Fig. 8.3) are areas with distinctive natural heritage characteristics, defined using information about species, habitats, landscape and geography (C. Mitchell, 2001). Within each zone a management vision is being developed that is congruent with national policy objectives but which grows out of consultation with key stakeholders. NHZs are thus a further example of the

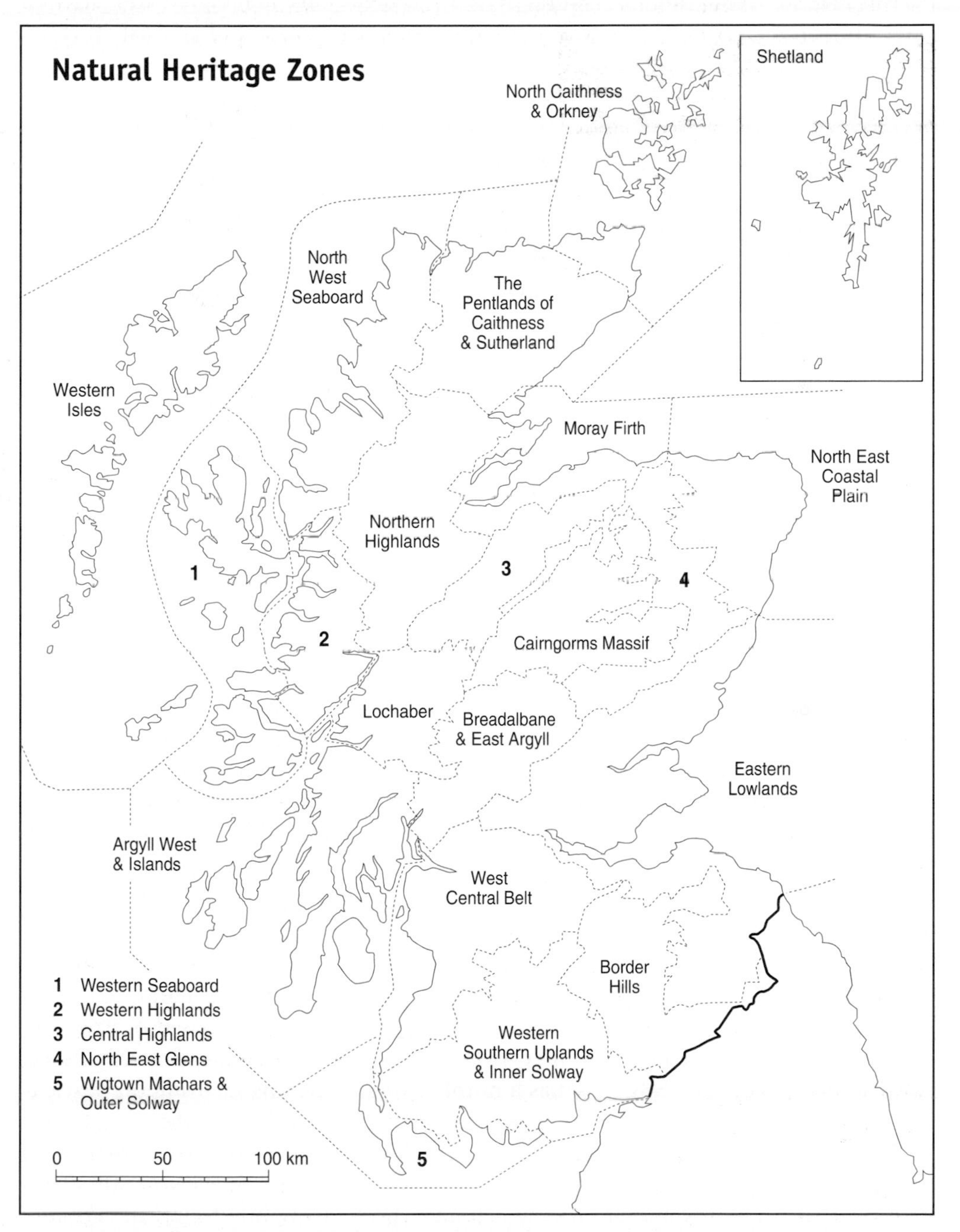

FIGURE 8.3 *Natural Heritage Zones in Scotland (after C. Mitchell, 2001). Used with the permission of SNH.*

participatory ethos.[5] To summarise, protected areas will always have an important part to play in conserving the natural heritage, but they are inadequate in isolation. They are a necessary but not a sufficient aspect of conservation policy (Bishop *et al.*, 1995; Cooper, 2000).

8.2.3 Biodiversity: BAPs, HAPs and SAPs

Although the term 'biodiversity' was only coined in 1985 (Scott, 1997), it is already established as a key word in the conservation lexicon. It refers to the variety and richness of the natural world which surrounds and sustains us. Although the word – a contraction of the phrase 'biological diversity' – is new, many of the ideas and activities represented by it are not. Holdgate (2000b) describes it as 'a gift-wrapped synonym for "nature"', but it is a more objective term; it bypasses the value-laden overtones of 'nature' and goes further by stressing the interconnectedness of the web of life (Scott, 1997). Biodiversity leapt to international prominence in 1992 when the UN Convention on Biological Diversity was signed at the Earth Summit. As defined within the Convention, it has three dimensions (Anon., 1994b): diversity between and within ecosystems and habitats, diversity of species, and genetic variation within individual species. It links the genetic and ecosystem levels of conservation with the more familiar ideas of species and habitat conservation (Kirby, 1999).

Biodiversity is now firmly established in science, politics and popular culture, but there are no satisfactory criteria for measuring it because familiar concepts such as species and race turn out to be elusive in practice (Balharry *et al.*, 1994). Moreover, the term has several shades of meaning (Jeffries, 1999). It refers to biological diversity itself, to the associated area of scientific research, and has also increasingly been used as a conceptual focus for conservation policy and practice in response to the twin threats of species extinctions and ecosystem loss. Clearly, then, biodiversity is not simply a branch of biology. It has political, ethical, cultural and socio-economic implications which give it powerful contemporary resonance (Dolman, 2000; Tilman, 2000).

Following the Earth Summit, the UK was the first country to publish its own Biodiversity Action Plan (Anon., 1994b), pledging to conserve and enhance biodiversity, to use its components sustainably, and to ensure that the benefits of biodiversity are equitably shared. Responsibility for implementing this vision in Scotland has rested since 1996 with the Scottish Biodiversity Group (SBG), a broadly-based partnership of those who work in, depend on and care for Scotland's countryside (Scottish Office, 1997; Usher, 2000). The SBG also has a number of sub-groups focusing on particularly important sectors (such as agriculture) and activities (such as public education).[6]

The SBG is overseeing the production and implementation of the government's list of Biodiversity Action Plans (BAPs). These are of two types. Species Action Plans (SAPs) focus on plants and animals which are wholly or mainly found in Scotland (such as red squirrels, corncrakes and chequered skipper butterflies), while Habitat Action Plans (HAPs) address special and/or threatened habitats such as Caledonian pinewoods, blanket bogs and machair. BAPs describe the current status of the species or habitat, outline the main factors which have caused loss or decline, and describe the conservation action which is currently underway. They then specify clear, costed targets for maintaining or increasing either the population and range of species or the size of habitats, and conclude with a list of actions which are needed to support those targets. There are now published action plans for forty-one habitats (Table 8.5) and 226 species (Usher, 2000). The SBG also supports local authorities, a majority of which are producing Local Biodiversity Action Plans (LBAPs) for their areas with much local participation. LBAPs aim to translate national objectives into effective local action, allowing local aspirations to drive the agenda. LBAP partnerships now cover most of Scotland (Abernethy, 2000). Important facets of Scotland's biodiversity are protected within designated conservation sites, but the main reservoir for biodiversity is the general countryside, and this is where BAPs are likely to produce the greatest gains (Usher, 2000).

TABLE 8.5 Scottish habitats for which a Habitat Action Plan has been published. After Usher (2000).

Ancient and/or species rich hedgerows	*Modiolus modiolus* beds
Blanket bog	Mudflats
Cereal field margins	Mud habitats in deep water
Coastal and floodplain grazing marsh	Native pine woodlands
Coastal saltmarsh	Purple moor grass and rush pastures
Coastal sand dunes	Reedbeds
Coastal vegetated shingle	*Sabellaria alveolata* reefs
Eutrophic standing waters	*Sabellaria spinulosa* reefs
Fens	Saline lagoons
Limestone pavements	Seagrass beds
Lophelia pertusa reefs	Serpulid reefs
Lowland calcareous grassland	Sheltered muddy gravels
Lowland dry acid grassland	Sublittoral sands and gravels
Lowland heathland	Tidal rapids
Lowland meadows	Upland calcareous grasslands
Lowland raised bog	Upland hay meadows
Lowland wood-pasture and parkland	Upland heathland
Machair	Upland mixed ashwoods
Maerl beds	Upland oakwood
Maritime cliff and slopes	Wet woodland
Mesotrophic lakes	

The UK's biodiversity programme has been both praised and criticised (Knightbridge, 2000).[7] On the plus side, it has facilitated the formation of effective partnerships with clear agendas and targets. It has completed all the SAPs and HAPs, released (or re-targeted) resources, and initiated much activity and research. Biodiversity has been successfully integrated into many policies and programmes, such as agri-environmental schemes. These are no mean achievements. However, on the negative side, it suffers from top-heavy bureaucracy ('pattern book conservation' (Marren, 2000: 44)) and from conflicting, incompatible targets, whereby action to save one species or habitat may jeopardise another. It is attacked for making little tangible progress – for being all plans and no action – and for not engaging local people sufficiently. Moreover, the government has been accused of failing to back up its brave words with adequate resources, and of garnering political credit while devolving the actual responsibility to local level by adopting the partnership approach. Most of the local organisations involved do not have the resources to implement the policy (as Lawson (1998: 82) puts it, 'no dough means BAPs fail to rise'), while much of the 'new' spending on biodiversity targets does not consist of new money but of existing resources re-targeted to reflect the new priorities. In other words, the conservation budget is being spread more thinly.

The government had hoped that industry and wealthy individuals would step forward to take up the slack by becoming 'biodiversity champions', sponsoring particular species, but only a handful of companies have made commitments, and most have chosen glamorous species such as otters, red squirrels or eye-catching butterflies. Esoteric species attract little funding, even though they may have considerable significance for biodiversity. A further criticism of the biodiversity project is that its rigid, static approach fails to take adequate account of either the dynamic nature of ecoystems or of the profound and inevitable changes to human and physical systems (including the climate) that are forecast for this century (Holdgate, 2000b; Ledoux *et al.*, 2000). As Adams (2000: 3) comments, 'our chief relationship with nature is through our consumption, not our conservation'. So although the first practical stage of the biodiversity programme is now complete – the action plans are written – the enormous challenge of implementing them stretches ahead.

8.2.4 Lottery funding for conservation

From a standing start in 1994, the National Lottery has rapidly become a major influence on conservation throughout the UK. By the start of 2000, the Heritage Lottery Fund (HLF) had awarded a remarkable £1.37 billion to conservation and access projects (Gay, 2000), and in the first four years of the Lottery the amount awarded to conservation was roughly equal to the combined grant-in-aid to all the government conservation agencies (Bishop *et al.*, 2000b). Lottery funds are awarded via a number of distributing bodies, but the most significant for conservation projects have been the HLF and the Millennium Commission. The key theme for the latter has been woodland creation, the Millennium Forest for Scotland Trust (MFST) being a prime example (Section 4.5.5).

In terms of its UK-wide significance, Bishop *et al.* (2000c) argue that Lottery funding has not only accelerated the quantity of conservation work being undertaken but is changing its quality and direction as well. Because of the sheer scale of funding suddenly available, it is allowing bigger dreams to be dreamed and then realised. It is also enabling site-based conservation to be linked to the wider countryside (Section 8.2.2.iii), and is resourcing community-led initiatives. Like a well known beer, it seems that Lottery funding can reach parts that other funding sources cannot reach (Bishop, 2000). One notable way that it has facilitated this is by enabling conservation organisations to purchase land; the HLF is the only major source of finance for land acquisition, and its grants have enabled the purchase of over 200 sites covering 50,000ha (Gay, 2000).

Scotland was one of the earliest jackpot winners when in 1995 the NTS was awarded £10.2 million for the acquisition of the 35,000ha Mar Lodge Estate, but since then England and Wales have outperformed Scotland in attracting Lottery funds (Dalton, 2000). Nevertheless, Lottery funding averages out at £12.79 per capita and £8.33 per hectare across Scotland (Bishop *et al.*, 2000b). Whatever one's views on charitable funds derived from gambling, it is clear that Lottery funding is now a major facilitator of and influence on conservation. There is a serious problem of sustainability, however, because conservation projects are inherently long term yet the Lottery only provides partial start-up capital. Projects may fail for lack of follow-on funding, the demise of the MFST being a sad example (Section 4.5.5). Nevertheless, Bishop's (2000: 1) view is one of tentative optimism:

> The Lottery is a powerful force for 'joined-up conservation'. . . . Whilst key questions on the additionality and sustainability of funding . . . remain, may be, just may be, the Lottery has helped us move into a new era of positive rather than defensive conservation.

8.3 CURRENT ISSUES AND DEBATES

8.3.1 What is conservation, and what should its aims be?

Conservation is a particular philosophy of management based on a view of limited resources. It is a response to loss or to the threat of loss. Because it involves sacrifice, usually by one group for a group elsewhere (in space, in time, or in the economy), it is inherently and unavoidably political. Historically, it has incorporated at least three main strands, the cultural, the economic and the ecological. Cultural conservation is concerned with preservation for aesthetic reasons, for science and education, for spiritual renewal, and in order to preserve our historical and cultural identity. The economic strand focuses on conserving society's material base. Ecological conservation addresses the preservation of biodiversity, highlighting the uniqueness and irreplaceability of species. The first two of these are anthropocentric and have a relatively long history, whereas the third, focusing on the intrinsic worth of nature itself, is ecocentric and recent (Section 8.1.1).

Conservation is a multi-faceted and slippery term, used in numerous different senses.[8] Because of its wide range of associations, attempts to define it tend to be eclectic and personal (Table 8.6). Moreover, because conservationists have diverse interests, priorities, beliefs and motivations, there has always been intense discussion about the aims, objectives and ethical basis of conservation. What should we be conserving, who for, and why? What mix of scientific, utilitarian, moral and aesthetic arguments should conservation rely on? Is naturalness next to godliness? What is the place of human beings in the 'natural' world (Section 13.3.1)? What are the appropriate spatial and temporal scales of operation? Whose values should count, and when? Should conservation rely on the past as a guide to the present, or should it be future-orientated? On this last point, Rawles and Holland (1994:37) suggest that 'conservation has as much to do with conserving the future as with conserving the

TABLE 8.6 Word associations and definitions of conservation:
A. Words most frequently associated with conservation during brainstorming sessions with undergraduate students between 1995 and 2000.
B. A selection of more formal definitions of conservation. For discussion of a wide selection of definitions, see Evans (1997).

A. Conservation: word associations		
Preservation	Restoration	Maintenance
Protection	Repair	Stability
Balance	Enhancement of environment	Care (for environment; for future)
Sustainability	Integration	Restrictions
Minimising impacts	Prevention	

B. Conservation: definitions	
Conservation is foresighted utilisation, preservation, and/or renewal of [resources] for the greatest good of the greatest number for the longest time.	Gifford Pinchot, 1910
Health is the capacity of the land for self-renewal. Conservation is our effort to understand and preserve this capacity. . . . Conservation is a state of harmony between men and land.	Aldo Leopold, 1949
[Conservation is] the rational use of the environment to achieve the highest quality of living for mankind.	IUCN, 1969
[Conservation is] the transfer of maximum significance from the past to the future.	Paul Evans, 1999b
Conservation involves making decisions about human actions in the light of their implications for non-human nature. It is about the terms and forms of human engagement with non-human nature.	Bill Adams, 2000

past', and Adams (1996), in *Future Nature*, agrees. In fact the very term 'natural heritage' embraces not just our inheritance but that which we bequeath to future generations (SNH, 1993).

In addition to all these fundamental ethical or philosophical questions (which are discussed in Section 13.3.3), there are perhaps four especially pressing practical dilemmas faced by conservation managers. Firstly, should conservation aim to preserve things as they are today, or endeavour to restore the environment to something like its diverse potential? Secondly, how closely should nature be managed? A third question concerns the exclusion or inclusion of people. Finally there is the dilemma of conservation's spatial frame of reference – how feasible is it to act on the slogan 'think globally, act locally'?

8.3.1.i Preservation versus *restoration*

Clearly preservation will always be a facet of conservation, but because ecosystems are highly dynamic they cannot be protected in the same way that our cultural heritage may (Brown, 1997). Nature (and hence its conservation) is about the continuity of dynamic processes, and so maintaining the *status quo* is neither possible nor even desirable (Rawles and Holland, 1994; Bachell, 2000). Out of this recognition has sprung the desire to replace defensive conservation with creative conservation (Adams, 1996) – the proactive re-creation of lost or damaged habitats (Section 10.2), and the reintroduction of lost species (Section 10.4). This progression from protection to enhancement has been dubbed restoration ecology (Holdgate, 1997). Conservation is thus increasingly seen as the management of change for the future.

This shift comes into sharp focus in the light of the predictions of rapid climate change during the present century (Hossell *et al.*, 2000; Holdgate, 2001; Section 1.2.2.iii). As climate and vegetation zones migrate northwards and upwards, so the characteristics of protected areas will change until, conceivably, they are no longer suitable for protecting the species or environments that they were created to preserve. Climate change therefore threatens to make the existing distribution of protected areas redundant, especially for species close to the edge of their range (Watt *et al.*, 1997). Consequently, there is growing interest in the establishment of habitat networks (Hampson and Peterken, 1998). These aim to establish connectivity between isolated habitats using linkages and wildlife corridors as stepping stones that will allow species to migrate in response to environmental changes. It is recognised, of course, that increased connectivity could have negative as well as positive impacts by facilitating the spread of undesirable species such as grey squirrels (Section 7.7.1) or sika deer (Section 10.3).

8.3.1.ii Intervention versus *'hands off'*

In the early days of statutory conservation there was considerable confidence amongst practictioners that they knew what they were doing and why, often resulting in interventionist styles of management. Subsequently, there was a shift away from this 'man knows best' approach towards natural, non-intrusive methods of environmental management. Non-intervention and working *with* nature are in vogue. Most recently, this 'bias towards the natural' has itself come into question (Budiansky, 1995). Does nature always know best? Given the extent to which the environment has been altered (Section 1.2.2), how sensible is it to let habitats evolve naturally from their current, 'artificial' state? Left to its own (human-altered) devices, the natural world will not always do the 'right' thing (in the sense of producing the results that society or ecologists want) because it has been so modified (Bachell, 2000). As a result, 'there are few cases where we can just let things run and expect that non-human processes will lead to an optimum balance' (Manning, 1997: 292). 'Hands off' management, as Aldo Leopold realised in 1933, shows 'good taste but poor insight' (in Budiansky, 1995: 135). In a woodland management context, for example, policies of non- or minimum intervention permit invasions by exotic species such as Sitka spruce and grey squirrels, as well as leading to the loss of certain valued cultural elements such as coppice woodland (Ratcliffe, 1995). In fact, in an environment with as long a human history as Scotland's, 'hands on' *versus* 'hands off' is probably a false choice; 'hands lightly on' may be an appropriate middle way.

8.3.1.iii *People* versus *no people*

Conservation has frequently adopted a 'keep people out' mentality, but this has increasingly been questioned. Should the natural heritage be protected *from* people or *for* people? The former has contributed to the alienation problem (Section 8.2.2.i) and is now hard to defend politically, especially in view of the new legal rights of access (Section 9.4.3). On the other hand, welcoming people onto protected, vulnerable sites runs the risk of compromising their specialness. Lister-Kaye (2001) for one argues unapologetically that there must be times and places for excluding people if the rising tide of natural heritage damage is to be halted.

Answers to these three dilemmas are usually site-specific. For example, if a particular habitat or species is the last surviving example in existence, a persuasive case can be made for vigilant protection, the exclusion of people, and manipulation of the surroundings to maximise the chances of survival. In more robust, less critical settings, it may be appropriate to experiment with non-intervention and to welcome the public, not least because environmental education is crucial for the cause of conservation (Section 14.1.4). The site-specific nature of these dilemmas is exemplified on NTS countryside properties where identical problems are being tackled in contrasting ways at different sites (Johnston, 2000). The overall ethos and objectives of conservation management at a particular site will usually determine what is 'right'; there is rarely an absolute, given answer.

8.3.1.iv *Local* versus *global*

A final dilemma is spatial in nature. Many practical decisions turn on the spatial scale of reference adopted, from local to global. For example, considerable resources are invested in protecting species which are rare and threatened within Scotland but which have large and robust populations worldwide. This applies, for example, in the case of corncrake (*Crex crex*) and dunlin (*Calidris alpina*). Mitchell (1999) highlights the artificiality of a system in which the apparent importance of the Scottish population of dotterel (*Charadrius morinellus*) was transformed at a stroke in 1992 by the accession of Sweden and Finland to the EU; having represented 95 per cent of the EU population, it instantly dropped to just 5 per cent. Globally, none of these three bird species is under threat, but in Scotland their numbers are small and declining. Should conservationists adopt a global view and relax, or work hard to preserve a place for these species within Scotland? Holdgate (2001: 20) advocates the former ('it is the survival of biodiversity at the global scale that counts, not the perpetuation of a "snapshot" ecosystem pattern in any one place') but many conservationists are motivated by more local concerns. How should emotive concepts like 'rare' and 'threatened' (with their powerful political and financial ramifications) be defined geographically?

Conservation is not alone in wrestling with spatial questions like these. Indeed, they crop up in many sectors of environmental management. For example, forestry plantations of exotic species, though widely reviled in Scotland (Section 4.3), actually ease the pressure on more fragile native forests elsewhere in the world (FoES, 1996). Equally, intensive agriculture has caused much environmental damage (Section

5.2.2), but without it a far larger land area worldwide would be needed to feed the world's burgeoning population, leaving very little (if any) space for nature. In all of these cases, the selection of the spatial frame of reference is profoundly significant for management choices.

8.3.2 *Threatened nature* versus *threatened communities*

As conservation values came to the fore during the later twentieth century, a 'jobs *versus* conservation' debate developed, polarised by the relatively depressed economy and the limited opportunities in much of rural Scotland. Most conservation controversies in Scotland – from the Flow Country (Warren, 2000a) to the Cairngorms funicular (Warren, 1997; Section 12.1) to the management of migratory geese (Section 7.7.3) – involve a tug-of-war between local socio-economic realities and the desire of society at large to protect landscapes and species. This tension is not uniquely Scottish. It afflicts many of Europe's peripheral regions (Mather, 1994). Indeed, conflict is a characteristic of environmental issues worldwide, the inevitable consequences of the diverse values and interests within a pluralistic society (Mitchell, 1997) and of the fact that halting biodiversity loss requires profound economic and political change (Dolman, 2000).

Nevertheless, solutions are especially elusive in Scotland because the areas with the greatest economic needs and those with the highest conservation value frequently coincide. This is because lack of opportunity for economic development over long periods has frequently meant that such areas are least modified by human activity. Their resulting high priority for nature conservation tends to lead to a disproportionate concentration of protected areas. But this same lack of opportunity also means that these areas are those most urgently in need of socio-economic assistance. Perhaps this is one reason why, when push comes to shove, conservation is rarely given priority in official decision making (the recent rejection of the Lingarabay superquarry proposal (Section 12.2) being the exception that proves the rule). The spatial coincidence of often mutually exclusive priorities greatly intensifies the problem. In the words of Selman (1988b: xvi):

> It will be the resolution of this conflict – between protection and development, between wildlife and people – which will doubtless pose a central problem for the future of countryside planning in Scotland.

That this conflict remains a live issue fifteen years after these words were written simply emphasises its intractability.

Smout (2000a: 4) characterises this ancient debate as 'the tension between use and delight'. Because of the enormous difficulty of the questions that it raises, both practical and ethical, it will probably always be with us. How to strike the right balance? Who judges what is 'right', and by what criteria? How should local, national and international priorities be weighed against each other? How to balance the rights of people against the rights of non-human nature, and the rights of this generation against those of future generations? These questions cause severe headaches for politicians and planners, and lead to contradictions in policy objectives at national and international level.

An important facet of this debate in Scotland has been the sometimes acute conflict between 'insiders' and 'outsiders' (Section 12.3), a friction which has greatly harmed the cause of conservation (Mather, 1993c, 1994). It was especially severe prior to the replacement of the NCC with SNH because the driving force for conservation in Scotland came from the urbanised south of England; consequently, the official voice of conservation frequently spoke with an English accent (Mather, 1993c). People in remote rural areas often feel that they are being dictated to by distant bureaucrats in urban heartlands. They resent conservation scientists' assumed authority to speak 'on behalf of nature', an assumption which marginalises local knowledge (Harrison and Burgess, 2000). When added to the widespread perception that designations blight development prospects, it is not surprising that a host of pejorative phrases have been used to describe conservationists and their activities – 'ecological imperialism', 'conservation sterilisation', 'green fascists'. Rare species generate little sympathy in communities which are themselves threatened with extinction.

At the heart of the insider/outsider conflict is the question of decision making. At what level along the spectrum from local to international should decisions be made? In arguing for high levels of protection, conservation professionals typically adopt national and international perspectives, whereas locals tend to give greater weight to their immediate concerns. On the one hand it seems patently unjust for the strong wishes of local people to be trumped by the views of those living elsewhere who may never have visited the place and whose lives may be unaffected by the decision. For example, commenting on the proposed Assynt Lochs SPA, a leading crofter said:

> The Assynt crofters feel that land management should be by negotiation and not by diktat, and that people who live and work in the area should have a far greater say in what happens in their area. (Macrae, 2000 in *SCENES* 151: 2)

hard to Promote Global to Small popula

On the other hand, should the legitimate self-interest of the people of one region determine the fate of a national asset of international significance? This impasse has become 'a severe, endemic and destructive element in modern Highland history' (Smout, 1993c: 28). In practical terms it becomes a choice between different levels of government. It seems logical to argue that assets of international value should be regulated at international level, but it is also widely recognised that conservation simply does not work unless local people have a sense of 'ownership' (Section 8.2.2.ii). Because local involvement is now held to be a key aspect of sustainable management, there is a trend towards handing more decision making down to the local level, even though this risks damaging environmental assets of national or international importance (Raemaekers and Boyack, 1999). The national/local tension is also a feature of the current debate over the management of SSSIs (Section 8.3.3).

The creation of SNH ameliorated these frictions somewhat, but it has by no means removed them. Despite the broadening criteria and a more inclusive ethos (Section 8.2.2), conservation is still far from being 'flavour of the month' in many rural areas, especially amongst crofters in the Gaidhealtachd (Gaelic-speaking areas) (MacDonald, 1998). Mitchell (1999: 126; 2000) describes SNH's work on the west coast islands as 'an essentially imperial operation' and is fiercely critical of the *modus operandi* of conservation charities such as the RSPB. The hostility and conflicting perceptions remain. 'Vermin' to a crofter may represent valued species to conservationists; designations which are justified in terms of the value of the natural heritage make resources appear valueless to locals. Consequently, the creation of new SSSIs and Natura 2000 sites continues to be highly controversial. For example, SNH has recently been forced to withdraw SSSI proposals on Islay (for seals) and Arran (for hen harriers) in the face of concerted local opposition and accusations of high-handed insensitivity and secretive bureaucracy (I. Mitchell, 2001). The clinching argument in Arran was that, in both notifying and adjudicating on new SSSIs, SNH acts as both judge and jury. If upheld in court, this critique could force a review of the entire designations procedure.

There have been some recent attempts to get beyond the negative inheritance of these long-standing 'jobs *v.* conservation' tensions, either by emphasising the positive spin-offs of conservation for communities, or by arguing that the friction is not inevitable. The first line has been taken, amongst others, by the government (SODD, 1999a: 6) which argues that 'conservation and enjoyment of the natural heritage [should] bring benefits to local communities and provide opportunities for social and economic progress'. Designations could therefore come to be viewed as beneficial accolades rather than constraints (Scottish Office, 1998b). Certainly there can be considerable employment opportunities associated with wildlife tourism at designated sites, with perhaps as many as 1,500 full-time equivalent (fte) jobs supported across Scotland (Crabtree *et al.*, 1994b). Natural heritage-related activities provide over 6,700 fte jobs and account for 2 per cent of all jobs in rural Scotland (SNH, 1998c).

The second approach has been championed by Hunter (1995: 17) who argues passionately that ecological restoration and social rehabilitation are not mutually exclusive but can and should go hand in hand, that 'people and nature can . . . co-exist in ways which will benefit both'. Taking the long historical view, both Hunter and MacAskill (1999) see a Scottish countryside without people as fundamentally unnatural. Thus conservationists who argue for the creation of people-free wilderness areas in the Highlands (c.f. Cairns, 2000) and those like Hunter who campaign for the repopulation of rural Scotland base their claims on starkly contrasting visions of naturalness. Most conservationists equate naturalness with an absence of human beings and their impacts, and regard ecological restoration (for example by reintroducing native species (Section 10.4)) as a goal worth agitating for. Hunter (1995: 163) points out the irony that:

> . . . there is one species which environmentalists seem curiously reluctant to have re-established in the numerous Highland localities from which this species was expelled . . . The species in question is man.

Use and delight, profit and beauty, will always be in tension, but there are some signs that this tension is becoming less destructive and more creative. The controversial topic of the place of *Homo sapiens* in the world is discussed in Section 13.3.1.

8.3.3 Rationalisation of conservation

The complex, overlapping system of landscape and nature designations has been much criticised (Section 8.2.1), and is widely seen as 'eclectic, discredited and increasingly ramshackle' (Stuart-Murray, 1999: 182). Reform has been called for for many years (Lowe *et al.*, 1986). Consequently, during the later 1990s a series of consultations explored ways of rationalising the designations morass, both at a general level (Scottish Office, 1996) and specifically with regard to NSAs (SNH, 1998b) and to SSSIs (Scottish Office, 1998b). Although SNH (1999c) made formal recommendations to government for making NSAs more effective, few substantive changes have yet resulted from these consultations, largely because of the upheavals surrounding the establishment of the Scottish Parliament in 1999. However, there may soon be scope for reducing the number of overlapping designations within national parks (Section 8.3.4), and a further consultation about reform of the SSSI system took place during 2001 (Scottish Executive, 2001a).

SSSIs have long been regarded as inadequate because, as a notification not a true designation, they rest on the controversial voluntary principle (Section 13.3.5). They are discretionary, not offering absolute protection. In the final analysis, the nature conservation interest of even an internationally significant site can be destroyed if the government considers that that is where the balance of advantage lies (Sharp, 1998). By contrast, Natura 2000 designations introduce a rigorous protective regime, with strong duties and penalties attached. Conservationists therefore argue that SSSIs need to be strengthened, and were horrified at the apparent recent drift in the opposite direction (Ratcliffe, 1998).

The reforms now proposed by the Scottish Executive (2001a) envisage that SSSIs will remain at the heart of natural heritage protection, but that their management needs to balance nature conservation with the sustainable development of rural communities (a balancing act to be echoed in the new national parks (Section 8.3.4)). The proposals for SSSIs include a stronger local voice in their management, an improved system of consultation, an end to compensation payments to landowners who withdraw plans for damaging new schemes, and more resources to help fund positive management. Also proposed are stiffer penalties for deliberate damage to SSSIs and for wildlife crime. SNH's increased budget of £48.5 million for 2001/2 should allow more resources to be devoted to the management of protected areas (Scottish Executive, 2001a). SNH has already been moving away from relying on the infamous lists of Potentially Damaging Operations that accompany SSSIs (and which have so often antagonised farmers and land managers) and has begun to adopt more positive management regimes (McLavin, 2000). The best scenario for the natural heritage is that a reformed, streamlined and less confrontational SSSI system could improve conservation's image sufficiently to allow the conservation ethic to escape from its 'key area' ghetto and pervade the whole countryside. The fear, however, is that the reforms will weaken SSSIs and permit further damaging development.

Finally, it has been recognised for some time that conservation policies need to be better integrated with other countryside policies (Lowe *et al.*, 1986; Bishop *et al.*, 1995). In the words of the Scottish Executive (2001a: 10), 'the protection of Scotland's nature is too important to be left to policies on nature conservation alone'. The universal adoption of sustainable development as a guiding principle now requires such integration (Henton and Crofts, 2000). It can be argued that merely tweaking the internal operations of conservation practice is wholly inadequate. As Mackay (1995: 215) says, 'there is little hope of the fundamental issues in Scottish land use being tackled squarely until they are tackled comprehensively'. He advocates a radical overhaul of Scottish environmental management whereby all sectors would fall within the remit of one government department, a recommendation now substantially realised in the creation of SEERAD. Such wholesale integration has strong pros but also strong cons (Section 2.4.3). Arguably the greatest need is for real integration between conservation and farming. With the focus shifting from identifying protected sites to securing the right kind of management for them, it has become ever more apparent that the systems for conserving nature and for supporting agriculture need to be harmonised (McLavin, 2000). Given that the agricultural area is about four times greater than that covered by conservation sites, making farming practices more environmentally sensitive would have a far greater beneficial impact for conservation than any adjustments to conservation policy *per se*. On this basis Macdonald and Johnson (2000) argue that the future of agriculture is the single most important factor determining the future of conservation. Given that the Scottish Executive (2001a) appears to be sympathetic to this argument, perhaps the talk about conserving the wider countryside will soon start bearing tangible fruit.

8.3.4 National parks

A bold step towards integrated, comprehensive management of sensitive upland areas is about to be taken in the creation of Scotland's first two national parks. The road to this point has been long and tortuous (Table 8.7),[9] but by 2003 national parks should exist in the two areas that have always been regarded as most deserving of national park status, namely Loch Lomond and the Trossachs, and the Cairngorms.[10]

8.3.4.i Background

There is rich irony in the fact that the birthplace of John Muir, the so-called father of national parks, has hitherto been one of the few nations on earth without any. Calls for national parks (NPs) in Scotland go back to the 1920s, but the 1949 legislation which introduced them to England and Wales excluded Scotland, even though the groundwork had been laid by the Ramsay Reports. Why was this? It was partly due to opposition from landowners and local authorities, partly because the Scottish Office did not want to obstruct the economic rehabilitation of rural areas (especially through HEP schemes (Section 6.2.1.iii)), and partly because the FC and the NTS

TABLE 8.7 A chronology of the key events in the Scottish national park debate. Compiled from numerous sources including Moir (1991), Mackay (1995), media coverage and *SCENES*.

Date	*Event*
1928	The *Scots Magazine* campaigns for NPs
1931	Addison Committee's recommendations for UK NPs include Scotland
1945, 1947	Ramsay Reports recommend five publicly-owned NPs; objectives included scenic & wildlife protection; development of public access & recreation; enhancement of rural life & industries
1948	National Park Direction Areas introduced to provide development controls in Ramsay's five areas; existed in planning system until 1980 when replaced by NSAs
1949	National Parks and Access to the Countryside Act excludes Scotland
1974	CCS publishes *A Park System for Scotland*, recommending 'Special Parks' not NPs
1988	Loch Lomond Regional Park created, administered by a Park Authority
1990	CCS publishes *The Mountain Areas of Scotland*, identifying pressures in the uplands and recommending wide-ranging solutions including NPs (CCS, 1990, 1991)
1991	Natural Heritage (Scotland) Act introduces Natural Heritage Areas (NHAs). Designation never used, but seen by some as a uniquely Scottish (and preferable) alternative to NPs
1992	Scottish Office rejects CCS's call for NPs, but establishes working parties to identify solutions for LLT and Cairngorms areas; NPs specifically excluded as an option
1993	Working parties publish their reports (LLTWP, 1993; CWP, 1993)
1994–5	Secretary of State responds to working party reports; establishes the Cairngorms Partnership and the Loch Lomond Joint Committee
1994	IUCN publishes *Parks for Life*; identifies Scotland as an area where 'action is now urgently required'
1996	Scottish Office review of natural heritage designations (Scottish Office, 1996) skirts round NP issue; makes positive comments about NHAs
1997	SWCL publishes *Protecting Scotland's Finest Landscapes* (SWCL, 1997), a powerful case for a new 'top-tier' designation which should be entitled 'national park'
1997	Secretary of State announces that the government is committed to a NP for LLT, and probably for the Cairngorms
1998	SNH publishes a consultation document on the principle of NPs in Scotland (SNH, 1998d)
1999	SNH publishes its Advice to Government, with detailed recommendations for NPs (SNH, 1999d)
2000	Scottish Executive consults on the enabling legislation for NPs (Scottish Executive, 2000c)
2000	The National Parks (Scotland) Bill is passed unanimously by the Scottish Parliament
2000–1	Detailed consultations on the names, boundaries, functions, powers, authority, representation and operation of proposed NPs in LLT and the Cairngorms (SNH, 2000b, 2000c). Boundary for LLT decided in June 2001 (Fig. 8.4)
2002–3	Creation of first NPs in LLT and the Cairngorms

TABLE 8.8 The main pressures and problems afflicting upland areas. The severity of these problems varies greatly from region to region. Collated from CCS (1990), Wightman (1996b), SWCL (1997) and Cairngorms Campaign (1997). For discussions of the particular issues and opportunities in the Loch Lomond and Cairngorms areas, see SNH (1999d).

Visitor and recreation pressure in areas of high conservation value, poor visitor facilities, and insufficient provision of information
Damaging land uses, such as overgrazing and inappropriate afforestation
Ecological change; loss and fragmentation of wildlife habitat
Unsympathetic developments. e.g. 'eyesore' housing & construction; bulldozed tracks high in the hills; quarrying & mining; energy generation
The ineffectiveness and lack of coordination of existing conservation designations
The absence of an integrated, strategic planning framework with adequate powers and resources: responsibility for most large upland areas is shared amongst several local authorities, so it is hard to tackle pressures in a coherent fashion
Core mountain properties (e.g. Glenfeshie) are traded on the international land market, creating uncertainty and prohibiting continuity of management for long-term objectives

believed that, between them, they could adequately provide scenic conservation and public access. There was also a perception that the need for recreational access was much less urgent than south of the Border (Moir, 1997).

During the four decades when NPs only existed in England and Wales there were several unsuccessful attempts to introduce them in Scotland (Crabtree, 1991; Moir, 1991), the last of which comprised the valedictory salvo fired by the CCS (1990, 1991). This consisted of powerfully and cogently argued proposals for integrated management of all of Scotland's mountain areas, and an unequivocal call for the creation of four NPs. Unfortunately the former were sidelined by the acrimonious debate over the latter. The four areas proposed were those which have always been the prime candidates: Loch Lomond and the Trossachs, the Cairngorms, the Ben Nevis/Glen Coe/Black Mount area, and Wester Ross. (Glen Affric, the fifth area recommended by the Ramsay Reports, was excluded.) Despite the strength of the CCS case, the overwhelming support for NPs amongst the general public, and the absence of effective alternatives for addressing the catalogue of management problems afflicting the uplands (Table 8.8), this proposal went the way of all its predecessors, sunk by a combination of lack of political will and the distinctly patchy support for NPs in the proposed areas (Moir, 1997). Instead, the Scottish Office put its faith in the voluntary approach epitomised by the Cairngorms Partnership. Those calling for NPs seemed perpetually doomed to be voices crying in (or for?) the wilderness.

All this changed in 1997 when the newly-elected government made a commitment to the creation of Scottish NPs. It is interesting to reflect on the politics of this. Instead of proceeding logically from problem to diagnosis to prescription, the political decision that Scotland must have national parks came first, and only then was attention given to what their form and functions should be. *Post hoc* justifications were clearly driven by a political, nationalistic agenda. Events have subsequently moved swiftly (Table 8.7), with extensive consultation at every stage.[11] As recommended by SNH (1999d), a two-stage process has been adopted in which enabling legislation sets a framework for all future NPs, and then secondary legislation establishes each park through individual designation orders (Scottish Executive, 2000c). This makes a 'menu approach' possible whereby the over-arching NP system can be fine-tuned for each park, and different objectives can be achieved in different parts of the parks through zoning. This distinctively Scottish, tailor-made approach is exemplified by the differing proposals for Loch Lomond and for the Cairngorms (SNH, 2000b, 2000c); a lighter touch is suggested for the latter reflecting the different circumstances.

Intriguingly, unlike every previous round of debate, this time there has been no strong 'no' campaign. Although many bodies and individuals remained unconvinced, and debate has been lively, only 5.5 per cent of the consultation responses specifically opposed the concept (SNH, 1999d). The explanation for this must lie in the combination of an extensive consultation process, the emergence of strong political will in favour of creating NPs, and the stress on a consensus-building, inclusive approach. The inclusion of social and economic promotion as primary aims, and the emphasis on involving local people, may also have won over some sceptics. With the enabling legislation unanimously passed in 2000, the details of Scotland's first two parks are now being hammered out (SNH, 2000b, 2000c, 2001). Additional parks, both terrestrial and marine, may be created in years to come in areas such as Ben Nevis/Glen Coe, the Southern Uplands and the Small Isles, but only after a similarly extensive process of consensus-building in the areas concerned.

8.3.4.ii The old debate: should Scotland have national parks?

Clearly this question has been overtaken by events, but it is instructive to review briefly the sets of arguments that were employed to support or oppose the creation of NPs in Scotland (Table 8.9).[12] Broadly speaking, support for NPs came from access, recreation and landscape bodies, and from some 85–90 per cent of the general public. Opposition came from land management bodies, development interests, local authorities in the Highlands, and some wildlife conservation groups, as well as from many of those living in the proposed areas.

Almost everyone agreed that upland areas were under increasing pressure and suffering from inadequate management (Table 8.8). But despite considerable consensus about the diagnosis of the problem, there was no agreement about the appropriate remedy. Conservationists saw NPs as the obvious solution, arguing that without them 'the processes of attrition, decline and conflict will continue' (SWCL, 1997: 10). Opponents, however, believed that the key issue was integrated management, and that NPs were neither the only nor the best means of achieving this. Pointing to experience in America (Runte, 1997), they argued that national parks are no 'magic bullet' for solving the problems of visitor pressure. Moreover, the problems are extensive in character, not limited to certain prime areas (Garner, 1989), so NPs (they argued) could only ever be a partial solution. In addition to the core debate (Table 8.9),

TABLE 8.9 The main arguments for and against national parks in Scotland that were made during the 1990s. These have been collated from various sources, but primarily from SWCL (1997), McDermott (1997) and Warren (1999b) in which these and other arguments are discussed in detail.

The case in favour of national parks
Negatives of the *status quo*
Existing designations are inadequate; do not provide effective protection from the wide range of pressures on upland areas (Table 8.8)
Voluntary principle is ineffective; partnership approach is good, but insecure without statutory remit
Powers and resources are inadequate and insufficiently integrated to achieve positive conservation and management of large upland areas
Absence of NPs is a national reproach
Benefits of NPs
Unified, single NP authorities permit integrated, strategic management in place of sectoralism
Improved visitor experience, and enhanced conservation
Economic benefits for rural communities: job-creation, inward investment, enhanced tourism, and a stronger identity for the marketing of areas and their products
Economic benefits for farmers: extra payments for access & conservation measures
Provide forums for conflict resolution; exemplars of good practice in integrated management
Enable Scotland to meet its national & international conservation obligations more effectively
The only plausible way to obtain extra funding for conservation (nationally & internationally)
The case against national parks
Arguments against national parks
Exacerbation of the 'honeypot' syndrome, leading to tourist overload
Adds another layer of bureaucracy to the already complex designations system
Opposition from locals, landowners or farmers would undermine the effectiveness of NPs
Erodes local democracy by giving national interests too much say; could lead to land nationalisation by back door (e.g. CCS (1991) and Wightman (1996b) both argue for public ownership of key areas)
Funding arguments: 1. Adequate funding hard to obtain. 2. Funding for NPs would divert resources from existing initiatives. 3. Money would be better spent strengthening existing designations
NPs would go against the new focus on the wider countryside (c.f. Section 8.2.2.iii)
Would encourage the public to see the countryside as a 'park' for their use, rather than as a working landscape, so could exacerbate conflicts
Better alternatives
Integrate, resource and strengthen existing designations; make them fulfil their potential
Give partnership initiatives (e.g. Cairngorms Partnership) extra resources and time to work
Put NHAs into practice

subsidiary questions included the desirability of the term 'national park', and the relationship between Scottish NPs and IUCN norms (Warren, 1999b).

Although many have bewailed the lack of NPs in Scotland, there are those who argue that the failure to create them in the 1940s proved to be a blessing in disguise. For instance, the Ramsay Committee's vision for publicly owned NPs, with its central planning ethos and its emphasis on recreation and development, might actually have been detrimental for conservation in Scotland (R. Watson, 1990; Mackay, 1995). So 'maybe Scotland did not miss much' (Smout, 2000a: 163). The delay has given Scotland a unique chance to learn from a wealth of international experience, to avoid the pitfalls of the English/Welsh model, and to design a park system tailor-made to address the needs of the Scottish uplands.

8.3.4.iii The current debate: what should Scottish national parks be like?

The government's decision that 'national parks are the right way forward for Scotland' (Dewar, 1997) immediately moved the entrenched debate into an entirely new phase. Subsequent years have seen a spate of consultations[13] and wide-ranging discussion of the pressing new question: what form should Scottish NPs take? The four-fold vision developed by SNH (1999d) is that NPs should:

- provide a greater clarity of national purpose for some of Scotland's most special areas
- secure higher standards of environmental stewardship
- engender trust between national and local interests in the delivery of conservation and community objectives
- pioneer techniques for achieving sustainable development.

This positive vision has been adopted, but many detailed questions have needed thrashing out. The key ones have been these:

1. **What should the purposes of Scottish national parks be?** The four aims of NPs enshrined in the primary legislation are these:
   - to conserve and enhance the natural and cultural heritage
   - to promote sustainable use of natural resources
   - to promote understanding and enjoyment (including recreation) of the special qualities of the areas by the public
   - to promote the sustainable economic and social development of communities.

   These aims are unexceptionable. Arguably, they should apply to all land management everywhere. However, whereas many of the world's national parks (including those in England and Wales) give nature conservation and recreation unchallenged primacy, within Scottish NPs social and economic objectives have officially been given co-equal status alongside environmental aims. Given that large parts of the two parks are working landscapes with rich human histories, and given the need for local support, the inclusion of rural development and protection of the cultural heritage are necessary innovations. Both the land and the people need to be sustained, hence the broad remit. As the Environment Minister stated at the outset, 'I have no interest in creating living museums where the emphasis is only on conservation' (Sewel, 1997). Scottish NPs are therefore setting out to deliver a new form of integrated rural management.

However, SNH and others have argued that in the event of conflict between conservation and the other three aims, conservation should win the day, so that in practice it will turn out to be first amongst equals. This, the so-called Sandford principle, was enshrined in the primary legislation despite attempts to remove it. Applying this principle is unlikely to be straightforward. It is easy to say that national park designation should become an accolade, not a constraint (Randall, 1998), but that is hard to reconcile with the need for more effective protection of biodiversity and the natural heritage. Striking an exemplary, integrated balance between the four aims will be a tricky tightrope to negotiate. The intention is that different needs will be reconciled through partnership working, and through a NP Plan based on zonal policies.

2. **What powers and resources should the park authorities have?** Conservationists have always argued that meaningful powers (including a town and country planning function) and significant resources are essential requirements of NPs (Aitken, 1997). They must have 'teeth'. Both of these are politically charged demands, resisted by local authorities (which want to retain their existing powers) and by central government (which is wary of entering into expensive long-term commitments). Annual running costs for the two new parks are likely to be around £4–6 million each. The primary legislation makes it clear that the parks will be centrally funded, but the issue of whether park authorities will have planning powers is to be decided on a park-by-park basis. It is likely, for example, that the Loch Lomond NP Authority will become the statutory planning authority for the area, whereas in the Cairngorms many planning issues may well remain the preserve of local authorities (SNH, 2001), a suggestion which is proving deeply controversial.
3. **What are the appropriate constitutional structures?** It would seem natural that the composition of the independent park authorities should reflect the range of interests in their area (e.g. owners, residents, agencies, local authorities, NGOs) but it also needs to strike a balance between local and national interests. These are highly sensitive issues (Jardine, 1998). The two new NPs both contain assets of national significance supported from central funds, but they will rely on local management and support; locals will resent constraints imposed in the national interest unless there are matching local benefits and unless locals are seen to have a voice in decision making. Recognising the challenge of reconciling national and local aspirations within the new parks, SNH (1999d) advocated a long-term 'contract' between national and local interests whereby locals are empowered and resourced to deliver national objectives. Without such a contract, NPs are unlikely to succeed because a condition of success is that they must first of all earn, and then sustain, the positive support of the local population

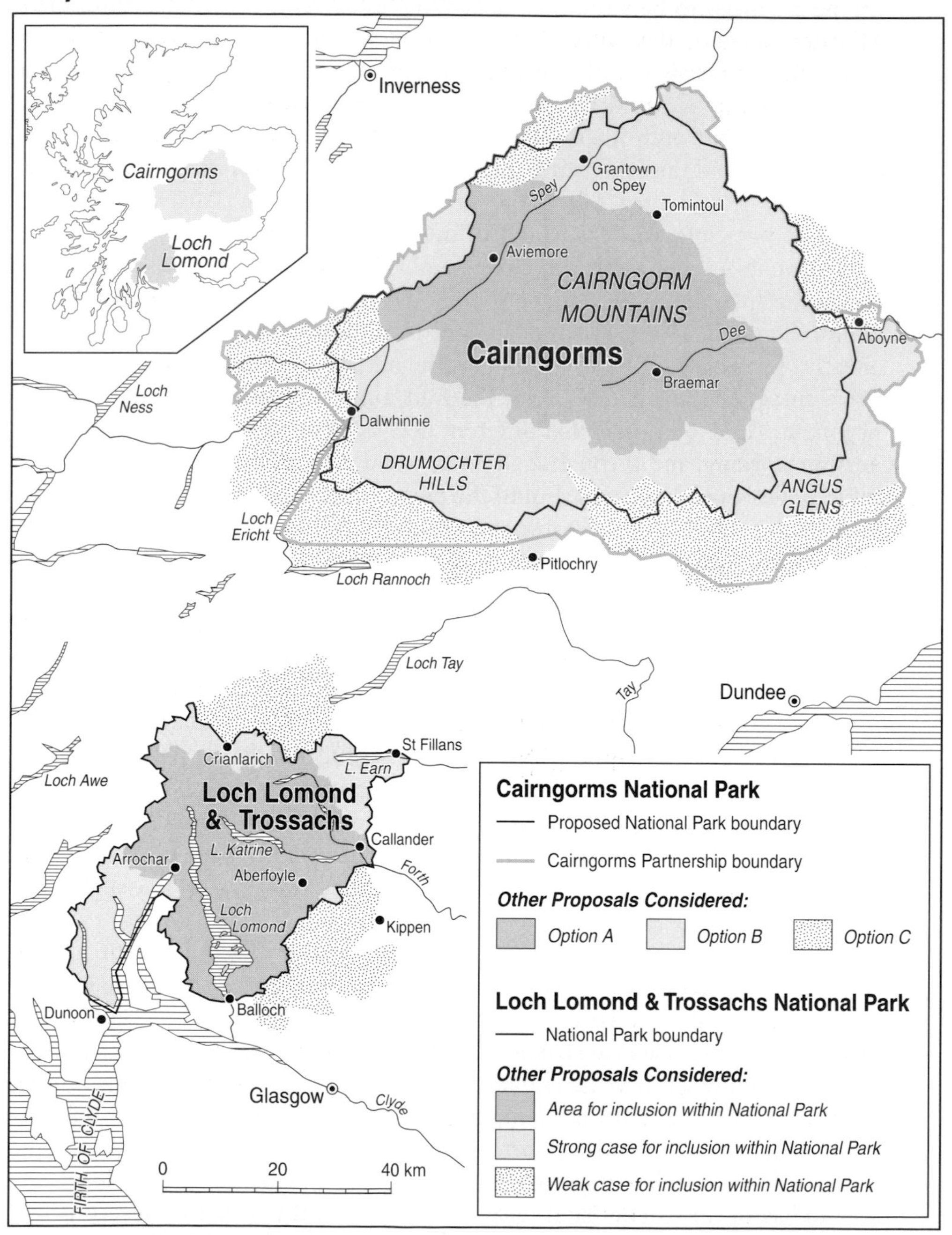

FIGURE 8.4 *Possible boundaries considered for new national parks in the Cairngorms and for Loch Lomond and the Trossachs. Also shown are the final choice of boundary for the latter, and the Cairngorms boundary proposed by SNH in August 2001. The inset shows the extent of mainland Scotland covered by the two largest options. Boundary options after SNH (2000b, 2000c), used with permission.*

(Crofts, 1999). In the event, the primary legislation stipulates that majorities of the twenty-five-strong park authority boards will consist of local interests (part elected, part appointed), and that the national perspective and appropriate expert knowledge will also be represented. Although the formalised voluntary approach which was demonstrated so effectively during the brief life of the Cairngorms Partnership has been jettisoned in favour of NPs, its principles of coordinated working have been built into the NP proposals. The Partnership is also acting as the midwife for the birth of the Cairngorms NP.

4. **Where should the boundaries of the two parks lie?** The maximum and minimum options for the parks (Fig. 8.4) reflect different visions of their functions. For the Cairngorms, the smallest and largest options would cover 2,100km^2 and 7,340km^2 respectively and include resident populations of 10,470 or 29,120 people (SNH, 2000c). The boundary decisions turn on a complex mix of environmental, cultural, economic and administrative considerations. Too small, and the natural heritage cannot be managed effectively; too large, and resources become stretched and administration unwieldy. The latter concern was immediately voiced when the Scottish Executive opted in 2001 for an extensive area for Loch Lomond and the Trossachs NP, likely to cost up to £5.8 million per annum to run. Under the enabling legislation, the areas chosen must conform to three conditions. These are, firstly, that they are of outstanding national importance for their natural and/or cultural heritage; secondly that they have distinctive characters and coherent identities; and thirdly that creating NPs will be the best means of meeting their particular needs. None of these criteria are definable in any absolute sense, but the first two are much more testable than the last, which (since it concerns hypothetical futures) is essentially unknowable.

SNH (1999d: 45) believes that NPs are not the whole solution to the problems of managing Scotland's natural heritage, but that they may 'help to pave the way to a more integrated and sustainable use of land, as part of their primary role in safeguarding and providing for the enjoyment of Scotland's most special places'. The creation of Scotland's first NPs is an exciting development, but of course their existence will not cause the long-standing tensions over upland management to evaporate overnight. The new NP authorities inherit an inspiring but unenviable challenge as they set out along uncharted tightropes amidst the crosswinds of high expectations, multiple objectives and conflicting demands. The battle for NPs in Scotland has been won. Ahead now lies the challenge of making them realise their positive potential.

NOTES

1. Smout (1993c, 2000a) provides enlightening discussions of the roots of 'green consciousness' in Scotland. More generally, the contested and culturally conditioned nature of 'nature' has spawned a plethora of texts. See, for example, the uniquely fascinating and monumental work by Glacken (1967), the books by Black (1970), Passmore (1974), Thomas (1983)

and Oelschlaeger (1991), the volume edited by Cronon (1996a), and the discussions by Simmons (1993, 1996), Budiansky (1995), Adams (1996), Coates (1998), Macnaghten and Urry (1998), Phillips and Mighall (2000), van den Born *et al.* (2001) and Nash (2001).

2. For discussions and critiques of the UK's response to the 1992 Earth Summit, see Carter and Lowe (1998) and Connelly and Smith (1999).
3. The history of nature conservation in the UK is described in detail by Evans (1997). Adams (1996), Sheail (1998) and Dixon (1998) also provide interesting discussions. A specifically Scottish focus is adopted by Davidson (1994), Bishop (1997), Moir (1997), Sommerville (1997), Arnott (1997) and Smout (2000a). Mackay (1995) provides in-depth descriptions and critiques of the records of the CCS and the NCC in Scotland, and the birth pangs of SNH.
4. Since 1990, Regionally Important Geological/Geomorphological Sites (RIGS) have complemented the SSSI system. See JNCC (1996b), Gordon and McKirdy (1997) and Gordon and Leys (2001) for information and discussion of earth heritage conservation.
5. For details of the objectives and operation of NHZs, see C. Mitchell (2001).
6. Existing knowledge of the biodiversity of Scotland is brought together by Fleming *et al.* (1997), and the current status of those habitats and species with action plans is reviewed by Usher (2000).
7. For a collection of articles discussing the pros and cons of biodiversity work throughout the UK, with numerous case studies, see *ECOS* 21(2), 2000.
8. The issues raised in this section are global in scope and can only be sketched in outline here; some are explored further in Chapter 13. For discussions of the practical and ethical meanings of conservation, see Holland and Rawles (1993), Rawles and Holland (1994) and Evans (1997). For thought-provoking discussions of the difficult choices facing conservationists, see Budiansky (1995), Holdgate (2000b) and Lister-Kaye (2001).
9. The 70-year-long history of the Scottish national park debate has been discussed by, *inter alia*, Ferguson (1988), Crabtree (1991), Mackay (1995), Evans (1997), Moir (1991, 1997) and Lambert (2001b).
10. For detailed descriptions of these two areas and their distinctive needs and pressures, see the respective working party reports (LLTWP, 1993; CWP, 1993). The speciality of Loch Lomond both nationally and internationally is described by Maitland *et al.* (2000), and its great popularity for recreation is discussed by Dickinson (2000b). For information and references on the international significance of the Cairngorms, see Section 12.1.
11. After SNH's initial 'invitation to contribute' in 1998, a formal consultation paper (SNH, 1998d) led to its Advice to Government (SNH, 1999d). The Scottish Executive (2000c) consulted on the enabling legislation, and on elections & appointments to NP Authorities, and SNH (2000b, 2000c, 2001) sought views on the specific proposals for the two new parks. The designation orders will also be consulted on.
12. The appropriateness of national parks in Scotland has been debated by numerous authors. In particular, see Garner (1989), Crabtree (1991), Dickinson (1991), Moir (1991, 1997), SWCL (1997) and Warren (1999b).
13. See note 11.

PART THREE:
Interactions and controversies

CHAPTER NINE

Access: whose rights and whose responsibility?

9.1 Introduction and background

Lying at the heart of the access debate is a clash of competing rights – the 'right to roam' *versus* property rights. A pair of quotations from *The Times* in the late nineteenth and late twentieth centuries encapsulates the nature of this head-on collision and its enduring power to provoke strong emotions. The first comes from a Leader column commenting on the first Access to Mountains (Scotland) Bill, and the second from a commentary on the contemporary debates surrounding the 'right to roam' and land reform. The uncanny similarities reveal how little the core arguments have changed.

> Is it not a matter of compromise? Surely the lords of the soil cannot claim so absolute a monopoly of earth's surface . . . as wholly to shut out the poor holiday folk, the artist and the naturalist. Surely the many have rights as well as the few. . . . On the other hand, numbers cannot claim utterly to destroy the rights of property; that is, the right to some exclusive use of it. The problem cannot be insoluble. (*The Times*, 25th March 1884 in Stephenson, 1989: 133)

> The conflict over common land and its enclosure has been at the heart of British politics since the Norman Conquest. . . . The correct policy is one of compromise. We must balance the needs of present generations against those of their descendants, and the demands of the excluded against those of the excluders. . . . We should settle our land grievances in the long-term interests of our nation, not just for the pleasure of present generations. . . . This great grievance, the running sore of British politics for a thousand years, has again become inflamed, and who knows what can be done to soothe it? (Roger Scruton, *The Times*, 23rd January, 1999)

Objectivity in this vexed arena is hard to come by: access to one may be disturbance or intrusion to another. It is not only emotionally but politically charged. This is inevitable for two reasons, firstly because 'property and ownership raise virtually all the questions which political philosophy can pose' (Cox, 1993: 275), and secondly because access issues go to the heart of what Smout (2000a: 170) dubs 'the quarrel over the countryside', a debate which he characterises as 'an argument over the limits and rights of property'. One reason why this quarrel defies easy resolution is the public nature of private landed property. Owning hills and glens can never be as exclusive as owning a house or a car, so the problem becomes one of drawing a line through the middle of the 'everyman's land' where private rights and public interest overlap. Although rights are commonly perceived to be absolute, Munton (1995) shows that they are simply interest claims that enjoy some legal, moral or social sanction. Because their continued existence depends on the evolving socio-political context, rights are properly seen as socially constructed, contested and negotiated concepts.

Many protagonists in this debate have views which have become entrenched during the long history of access controversy in Britain (Williamson, 2001).[1] In Scotland, that history stretches back to the so-called 'battle of Glen Tilt' in 1847 and the founding of the Scottish Rights of Way Society in 1843. Between 1884 and 1939 there were numerous attempts to pass an Access Bill which, had it become law, would have made it illegal to exclude the public from uncultivated mountain and moor – what would now be called a 'right to roam'. The growing popularity of walking in the countryside led to the formation in 1905 of the Ramblers Association (RA), which campaigns tirelessly for improved access, and whose membership has now grown to 130,000 in the UK and 6,100 in Scotland (RA, 2000). The 1949 National Parks and Access to the Countryside Act made some provision for improved public access in England and Wales, but it had no relevance for access in Scotland. Public rights of way are relatively short or fragmentary routes and do not constitute a network. There are just 15,000km of rights of way in Scotland, compared with 160,000km in England, and most of these have little or no legal status (Davison, 2001).

Although the issue has a long and controversial history, it has been pushed steadily up the agenda in recent years by the huge growth in outdoor recreation (Fig. 9.1). The area of land devoted to recreational use has increased by 138 per cent since the 1940s (Mackey *et al.*, 1998), and outdoor recreation now ranks as a major land use in its own right, even dominating in much of the uplands (Balfour, 1994; Mackay, 2001). For example, a survey by HIE (1996) suggested that 767,000 mountaineers (including walkers, climbers and cross-country skiers) notched up 8.1 million days in the hills during the previous year in the Highlands, contributing £162 million to the economy and supporting 6,100 jobs. Such mountain pursuits extend the tourist season and help to reduce seasonal unemployment in rural areas.

High percentages of Scots now visit the countryside for leisure pursuits (Greene, 1996). For example, the 1998 UK Day Visits Survey showed that 60 per cent of Scottish adults visited the countryside that year, taking over 32 million walks longer than two miles, 4.5 million cycle rides, and generating £545 million of expenditure (NCSR, 1999). In terms of trends in recreation, reliable runs of data are surprisingly sparse, but Hunt (2001b) presents some revealing statistics. The proportion of the population who spend at least one day in the countryside each month has risen from 25 per cent in 1987 to 40 per cent in 1998, and 68 per cent of people now visit the countryside at least once a year. The numbers involved in outdoor recreation showed continuous growth throughout the 1990s, increasing at 3–4 per cent per annum, with

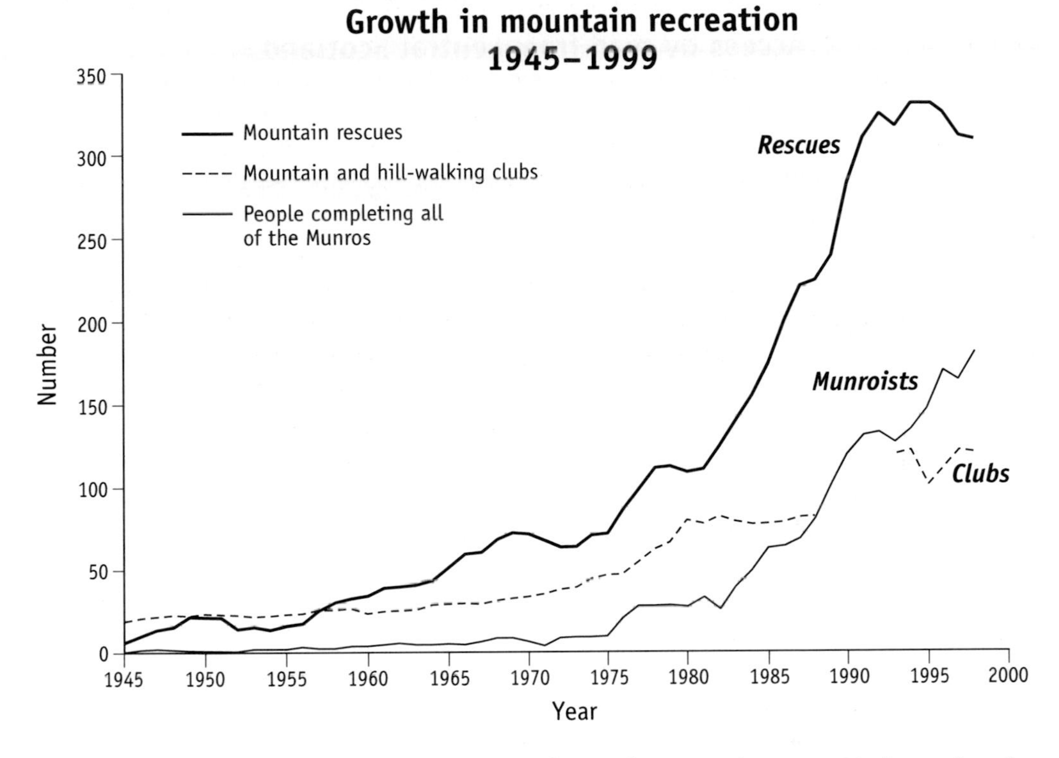

FIGURE 9.1 *Three indirect measures of the growth in outdoor recreation, 1945–99: the number of mountain rescues, the number of people completing all the Munros, and the number of walking/climbing clubs. The first two of these are shown as three-year running means. Data compiled by Bob Aitken and used with permission.*

walking being the most popular and fastest growing pursuit, rising at 8 per cent per annum. Important though such national trends and statistics are, what really counts is what happens at the local level, and the development of local solutions (Sidaway, 2001).

The explosive growth of outdoor recreation is a subset of the development of tourism more generally, growth which has been driven by three main factors:

- increasing leisure time
- increasing disposable income
- increasing physical mobility.

This last has resulted from the steady improvement of transport infrastructure, beginning with the arrival of the railways in the nineteenth century and continuing through to the post-war growth of car ownership and the great improvements in the road network which have made remote areas so much more accessible (Fig. 9.2). Moreover, access into and within the countryside has been facilitated by the construction of forest roads and tracks on sporting estates, some 10,000km of which were built between the 1940s and the 1980s (Mackey *et al.*, 1998).

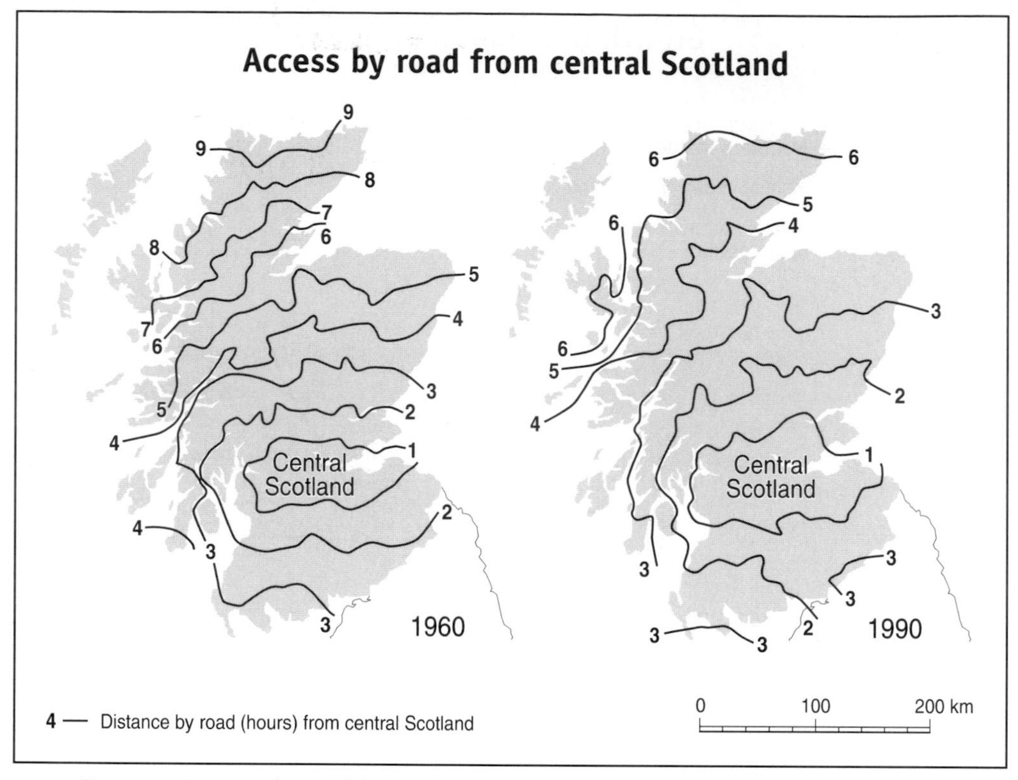

FIGURE 9.2 *Access by road from central Scotland in 1960 and 1990. After CCS (1990).*

Tourism and recreation are now the biggest sectors of the economy in a majority of rural Scotland, having overtaken the established land uses of farming, forestry, fishing and field sports both in terms of employment and GDP (Slee, 1998; Snowdon and Thomson, 1998). Tourism supports around 8 per cent of all employment in Scotland and over 13 per cent in the Highlands (Miles, 2001). But it is the traditional land uses which shape the appearance and determine the quality of the landscape which visitors come to enjoy. The fact that the environment is the single most important factor attracting tourists to Scotland puts the role of the traditional land uses in a new and rather interesting light (MacLellan, 1998). As Slee (1998: 94) comments, 'the land managers of the scenic resources that provide the setting and backcloth for tourist activity are more often bystanders than stakeholders in the tourism industry'. With a few notable exceptions in prime tourist areas, such as Rothiemurchus Estate by Aviemore, land managers tend to view the rising influx of visitors as a cause of cost and nuisance, not as an opportunity. Because of the heavily urbanised nature of Scottish society (and despite the blurring of urban-rural distinctions), many recreationists have a limited understanding of the nature and needs of rural life.

> The Scottish countryside is now accessible to everyone yet misunderstood by most. The stereotypes of rural life, and particularly Highland life, are powerful, and most townspeople cling tenaciously to them, perhaps because they want the

countryside to appear archaic, simple, picturesque and, above all, different. (Clark, 1983: 236)

Most informal recreation is heavily dependent on 'by-product access' – that is, it takes place on privately owned land which is primarily used for farming, forestry or field sports (Crabtree *et al.*, 1994c). In the days when the scale of public access was limited, many landowners remained unaffected or turned a blind eye, contributing to the widespread perception of a long-established right of access to the countryside. But with increasing numbers has come more disturbance and damage (mostly through ignorance but sometimes by wilful intent), affecting both the business activities and the private benefits enjoyed by rural people, and creating friction between the public and land managers. When such confrontations make the news, they are usually presented in stereotypical terms (the titled, tweedy laird pitted against the innocent, cagoule-clad backpacker), exacerbating a perception of rural conflict with overtones of class warfare. In all but the most remote areas, it is no longer possible to overlook the problem, nor for private estates to attempt to prohibit access to large parts of their ground for long periods of the year as they once did. Consequently, many access providers are now attempting 'access management' of various types, especially by providing information to potential users (Section 9.3.2). While private owners rarely encourage public access, the main landholding public agencies have long taken a proactive stance, Forest Enterprise being a prime example. Local authorities also have a role to play in managing and providing for open-air recreation in the countryside, but most have focused more on urban-based sport and recreation facilities.

There are some recent indications that the rate of growth of outdoor recreation is beginning to slow or even plateau (Greene, 1997; Aitken, 2000). Certainly, the UK Day Visitor Surveys in 1994, 1996 and 1998 showed little change in the percentages of Scottish adults visiting the countryside. However, this is no cause for complacency because the numbers involved are high (Table 9.1) and the average number of trips per individual appears to be increasing. However, it is notoriously difficult to assess trends in informal outdoor recreation, whether at local or national scales. Factors as varied as weather patterns, petrol prices and the strength of sterling all affect people's decisions, so one is frequently not comparing like with like. Moreover, direct datasets with which to assess such trends are few, short and rather inaccessible, making a reliance on proxy data such as those in Figure 9.1 unavoidable. Some organisations have operated automatic people counters on popular routes such as Ben Lomond (NTS) and in the Cairngorms (SNH; cf. Gardiner, 2000), and this information is of interest (Table 9.1), but there has been no systematic effort to acquire long-term, national datasets. Given the high and rising profile of access issues in recent years, this is a strange omission.

9.2 TYPES AND IMPACTS OF PUBLIC ACCESS

Outdoor recreation has diversified greatly in recent years. It is not just the number of people that has multiplied but the range of pursuits for which they desire access to

TABLE 9.1 Numbers of annual visits to some popular mountains and mountain areas. These figures are based on diverse methodologies and so are not directly comparable.

Location	*Usage* *Walkers per year*	*Source*
Cairngorm Mountains	123,000	Taylor & MacGregor, 1999
Cairn Gorm	59,000	Gardiner, 2000
Ben Macdui	11,600	Taylor & MacGregor, 1999
Braeriach	4,900	Taylor & MacGregor, 1999
Sgor an Lochain Uaine	4,300	Taylor & MacGregor, 1999
Lairig Ghru	4,200	Gardiner, 2000
Cairn Toul	3,800	Taylor & MacGregor, 1999
Glencoe	80,000	HIE, 1996
Ben Nevis	60,000–85,000	Mackay, 2001
E. Grampians & Lochnagar	50,000–60,000	Mather, 1998
Goat Fell	33,000	SNH, 1997a
Ben Lomond	31,000	SNH, 1997a
Ben Lawers	>22,000	Johnston, 2000
East Torridon	20,000	SNH, 1997a
Ben Vorlich	15,000–20,000	SNH, 1997a
Coire Dubh, Torridon	15,500	SNH, 1997a
Stac Pollaidh	15,000	SNH, 1997a
Ben Wyvis	8,000	SNH, 1997a
Slioch	6,500	SNH, 1997a
Ben Vane	5,200	SNH, 1997a
Glen Shiel:		Herries, 1998
South Cluanie ridge	2,500	
Five Sisters of Kintail	1,800–2,000	

rural and upland areas (Table 9.2). One pastime that has come to prominence and which has caused people to seek out previously little-visited mountain areas is 'Munro-bagging', the challenge of climbing some or all of the 277 Scottish mountains over 3,000 feet (914m). Areas with concentrations of Munros, such as the Cairngorms and the West Highlands, saw huge increases in the numbers of hillwalkers in the 1980s and 1990s. Even inaccessible Munros are climbed by almost a thousand people each year, and the easily reached ones by many thousands (Mackay, 2001; Table 9.1). A few of the Corbetts, hills between 2,500 and 3,000 feet in height, have become almost as popular. Surveys show that the average walker is young to middle aged, middle class, highly educated, professional and Scottish, although a growing number come from overseas (Mather, 1997; Mackay, 2001). Mountain biking has also experienced prodigious growth in the last fifteen years, and there has been a rise in the amount of group access for recreational, educational and charitable purposes.

TABLE 9.2 The main types of recreation for which people desire access to or through open country (excluding game sports). For details of water-related recreational pursuits and their impacts, see Table 6.5 and Section 6.5.4.

Walking	Short-distance, low level, e.g. walking the dog Long-distance, high level, e.g. hill walking & 'Munro-bagging' Orienteering
Climbing	Rock climbing; ice climbing; abseiling
Bicycling	Cycle touring; mountain biking
Air sports	Paragliding; hang-gliding; microlights; gliding; ballooning
Water sports	Boating; canoeing; sailing; rafting; windsurfing; sub-aqua
Fishing	Land-based or boat-based; fly fishing or coarse fishing
Camping & mountain bothying	Camping at official campsites; wild camping; bothy-to-bothy walking
Skiing	Downhill; cross-country ski touring; ski mountaineering
Nature study	Bird watching; wildlife photography; scientific research
Running & jogging	Low level; hill running; mountain marathoning
Motor sports	Trial biking; off-roading (4WD)
Pony trekking	
Outdoor education & training	Courses at outdoor centres for recreation, and for certification (e.g. sports instructors; mountain leadership)

All this fresh air and exercise is undoubtedly good for the physical, mental and spiritual health of the population; enjoyment of the outdoors can be a 'safety valve' releasing the pressures of daily life. Sadly, however, the huge popularity of outdoor recreation is exacting a toll on the natural environment, as well as creating tensions between different groups. These tensions may arise between visitors and residents but also between incompatible pursuits (the peace sought by a lochside walker being shattered by a jetskier, for example). The primary impacts of recreation can be broadly divided into quantifiable (though not always costable) and unquantifiable (Table 9.3), even though the categorisation of some impacts is debatable. The actual costs to landowners from public access are highly variable and site-specific. They include, for example, footpath and bridge repair, staff time, and vandalism. In a survey by Crabtree *et al.* (1994c), most farms and half the estates reported no costs, but 20 per cent of estates suffered annual costs in excess of £1,000, and only a small minority (in the main tourist areas) can offset such costs with access-related income. Apart from the criminal fraternity, few individuals visiting the countryside act with malicious intent. It is an observed fact that the vast majority wish to behave responsibly and already do so (Scottish Executive, 2001e). Nevertheless, the cumulative impact of the increasing numbers partaking in outdoor recreation can cause problems.

TABLE 9.3 The primary impacts of informal recreation. See text for elaboration. Vandalism and theft are very rarely associated with recreation; they are criminal activities mostly confined to urban peripheries.

Quantifiable impacts	*Unquantifiable impacts*
Footpath erosion	Natural heritage damage
Disruption of field sports	Invasion of privacy; annoyance
Litter	Effect on amenity value
Damage to agricultural crops, walls, fences and gates	Livestock disturbance and worrying by dogs
Forest fires	
(Vandalism and theft)	

9.2.1 Quantifiable impacts

The disruption of sporting activities (notably stalking and grouse shooting) has been a significant facet of the access debate, with walkers and (to a lesser extent) mountain bikers causing most disturbance (Cox, 1993; Staines and Scott, 1994). In the days when walkers were few and stalkers were numerous, friction was rare, but as that balance has reversed so conflict has increased. In the worst cases, disruption of stalks makes it difficult to achieve the increased annual culls for which the DCS is pressing (Section 7.6.1), and estates may lose stalking clients. Animal welfare considerations have also been raised, such as the possibility of disturbance of deer at calving time. Although hillwalking and stalking co-exist with little friction throughout much of the deer range, in popular areas with concentrations of Munros many estates have written off whole areas as effectively unstalkable (ADMG, 1997). Antagonism can develop between walkers and stalkers as both can have days 'ruined' – walkers if they are prevented from climbing their chosen hill, stalkers if they have a blank day due to human disturbance. This is regrettable because hillwalkers and stalkers have a considerable amount in common. Both share a love of the hills in all their moods, possess necessary mountain-craft skills and fitness, and are 'bagging types', whether it is a Munro or a stag that is in the bag at the end of the day. Most interactions are entirely amicable and considerate (Herries, 1998; Taylor and MacGregor, 1999). However, as is so often the case, an errant (and vocal) minority greatly aggravate the situation and polarise the debate. It is probable that both sides exaggerate the degree of conflict and paint the 'opposition' in a bad light to further their own ends, choosing to ignore the fact that it is certainly possible (if not always easy) for stalking to be successfully combined with a policy of open access. This has now been demonstrated, for example, at Creag Meagaidh, Abernethy, Mar Lodge, Kintail and Affric (Johnston, 2000).

Another concern is footpath erosion. Paths up popular mountains can become badly eroded, in a few cases creating scars which are visible for miles (Fig. 9.3), while the low ground approaches to many hills have become unsightly, unpleasant morasses. This is the inevitable result of the passage of thousands of boots and mountain bikes

FIGURE 9.3 *An aerial view of footpath erosion scars on Stac Pollaidh, Assynt. Photo: © Patricia & Angus Macdonald/SNH.*

across ground which frequently consists of saturated, erodible soils. As numbers of walkers have increased, so footpaths have grown in width and number, a pattern seen in the Cairngorms particularly (Lance *et al.*, 1991; Watson, 1984, 1991). The result is that large amounts of money now have to be spent on path construction, repair and maintenance to counteract the effects of decades of neglect. Costs vary widely, but repairing a high-level footpath costs about £100 per metre, while each metre of a steep, stone-pitched path can cost nearly £200 (Johnston, 2000). Much of this public-spirited work is done by charities. In the five years to 2000, for example, the Footpath Trust completed 150km of mountain paths at an average cost of £17,000 per kilometre,[2] and during the 1990s the NTS spent £350,000 on the Ben Lomond paths (Johnston, 2000). In 1999 the John Muir Trust estimated that restoring the badly eroded path on Schiehallion would cost around £150,000. Given such daunting costs and the limited funds available, it is not surprising that many paths continue to deteriorate.

9.2.2 Unquantifiable impacts

These are no less serious for being hard to measure precisely (Table 9.3). Damage to the natural heritage interest is a significant example which can take many forms (Sidaway, 1994). The trampling of sensitive vegetation and soils, especially on fragile montane heaths or wetlands, can lead to habitat loss and erosion. Persistent disturbance of fauna (especially by dogs) during breeding can impact on breeding success and lead to the redistribution and/or reduction of populations; ground-nesting birds like dotterel (*Charadrius morinellus*) are particularly vulnerable. Increasing access on upland moorlands affects many moorland breeding birds (notably raptors) which are easily put off their nests (Phillips, 2001). Natural heritage damage can also include pollution and a variety of hydrological and geomorphological effects (Sidaway, 1994). Although recreational impacts on the natural heritage are relatively insignificant compared, for example, to those resulting from land use change, they can cause serious and persistent problems locally. Also of concern are illegal activities such as the removal of biota (by egg collectors for example) which can reduce species diversity and/or population levels.

A further intangible impact of increasing use of the countryside is that the highly-prized qualities of wildness and solitude – the very qualities which bring so many to remote places to get away from it all and 're-create' – may be compromised by the presence of other people. This loss of amenity affects owners and the public alike, but few see it as a major problem (Mather, 1998; Taylor and MacGregor, 1999). There seems to be 'a tacit acceptance that some loss of wildness is an inevitable accompaniment of freedom of access' (Mather, 1997: 197). Other impacts include annoyance arising from obstructive car parking, gates being left open, and public use of private roads. These can drive farmers to distraction, especially when added to the more serious problems of vandalism, out-of-control dogs, theft, poaching, fly-tipping and litter (Costley, 2001). Most affected are those living in the vicinity of the major urban areas, and those who manage land in the more popular upland regions.

9.3 RECENT DEVELOPMENTS

9.3.1 Access initiatives by SNH

Since its creation in 1992, SNH has actively sought solutions to the problems associated with access for informal recreation (SNH, 1992, 1994c, 1998e). This has involved working on three fronts: facilitating improved access on the ground, brokering agreements between hitherto warring parties, and developing recommendations for new access legislation (Section 9.4.3). An example of the first is the 'Paths for All' initiative, and the Access Concordat is a shining example of the second. Both are described below. Another facet of SNH's practical access work is its oversight of Scotland's 731km of long distance walking routes, such as the West Highland Way, the Southern Upland Way and the new Great Glen Way (SNH, 1997b, 2000a). These mostly follow the old drove roads and other traditional routes through the uplands.

9.3.1.i Low ground access: the Paths for All Partnership

Launched in 1996, the Paths for All Partnership (PFAP) aims to improve public access to the countryside around towns and villages, not only for walkers but also for cyclists and horse riders. For many years, much of the media coverage and rhetoric surrounding access issues has focused on the uplands, and this has fostered the misperception that access problems are restricted to upland areas. More recently, it has been realised that less than 10 per cent of all walks are taken in the hills and that most of the demand for access is in fact in the farmland and woods close to where people live (SNH, 1994c; Mackay, 2001). The PFAP was developed in response to this realisation, to focus resources where the needs are. The Partnership, promoted and substantially funded by SNH, consists of eighteen diverse organisations representing land managers, recreational users and national agencies. The ambitious aim of the PFAP is to have established 100 well managed path networks by 2006; the total number of planned networks in all thirty-two local authority areas already exceeds this (PFAP, 2000). In 1999/2000 alone it spent £404,000 on footpath work (PFAP, 2001). In addition to the physical provision of paths, it works hard in promoting and facilitating their use, in obtaining funding, and in education and training. If it manages to maintain its momentum and realise its vision of 'path networks criss-crossing the country like gossamer webs' (Magnusson *in* PFAP, 2000:2), people and communities across the country will benefit substantially.

9.3.1.ii High ground access: the Access Concordat

Arguably one of SNH's most remarkable achievements in this field was its brokering of the Concordat on Access to the Hills, launched in 1996 (Table 9.4). Achieved through the work of the Access Forum, it was a bold attempt to establish tolerant communication between the interested parties, and to build a consensus allowing them to sign a general statement of access principles by which all would agree to abide. Given the diverse (or even diametrically opposed) perspectives and objectives of the ten organisations comprising the Access Forum, it was something of a miracle that any form of words was agreed. Much hard bargaining and no less than eight drafts were required, but the final product was widely welcomed as a breakthrough after years of acrimonious stalemate. Significantly, in explicitly accepting that with rights come responsibilites, it constituted a blow against the prevailing culture of dutiless rights. Nevertheless, though lauded by its proponents as a triumph for the voluntary principle (Section 13.3.5), it was criticised by its detractors as a spineless compromise. The Access Forum, created in 1994, continues as a standing 'liaison committee' to handle issues as they arise, and to produce codes of good practice and proposals for new legislation (SNH, 1998e). Commenting on the launch of the Concordat, Warren (1999c) observed that peace had apparently broken out on the centuries-old access battlefield, but this proved to be prematurely optimistic. Trailblazing though the agreement was, it is shortly to be superseded by new legislation which has stirred up renewed controversy (Section 9.4.3).

Table 9.4 The key elements of the 1996 *Concordat on Access to Scotland's Hills and Mountains*. Access Forum membership has broadened since 1996 to become more representative of lower ground access. These newer members are in bold type. Operating under the same chairman is the linked Access Forum for Inland Water; it includes a similar range and balance of organisations but with an emphasis on water resource interests (cf. Davison, 2001).

The Concordat	Access to the hills for informal recreation should be based on: • free access exercised responsibly and subject to reasonable constraints for management and conservation • acceptance by visitors of the needs of land management • acceptance by land managers of the public's access expectations • acknowledgement by all of a common interest in protecting and enhancing the special qualities of the ulands
Making the Concordat work	Good practice should involve the following: • courtesy and consideration of others • welcoming visitors • making good advice and information readily available • respect by visitors for the welfare needs of livestock and wildlife • adherence to codes of good practice by all • restrictions on access should be for the minimum period and area, only when essential, and should be fully explained
Members of the Access Forum	ADMG, COSLA, **FC**, MCoS, NFUS, RA, SCAC, **SCU**, SLF, SNH, **SRWS**, SSA, SPC, **STB**

9.3.2 Information provision and grant availability

Although very few walkers consciously set out to disrupt rural activities, lack of information often leads to conflict which could have been avoided. Uncertainty about where and when access is permitted is also known to be a factor limiting recreation by the general public. Land managers and interested organisations have worked hard in recent years to bridge this information gap. A number of approaches have been adopted, including explanatory signs, maps, leaflets and guidebooks, all of which are tried and tested means of informing and educating recreationists. However, while these offer general guidance, sometimes information is needed on a daily basis. For example, unless hillwalkers know where stalking is taking place that day they cannot choose to avoid it; they are unlikely to react positively to a sign asking (or telling) them to keep off the hills throughout the stalking season.

This specific problem has been tackled with the Hillphone scheme, coordinated and partly funded by SNH. This provides contact telephone numbers for estates or Deer Management Groups so that walkers can easily obtain information about estates' stalking intentions, enabling them to plan routes which will not conflict with deer management. It was trialled during the 1996 stalking season, and has since been extended to twelve areas. Of course, making information available does not ensure that it will reach all those who need to know, but it is a helpful and necessary step. With the explosion of information technology, it may not be long before the internet becomes the key way of making timely information available.

If lack of information has afflicted the walking public, lack of funds for necessary footpath maintenance has been a long-standing problem for land managers. Many estates provide, in effect, a public service (albeit unintentionally and often reluctantly) by allowing the public to use their networks of paths and tracks. Recognising that this public use imposes significant extra costs on landowners, grants have begun to be made available for access-related expenditure such as footpath repairs and signposting. Such payments, mostly on a 'matching funding' basis, can now be obtained from SNH and the Forestry Commission (FC), as well as from the likes of the Scottish Mountaineering Club and the Brasher Trust.

9.4 MAIN DEBATES

9.4.1 How to preserve forest access?

In contrast to the fragility of much of the uplands, forests are robust, able to withstand frequent use and to absorb large numbers of visitors (SNH, 1994c). Demand for public access to state forests arose early in the life of the FC, leading to the creation of six National Forest Parks since 1935 (Pringle, 1995) and, more recently, nine Woodland Parks. The popularity of forest recreation is high and growing, ranking second only to seaside walks (STS, 1996). About two million people who live in Scotland visit woodlands as part of a recreational trip every year and many visit frequently (FC, 1999a).

Public access had long been guaranteed in FC forests, subject only to restrictions during forest operations. However, in 1981 the Thatcher administration obliged the FC to begin selling off land to the private sector, setting it a target of raising £20 million annually from woodland sales. Since public access is rarely encouraged in private forests, access rights were mostly lost on sale. Public access brings with it many negatives for private owners, including risks of fire, open gates, litter, vandalism, liability during forest operations and interference with sporting interests (Christie-Miller, 2000), and since there are usually no tangible compensatory benefits, they have little incentive to facilitate public access. Nevertheless, the larger, traditional private estates welcome large numbers of people, and the Timber Growers Association has a 'walkers welcome policy', so almost as many people visit private forests as state forests (SNH, 1994c).

Because of the FC's continuing programme of disposals, there was mounting concern throughout the UK in the 1980s and 1990s that forest recreation was becoming increasingly restricted. The challenge was to find ways of safeguarding forest access. There are at least four possible ways of doing this:

1. **Access agreements**. During the 1980s the FC made no attempt to safeguard access in the forests that it sold to the private sector, but it was then empowered by the government to enter into access agreements with local councils. The government hoped that this would offer a solution, but this proved a vain hope. Councils were reluctant because of the costs involved and because such agreements were largely unenforceable, while owners were reluctant because access agreements reduced the market value of the wood. Between 1981 and 1996 the FC sold 75,500ha (10 per cent) of the area of its forest estate and 48 per cent of its woodlands in Scotland (RA in Lean, 1996). Of the 820 woodlands sold, only a handful had access agreements, and since they only applied to the first time buyer anyway they hardly inspired confidence that access would remain guaranteed. This approach was palpably failing.
2. **Grants**. Stung by such accusations, from the mid-1990s the FC introduced a series of woodland grants which were conditional on permitting access. Some, such as the Woodland Improvement Grant, have public recreation as a main aim. While helpful in some localities, the overall trend of lost woodland access through on-going FC sales continued. Despite the incentives on offer, it was perhaps unrealistic to expect large scale new recreational opportunities to develop in the private sector (HGTAC, 1998).
3. **Sale conditional on guaranteed access**. This option is partly in force already. Before the FC can sell a woodland which is currently used for recreation, it is now obliged to ensure continuity of access through agreements with local authorities.
4. **Public rights of access to private forests**. In many European countries, the law provides for a general right of access to all forests, whether state-owned or private, balanced by safeguards and rights for private owners (PSPS, 1991). Hitherto, this has not been the case here, but the proposed new right of access (Section 9.4.3) includes private woodlands (except newly-established plantations).

The first two of these options, practised into the late 1990s, were considered to be entirely inadequate by the ever more vocal access lobby. In 1998 the newly elected Labour Government responded to the mounting chorus of criticism by halting sales of state forests. The new policy not only ended the large-scale disposals of forest land, thereby safeguarding public access, but freed the FC to purchase land for community use. With the combination of this policy turnaround, the FC's emphasis on access, and the inclusion of woodlands in the new general right of access, forest access looks set to be much more secure in the present century than during the closing decades of the last.

9.4.2 How should access be managed?

Before addressing this question it should be recognised that in some quarters divisions remain over the more fundamental issue of whether any restrictions on public access are acceptable. At one end of this polarised debate stand the old-school landowners. In their view, land ownership should confer absolute freedom to manage access, including the option of complete prohibition. At the other extreme are those who believe that the 'right to roam' is an absolute right because the land belongs to the people; any infringement of this right is unacceptable.

> Just as the right to free speech is regarded as a fundamental right, so should access. . . . Freedom to roam is the natural condition of humanity. Like many other freedoms, it has been sacrificed on the altar of exclusive property rights. (Wightman, 1996a: 198, 199)

Such an attitude is not new. In 1890 the president of the Scottish Mountaineering Club stated that 'the love of scenery and the hills is implanted in the heart of every Scot as part of his very birthright' (in Smout, 1993c: 14). The incompatibility of such views illustrates why the Access Concordat was essential, why it was difficult to achieve, and why its launch represented unprecedented and remarkable consensus. Thankfully, such extreme positions are now held by very few; it was the views of the moderate majority which prevailed. Most now realise that an unlimited right to roam could have no reality. After all, even the right to free speech has its limits (based on content). The Concordat enshrines the principle that it is acceptable to restrict access for certain essential land management and conservation purposes. Once this principle is accepted, the next question is the practical one of how such restrictions should be operated.

Several options exist. The first is to limit the times when access is permitted. This is routinely done by forest owners during felling operations, by sporting estates during the stalking season, and by nature reserves during vulnerable periods such as the nesting season. An additional or alternative approach is to manage the routes which walkers adopt. This can be done by forbidding or discouraging access to certain areas and recommending alternative routes or destinations. For example, walkers can be asked to keep to paths to avoid disturbing wildlife or damaging fragile vegetation. One dimension of this approach, and one supported by many walkers, is the use of 'the long walk in' (Kempe, 1994; Mather, 1997; Taylor and MacGregor, 1999). The thinking behind this is that an effective way to protect a vulnerable upland area is simply to make it hard to reach. If the only means of getting there is a long, arduous trek then the 'attrition of distance' will mean that only a small number of keen, fit walkers will reach it. This approach has been adopted in the Cairngorms, for example, as a tool to manage visitor pressure (Cairngorms Partnership, 1997). Both these options – restrictions in time and restrictions in space – are established and accepted by most (Mackay, 2001).

Two further possibilities stir up much greater controversy. The first is to charge for access. This could be described, very literally, as the *'quid' pro quo* option. Many recreationists are prepared to pay for services provided, such as toilets and car parks (although walkers and climbers are notoriously reluctant to pay even for these), but asking people to pay simply to gain access to the hills would be vociferously resisted. Voluntary donation boxes in the English Lake District are helping to defray path maintenance costs, but a proposal in 1997 to charge walkers one pound to ascend Ben Nevis created a storm of protest because many regard the long-standing tradition of free access as a fundamental right.

The issue at stake here is essentially one of equity. Providing the four essential components of access infrastructure – paths, parking, signage and litter removal – is expensive. For example, in 1998/99, major landowners in the Cairngorms (including some public bodies) spent £1.27 million on facilitating outdoor recreation and access, only a quarter of which was grant-aided (Cairngorms Partnership, 2000). Who should be liable for these costs? Landowners cannot benefit from most forms of access, yet they suffer the impacts. Should they bear the brunt of the expense on their land, accepting it as one of the many unavoidable and costly responsibilities of landownership? Or, mirroring the 'polluter pays' principle, should there perhaps be an 'eroder pays' principle (sometimes dubbed a 'boot tax')? If such a principle were to be accepted, the practical problem would be how to collect such payments – directly from walkers (somehow) or indirectly from general taxation? Both have obvious pros and cons, but there is a general belief that the costs of access should be met from the public purse (STS, 1991). Similarly, SNH (1998e) believes that directly charging walkers for access per se is wrong in principle, and recommends that access provision by landowners should be substantially assisted by public bodies and Lottery funding. But the day may yet come when there is no such thing as a free walk.

The final and most controversial concept – one which lurks abhorrently in a possible future – is the introduction of a quota and permit system for the most popular areas. In popular American national parks such as Yosemite and Yellowstone such rigorous control became necessary years ago, with requirements for advance booking and permits for wild camping and off-trail walking. This is the 'nuclear option' for access management, and one which no one wants to see operating in Scotland. The threat of it is a strong incentive to make existing arrangements work effectively for people and the environment. As Thomson (2001) explains, the challenge for access managers is to tread the line between inadequate management and heavy-handed approaches which destroy the sense of freedom that people seek in the outdoors. In a rapidly changing and highly diverse country, striking the right balance between under-managing and over-managing will be far from easy.

9.4.3 Should there be a legal right to roam?

This is the fundamental issue that has always been at the heart of the access debate – the clash alluded to at the start of the chapter between property rights and the rights of the general populace. It is often said that there is no law of trespass in Scotland, but although this is true *de facto* it is not true *de jure*. Although there is a long-standing *tradition* of access to open country, some forms of trespass are civil wrongs, and wild camping and lighting fires on someone else's land remain criminal offences under the 1865 Trespass Act (Scottish Executive, 2001b). However, the current situation is a confusing grey area. Although no one has a general right to be on someone else's property, they are usually committing no wrong just by being there; trespassers cannot be prosecuted. The situation regarding access to inland water is also arcane and confusing (SNH, 1994c).

The access lobby, spearheaded by the RA and given passionate and articulate voice by Marion Shoard (1997, 1999), has long campaigned for a legal right of access to open land throughout the UK, subject only to restrictions to protect privacy and essential land use needs. Scottish landowners have always argued that this would be a fundamental encroachment on property rights. As Lyddon (1994: 197) presciently observes, 'rural property is increasingly being seen as common property', a trend which landowners understandably resent and resist. The SLF (1993, 1995) has long believed that the law has no place in access management, arguing that legal rights generate confrontation and create more problems than they solve. In its view, access management needs a sensitive touch, not the heavy hand of the law. There has been considerable sympathy for this argument amongst the interested organisations, and SNH (1994c: 41) initially eschewed a legislative solution in favour of 'a vigorous effort to promote a voluntary approach'. More recently, many argued that legislation should only be considered as a last resort after the hard-won Access Concordat had been given time to fulfil its very positive potential (SPRM, 1998). The reality, however, is that the Concordat has (regrettably perhaps) been overtaken by political events, and by a growing conviction that new access arrangements are needed for *all* of rural Scotland, rather than just for the hill ground issues addressed by the Concordat.

There are some clear arguments in favour of access legislation (SNH, 1998e; Mackay, 2001; Davison, 2001).[3] These include:

- The confusing, unsatisfactory and outdated nature of the current legal position (whereby the public have no clear rights and landowners have no powers of redress). The lack of clear and secure rights makes it difficult to promote responsible behaviour or to justify public investment.
- The use of land for recreation has no clear basis in law, and existing legal powers to promote access for this purpose (through access agreements, for example) have been little used.
- The inadequacy of the framework for managing and meeting the costs of public access.
- The lack of permanence of most access provision (typically dependent on landowners' consent), and the lack of information and a sense of welcome in much of the countryside.
- The congruence between encouragement of access and the wider public policy agenda (such as social inclusion, public health and quality of life).

- The widespread perception that the voluntary principle (Section 13.3.5) has not worked effectively in recent decades.

For these and other reasons, and after wide-ranging consultation, the Access Forum concluded that new legislation was, after all, the best solution (SNH, 1998e). Accordingly the Scottish Executive is fulfilling one of Labour's manifesto pledges by enacting access legislation, incorporating it within the Land Reform Bill which is expected to be enacted in 2002 (Scottish Executive, 2001b; Section 3.4). The inclusion of access proposals within the Land Reform Bill was regarded as a mistake by many, not least the Access Forum. Land reform and access are both significant and extremely complex matters, arguably deserving separate legislation, but political agendas prevailed. The new law will avoid the contentious and unqualified phrase 'right to roam'. Instead, it will introduce a general right of access to land and inland water, exercised responsibly, for recreation and passage. This right will apply to enclosed land as well as open and hill ground, but will be qualified by the new Scottish Outdoor Access Code developed by the Access Forum (Table 9.5). This code stresses that access should be taken responsibly, with respect for the privacy of those who live and work in the countryside, and it specifies just what comprises 'responsible' behaviour. The legislation will place obligations on local authorities and land managers to facilitate access, but will provide safeguards for the operational needs of land managers, and allow for constraints to satisfy conservation needs. The Bill provides for the creation of core path networks, for the establishment of local access forums, and promises extra funding for access provision.

It seems that the new access legislation will address many of the deficiencies of the current situation. Unsurprisingly, however, the far-reaching nature of the proposals has provoked considerable controversy. Landowning and farming interests in particular have expressed great concern over the potentially damaging implications for rural land use (ADMG, 2001; Ross, 2001), and the NFUS was so dissatisfied with the draft Bill that it withdrew from the Access Forum. Specific worries relate to non-pedestrian access, liability, access by commercial groups, night access and wild camping, but perhaps of most concern is the apparent lack of any effective means of ensuring that 'responsible access' is what actually takes place, or any effective deterrents or sanctions if it does not. The perception in these quarters is that an injustice is being perpetrated, whereby walkers are to be given legal rights, whereas the interests of land owners and managers are to be protected merely by an advisory code which many fear will be known by few and flouted with impunity. A right of access is an easy concept to grasp and exercise, whereas a duty of responsibility is far harder to define and police, especially since irresponsibility is in the eye of the beholder to some degree. If the new arrangements are to work, it will indeed be crucial to ensure that actual behaviour on the ground reflects the precepts of the Code, yet no one thinks that this will be easy to achieve. Quite the contrary, in fact, not least because all parties to the Access Forum are adamantly opposed to any new criminal sanctions (Thomson, 2001). From the other end of the spectrum, the access lobby perceives the new law as an overdue (and inadequate) correction to the long-standing injustice of the exclusion of people from

TABLE 9.5 Key elements of the Draft Scottish Outdoor Access Code. For full details, see Scottish Executive (2001c).

Key principles	• The exercise of the right must be integrated with farming, forestry, field sports, conservation and recreation • All forms of damage are to be avoided, and disturbance to be minimised • People exercise the right at their own risk, and must respect others' privacy
The extent of the right of access	• The right includes all land and inland water • It is for the purpose of open air recreation and passage • It can be exercised by individuals, groups or clubs, day or night
Qualifications on the right of access	The right does not apply to: • buildings, curtilages or restricted areas (e.g. military areas; airfields) • field sports (e.g. stalking; shooting; fishing) • motorised recreation (e.g. off-road driving; water skiing; jet skiing) • picking, uprooting or destroying plants (e.g. fungi; wild fruits) • provision of facilities for large, organised group events • irresponsible and criminal behaviour
Responsibilities of the public	The right must always be exercised responsibly. This means: • thinking ahead about how to minimise your impact and being prepared to change your behaviour • respecting people's privacy and peace • taking personal responsibility for your own actions • respecting the needs of others enjoying the outdoors • helping land managers to work safely and effectively • keeping dogs under close control • not damaging the environment, disturbing wildlife or leaving litter • putting something back into the outdoors • following the Scottish Outdoor Access Code and any local guidance
Responsibilities of land managers	• to influence, sensitively, how people exercise the right of access • to be more aware of the impacts of operations on the public • to facilitate access whenever and wherever it is reasonable & practicable • to work in partnership with public bodies • to follow the Scottish Outdoor Access Code and any local guidance
Responsibilities of public bodies	• to make full use of their duties and powers • to meet their responsibilities as land managers • to secure adequate funding for access work • to set standards and monitor their performance • to show courtesy and consideration through their staff • to follow the Scottish Outdoor Access Code

large parts of their own country. Also concerned about the likely effects of the proposals are some nature conservationists. Lister-Kaye (2001), for example, argues passionately that there are times and places where nature conservation and public access are incompatible, and that effective environmental protection necessitates some no-go areas. There is clearly a danger that if new access arrangements reduce the incentive of those who manage the land to conserve it, a new version of the tragedy of the commons could result (Holdgate, 2000b).

The current proposals also raise serious funding issues. Significant financial and human resources are required if the existing access infrastructure is even to be maintained, let alone developed. SNH estimates that making the proposed arrangements work effectively will cost around £16 million per annum, partly in the provision of a 165-strong ranger service, but mainly in improving the provision and management of path networks (including the necessary signage and information). Crucially, there will also be the challenge of carrying out public promotion and education on a hitherto undreamt of scale. Complementing the work of the rangers, the role of estate stalkers could well evolve from an exclusive focus on deer management to incorporate some rangering functions. The Access Forum and SNH are clear that insufficient funding could fatally undermine the new arrangements (SNH, 1998e), but finding the money to put them in place and maintain them may not be easy. Local Authorities will also need significant funds to fulfil the access obligations which the new law will lay upon them, a need which the Scottish Executive has recognised in providing them with an extra £13 million over the period 2001–4 for improving access management. Local Authorities are also considering other ways of raising funds such as the reintroduction of sporting rates, a proposal which horrifies private owners.

The compromise called for since 1884 is finally being hammered into legal existence, as rights are being balanced by responsibilities on both sides of the (literal and metaphorical) fence. That access legislation is coming is now beyond doubt, but as Scott (2001: 212) comments, 'legislation could either happen well or happen disastrously'. It must be hoped that the balancing act proves to be sustainable, both sociopolitically and environmentally, and that the remarkable but fragile consensus built by the Access Forum turns out to be durable and fruitful. Williamson (2001: 68) suggests that 'our boxed, stereotypical, and polarised thinking on access has not produced any real progress in the last seventy years or more'. Some would bridle at this pessimistic verdict, but few would disagree with his conclusion that what is now needed is new and properly integrated thinking in place of the compartmentalised, reductionist approach that has hindered conflict resolution for so long. Certainly, the new access arrangements amount to a change of culture (Mackay, 2001). Making them work on the ground will require responsible, considerate and generous behaviour by all involved.

NOTES

1. Smith (1997) outlines the history of access issues in Scotland and discusses recent developments. Detailed and colourful accounts of the history of access struggles throughout Britain are provided by Stephenson (1989) and Shoard (1997, 1999). As well as being very well informed, Marion Shoard's books constitute powerfully argued and impassioned pleas for a right to roam. Current access issues in Scotland are discussed in the volume edited by Usher (2001).
2. Data reported in *SCENES* 147: 8 (2000), announcing the cessation of the charity's work.
3. Davison (2001) gives a detailed account of the development of the new legislative proposals.

CHAPTER TEN

Natives, exotics and reintroductions: what species where?

For most of history, the only criterion by which human beings judged other species was their usefulness, but in recent centuries other dimensions became important, so that certain species came to be valued for their attractiveness, their novelty or their potential for game sport. European explorers and colonisers took animals and plants to distant corners of the globe, and collectors brought exotic species back, with no thought to the ecological ramifications of this artificial mixing. Even extinctions were not widely condemned, being regarded as an unfortunate by-product of the inexorable march of human progress, or even (in the case of certain predators) as a desirable objective. Only in the second half of the twentieth century, long after the process of global species mixing had passed the point of no return, did a concern for the integrity of native species and habitats start to become widely shared.

In Scotland, native species are usually taken to be those which have colonised naturally during post-glacial time (the last eleven millennia), and exotic (or alien) species as those which have been introduced by human agency. The two terms are often used as polar opposites, but they actually represent opposite ends of a continuum (Table 10.1).[1] Exotic species are commonly subdivided into archaeophytes and neophytes,

TABLE 10.1 Definitions of the terms 'native', 'exotic', and four intermediate categories. Based on Usher (1999b).

Classification	*Definition*
Native	Species which have arrived since the last glacial period without human aid
Formerly native	Species no longer present but which occurred naturally during post-glacial time. Subdivisions: 1. Those which could survive if reintroduced. 2. Those which could no longer survive in today's environment & climate
Locally non-native	Species introduced by humans beyond their natural geographical range
Long-established (or 'naturalised')	Species introduced by humans long ago which are now part of the food webs of native species
Recently arrived	Species colonising as a result of human activities (eg. land use practices; human-induced climate change)
Exotic	Species introduced by humans, either deliberately or accidentally

those introduced before or after AD 1500 respectively, a date selected to represent the beginning of the modern era (Dickson, 1998). The often emotive debate surrounding the rights and wrongs of native *versus* exotic species touches on almost all aspects of environmental management. It is one of the most complex of all conservation issues. After a general discussion of this debate, this chapter explores three aspects of it in the Scottish context: the place of native tree species and native woodland (Section 10.2); the threat to the genetic integrity of native red deer from hybridisation with exotic sika deer (Section 10.3); and the controversies concerning reintroductions of extinct native species (Section 10.4).

10.1 THE RIGHTS AND WRONGS OF NATIVE AND EXOTIC SPECIES

There is great power in the twin ideas of righting past wrongs and of recreating a 'paradise lost'. Given the pervasiveness of the human impact on the Scottish environment (Section 1.2.2), many argue that there is both an ecological and a moral case to be made for putting the clock back and reversing some of those effects. At the simplest level, the ecological case rests on the belief that 'nature knows best' – that native species are best adapted to the environment and exist in a natural balance with each other. The moral argument is that since the blame for environmental disruption lies at our door, we are duty-bound to do something about it. Taken together, such views underlie the pervasive 'bias in favour of the natural' in conservation circles, and the widespread desire to see native species conserved, lost environments restored, lost species reintroduced and exotic species removed. Increased 'naturalness' is the explicit or implicit rationale for much conservation management, and is the imperative which now sets the agenda in many sectors of environmental management. The often unquestioned assumption is that native and natural are good, while exotic and artificial are bad. However, this dualistic scheme is simplistic and hard to sustain. It ignores a raft of ethical, philosophical and practical dilemmas about the place and influence of *Homo sapiens* in the world (Budiansky, 1995; Worrell, 1997) (Section 13.3.1). It also fails to recognise that management decisions very rarely consist of easy black and white choices. Reality consists of many shades of grey.

This is not to imply that differentiating between native and exotic species is unimportant. On the contrary, there are strong arguments supporting a preference for natives, the authentic 'locals' who 'belong' in an area. Because of their long-standing presence, native species are well adapted to the environment and exist in a dynamic (not static) equilibrium with the other components of the ecosystem. Native plant species typically host a greater number and diversity of fauna than introduced species. Equally, there are many dangers associated with exotics, quite apart from a principled objection to their alien origins. Indeed, the UN Environment Programme regards invasive aliens as one of the greatest threats to the planet's biodiversity, second only to habitat loss (MacKenzie, 2001). The IUCN states uncompromisingly that invasive alien species are now recognised as 'one of the greatest threats to the ecological and economic well being of the planet' (McNeely *et al.*, 2001: viii). The unstoppable wave of global species mixing 'threatens to homogenise the world's ecological assemblages into one giant mongrel ecology', argues Hettinger (2001: 216), replacing biodiversity with biosimilarity. He suggests that an important rationale for excluding exotics is in order to preserve difference, in the same way that people resist the homogenisation of human cultures through creeping 'McDonaldisation'.

By no means all exotic species are damaging. In fact, their presence can sometimes have ecological benefits. Nevertheless, there are innumerable instances worldwide in which introduced species are causing severe environmental and economic problems (Hettinger, 2001; McNeely, 2001; Woods and Moriarty, 2001). In particular, impoverished island biotas such as those of the UK are notoriously vulnerable to invasion by exotic species which frequently damage the nature conservation interest, not least by posing a threat to native populations (Kirby *et al.*, 1997; Holdgate, 2000b; Peterken, 2001). Contemporary Scottish examples include the dwindling range of the red squirrel following the arrival of its grey cousin (Section 7.7.1), the damage being caused by mink on the Western Isles (Section 7.7.2) and the threat to powan in Loch Lomond from introduced ruffe (Section 6.5.5.ii). Other damaging exotic species in Scotland include giant hogweed, New Zealand flatworms and rhododendron, the last of which is spreading rapidly through western woodlands, obliterating ground vegetation and preventing tree regeneration. Epidemics of introduced diseases can make short work of native species which have no resistance to them. This is exemplified by the outbreak of infectious salmon anaemia in 1998 and the newly evolved hybrid blight afflicting Scotland's native alder populations. An audit by SNH found that at least 992 alien species now exist in Scotland (Welch *et al.*, 2001). Most of these are vascular plants introduced as garden outcasts and not widely distributed, but there are 49 bird species, 16 fish species and 13 mammal species of alien origin. Of the introduced species, 76 are regarded as 'potential problem species', and no less than 10 of the 13 mammal species are judged to have damaging impacts. As a consequence of the degrading (and often hugely expensive) impacts that exotic species can have, they are almost universally regarded as 'one of the major ecological evils that environmentalists are called upon to resist' (Hettinger, 2001: 194). Because of their potential to wreak havoc, ecological wisdom dictates that native species should be the much preferred 'default option'.

But that said, any hard and fast scheme which damns all exotics and sanctifies all natives faces a phalanx of hard questions. To try to base management decisions on such a scheme is to run immediately into a thicket of issues with no easy answers.

- **How far back to go?** Native or exotic status is not set in stone for all time. All native species were alien once, in the sense that they initially arrived as colonisers, and over geological timescales the floral and faunal composition of ecosystems is in a state of constant flux. Several species that are known to have been present in Britain during previous interglacial periods did not happen to get here at the start of the present one. Examples include fallow deer, rabbits, Norway spruce, hemlock and rhododendron. To choose the particular species composition of the present interglacial as the only arbiter of nativeness is an arbitrary choice, however convenient. This choice is arbitrary taking the long geological view, but it can be justified on the grounds that the most recent glaciation wiped the ecological slate almost completely clean. The complement of native species that we have is the hand that nature has dealt us in the current interglacial. In this sense the current composition is an accident of nature (Kitchener, 1998). Many long-established (or naturalised) species have now become 'functionally native' by becoming a part of the 'natural' food chain (Usher, 1999b); the rabbit, a major prey species for many predators, is a familiar example. So how long does a species have to be resident before it is given a passport?

- **What baseline to choose?** A closely related question is what moment in history or pre-history should be selected as the authentic baseline to guide our species choices today. Until when was nature natural? It seems intuitively obvious that this moment should be before the era of significant human modification, perhaps five or six millennia ago before the Mesolithic. But at that time, the Climatic Optimum (Section 1.2.1), both the climate and the floral and faunal composition of Scotland were very different, so the environment then may not be an appropriate guide for today. On the other hand, to choose any subsequent moment along the continuum from less to more modified is entirely arbitrary. Why freeze-frame one moment of history as opposed to any other? Complicating the picture still further is the fact that some species have colonised naturally in some places but have been introduced by humans elsewhere, and thus may be both native and exotic (Dickson, 1998). An example is the hedgehog, native in mainland Scotland but exotic in the Outer Hebrides where its recent introduction has proved devastating for ground nesting birds (Usher, 1999b). Considerations like this lead Bachell (2000:26) to declare that 'there really is no measure of authenticity in nature conservation'. It is based on speculation about unknowables. Equally critical is Brown (1997:196), who argues that 'the term native is misleading as it creates arbitrary and ecologically unsound distinctions'. He believes that it is time to redefine it because the 'post-glacial pre-Mesolithic' definition effectively fossilises nature, failing to take account of its dynamic attributes.
- **What scale and genetic level to operate on?** To qualify as native, should a species be naturally present within the UK, within Scotland, within a parish, or should it be (or have been) present on the specific site in question? For example, Scots pine is native to Scotland but there are places where it never grew. Does this make it an exotic in such places? To go a step further, distinct sub-populations of Scots pine with different origins exist. Should such precise genotypic characteristics be used in deciding native or exotic status? Of course, when it comes to reintroducing extinct species (Section 10.4), such genetic precision is an unattainable counsel of perfection; it is a case of finding a donor population elsewhere that is as genetically similar as possible to the extinct population. To avoid classifying such reintroduced species as exotics requires a relaxation of the genetic criteria. It comes as a shock to discover that even animals as quintessentially native to Scotland as red squirrels (Section 7.7.1) and capercaillie (Section 7.3.1), both of high conservation importance, are genetically exotic, having been reintroduced

in the late eighteenth and early nineteenth centuries from Scandinavia (Kitchener, 1998). A further example of a genetic dilemma comes from Ben Lawers where the NTS wish to conserve two remaining woolly willows which will not reproduce without intervention (Johnston, 2000). Does the inevitability of their death justify the introduction of genetically different material from elsewhere?

- **What about climate change?** At a time when climate change is set to bring profound changes to Scotland's environment and biodiversity in the coming decades (Kerr *et al.*, 1999; Section 1.2.2.iii), it seems illogical to perpetuate the concept of a fixed endowment of native species (Moffatt, 1999). Holdgate (2001:20) argues that because we live in an age of 'recombinant biogeography', facing inevitable changes in biodiversity, 'conservationists will need to stop being 'purist' and accept some invasive species we now regard as alien'. An ironic implication of the prospect of climate change is that it could return Scotland's climate to something approaching that of the Climatic Optimum (Section 1.2.1). Consequently, using the species present at that time could, after all, turn out to be both appropriate and successful (Ashmole, 2000).
- **How sure is our knowledge?** Implicit in all the above questions is an assumption that our understanding of the past is reliable, and that we know whether particular species arrived here by natural or human means. Notwithstanding the great advances in palynological, archaeological and historical research, the reality is that many uncertainties remain and that the status of certain species is debatable (Dickson, 1998; Peterken, 2001). Our understanding is full of caveats and in a state of constant flux as new discoveries challenge past certainties. The continual emergence of new information argues for a flexible, reversible approach that keeps options open, rather than the dogmatic application of hard-and-fast distinctions. In this context, the burgeoning interest in environmental history in Scotland is an important development.[2]
- **What about human preferences?** Because introduced species contribute to the overall diversity of landscapes and wildlife they are often widely welcomed by the public. Fallow deer, grey squirrels, rabbits and rhododendron are all popular in certain quarters. Exotic trees such as beech,[3] larch and the controversial sycamore have been here for centuries, support a rich flora and fauna, and are much loved by people (Fig. 10.1). Urban floras are dominated by introduced plants, mostly garden escapes (Dickson, 1998). Some introductions were very ancient; pheasants, much valued for sport, have been here since Roman times, and voles were introduced to the Orkneys by Neolithic people (Kitchener, 1998). Should such species be eradicated and human preferences set aside simply because they did not arrive on these shores naturally? Should Neolithic cultural artefacts be carefully protected but Neolithic ecological effects be destroyed? Smout (2000b) comments that the general public care more about aesthetics than ecology. How should that 'democratic preference' be weighed against the ecological bias towards the natural? Should even invasive aliens like rhododendron and giant hogweed be regarded as interesting additions to our native flora (c.f. Dickson, 1998)?

FIGURE 10.1 *Mature beech woods beside Loch Insh: an exotic species to be deplored or a welcome addition to the landscape? Photo: © the author.*

- **Is *Homo sapiens* native or exotic?** Mesolithic people arrived in Scotland around the same time as Scots pine, soon after the glaciers departed, and so could be regarded as native. On the other hand, the Scottish people are far from being genetically pure, a consequence of the repeated invasions by Scotti, Angles,

> Vikings, Danes, Jutes and English, and the many other peoples who have 'introduced' themselves from around the world. If we are native, then are not our actions as 'natural' as any other process (c.f. Section 13.3.1)? Or are we perhaps naturally alien (Woods and Moriarty, 2001)? If we are exotic, there can be little justification for persecuting other exotic species; the logical end point of a drive for naturalness would then be mass emigration. There are some who interpret calls to extirpate exotic species as an expression of racist, xenophobic eco-imperialism tantamount to ethnic cleansing (Rickman, 1994; Brown, 1997). Discussing this critique, Peretti (1998: 190) asks provocatively: 'If peaceful coexistence in a multicultural society is a good goal for humans, why not for other species?' Others argue that exotics have no right to be here at all, even suggesting that native animal species have *more* right to be here than we have (Yalden, 1999).

To many of these questions, the pragmatic answer has to be, 'it all depends'. For example, using genetically authentic native species will have maximum importance in native woodlands where conservation and biodiversity are paramount, and least significance where timber production is dominant. This place specific approach is mooted by Cooper (2000: 1148): 'In a biodiversity reserve perhaps aliens are genuinely out of place, while their freedom to expand and our loss of control over them may be just what wilderness areas are all about'. Native species are not invariably the best management choice, and neither should all exotic species be rejected out of hand in an undifferentiated fashion simply because of their alien status. A geographical distinction can be drawn between three types of exotic species (Holdgate, 2000b; Peterken, 2001): those extending their natural European range as a result of climate change ('recently arrived' in Table 10.1), introduced species which were present in previous interglacials, and introductions which were trans-continental or even trans-hemispheric (such as the wallaby population on the island of Inchconnacon in Loch Lomond). The first two may be seen as more acceptable than the third, but it is all a matter of degree. Woods and Moriarty (2001) argue that because 'native' and 'exotic' are cluster concepts with overlapping boundaries, absolutist conservation policies are misplaced.

In applying a bias towards naturalness, it is therefore quite clear that balances need to be struck between ecological and cultural criteria, between purism and pragmatism. In trying to undo what we regard as the errors of our forbears, it would surely be a mistake to allow the pendulum to swing to an equally damaging extreme of 'eco-fascism' which only welcomes those species which are somehow certified as genetically pure and native. As Smout (1999b: 11) observes, 'we should learn from history but never be its prisoner'. Such pragmatism often prevails in practice. For example, at the RSFS's millennium forest at Cashel, Scots pine has been included as a native, even though the site is outwith the accepted pine zone (Hunt, 2000). The reality, however unpalatable, is that the majority of exotic species are here to stay (and to spread), so in most contexts we should work with them, not against them because large scale 'grandslam schemes of extermination' against well established aliens are futile (Dickson, 1998: 41). As Peretti (1998: 185) says, 'attempting to keep nature 'pure', 'wild' and alien-free may be impractical, impossible, or even undesirable'. Jamieson (1995:340) even suggests that 'a celebration of alien plants and surprising biological juxtapositions may be more in tune with the postmodern world than attempts to protect native species'. Except in small areas of high conservation value, limited resources should not be wasted on doomed attempts to rid Scotland of exotics. Nevertheless, the damage that non-native species can do suggests that they should be managed in a well informed, vigilant and precautionary way.

10.2 NATURALNESS IN WOODLAND MANAGEMENT

10.2.1 'The Great Wood of Caledon': the return of the native

Scotland's native woodlands, a unique oceanic version of boreal forests, have been decimated more extensively than perhaps anywhere else in Europe except Iceland; they 'survive precariously as tattered fragments' (Worrell, 1996: 33). Woodlands of native species (Table 10.2) cover just 2 per cent of Scotland and less than 1 per cent of lowland Scotland, the total area being 320,938ha (N. Mackenzie, 1999). Of the patches which survived into the twentieth century, significant areas were subsequently cleared or underplanted for plantation forestry (MacKenzie and Callander, 1995), and this largely explains why the area of semi-natural woodland declined by about 20 per cent between the 1940s and the 1970s (Tudor *et al.*, 1994). Although felling and underplanting have not been grant-aided since 1985, many native woodlands are small and in poor condition, failing to regenerate due to grazing pressure from deer and sheep. They are dying on their feet.

TABLE 10.2 Trees and shrubs native to Scotland. Some of these species are rare or have a restricted natural distribution. After N. Mackenzie (1999).

Alder	*Alnus glutinosa*	Oak, pedunculate	*Quercus robur*
Ash	*Fraxinus excelsior*	Oak, sessile	*Quercus petraea*
Aspen	*Populus tremula*	Pine, Scots	*Pinus sylvestris*
Birch, silver	*Betula pendula*	Rose, dog	*Rosa canina*
Birch, downy	*Betula pubescens*	Rose, guelder	*Viburnum opulus*
Blackthorn	*Prunus spinosa*	Rowan	*Sorbus aucuparia*
Cherry, bird	*Prunus padus*	Whitebeam	*Sorbus rupicola*
Cherry, wild (gean)	*Prunus avium*		*Sorbus arranensis*
Elder	*Sambucus nigra*		*Sorbus pseudofennica*
Elm, wych,	*Ulmus glabra*	Willow, goat	*Salix caprea*
Hawthorn	*Crataegus monogyna*	Willow, grey	*Salix cinerea*
Hazel	*Corylus avellana*	Willow, eared	*Salix aurita*
Holly	*Ilex aquifolium*	Yew	*Taxus baccata*
Juniper	*Juniperus communis*		

Semi-natural woodlands consist predominantly of trees and shrubs which are native to the site and growing by natural regeneration (Rodwell and Patterson, 1994). The term *semi*-natural is used because none have entirely escaped the influence of human beings and their domestic stock over the millennia. Such woodlands are classified either as primary (ancient semi-natural) woodlands, sites which have always been wooded, or secondary (semi-natural), those which have become established in historical times by natural or non-natural means. These woodlands are now widely regarded as precious components of Scotland's natural and cultural heritage, especially for conserving biodiversity (Newton and Humphrey, 1997), but prior to the 1980s this was not generally recognised or translated into policy. Since then considerable momentum has built up for bringing the struggling fragments into positive management, and expanding their area either by planting native species or by encouraging natural regeneration. While timber production is sometimes included as a management objective, the enthusiasm springs from environmental concerns, the aim commonly being the re-creation of a site's natural ecosystem. The focus is thus on ecological communities, both floral and faunal, not simply on the trees themselves.

One of the first groups to advocate extensive re-creation of lost woodland was Reforesting Scotland, but during the 1990s a wide range of native woodland initiatives sprang up all over the country, operating at many scales and adopting diverse approaches (Bachell, 2000; Begg and Watson, 2000; Ashmole, 2000). Trees for Life even advocate the creation of a 2,238km^2 wild forest in northern Scotland, complete with a core wilderness zone and reintroduced mammals such as the wolf (Featherstone, 1997). As a result of such initiatives, recent years have seen the long-standing trend of fragmentation and decline being gradually reversed. In 1994 the FC published guidelines on creating new native woodlands (Rodwell and Patterson, 1994), and there is now a wide range of grant-aided mechanisms to support their creation and management (House, 2000). Since 1993 over half of all planting, restocking and regeneration schemes have used native species (N. Mackenzie, 1999). During the 1990s no less than 50,000ha of new native woodlands were created, and the Scottish Executive (2000b) is committed to a further 15,000 by 2003.

The highest profile aspect of the native woodland movement has been the drive to safeguard and restore the Caledonian pinewoods. The Scots pine (Fig. 10.2) has an almost totemic significance, symbolising natural Scotland. Ancient native pinewoods have a uniquely special aura to them. In the oft-quoted words of Steven and Carlisle (1959: v), 'to stand in them is to feel the past'. If current efforts bear fruit, it is also, in all probability, to feel the future. Scotland's Scots pine woods represent the western outpost of the boreal forest which stretches across northern Eurasia from Siberia (Fig. 10.3). They once covered perhaps 1.5 million ha of the country (Tuley, 1995) and were a key element of the celebrated (and controversial) 'Great Wood of Caledon' (Dickson, 1993; Smout, 2000a). Now the total area of Caledonian pinewood is just 17,882ha, comprising eighty-five small fragments which frequently consist of old and dying trees (Jones, 1999b) (Fig. 10.4).[4] The scattered, gnarled remnants at Glenfeshie are often held up as examples of their degenerate condition. Native pinewoods are important not just for cultural, aesthetic and recreation reasons, but also ecologically; they support many uncommon or rare species such as red squirrel, pine marten and capercaillie, and also Scottish crossbill, Scotland's only endemic bird.

The rise of concern for native pinewoods is traced by Callander (1995) from Steven and Carlisle's (1959) seminal text, to flagship conferences in 1975 and 1994 (Bunce

FIGURE 10.2 *An old Scots pine on Mar Lodge Estate. Photo: © the author.*

and Jeffers, 1977; Aldhous, 1995), and through to the publication of the Caledonian Pinewood Inventory in 1994. Grants for native pinewoods have been available from the FC since 1978, and by 1994 half of all planting and regeneration schemes in the Highlands were for native pinewoods (Callander, 1995). Much effort has been focused around the flanks of the Cairngorms where the largest remnants of native pinewood exist (Fig. 10.4). For example, the RSPB has worked hard to protect and enhance its forests at Abernethy (Taylor, 1995), and the Cairngorms Partnership has

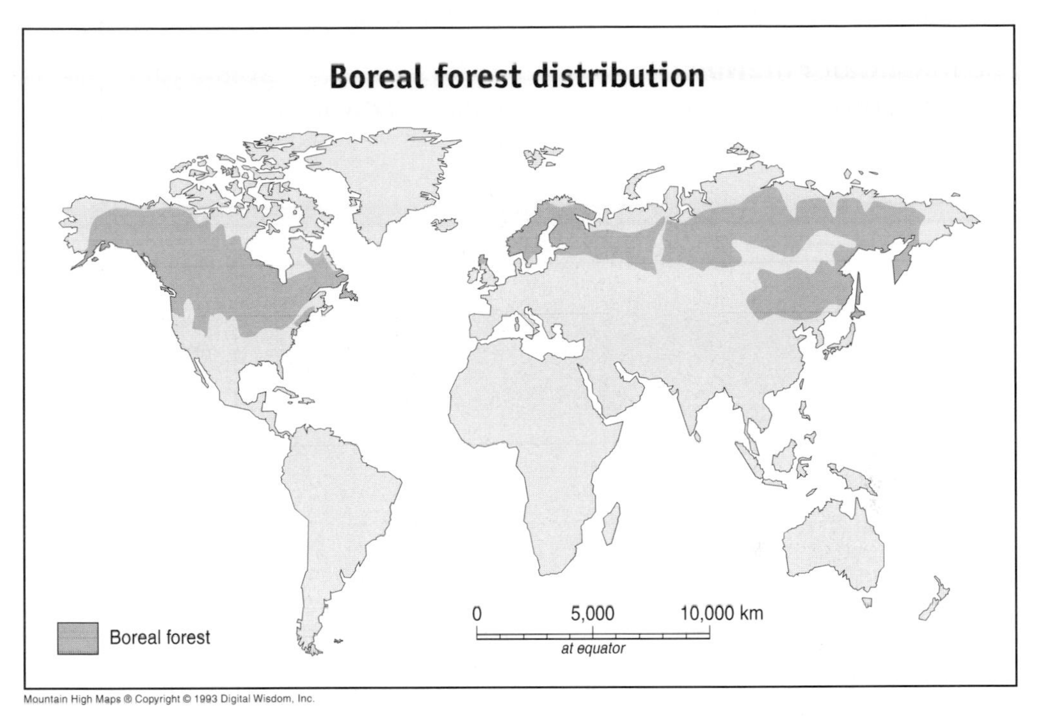

FIGURE 10.3 *Global distribution of boreal forest. After Worrell (1996).*

placed the protection and regeneration of native woodland at the heart of its forest strategy (Hampson, 1999).

It was concern for the native pinewoods which first sparked widespread interest in native woodlands, and several native pinewoods have now been designated as Special Areas of Conservation (Jones, 1999b). However, it can be argued that the focus on Caledonian pinewoods, which comprise less than 8 per cent of Scotland's total native woodland area, has distracted attention from the majority, many of which are in poor condition and in need of constructive management (MacKenzie and Callander, 1995, 1996). Recent years have seen international recognition of the need to conserve and enhance the whole spectrum of native woodland types which exist in Scotland (Hampson and Peterken, 1998), and Habitat Action Plan targets have been set for some of these (Table 10.3).

The native woodland movement has gathered considerable momentum, but it has not taken root to the same extent everywhere. The increase in the use of natives has been much less marked in the Lowlands than in the Highlands where 88 per cent of native woodlands are found (N. MacKenzie, 1999); natives comprise 75 per cent of new planting in the latter as against 25 per cent in the former, and the proportion of native broadleaves used in woodland schemes in the Lowlands actually declined in the late 1990s (MacKenzie and Callander, 1995, 1996; N. MacKenzie, 1999). This is despite the fact that losses of native woodland have been even more pronounced in south Scotland (Newton and Ashmole, 1998). In the Borders, native woodland now

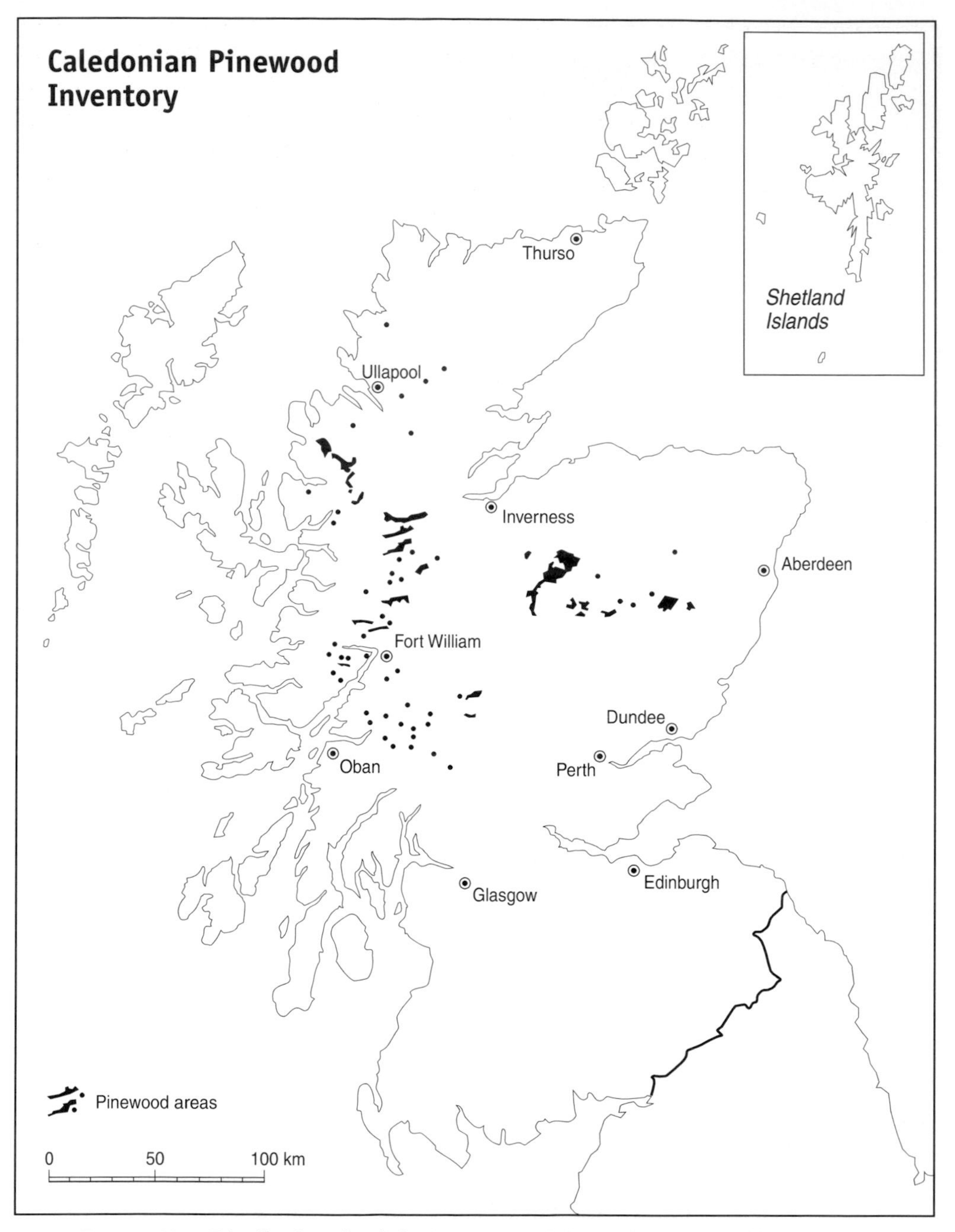

FIGURE 10.4 *Distribution of Caledonian pinewoods in Scotland. After Aldhous (1995).*

covers a mere 0.4 per cent, and the *largest* native woodland is just 24ha. The part of the country with the least native woodland is recovering slowest.

Nor is the native woodland movement free of detractors. Those advocating the recreation of pristine wildwood are sometimes accused by archaeologists and social historians of 'ecological fundamentalism' because woodlands can damage and obscure important elements of Scotland's cultural heritage (c.f. Smout, 1999a). Others criticise the focus on trees to the exclusion of other species; 'the single-minded obsession with trees that is stalking the country is very worrying' (Fenton, 1999: 69). There are also fears that the stress on native woodlands will lead to a future shortfall in timber production, and that it exacerbates the unhelpful distinction between production forests and forests for conservation and amenity (Brown, 1997). Given the fact that public preferences and market needs change far more rapidly than forestry ever can (Section 4.1), is it wise to put so many of our eggs in one basket? For these and other reasons, Wigan (1999c) trenchantly criticises what he calls the 'native theology', arguing that 'we should not be trying to force nature into a preconceived antediluvian state conjured up by inspired guesswork'. Moreover, there are a wide range of ecological, philosophical and socio-economic questions surrounding the extension of native woodland to which there are no easy answers. It is to these questions that we now return.

TABLE 10.3 The total areas of the different types of native woodland in Scotland, and restoration and expansion targets specified in Habitat Action Plans (HAPs). The need for HAPs for upland birch and lowland mixed broadleaves is being considered. After Scottish Executive (2000b).

Habitat	*Area (ha)*	*% of native woodland area*	*Restoration target (ha)*	*Expansion/ creation target (ha)*	*Completion date*
Upland birchwoods	64,000	42	–	–	–
Upland oakwoods	30,000	20	3,000	3,000	2005
Native pinewoods	30,000	20	5,500	30,500	2005
Upland mixed ash woods	12,000	8	1,600	2,200	2015
Wet woodlands	12,000	8	1,600	2,200	2015
Lowland wood pasture	–	–	150	–	2010
Lowland mixed broadleaved woods	3,000	2	–	–	–

10.2.2 Species choices: is native always best?

In the early days, the native woodland movement was motivated by the conviction that because humans had removed forest from much of Scotland in recent centuries, restoring native woodlands must be the right thing to do. Those simple certainties have now given way to a much more complex picture, partly arising from the realisation that woodland cover has been steadily reduced by the *combined* actions of people and nature for at least six millennia (Smout, 1999b, 2000b).[5] One result of this is that Fraser Darling's long-accepted notions of the Highlands being a 'wet desert', a terrain ecologically devasted by human action since medieval times, are being substantially modified (Tipping, 1993; Smout, 1997c). Nevertheless, there are many good reasons for prizing and restoring the ancient woods that mean so much to so many. Peterken (1996) sets out many of them, an interconnected range of scientific, cultural, conservation and material reasons why natural woodland is important today. But 'putting back what's been lost' is now seen as an inadequate and sometimes inaccurate rationale. Equally, the importance of distinguishing between the closely related but distinct concepts of nativeness and naturalness is also being acknowledged; exotics can be made to mimic natural ecosystems, while native species can look out of place if used insensitively. Blanket planting with Scots pine could be just as unpopular as wall-to-wall Sitka spruce plantations (Lister-Kaye, 1995).

Many of the questions identified in Section 10.1 come into sharp focus in the context of native woodland management. For example:

- **What baseline to choose?** What period of the past should we take as our guide for the future? Very different choices have been made by the diverse native woodland groups. At Glen Finglas the Woodland Trust is looking back to medieval times (Begg and Watson, 2000), whereas the objective of the Borders Forest Trust's 'Wildwood Project' is to re-establish a woodland wilderness at Carrifran based on the species present around 6,000 years ago (Ashmole, 2000). Recreating the 'natural forest cover' is problematic, requiring careful consideration of period, place and plants (Dickson, 1993; Tipping *et al.*, 2000). If, as is the case around Loch Garten on Speyside, ancient woodland is underlain by traces of even more ancient Neolithic cultivation (Smout, 1997c), should native woodland give way to native archaeology? Looking forward, should native woodland be allowed to regenerate across cultural landscapes rich in human artefacts?
- **How sure is our knowledge?** It is impossible to know which additional European tree species would have colonised naturally in the absence of humans (Brown, 1997), so the past provides an equivocal guide to what the 'natural present' might have been, or to what the future should look like. Furthermore, it is not just the details of environmental history which remain unresolved. Controversy surrounds aspects of the big picture as well, making it hard to set locational objectives for native woodland expansion. J. Fenton (1997, 1999), for example, challenges several of the accepted beliefs about ecosystem restoration. He argues that large parts of the Highlands have been *naturally* treeless for over 4,000 years, and that much of the vegetation cover, being relatively natural, is not in need of restoration. The idealised 'Great Wood of Caledon', he believes, does not need to be put back because it is a myth (cf. Breeze, 1997; Smout, 1997c).
- **What about human needs and preferences?** Many of the 1,700 tree species which have been introduced to Britain since Roman times (White, 1997) are valued parts of our landscapes, performing important aesthetic and cultural functions (Mabey, 1996). For instance, alien trees like beech and larch contribute

much to the landscape (as well as being economically valuable). Examples of non-native trees which have been culturally adopted include chestnut (the 'conker tree'), Norway spruce (the 'Christmas tree') and sycamore (with its 'helicopter seeds' beloved of children). In Scotland, the exotic Sitka spruce is widely disliked because of its association with 'mass production' forestry. Yet in its natural habitat (and when allowed to grow to maturity here) it is a magnificent tree; in its native British Columbia it has iconic status for the conservation movement (Edwards, 1999). Exotic conifers are also a *sine qua non* of an internationally competitive homegrown timber industry. The real issue is the way that introduced species are used (Bell, 1997). Equally, replacing single-purpose management for timber production with single-purpose management for conservation is unlikely to prove wise. While accepting that there will always be special exceptions, the multi-purpose ethic should probably be applied just as much to native woodland as to any other kind of forestry (House, 2000). Importantly, this is likely to be the best means of guaranteeing the long-term survival of native woods because in future, as in history, woodlands are only likely to survive if they have a use; mere historical accuracy guarantees nothing (Smout, 1999b).

- **What scale and genetic level to operate on?** Although a concern for genetic and site-specific authenticity has arisen only recently (Bachell, 2000), it has been officially adopted by the FC in its recommendation of the use of planting stock from 'local seed zones' (Herbert *et al.*, 1999). But given the extent to which human and natural influences are interwoven in today's cultural landscapes, is such a high standard appropriate? 'If there is no real purity, why be purist?' (Taylor, 1996: 13). Again, the answer will depend on management objectives, but also on further research. For example, although the distribution and population genetics of Scots pine are reasonably well understood, knowledge concerning other native species remains inadequate for genetic authenticity to be confidently pursued. Aiming high is always laudable, but in practice these fine genetic distinctions represent a counsel of perfection which is being overwhelmed by genetic pollution from the 50 million or so trees that are imported to Britain every year, many of which are overseas genotypes of native species (White, 1997).

Although 'native', 'natural' and 'semi-natural' cannot always be unequivocally defined or quantified and have many shades of (evolving) meaning, they are nevertheless useful concepts when it comes to describing woodlands and setting management objectives, even if it is simply adopting a trajectory from less to more natural. The principle that native species and local provenances should be encouraged is widely accepted nationally and internationally. It is thus likely that the importance of native species in Scottish forestry will continue to grow. It is clear, however, that natural and anthropogenic are not black-and-white alternatives but the end points of an extended continuum, a continuum along which many complex combinations exist (Table 10.1). Brown (1997: 197) believes that there is 'no rational, objective criterion by which native woodland . . . can be distinguished from non-native woodland'. The choice facing woodland managers is thus between gradations of naturalness, a spectrum which has been thoughtfully investigated by Peterken (1996). In practical terms, for *existing* semi-natural woodlands in Scotland the options are twofold (Peterken, 2000): to maintain the existing mix of native species or to bring back elements of the past. For *new* native woodlands the choice is between regulated or unregulated succession. For any particular site, it is arguable that appropriateness is a more practicable and acceptable aim than naturalness.

A simple notion of naturalness has often been employed as a trump card with moral authority, but given the tangled web of uncertainty which surrounds the concept, this will no longer suffice. Perhaps partly because of this realisation, the high tide of 'anti-exotic' feeling is beginning to recede. One example of this is that attitudes towards exotic conifers are changing. It is being recognised that some new conifer habitats have considerable conservation value, such that several SSSIs now exist (Hodge *et al.*, 1998). Ratcliffe (1995a: 5) suggests that 'extended rotations of exotic conifers . . . can be correctly seen as interesting, valued and diverse ecosystems'. Recent work reveals that long-maintained artificial ecosystems may have high biodiversity, that plantation forestry may well be sustainable (Evans, 1999), that conifer plantations are important habitats for native fungi (Humphrey *et al.*, 2000) and that introduced conifer and broadleaved species can contribute to nature conservation objectives (Kirby *et al.*, 1997; Petty, 2000; Peterken, 2001). Even Friends of the Earth concede that 'exotics do appear to be a commercial necessity' (FoES, 1996:60), although their preference is for Norway spruce, 'native just across the North Sea'.

Although the vision of restoring the 'Great Wood of Caledon' to its former glory is an inspiring one, it needs to be tempered with pragmatism, and with the realisation that 'the woods are an expression of mankind as well as nature' (Peterken, 1996: 7). Perhaps we should simply acknowledge that nativeness and naturalness are the preferences of our time. Although these concepts have clear advantages in terms of ecology, conservation and landscape, essentially their adoption springs from ethical and value judgements (Worrell, 1997). It is part of the wider trend away from anthropocentric viewpoints towards ecocentric beliefs (Section 13.1). Maybe we should stop casting around for some 'objective' rationale and simply do what we believe is best (Smout, 1999b) because 'in the end perhaps, restoration is just as much about aesthetics as science!' (Johnston, 2000: 40).

10.3 THE SIKA THREAT: LOSING A NATIVE?

Sika deer (*Cervus nippon*) (Fig. 10.5) are native to eastern Asia and are distant relatives of red deer (Putman, 2000). They were introduced to thirteen sites in Scotland in the late nineteenth and early twentieth centuries (Ratcliffe, 1995b). By the 1920s, as a consequence of escapes and intentional releases, they were breeding in the wild. The story since then has been one of steady range expansion, and of increasing population size and density, so that they are now thought to be present in more than half of the Scottish mainland (Pemberton *et al.*, 2000) (Fig. 10.6). This expansion has been facilitated by the rapid extension of young coniferous forest in which sika thrive

FIGURE 10.5 *A sika stag. Photo: © Steve Smith.*

(Ratcliffe, 1987b), and by the notable fecundity of the species; even one-third of calves may be fertile (Staines, 1999a). Sika are smaller bodied than red deer and the stags have simple antlers and a distinctive whistling call which contrasts with the red stag's impressive antlers and rutting roar. They can maintain higher densities than red deer with no suppression of reproductive performance, and may be able to sustain a 25 per cent cull, compared to about 20 per cent for woodland red deer (Chadwick *et al.*, 1996). Recent estimates suggest that sika numbers are growing rapidly despite the fact that annual culls have trebled since the late 1980s (Table 10.4). Population estimates are necessarily very approximate because of the difficulty of counting the numbers of a secretive and often nocturnal woodland creature (although indirect methods of estimating sika abundance exist (Fernanda *et al.*, 2001)). There is little doubt, though, that numbers are growing apace, so the current 'best guess' of 20,000 in Scotland could easily be a substantial under-estimate (McLean, 2001a).

The expansion of an introduced species would be a matter of concern even if there were no ill effects on the flora and fauna. However, the apparently inexorable spread of sika is worrying for three reasons. The first is that sika cause at least as much damage to commercial forestry and agriculture as red and roe deer (Youngson, 1997) (Section 11.1.1). The second is that they are thought to displace roe deer from young forests (Chadwick *et al.*, 1996). The third, the focus here, is that sika deer hybridise freely with red deer and produce fertile offspring. Although both species are much

TABLE 10.4 Sika population estimates in Scotland and cull levels since 1986. Cull data from Deer Commission (RDC/DCS) Annual Reports. Population estimate for 1993/94 from Harris *et al.* (1995) and for 1999/00 from McLean (2001a).

Year	*Population estimate*	*Stags*	*Hinds*	*Calves*	*Cull Total*
1986/7		537	549	98	1,184
1987/8		553	506	166	1,225
1988/9		635	503	138	1,276
1989/90					
1990/1		840	703	197	1,740
1991/2		850	733	257	1,840
1992/3		1,078	934	279	2,291
1993/4	9,000	1,029	1,127	398	2,554
1994/5		1,195	1,138	384	2,717
1995/6		1,116	1,070	434	2,670
1996/7		1,513	1,492	625	3,630
1997/8		1,566	1,466	668	3,700
1998/9		1,804	1,751	781	4,336
1999/2000	>20,000	1,818	1,778	781	4,377

more likely to true breed, some hybrids are possible in each generation, especially when sika stags are actively colonising, resulting in introgression of sika genes to the red deer genome.

Despite the ease with which red and sika deer interbred in nineteenth-century deer parks, concern over hybridisation in the wild was slow to arise. Red-sika hybrids were reported from the late 1960s, but not until twenty years later were the serious consequences of hybridisation highlighted by Ratcliffe (1987b). Subsequent research has confirmed that the genetic integrity of red deer is at risk (Abernethy, 1994a, 1994b). In fact, introgression is now widespread, and in some populations it is well advanced, notably in the Borders, north Kintyre, Ross-shire and the Cowal peninsula (Swanson, 2000). Large numbers of deer have a trace of the other species in them. A genetic study revealed that one-third of 'red deer' samples in west Argyll contained sika DNA, and in Sutherland and Ross-shire over 75 per cent of 'sika deer' showed evidence of hybridisation sometime in the past (Swanson, 2000). Unchecked, this trend would ultimately lead to the demise of genetically pure Scottish red deer, just as Scotland's 3,500 wildcats (*Felis silvestris*) have lost their genetic uniqueness through interbreeding with feral and domestic cats (*F. catus*) (Balharry *et al.*, 1994; Harris *et al.*, 1995; Kitchener, 1995). Such a saga has already unfolded in the English Lake District and in Ireland's Wicklow Mountains where pure red deer no longer exist (DCS, 1998a). It was long thought that railways and major roads would act as barriers, helping to keep populations such as the Peebles sika and Galloway red deer separate, but by the late 1990s the range of these two populations overlapped, and hybrids were found in mid-Galloway (DCS, 1998b).

Hybridisation is of concern not only to conservationists and ecologists but also to the deer stalking industry by threatening the fabled majesty of red deer stags. If 'the

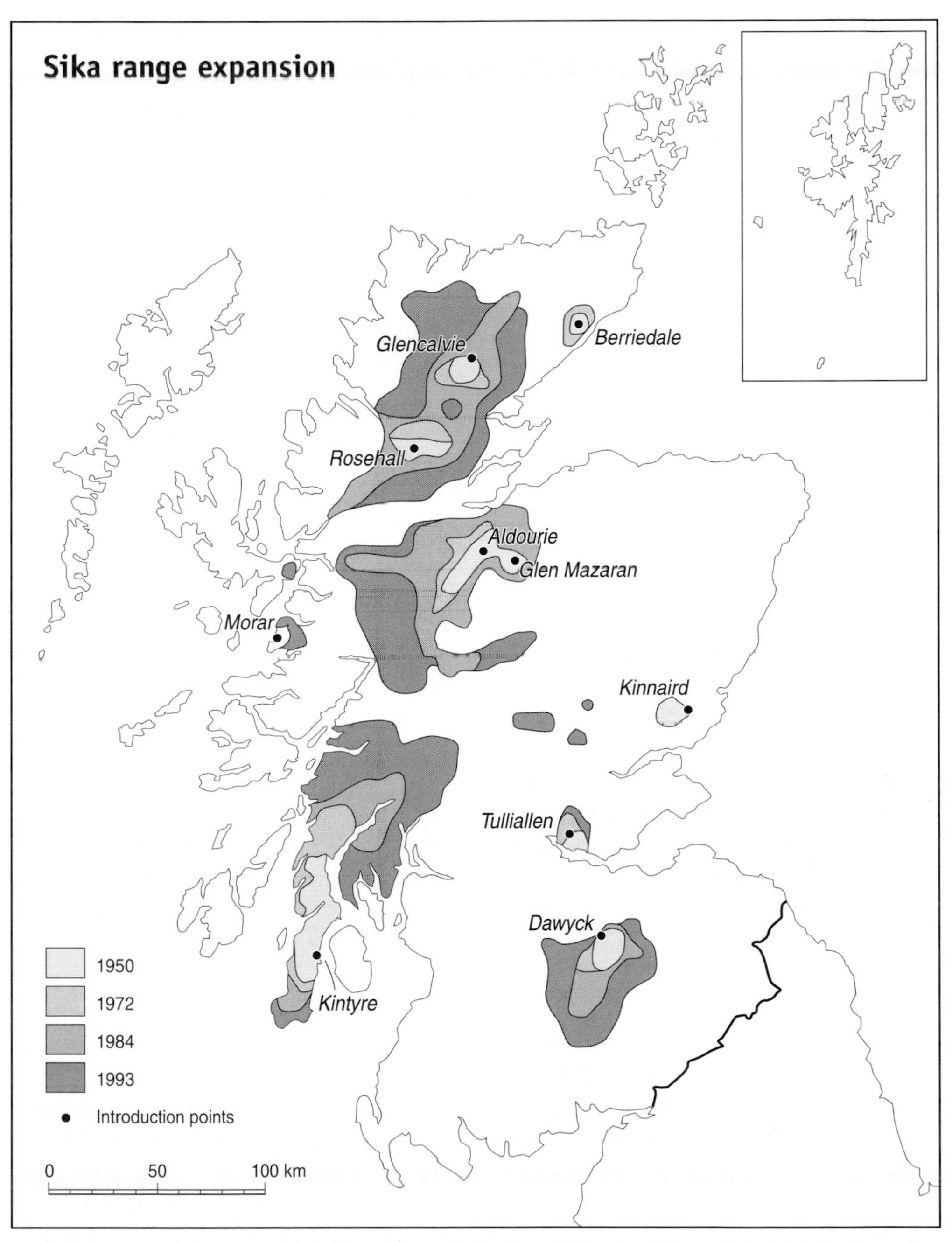

FIGURE 10.6 *The expansion of the range of sika deer in Scotland from their initial points of release or escape. After Abernethy, 1994a. See Fig. 7.2 for their current distribution.*

monarch of the glen' was replaced by 'the mongrel of the glen' (Ratcliffe, 1995b), there could be damaging implications for stalking and venison income, as well as for land values and employment, all of which could critically undermine the shaky finances of many sporting estates. Genetic mixing could thus have economic consequences. For all these reasons, the consensus view is that the genetic integrity of red deer must be conserved, and that sika and hybrids should be shot on sight. The sika policy worked out jointly by the DCS, SNH and the FC (DCS, 1998a) recommends that the overriding aim of sika management should be population control in order to protect woodlands, agriculture and the natural heritage. It calls for rigorous control of all sika populations, a major effort to remove all sika from the open range, and the establishment of 'refugia' for red deer on some Hebridean islands. The DCS does not want sika to be treated as a sporting resource (McLean, 2001b), arguing that any sporting gain is greatly outweighed by their damaging potential. The hope is that, in the long term, sika will not spread throughout mainland Scotland and will be largely confined to woodlands (DCS, 2000a).

As in the case of native woodland, however, the situation regarding sika is not as clear cut as it might at first appear (Ratcliffe, 1995b). Several factors muddy the waters. Some naturalists consider sika to be an interesting addition to Scotland's depleted large mammalian fauna, while sporting interests (especially in the Borders) regard their presence as a lucrative and welcome asset. More fundamentally, the genetic discreteness of the native red deer population turns out to be doubtful (Balharry *et al.*, 1994; Swanson, 2000). This should not really come as a surprise given the numerous introductions to deer parks of non-Scottish red deer and wapiti (*Cervus canadensis*) since Victorian times to 'improve' stock, and that many of these later escaped and bred with wild deer. We are already a century too late to preserve the genetic integrity of mainland red deer. Rose (1995:69) comments that 'genetic threats, like racial threats, tend to generate strong emotions and irrational responses'. Why, he asks, should we intervene, when the naturalised sika population is clearly thriving here? Taking a global perspective, Scotland may now have one of the largest and healthiest populations of sika deer in the world because many former populations in Japan and eastern Asia are now extinct (Swanson, 1997). Does this argue for conservation rather than eradication, at least for the Japanese-origin sika around Peebles which have lived largely in isolation from red deer and so may be genetically pure (Harris *et al.*, 1995; Putman, 2000)? Ironically, Scottish expertise in sika management is reflected by the recent request from the Japanese government for advice on the management of *Cervus nippon* (DCS, 1999)!

From a practical standpoint it has also become clear that controlling sika, let alone eradicating them, is extremely difficult, costly and time-consuming (McLean, 2001c). Although sika densities are often greater than red deer densities in mature woodland, commonly reaching 20–40/km^2, sika are much harder to keep track of, are more nocturnal, and emerge from cover more reluctantly, especially when heavily culled (Chadwick *et al.*, 1996). Research in Sutherland showed that it required thirty man hours to cull each animal, and suggested that in order to reduce a sika population, at least one stalker is needed per 200–300ha (McLean, 1993); the site actually had one stalker per 8,000ha. 'Shoot on sight' is easy to say but hard to do. Positive identification of hybrids is difficult even with close examination of carcases; it is impossible when an animal is glimpsed fleetingly in woodland twilight. Swanson (2000) reports that only 2 per cent of genetically identified hybrids in Argyll showed any visual sign of hybridisation, and that stalkers almost universally failed to identify hybrids. This is because the frequency of backcrossing dilutes distinctive sika characteristics such as the white rump and metatarsal gland on the hock. It is clear, then, that hybrids cannot be selectively culled (Pemberton *et al.*, 1999). An alternative would be heavy shooting in buffer zones around defined sanctuaries of each species. Such a policy, advocated by Ratcliffe (1995b), would also require active woodland management to facilitate deer control (Section 11.1.3).

Consequently, a pragmatic case can be made that we should 'live and let live'. Any attempt to extirpate sika would incur enormous costs and would almost undoubtedly fail. Even if successful it would not eradicate the genetic legacy. In terms of appearance, most hybrids look like either red deer or sika deer. Why resist the inevitable? As Rose (1995: 69) laconically comments:

> I believe my great-grandmother was French – rumour has it she was also pretty. I can live with that.

Ratcliffe (1998: 24) agrees: 'the battle is almost lost, so why not treat them as a naturalised species?' Only a small minority subscribe to this relaxed, inclusive view (at least publicly), despite an impotent acceptance that further expansion of their range and population is inevitable (Youngson, 1997; McLean, 1999). In fact, many believe that it is only a matter of time before the mainland Scottish deer population, like that of Co. Wicklow, consists entirely of hybrids (Abernethy, 1994b; Pemberton *et al.*, 1998), a destiny which many see as profoundly sad.

So should we worry about genetic purity or adopt a pragmatic approach? Recent genetic research provides ammunition for both purist and pragmatist. Swanson (2000) concludes that although continued spread of hybrid DNA is impossible to prevent by selective shooting, the *appearance* of mainland red and sika deer is unlikely to change significantly in the foreseeable future. In other words, they may remain phenotypically distinct despite genetic mixing. If so, and if aesthetics are the main concern, there seems little to worry about. But if conserving genetic integrity is indeed judged to be important and intrinsically valuable, then we are staring disaster in the face. In the longer term, the evidence from Wicklow suggests that even the pragmatists may have something to worry about; there, after many generations of hybridisation, recognisable hybrids have become common (Abernethy, 1994b). (Given the very different context, however, a similar outcome is not inevitable in Scotland (Swanson, 2000).) If realistic management objectives are to be defined for red and sika populations, achievable policies based on clear thinking are required. In particular, a clear distinction needs to be drawn between conserving the red deer *phenotype* (probably an achievable aim) and conserving the *genotype*, a goal which may well be out of reach already. Arguably, conservation effort should focus on the former rather than the latter.

Compared to the considerable time, effort and money that is being put into saving native woodlands and elements of the native fauna such as red squirrels and corncrakes, it is striking how little has been done to preserve the genetic integrity of Scotland's largest native land mammal. Sika have been rapidly colonising for decades, yet there has been (and remains) no concerted effort even to monitor the situation systematically, let alone control or reverse it. The DCS's Annual Reports have repeatedly talked of sika 'continuing to cause concern', but this has led to only limited research and even less action. With its restricted budget and its responsibility for overseeing the management of the much larger wild red deer population, the DCS has woefully inadequate resources with which to tackle the sika problem. Unless its budget is significantly increased, and/or an NGO takes up the challenge, the days of a mainland red deer population which is red in both appearance and ancestry may be numbered. Already, the only realistic hope for conserving pure red deer is to protect the populations on those Hebridean and west coast islands where hybridisation has not yet occurred (Pemberton *et al.*, 1998, 2000). Accordingly, introductions of deer to the islands were made illegal in 1999 (DCS, 1999). A genetically pure population of Scottish red deer exists in New Zealand, but it would be ironic indeed if eventually the only large population of true Scottish red deer were to exist on the other side of the world, leaving Scotland itself with a hybrid population. That irony will be compounded if the native woodlands that are being restored and kept pure with such zeal end up being inhabited by hybrid deer.

10.4 REINTRODUCING LOST NATIVES

As well as introducing many species to Scotland, human beings have driven many others to extinction, both intentionally and otherwise (Lambert, 1998; Yalden, 1999) (Tables 1.1 and 1.2). The composition of the mammalian fauna, in particular, has been radically altered (Kitchener, 1998), although climate change, not humans, was the cause of extinctions in the immediate post-glacial period. A natural outworking of a concern for naturalness and authenticity in the environment is a desire to see introduced species controlled or exterminated, and to see missing members of the cast written back into the story. This is not a new idea. Capercaillie were reintroduced in 1837 (Petty, 2000). Sea eagles and red kites, persecuted to extinction in Victorian times, were the subject of successful (if controversial) reintroduction programmes in the later twentieth century (Table 10.5), and otters have been returned to areas where they were locally extinct. Species have also been known to re-establish unaided, most famously the ospreys which returned to Loch Garten in 1954 and now occupy almost eighty sites in the Highlands (Dennis *et al.*, 1993).

More recently, however, there has been growing interest in reintroducing some of the larger mammals, a prospect that has proved considerably more controversial than rebuilding our avifauna (which anyway is largely complete (Yalden, 1986)). Not only do many of these species need large areas of suitable habitat in order to survive (Summers *et al.*, 1995), but the prospect of their presence excites strong reactions, both positive and negative. On the positive side, strong ecological arguments can be made that ecosystems can only function naturally and properly with the presence of such animals. Long-term campaigners such as Dennis (1998: 6) regard reintroductions of

TABLE 10.5 Recolonisations and reintroductions of birds to Scotland. Method: RC = recolonised, RI = reintroduced. Purpose: SR = species restoration, H = hunting. After Kitchener (1998).

Species	*Date of re-arrival*	*Method*	*Purpose*
Capercaillie (*Tetrao urogallus*)	1837	RI	H
Great spotted woodpecker (*Dendrocopos major*)	1887	RC	–
Osprey (*Pandion haliaetus*)	1954	RC	–
Goshawk (*Accipiter gentilis*)	Mid-1960s	RI	SR
Spotted crake (*Porzana porzana*)	?1966	RC	–
Sea eagle (*Haliaeetus albicilla*)	1975	RI	SR
Red kite (*Milvus milvus*)	1991	RI	SR

large mammals as both feasible and essential, arguing that 'it will be impossible to recreate an ecologically vibrant Caledonian Forest if the major players in the ecosystem are not present'. Similarly, Yalden (1999: 272) believes that restoring 'something like the full magnificence of the European fauna to at least a small part of these islands would be an appropriate target for the next millennium'. On the negative side there are strongly-held beliefs that reintroductions would place human livelihoods and even human safety in jeopardy. Peterken (1996: 374) comments witheringly on this public mind-set which sees wolves as 'marauding packs of ravening beasts terrorising farms, villages and ramblers', and in which 'beavers, once set free, are expected to fell all forests and to clog every stream and river'. Wrong these perceptions may be, but they represent a significant social barrier to reintroductions. Whether desirable or not, large scale ecological reconstruction is viewed by many as simply impracticable: 'resurrecting the primeval food web can only be a fireside dream' (Tapper, 1999: 66).

In reviewing the possibilities for restoring mammalian species to the UK, Yalden (1986) makes a strong ecological and moral case for the feasibility and rightness of reintroductions. His moral argument is that while British conservationists have taken the lead in global campaigns to save large mammals such as lions, tigers and pandas, they have done nothing to restore our own large mammal fauna. A more explicit moral tone is adopted by Morris (1986: 49) who sees reintroductions as 'atoning for past follies'. Yalden (1986) also argues that the recreation of a large mammal fauna could have lucrative potential for eco-tourism, and he disparagingly contrasts the 'unimaginative, timid' British attitude with the existence of successful reintroduction programmes in many other European countries.

The last twenty years have seen such programmes multiplying across Europe (Klaffke, 1999). The 1979 Bern Convention on the Conservation of European Wildlife and Natural Habitats encourages reintroductions of native species, and most significantly, under the 1992 Species and Habitats Directive, EU countries are obliged to explore the desirability of reintroducing formerly native species where this will aid species conservation. The IUCN has identified several motivations for reintroductions: to enhance long-term survival of a species, to re-establish a keystone species, to promote biodiversity, and to provide economic benefits (Howells and Edwards-Jones, 1997).

Within Scotland there has been speculative talk of reintroducing lynx, aurochs (wild cattle) and brown bear, but the three species to which serious consideration has been given are wolves, beavers, and wild boar.

10.4.1 Wolves

The last Scottish wolf was probably killed in the late seventeenth century, but there are now those who would like to hear the howl of the wolf return to the glens. Reintroduction proposals have recommended a two-stage process, with wolves (*Canis lupus*) being released experimentally on the island of Rum (a National Nature Reserve), to be followed, if successful, by releases on the mainland. Yalden (1986, 1999) argues strongly for such an experiment, estimating that Rum could support up to twenty wolves. The Affric area is also touted as a potential site (Spinney, 1995). The motivation behind such proposals is partly ecological, partly ethical, partly aesthetic or romantic. The ecological argument is that predation by wolves would be a more natural and efficient control on the expanding red deer population (Section 7.5.1) than current culling practices, and that their reintroduction would be a step towards the ecological restoration of the Highlands (Featherstone, 1997). Ethically it is suggested that reintroduction is a moral imperative because (as for other extinct mammals) it was human persecution which eradicated Scottish wolves. The romantic and emotional arguments for the return of an animal that epitomises the wild are obvious. They spill over into a wildlife tourism case – that wolf watching would attract tourists, as it has done very lucratively in North America (Isaacson, 1999). Tourist revenue could also offset lost income from venison sales and livestock losses.

Unsurprisingly, there has been outspoken opposition to the idea from the public, from land managers and from politicians. Amongst the public, there is a widespread perception that 'the big, bad wolf' would threaten human safety, even though healthy wolves almost never attack people. Farmers, landowners and crofters are all vigorous in their opposition (Spinney, 1995). Livestock farmers envisage substantial losses to wolves, pointing out that no intelligent wolf is going to target fleet red deer on the open hill when quantities of docile domestic prey are corralled on farms. Wolf reintroduction would probably not be an effective method of reducing deer numbers anyway (McLean, 2001b). The evidence from Norway and America is that low numbers of large predators have little effect on deer numbers (Holt, 2001; Reimers, 2001). To have a significant impact on deer populations, a very large number of wolves would be needed, and there is unlikely to be enough space in Scotland (either ecologically or socio-politically) for such large wolf populations. The problem with small populations, however, is that wolves might interbreed with dogs as is happening in Italy and Sweden (Balharry *et al.*, 1994). Recent evidence also casts doubt on the appropriateness and viability of Rum as a trial site. Wolves may never have been native to the island, and at 87km^2 it might only support four wolves (when twenty is the minimum possible population) (Holt, 2001).

Politicians, meanwhile, cast a nervous eye on the controversies that erupted in the late 1990s in the USA over wolf reintroductions to Yellowstone National Park (Isaacson, 1999), and in parts of southern Europe where wolves are colonising naturally

(Klaffke, 1999). They fear huge compensation claims on the public purse. In Norway, around 5 per cent of the summer sheep population is lost annually to predation from the protected and fast-growing populations of bears, wolverine, wolves and lynx, and the government pays out some £2.3 million per annum in compensation (Reimers, 2001). A wolf cull is now on the cards (Marchant, 2001).

Although wolves might one day roam Scotland again, for now it seems that the political odds are stacked conclusively against wolf reintroduction, a reality which even Yalden (1998) reluctantly conceeds. The Highland Wolf Fund, established in 1994 to promote the idea, has now been wound up, and SNH has no plans to investigate the concept. Holt (2001) suggests that reintroduction could create a permanent rift between conservation bodies and rural communities, a divide which could be devastating for conservation in the Highlands. The reintroduction spotlight is now trained on a smaller and less controversial species, the beaver.

10.4.2 Beavers

The Eurasian beaver (*Castor fiber*), Europe's largest indigenous rodent, is best known for its ability to fell trees in riparian woodlands and to construct dams and lodges, although it builds less extensively than its Canadian cousin (*C. canadensis*) (Conroy *et al.*, 1998). Debates about the wisdom or otherwise of reintroducing beavers to Scotland have aroused strong emotions since the idea was first mooted in the late 1970s (Lever, 1980). As for all prospective reintroductions, these debates divide into questions of desirability and feasibility:

1. **Desirability**. Many argue, firstly, that beavers are an element of our native biodiversity that is missing due to human activities, and that we therefore have a moral responsibility to consider their return (SNH, 1998f). To its advocates, the beaver is 'a harmless rodent with nothing more ominous in its nature than a talent for civil engineering and a predilection for willow bark' (Crumley, 1996). Secondly, beavers are an important keystone species in forest and riparian ecosystems, and their activities have a range of hydrological and conservation benefits (Macdonald *et al.*, 1995; SNH, 2000d). Thirdly, beaver-watching could contribute to local economies, as it does in several European countries (SNH, 1998f); beavers could generate income just as ospreys have on Speyside. Ultimately, if beaver populations thrived, they could become a game animal from which landowners could derive a revenue, as they do in Norway and Sweden (Reimers, 2001).
2. **Feasibility**. It is salutary to note that two attempts to reintroduce beavers to Scotland in the late nineteenth and early twentieth century ended in failure (Conroy *et al.*, 1998). On the other hand, beavers were successfully re-introduced to thirteen European countries during the twentieth century (Macdonald *et al.*, 1995). To be feasible, there is a need, firstly, for suitable habitat to enable a viable, self-sustaining population to become established in the wild. Secondly, the potential conflicts with other conservation and land use interests which could arise if beavers become successfully established must be potentially surmountable.

There is little doubt that reintroduction is ecologically feasible, and that viable (if fragmented) beaver populations could be successfully established in Scotland (Macdonald *et al.*, 1995; Parker *et al.*, 2000; South *et al.*, 2000). Doubts have been expressed about the availability of enough suitable habitat (Collen, 1995), but SNH (1998f) estimates that even without restoration of riparian woodland there is sufficient habitat to support a population of between 200 and 1,000. Beavers are also known to tolerate human disturbance well (Yalden, 1986). Concerning conflicts, the potential impacts of a thriving beaver population include:

- expansion of wetlands
- damming of culverts leading to damage to roads and railways
- raised water tables
- damage to flood protection levées
- alteration of the ecology of river beds used by spawning salmon
- changed fluvial sediment dynamics
- damage to forests and agricultural crops (both directly, and indirectly through flooding).

Anglers and farmers are strongly opposed to reintroduction, whereas foresters (with some outspoken exceptions) are less concerned because direct damage to valuable conifers would be slight due to beavers' preference for broadleaved species (Macdonald *et al.*, 1995). Parker *et al.* (2000), drawing on Norwegian experience, suggest that broadleaf riparian forests (especially of aspen, rare in Scotland) would suffer, but that anglers' fears are unfounded. While there is no doubt that a beaver population could cause significant localised damage which could rebound to threaten the population (Conroy *et al.*, 1998), there is a growing consensus that the overall impact of beavers would be benign, and that the benefits would outweigh any financial losses (Macdonald *et al.*, 1995; SNH, 1998f).

Following public consultation which revealed strong majority support for the idea (SNH, 2000d), a trial reintroduction was given the go-ahead in 2000, probably involving the introduction of twelve Polish beavers to the Knapdale area of Argyll in 2002. If successful, the seven-year pilot project will be followed by widespread introductions. Although SNH will only shoulder £250,000 of the projected cost of around £400,000, there is concern that other initiatives may suffer budget cuts as a result. Such *realpolitik* considerations temper the widespread enthusiasm for the return of the Eurasian beaver.

10.4.3 Wild boar

Unlike wolves and beavers, wild boar (*Sus scrofa*) are not threatened in Europe and SNH is under no legal obligation to consider their reintroduction. There is nevertheless an interest in exploring the feasibility of restoring an animal which some believe would be 'a magnificent presence' in our woodlands (Yalden, 1999: 269). Evidence from Europe provides mixed messages about the likely ecological impact of wild boar in Scotland. Leaper *et al.* (1999) review studies showing that wild boar can both promote and suppress regeneration in forests, and can either enhance or damage biodiversity. The FC is experimenting with using wild boar in place of mechanical scarifiers in a Morayshire forest, and the signs are encouraging.[6]

But despite having some beneficial effects, wild boar are generally considered a pest because of the extensive damage that they often cause to agricultural crops and fences. Soaring continental populations are now raising the spectre of interbreeding with domestic pigs, and the associated threat that Classical Swine Fever (endemic in wild boar) could spread amongst Europe's pig farms (D. MacKenzie, 1999). Recent outbreaks in the Netherlands (1997) and eastern England (2000) proved devastating for the pig industry. Wild boar are known to disturb ground-nesting game birds such as pheasant and partridge, behaviour which would also endanger the struggling Scottish populations of capercaillie and black grouse (Leaper *et al.*, 1999). Despite all this, they are tolerated in Europe because of their value as a game species and have, for example, been widely introduced in France for this reason. Studies examining the biological feasibility of restoring wild boar to Scotland have also come to conflicting conclusions. Focusing on reintroduction within woodlands, Howells and Edwards-Jones (1997) conclude that sufficient habitat does not exist and that a self-sustaining minimum viable population (MVP) of 300 animals would not survive. Leaper *et al.* (1999), using a broader range of habitat and suggesting an MVP of 25–50, come to the opposite conclusion. They identify four sites potentially able to support populations of 300–1,000 animals.

If the ecological viability is debatable, the desirability of restoring wild boar seems questionable at best. Farmers and foresters would oppose the idea, and even the conservation and game shooting fraternities would have mixed feelings because of the potential for damage to their other interests. Once established, populations could expand fast, prove hard to control, and be highly disruptive (Morris, 1986). Leaper *et al.* (1999) rest their case for reintroduction primarily on the moral argument of righting past wrongs, but this seems insufficient by itself. Although reintroduction of wild boar to the wild is perhaps ultimately feasible as part of a wider ecological restoration programme (Howells and Edwards-Jones, 1997), it is unlikely to happen in the foreseeable future.

10.4.4 Conclusion

Reintroductions are expensive, long-term operations, with no guarantee of success, but the main barrier is not economic or ecological but social. In a survey, Macmillan *et al.* (2001a) found that people are more in favour of reintroducing beavers than wolves, but that their top preference is for restored native woodland excluding both these creatures. Scotland has long been spared a dilemma which is common around the world: how to deal with endangered, protected wild animals which damage people's livelihoods. Reintroductions will require people to relearn the lost art of coexisting with such creatures. The vision of a future Scottish environment in which a restored faunal assemblage thrives in expanding native forests is undoubtedly an inspiring one (Featherstone, 1997). Dennis (1998: 7) argues that our choice is not between wild places and people but between 'a rich or an impoverished future for mankind', and his enthusiasm is infectious. There remain, however, a number of significant ecological, economic, political and psychological hurdles to overcome before such a vision could begin to take shape on any significant scale.

NOTES

1. The difficulty of precisely defining the terms 'native' and 'exotic' is explored by Webb (1985), Usher (1999b), Hettinger (2001) and Woods and Moriarty (2001). Building on Webb's work, Usher proposes nine indicative criteria for distinguishing between native and exotic species in Scotland.
2. The Centre for Environmental History and Policy, a joint initiative of the universities of Stirling and St Andrews, was established in 1999.
3. Beech is native to southern Britain but not to Scotland; it was introduced about three centuries ago.
4. A larger area for native pinewoods is given in Table 10.3. The figure quoted here relates specifically to woods in the Caledonian Pinewood Inventory, as opposed to native pinewoods more generally. See Jones (1999b).
5. Detailed critiques of the native/exotic debate in the context of woodland management are given by Brown (1997), Ratcliffe (1997) and Worrell (1997). Brennan (1993) discusses the criticism that restoration forestry amounts to 'faking nature' because it simply creates artefacts.
6. Reported in *SCENES* 158: 4, 2001.

CHAPTER ELEVEN

Integrating forestry: deer in forests, and woods on farms

For historical reasons (Section 4.2), forestry in Scotland for the last century or more has tended to be practised in isolation, rather than being integrated with other land uses as in most other parts of the world. Within locked, ring-fenced plantations, timber production has typically been the sole objective, all available land being devoted to trees to the exclusion (literally) of all else. For much of the twentieth century, land was used for single purposes: forestry *or* farming *or* nature conservation. The boundaries were precise, impermeable and jealously guarded. During the multi-purpose revolution of recent years, such hard and fast distinctions have become blurred as diverse objectives and activities are increasingly required from the same area of land. In every sense, the fences have been coming down; impermeability is giving way to porosity. Even in settings in which timber production remains the prime goal, forestry is now expected to yield multiple benefits (Section 4.5.1).

Traditionally, foresters have regarded deer as a threat to be kept at bay, while farmers have viewed forestry in much the same light. Now, however, deer are increasingly being permitted or encouraged to rediscover their natural forest habitat, and farmers are establishing woodlands on agricultural land. Both these trends have been rapidly developing during the last twenty years, and each constitutes a significant example of the wider moves towards integrated land use (Section 2.4.3). This chapter explores the opportunities and challenges that these trends present, and the reasons why, after decades of rivalry, foresters are increasingly accepting deer managers and farmers as friends (or at least as necessary acquaintances).

11.1 Deer and trees: can they co-exist?

Attitudes to deer range between two extremes (Prior, 1994): the 'wild west' approach, which operates on the maxim that 'the only good deer is a dead deer', and the 'Bambi Syndrome'. Foresters and farmers frequently have some sympathy with the first of these due to the damage that deer can have on their livelihoods, while the second is widely held by the urban public which abhors the violent death of beautiful creatures. During the twentieth century the forest area in Scotland more than tripled and deer numbers quadrupled. Conflict and damaging interactions have been the inevitable result. Foresters once believed that having cleared deer from an area and fenced it they could then practise silviculture in a deer-free environment. This fond notion has been resoundingly disproved by the deer themselves. Plantations throughout Scotland are now home to thriving and growing populations of red, roe and sika deer, and there are sometimes more deer inside the (remains of the) fences than outside. Unless these populations are brought under control they have the potential to compromise many of the current enlightened initiatives in forestry, agriculture and conservation. Woodland deer have long been regarded by sporting interests as the poor relation of the open range 'monarchs of the glen', and have been overlooked – out of sight, out of mind, and unmanaged – but it is now critical that far more time and resources are devoted to their management.[1]

Curiously, no one seems to have foreseen the magnitude of the woodland deer problem or its implications. By the mid-1990s, however, the Deer Commission perceived it as the most important that they had to address (RDC, 1996). Deer are present in productive forests in most parts of Europe, but the big difference here is the vast reservoir of wild deer on the surrounding open land (Mutch, 1987). Any weaknesses in forest fences (caused by rockfalls, washouts, snowdrifts or simply age) are rapidly found and exploited, especially in winter when deer persistently seek entry for food and shelter (Fig. 11.1). The result is much higher densities within Scottish forests than are tolerated in, for example, Germany. This is partly because the much shorter forest rotations here increase the percentage of forest areas at the pre-canopy closure stage which is optimal for deer, allowing the density of resident deer to exceed 40/km^2 in some plantations (Staines, 1999a). An extreme density of 81.5/km^2 is reported by McLean (1999). All deer species love woodlands. If it was just shelter that they desired, happy co-existence would be the order of the day. Unfortunately, their presence is accompanied by a range of negative impacts.

11.1.1 Impacts of deer on forestry

11.1.1.i Tree damage

Deer harm trees primarily by bark stripping and browsing but also through fraying, thrashing and root damage.[2]

- **Bark stripping** (Fig. 11.2a). Mainly carried out by red and sika deer, this can cause slower growth, infection, or (if the tree is ring-barked) death. The occurrence and severity of bark-stripping is extremely variable (Gill, 1992a), and although the impact on tree volume is minimal, the effect on timber quality can be considerable (Welch and Scott, 1998). Bark stripping remains something of a mystery. Why do deer do it? What explains the occurrence and timing of sudden outbreaks of widespread damage? It may be a desire for food and roughage, or a behavioural response to population pressure, disturbance, boredom or stress. There may be a critical deer density which triggers it, or it may serve no apparent function (the cervine equivalent of a boys' night out). Mitchell *et al.* (1977) and Staines and Welch (1987) review the many suggested explanations and both conclude that no single cause adequately accounts for all occurrences.

FIGURE 11.1 *Red deer stags in pinewood in winter, Glen Feshie. Photo: © Neil McIntyre.*

- **Browsing.** This refers to all forms of feeding damage other than bark stripping, and its intensity increases with deer density (Gill, 1992a). When it only affects lateral branches it is of little consequence, but when deer eat a tree's leading shoot the results include delayed growth, deformity (especially multi-trunking) and structural defects, all of which have economic implications (Staines and Welch, 1987). Trees are most vulnerable to leader-browsing at a height of about 0.4–0.6m (Staines and Welch, 1987); they are out of danger from roe deer at about 1m (Prior, 1995) and from red and sika deer at about 1.8m (SNH, 2000e). Sustained browsing of saplings can hold them in check indefinitely, creating close-cropped bushes instead of trees (Fig. 11.2b). The only benefit of browsing is that it can cause some rapid compensatory growth (Putman, 1996). Both bark-stripping and browsing occur mostly in late winter in coniferous forests, whereas broadleaves are more usually damaged in spring and summer (Gill, 1992a; Harmer and Gill, 2000).
- **Fraying and thrashing.** Deer also damage trees with their antlers, either fraying the main stem (to remove antler velvet or to mark territory) or thrashing the lower lateral branches. This causes unsightly damage and so may be unacceptable in amenity woodlands, but it has little impact on tree growth. Root damage through trampling can have an impact on shallow-rooting tree species such as Sitka spruce.

The degree of deer damage to trees is related to a wide range of factors (Miller *et al.*, 1998):

FIGURE 11.2 *Deer damage to trees.*
a. Bark stripping: a stripped section of trunk on a young Lodgepole pine.
Photo: © the author. (above)
b. Browsing: heavily browsed trees, an example of 'cervine topiary'.
Photo: © Laurie Campbell. (opposite)

- Tree species, age and density (Fig. 11.3; Table 11.1).
- Deer species and density.
- Site characteristics. These include soil quality, and the availability and quality of alternative food for the deer. For example, trees suffer less in the east than on the poor soils of the west.

TABLE 11.1 Relative preferences of red deer and sheep for the foliage of various native tree species (SNH, 2000e).

Preference ranking	*Tree species*
1 (most preferred)	Ash, aspen, willow, hazel, holly, rowan
2	Oak
3	Birch, pine
4 (least preferred)	Alder, juniper

The recognition of such factors, and of the general spatial and temporal patterns of deer damage, provides useful management information, but the reality is that if deer are hungry they will eat almost any kind of tree at any time. It is therefore very difficult to predict with any accuracy how much damage will be done by different deer species at different times within different woodlands (Harmer and Gill, 2000). Of the commercially grown conifers, Sitka spruce usually survives best and Lodgepole pine the worst, but the latter can prove undesirable to deer and even Sitka spruce can sometimes be severely damaged (Mitchell *et al.*, 1977; Welch *et al.*, 1991, 1992). Given that Sitka spruce is most resilient to deer damage, the current emphasis in forest management on other more vulnerable coniferous and broadleaved species (Section 4.5.1) can only exacerbate deer-related problems in woodlands.

Concern about deer damage to forestry has mostly centred on red deer. However, sika deer are thought to cause at least as much (if not more) damage to commercial crops as red deer, even damaging less susceptible species like Sitka spruce. Bole-scoring by stags can affect up to 90 per cent of trees in Sitka and Norway spruce forests, and

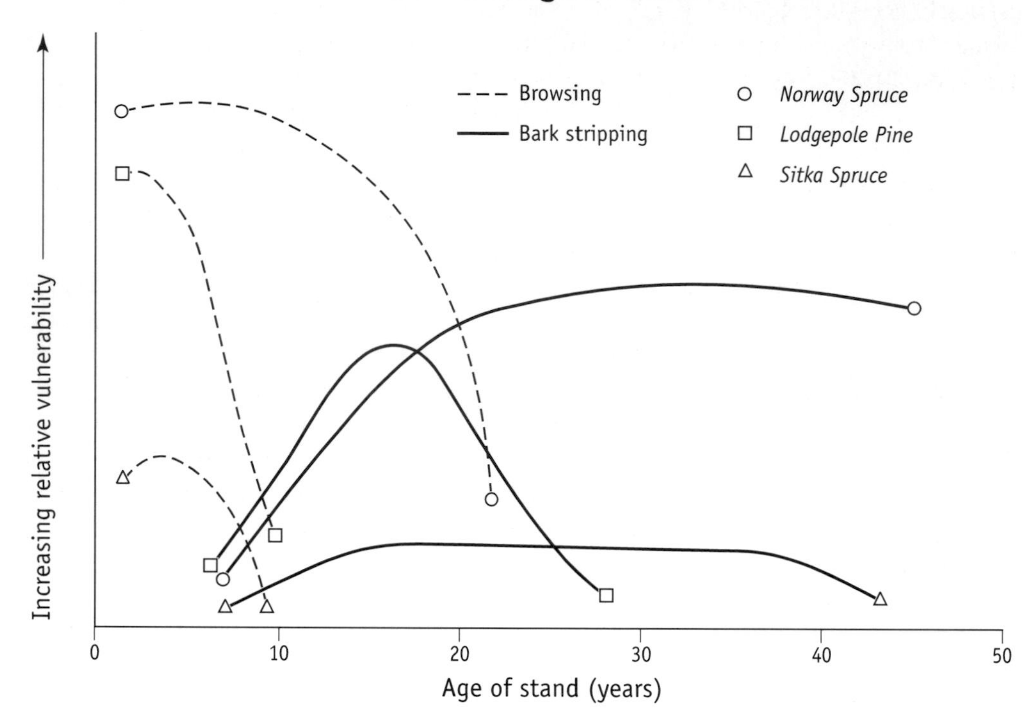

FIGURE 11.3 *A schematic representation of the vulnerability of three main commercial tree species to deer damage (bark stripping and browsing) in relation to age. After Staines and Welch (1987).*

in the Peebles area, damage by sika costs Forest Enterprise up to £4,000/ha (Putman, 2000). Although the extent of sika damage to native woodland regeneration is unknown (DCS, 1998a; McLean, 1999), recent evidence indicates that many woodland regeneration schemes are suffering (McLean, 2001c). Roe deer, too, can cause significant damage (Prior, 1995).

A critical unknown about deer damage concerns the economics (Gill *et al.*, 2000): what is its real economic significance for various end markets, and how does this relate to the costs of deer control and tree protection? Given the length of forestry rotations it is notoriously difficult to quantify the various elements of this equation, not least because damage does not necessarily equate to economic loss: damaged trees may recover completely. The economic impact of deer has two facets: the costs of deer control and the impact of tree damage on timber value. There are few available data about either aspect. The former includes fencing (erection and maintenance), employment of rangers, and the opportunity cost of leaving areas unplanted to facilitate control (especially for roe deer). The Forestry Commission (FC) spent about £1.5 million on deer control in Scotland in 1990 (Gill, 1992b). Bullock (1999) estimates that protecting trees from deer can cost £3/ha/year, while Gill *et al.* (2000) suggest that fencing costs can reduce profits by up to 30 per cent. Consequently, given that the costs of deer control can be partly offset against revenue from venison sales, culling is usually more cost-effective than fencing. Estimates of the impact of deer damage on the value of standing timber vary widely, from an 8–17 per cent reduction in value (Scott and Palmer, 2000) to Wigan's (1996b) suggestion of a sixteenfold reduction. A study by Callander and MacKenzie (1991) revealed losses of £2000/ha. Notwithstanding such startling figures, deer control may not be cost effective in many upland forests, especially if timber prices fail to recover from their current depressed state. Low timber prices could even make woodland deer a more valuable asset than the woodland itself in the long term.

11.1.1.ii Problems of re-establishment

Removing deer and ring-fencing is not difficult when afforesting unplanted land. However, at the end of a forestry rotation, restocking plantations that have a resident deer population is highly problematic. Unless deer densities are very low, young trees will have little chance of surviving or thriving and protecting them is expensive. It is rarely possible to remove all deer (especially roe) from a partly wooded area, and keeping woodland deer densities low enough during the establishment phase to allow trees to 'get away' is difficult, labour-intensive and expensive. Deer fencing is only economic for large areas, yet best forestry practice now discourages large scale clear-felling, promotes small-area re-establishment (preferably by natural regeneration), and advocates mixed age, mixed species woodlands (Section 4.5.1). How can such desirable silvicultural objectives be achieved in the presence of deer? Since all of them exacerbate the risk of deer damage (Harmer and Gill, 2000), the answer is 'only with great difficulty'. Unless deer densities are reduced to very low levels, only the hardiest of tree species such as Sitka spruce will survive. By determining the species of seedlings and saplings which reach maturity, deer can affect the composition of woodland canopies for decades or even centuries (Gill, 2000). This is perhaps the most pervasive impact that deer have on woodlands. Heavy culling is the only alternative to small exclosures, but either option adds greatly to the costs of woodland re-establishment. Regeneration of coppice is a particular problem; fencing costs often make coppice management uneconomic (Putman and Moore, 1998; Gill, 2000). With hindsight, it can be seen that twentieth-century foresters planting on the open hill had it easy. As the early plantations reach maturity, foresters now face the challenge of replacing them with new woodlands that satisfy today's multiple objectives, and of doing so in a deer-rich environment. This is proving anything but straightforward.

11.1.1.iii Neighbour and personnel conflicts

Forest owners look to their neighbours to control their open range deer. Equally, sporting estates expect forest interests to maintain their fences in order to prevent hill deer being 'lost' into the forests. Failures (whether perceived or real) to fulfil these obligations cause friction. Sometimes sporting rights are retained when woods are sold, so the new owner is dependent on others to respond to deer break-ins, a frustrating situation which can lead to damaging delays. Tensions can even arise between stalkers and foresters on the same estate because traditional stalkers tend not to be interested (or skilled) in woodland stalking and give a low priority to controlling woodland deer. These negative impacts of deer are counterbalanced to some degree by two positive attributes. Firstly, deer are a sporting asset. Commercial letting of woodland stalking, together with sales of venison and other by-products, can generate a steady income for forest owners. This applies to red, roe and sika. Secondly, deer are a natural part of Scottish forests, so their presence increases the naturalness of the woodland habitat. This is clearly a less tangible asset but it is increasingly valued by society.

11.1.2 Impacts of afforestation on deer

Similarly, afforestation has both positive and negative impacts on deer. On the positive side, increasing the area of woodland constitutes a potential range expansion, especially for roe and sika deer for which twentieth-century afforestation was a major factor increasing their population and range. Also, deer grow larger and healthier in woodland, and reach sexual maturity earlier, because it is their natural habitat. On the negative side, although afforestation has led to overall range increases, the early decades after the establishment of fenced forests see localised range contraction, especially for red deer. Frequently, the areas lost to deer are their prime wintering grounds. This increases deer densities on the open hill, putting deer under extra stress and at extra risk of winter mortality. In turn, this results in greater impact on the montane vegetation, increased likelihood of deer marauding onto agricultural land, and more pressure on forest fences. The loss of wintering ground has the greatest effect on stags. They therefore tend to be the first to break into plantations and onto arable land, resulting in far more stags being shot for these reasons than hinds. This exacerbates the problem of imbalanced sex ratios (Section 7.5.1).

11.1.3 Management responses

For many decades foresters have endeavoured to exclude all deer, but it is now accepted that sooner or later deer will inevitably gain access to forests and that damage will occur. It is thus essential that the traditionally reactive approach to deer damage is replaced by proactive, predictive management of this wildlife resource (Mayle, 1996). The questions for forest managers are, firstly, what is the acceptable deer density, and secondly, how to achieve it. Answers to the first question will vary widely according to overall management objectives, and the relative importance attached to timber production, woodland stalking, amenity, conservation and naturalness. Options for attaining the desired deer density include the following:

- **Forest design for wildlife management.** If deer are invisible in dense woodland it is impossible to cull or manage them intelligently (Cooper, 1987). Without reliable population data, culling simply becomes predation, not management. Prior (1994) argues that most deer problems in forests can be avoided by building deer into the planning stages of plantations as a site factor, a practice that was conspicuous by its absence until recently (Mutch, 1987). Forest design options to facilitate wildlife management include excluding deer wintering areas from plantations, leaving open corridors within woodlands, especially near water-courses, and creating deer lawns of palatable species to attract deer to known locations (Ratcliffe, 1985). It is also common practice to include deer leaps in forest fences – ramped gaps which allow deer to jump out but not in.

- **Adequate control personnel.** At present, many rangers are responsible for vast areas. In Shin Forest, for example, there is one stalker per 8,000ha (McLean, 1993). Large numbers of control personnel are needed not only because of the practical difficulties of managing forest deer but also because the enhanced productivity and lower mortality of woodland deer can necessitate culling rates as high as 25 per cent (Ratcliffe, 1998). Yet at present perhaps half of all Scottish forests have no wildlife management whatever, and the FC only employs sixty-four rangers (= one ranger per 59,000ha) (Wigan, 1996b). Given the areas that rangers are often expected to manage, it becomes effectively impossible to control deer populations at densities of 20 deer/km^2 or more (Latham *et al.*, 1998).
- **Habitat management.** Several options exist for managing deer by manipulating their habitat (Putman, 1998), including browse plots, diversionary feeding and substitution. Browse plots are areas planted with 'sacrificial' trees of species preferred by deer. Diversionary feeding works through the creation of attractive alternative forage. In both cases the hope is that the deer will satisfy their needs without needing to utilise the valuable commercial tree crop. Substitution involves allowing deer access to older plantations when they are fenced out of newly established ones. Given the behavioural unknowns concerning deer damage to trees, this is a potentially risky option; fence removal can sometimes result in serious damage (Scott, 1998). For managers with multiple objectives, such as estate owners and conservation organisations, it is nevertheless attractive because it can result in less deer pressure on fences, reduced fence maintenance costs, better quality deer, and enhanced naturalness. Forest owners aiming single-mindedly for timber production, however, have little incentive to risk their investment.
- **Individual tree guards and chemical repellents.** These strategies can be effective for small amenity woodlands, but are prohibitively expensive elsewhere.

Putman (1998) argues strongly that no single approach is likely to be effective, and therefore that a combination of population control and habitat manipulation will probably yield the most satisfactory results.

11.1.4 Restoring native woodlands: fencing versus *culling*

Of all the issues surrounding the impacts of red deer on the natural heritage, the one of greatest concern has been deer damage to native woodlands. This has been a long-standing worry. In 1892 Nairne (in Smout, 2000a: 58) observed that 'natural reproduction can never go on in . . . forests where deer are present, as they destroy the young trees with avidity'. It has long been recognised that high densities of red deer

FIGURE 11.4 *Mature Scots pine at Rothiemurchus near Aviemore. Note the absence of regenerating young trees. Photo: © Lorne Gill/SNH.*

can prevent the natural regeneration of native pinewoods and oakwoods, and that there has been none in some woods since the eighteenth century (Steven and Carlisle, 1959) (Fig. 11.4). There are several factors which will influence both the occurrence and type of regeneration (SNH, 1994b; Hester and Miller, 1995), but a high density of herbivores, notably red deer and sheep, is regarded as the paramount hindrance (SNH, 2000e). It can completely suppress regeneration. This is demonstrated by the frequent success of exclosure fencing in promoting forest re-establishment near to seed sources (Scott *et al.*, 2000) and by the presence of trees in places inaccessible to deer, such as crags and islands (Fig. 11.5).

What is the best strategy for addressing this issue? In most situations it comes down to a choice between fencing deer out of woodlands and heavy culling of deer around and within them. The pros and cons have often been debated (Staines *et al.*, 1995; Ratcliffe, 1998; Miller *et al.*, 1998). The salient advantage of fencing is that it works; exclosure fencing can 'kick start' regeneration during the critical early period. It may also be the only practical solution, especially for protecting woodland that is highly valued for aesthetic, economic or genetic reasons. Furthermore some ecological groups advocate fencing in preference to culling because it minimises on-going human intervention (Featherstone, 1997). Consequently, fencing out herbivores was, for many years, the main recommendation for encouraging natural regeneration.

However, deer fencing has been rapidly falling out of favour with the recognition of the many problems and disadvantages associated with it. There are at least eight:

FIGURE 11.5 *Where the deer cannot reach: dense vegetation (both trees and understorey) on an island in Loch Eilt contrasts with the bare hills beyond. Photo: © the author.*

1. Fencing simply moves the deer problem elsewhere, rather than solving it.
2. It restricts deer movement, concentrating the herd and sometimes cutting it off from lower, sheltered ground.
3. The visual impact is considerable, not only being unsightly in itself but also creating intrusive vegetation effects. Either side of fences, the juxtaposition of

grazed and ungrazed ground vegetation, and of woodland and open ground, creates hard, straight edges which are the antithesis of a natural-looking structure (Fig. 11.6). Such vegetation contrasts can sometimes be more visually intrusive than the fence itself.

4. Fences impact negatively on recreation by impeding human access.
5. They are expensive to erect, maintain and dismantle.
6. Exclosure fencing can result in a reduction of floral diversity as, in the absence of grazing pressure, the ground vegetation may become dominated by just a few species such as heather and blaeberry (Milne *et al.*, 1998).
7. Fences around and within woodlands cause significant mortality in certain bird species, primarily capercaillie, black grouse and red grouse (Baines and Summers, 1997). For the threatened capercaillie (Section 7.3.1), fence strikes are the single greatest cause of mortality (Petty, 2000). This creates a painful dilemma: the fences which help to recreate the lost habitats which will enable populations of capercaillie and black game to recover are lethal to those very species (McCall, 1998). Consequently, the National Trust for Scotland (NTS), the RSPB and the FC are all experimenting with ways of making fences more visible to woodland grouse (Summers and Dugan, 2001), experiments which have greatly reduced fence strikes (Petty, 2000). Inevitably, though, they maximise the visual intrusiveness of fences. Efforts to promote capercaillie by removing fencing and reducing deer populations are being made by the FC, the NTS at Mar Lodge, and by a few private estates such as Glen Tanar.
8. The final disadvantage of deer fencing is that, ironically, it may be counter-productive for natural regeneration. The sudden removal of grazing pressure following fencing allows established but previously suppressed seedlings to thrive, but the rapid development of prolific ground vegetation can sometimes restrict or prevent on-going regeneration (Miller *et al.*, 1998; Humphrey and Nixon, 1999). This dense, ungrazed vegetation may also work against the breeding success of capercaillie and black grouse (Dugan, 2000). A complete cessation of grazing can also increase the numbers of field voles which destroy seedlings (SNH, 2000e).

Given this litany of problems, an alternative approach has been attempted, namely the reduction of deer numbers by heavy culling. This is now widely regarded as the ideal ecological solution. The precise threshold population density which needs to be attained before regeneration is likely to succeed varies much between sites (Stewart and Hester, 1998), but in the uplands it is often around 4–8 red deer/km^2 (c.f. typical winter densities on low ground of more than 150/km^2) (SNH, 2000e). In the lowlands the limited data that exist suggest that regeneration can sometimes succeed at densities as high as 15–25 deer/km^2 (Harmer and Gill, 2000). At densities less than 4 deer/km^2 the presence of deer actually promotes regeneration without threatening the success of seedlings which establish. This is because herbivores reduce vegetation competition through grazing and create gaps in the field layer which act as microsites for seedling establishment (Hester and Miller, 1995; Miller *et al.*, 1998). Light grazing

FIGURE 11.6 *Vegetation contrasts either side of a deer fence in Glenfinnan: ungrazed rank heather to the left, close-cropped grass to the right. Photo: © the author.*

also promotes greater diversity in vegetation structure and species composition than either overgrazing or the complete absence of grazing that results from exclosure fencing (Humphrey and Nixon, 1999). Thus the best and most natural prescription for encouraging regeneration of woodland and scrub is not the complete exclusion of herbivores (which are important components of natural ecosystems) but reducing grazing pressure to low levels (Hester and Miller, 1995; Milne *et al.*, 1998; SNH, 2000e).

Some bold experiments with heavy culling and fence removal to promote natural regeneration have been underway since the late 1980s to test the feasibility of this approach (Fig. 11.7). The most notable and large-scale examples have been undertaken by SNH at the Creag Meagaidh NNR (Ramsay, 1996) and by the RSPB at Abernethy (Beaumont *et al.*, 1995; Taylor, 1995). At Abernethy, reduction of deer density from 12/km^2 to less than 5/km^2 and the removal of 40km of fencing has resulted in a wide array of ecological and human benefits (Dugan, 1997). Ecologically the results have included prolific regeneration of Scots pine, thriving ground vegetation, reduced deer mortality and a dramatic increase in numbers of black grouse. Human and economic benefits have included improved access and aesthetics for the 100,000 or more annual visitors, a reduced maintenance budget, and, surprisingly, sustained sport for stalking tenants.

Heavy culling is not without its own problems and challenges, however. Firstly, it is costly, although this can be largely offset by the sale of hunting permits and venison. Secondly, it is still resisted by many private landowners. This is partly due to the perceived financial risks involved and partly to the persistence of certain Victorian myths (Section 7.6.1). It is significant, for example, that neither of the examples given above were carried out by private owners. However, private estates are now under heavy pressure from the DCS to increase their culls (Section 7.6.1), and some are experimenting successfully with fence removal. Glenfinnan Estate, for example, removed fencing from a twenty-year-old plantation in 1996 and the forest has thus far suffered no economically significant deer damage (Warren and Gibson, 2002). Thirdly, it might prove hard to sustain the kind of early success demonstrated at Creag Meagaidh and Abernethy. Over time, the reproductive performance of the woodland deer populations is bound to increase and the newly regenerated seedlings will become young trees, making higher culls necessary in increasingly dense woodland (Ratcliffe, 1998). Dugan (2000), however, dismisses such nagging worries, arguing that, at Abernethy at least, long-term habitat objectives and the sustained provision of appropriate project resources will allow the current low deer densities (less than 3.5/km^2) to be maintained. Fourthly, because levels of deer damage are only weakly related to deer density, culling is only likely to prove successful if deer numbers are held at very low levels over a large area in the long term (Putman, 1998). In no sense, therefore, is population reduction a simple panacea.

Local factors and management objectives will determine whether fencing or culling is the most appropriate solution. Increasingly, however, it is being realised that a combination of both tools is likely to achieve the best results, using culling backed up with fencing where necessary. Ratcliffe (1998) argues that although the days when large-scale perimeter fencing was used automatically are over, fencing will probably always have a role to play, even if only in the form of small-scale temporary exclosures moved around woodlands over time (Humphrey and Nixon, 1999) or to protect valuable seed sources and stands of vulnerable trees like aspen.

FIGURE 11.7 *Naturally regenerated young Scots pine trees around a mature mother tree, Inshriach, Strathspey. Photo: © Neil McIntyre.*

In conclusion, it appears that the long decades of friction between foresters and deer managers may be drawing to a close as it is realised that each have assets to offer the other. Attempts to manage the two enterprises in watertight compartments have demonstrably failed, so even if cooperation is reluctant it is an unavoidable necessity.

One symbol of this dawning era of collaboration is the increasing involvement of woodland interests in deer management groups (Section 7.5.2) (Heggs, 1997). Having accepted that woodland has to be managed *with* deer, not against them, the emphasis now is on finding ways to maximise the positives and minimise the negative interactions. This new climate of cooperation has come none too soon. When combined with some of the recent trends in land use change, the rapidly increasing populations of most deer species are going to make deer management an increasingly significant concern not only for foresters but for farmers, crofters and conservationists alike. Putman and Moore (1998) identify several such land use trends which are exacerbating deer-related problems: increasing agricultural set-aside, increasing use of broadleaved tree species, the emphasis on continuous cover silvicultural systems, and, notably, the spread of farm woodland schemes which facilitates further increases in deer range and density. The last of these – the blurring of the once rigid demarcation between farming and forestry – is the focus of the second half of the chapter.

11.2 INTEGRATING FARMING AND FORESTRY

11.2.1 Overcoming a tradition of separation

Together, farming and forestry dominate rural Scotland. Yet although they operate on adjacent land, differences between their production cycles, their markets and their tax treatments have created largely separate (and sometimes rival) industries. Robinson (1990) even uses the term 'apartheid' to characterise the relationship between the two sectors. While most farmers appreciate the existing trees and woods in the landscape, few have a positive attitude towards woodland planting and management (Scambler, 1989; Lloyd *et al.*, 1996). Only on private estates is there an established tradition of integration. Why is this?

Part of the answer is financial. Forestry represents a long-term investment, generating sporadic income and tying up land which could be producing an annual return. Relative to agricultural crops, trees yield a poor return after decades of growth. Forestry frequently reduces the capital value of the land, while the long-term and (in effect) irreversible nature of woodland establishment reduces a farmer's ability to adapt in a rapidly changing world. A second factor explaining why no tradition of farm forestry exists in Scotland is that, unlike all other crops produced by tenant farmers, trees on tenanted farms remain the property of the landowner. Given that 35 per cent of Scottish farms are tenanted (Wightman, 1996a), this is significant.

A third group of reasons for farmers' traditional resistance to forestry are psychological or perceptual. The 'either/or' mentality which has grown up in recent times contrasts sharply with many other parts of Europe where farmers are familiar with woodland management and where silvicultural and agricultural skills go hand in hand (Hulbert *et al.*, 1999). The beliefs that 'good land is for food production' and that 'farmers are not foresters' are deeply imbedded in Britain's farming culture (Clark and Johnson, 1993; Lloyd *et al.*, 1996). Many farmers feel that planting trees on arable land is totally inappropriate (Mather, 1996a). Such 'in-grain-ed' attitudes are an inevitable result of a century in which national policies were single-mindedly orientated towards self-sufficiency in foodstuffs. This ensured that forestry was only practised on the poorer, marginal land not required for farming.

Furthermore, the treelessness of much of the Scottish landscape in recent times has fostered a subliminal perception amongst farmers that forestry is an alien activity, carried out by 'others' and to be resisted, a perception that was reinforced by the rapidity and aggressive style of much twentieth-century afforestation. In the period 1975–90, for example, the forested area expanded by 40 per cent, mostly at the expense of upland sheep farms (Mather and Thomson, 1995), doing nothing to endear forestry (or foresters) to farmers, and doing much to engender a siege mentality. When the forested area in a district exceeds 30 per cent, agriculture tends to go into decline (Mather and Thomson, 1995). Farmers often perceive forests negatively as a source of vermin and as a potential problem in relation to public access, vandalism and fly-tipping (Lloyd *et al.*, 1996; Edwards and Gemmel, 1999). Tenancy arrangements in which woods remain the realm of the landowner reinforce the 'otherness' of forest activities in the minds of many farmers.

Integrating agricultural employment with forestry was repeatedly recommended throughout the twentieth century, from the 1918 Acland Committee to the inter-war Forest Workers Holdings Programme to the FC's social policy of the 1950s (Mackay, 1995). During the 1970s and 1980s there was intense discussion amongst both farming and forestry circles of how best to diversify while maintaining farm incomes, with some arguing that farming faced a stark choice between integration or contraction (MacBrayne, 1982). Farm forestry featured prominently in these discussions, and some practical experimentation was carried out (Halhead, 1987). Despite all of this, little changed on the ground. Forestry grants were geared to foresters, not farmers, and although tax incentives proved irresistible to private forest companies they were irrelevant to struggling hill farmers (Mather and Murray, 1988).

By the mid-1980s, however, the established perception of forestry and farming as competing land uses had begun to change as the strength of the case for woodland expansion on farms became apparent. The main arguments, all of which have been strengthened by subsequent events, were that:

- most farmland could grow trees (and historically did grow trees)
- diversification is desirable and necessary for economic and environmental reasons
- there is an over-supply of agricultural products and a timber deficit
- woodlands offer opportunities for recreation, wildlife, amenity and game sport
- social and practical considerations argue in favour of farmers continuing to manage rural land.

During the last fifteen years the barriers (both perceived and real) preventing integration have been progressively reduced. A combination of environmental, social and political forces is now pushing rural land use away from monocultural production and towards multiple use, while the dire state of agriculture (Section 5.2.3), especially hill farming, is forcing farmers to investigate alternative land uses (Section 5.3.3.i). All this augurs well for both farm woodlands and agroforestry.

11.2.2 Farm woodlands

Until the mid-1980s the high value of agricultural land, CAP subsidies, and the absence of financial inducements militated heavily against farm woodlands in the UK. Since that time there has been a radical change in the climate for farm forestry as the importance of non-market, environmental criteria such as landscape enhancement and biodiversity has rapidly grown (Crabtree *et al.*, 1997; Kirby *et al.*, 1999). Recognising the economic disincentives, public money has been made available at ever more generous levels to help bridge the gap. The Broadleaved Woodland Grant Scheme of 1985 started the trend, followed by the 1988 Farm Woodland Scheme. This was replaced in 1992 by the Farm Woodland Premium Scheme (FWPS) with its 'better land supplement' which proved attractive to farmers throughout Scotland, primarily to improve amenity, shelter and wildlife habitat; timber production is a minority objective (Crabtree *et al.*, 1997). Grant levels have continued to rise since then, to a total spend of some £4.5 million in 1999/2000, and the FC's target for annual planting under the FWPS is now set at 4,000–5,000ha (FC, 2000a). There has been an increasing emphasis on environmental objectives, and grants are also now available for short rotation coppice or 'arable energy' schemes (Coates, 1999; Dawson, 1999).

The FWPS explicitly encourages forestry to come 'down the hill' onto good farming land. This is the inverse of twentieth-century policy priorities but is in keeping with the government's desire to see more of a mosaic of rural land uses in which forestry plays a prominent role (SODD, 1999c). There is no desire, however, to see widespread afforestation of prime agricultural land, given that it is such a scarce asset, nor to plant trees on the better quality in-bye land which is crucial for the maintainence of hill farming enterprises. Nevertheless, as forestry moves onto better land (including set-aside) it opens up opportunities for new tree species and new types of forestry that cannot be considered in the exposed, nutrient-poor uplands to which forestry has hitherto been confined.

Initially, the take-up on the various schemes was limited. Given that agricultural subsidies were typically running at levels some 100 times greater than the support available for forestry, the incentives proved inadequate to persuade farmers to take the 'quantum leap' now being expected of them. However attractive the incentives, many farmers simply have no interest in trees (Scambler, 1989), and deep-seated views established over generations rarely change overnight (Mather, 1996a). Furthermore, the probable high demand for public access offsets the positive attributes of farm woodlands in the eyes of many farmers (Sanders, 1999), as does the fear that increasing the woodland area will increase the numbers of deer and hence the severity of deer damage to crops (Scott and Palmer, 2000). However, the initial inertia is now being overcome. By the end of 1998, 30,722ha of woodland had been planted on 2,113 farms, mostly consisting of small (3ha or less) broadleaved woodlands, and these have led to substantial increases in biodiversity and enhancement of the local landscape (Coates, 1999; Egdell, 1999; Fig. 11.8). Most recently, as a result of 'modulation' under the EU's Rural Development Regulation (Section 5.2.1.ii), significant additional funds are being made available to farmers to establish woodlands on farmland.

FIGURE 11.8 *Farm woodlands in the landscape: arable farmland, Cargill, Perthshire. Photo: © SNH.*

Forestry on crofting lands is also now proving a quiet success and is generating widespread interest (Shucksmith *et al.*, 1996). Traditionally, trees have played little part in crofting, partly because trees belonged to the laird. Since 1992, however, crofters have been allowed to plant trees on their common grazings. Though cautious at first, an increasing number of crofters are embracing this opportunity of diversifying (Crofters Commission, 1999). Almost 1,000 have now established woodland schemes ranging in size from 1ha to 450ha (FC, 1999a). The extent to which these trends in farm woodland establishment and crofter forestry continue will depend crucially on the level of incentives available, and on the nature of CAP reform (Thomson, 1993b), but if large areas of countryside come out of farming in years to come, woodlands represent an obvious alternative land use, one which can contribute diverse, multiple benefits to rural development (Coates, 1999).

11.2.3 Agroforestry

Whereas farm woodlands represent integration of agriculture and forestry at the farm scale, agroforestry involves integration at the field scale (Jarvis, 1991).[3] There are two main approaches:

1. Silvopastoral systems combine trees with livestock, the animals grazing amongst trees planted at wide spacings until increasing tree shade prevents adequate pasture growth. Sheep grazing amongst sycamore is a frequent combination.

2. Silvoarable systems combine trees with arable crops. Fast-growing tree species such as poplars, or fruit- and nut-producing trees, are planted in widely spaced rows with an arable crop in between.

Both approaches have the potential to deliver greater yields and financial returns than either conventional agriculture or forestry in isolation (Knowles, 1991; Sinclair, 1999), especially in the lowlands, although economic appraisals are sensitive to the assumptions made (Thomas and Willis, 2000). Shelterbelts, being variously classed as farm woodlands or agroforestry, constitute an intermediate approach between silvo-pastoral and silvoarable systems. Biomass forestry (Section 4.5.5) can also be regarded as a form of agroforestry. Historically agroforestry was common in Britain, and it is widespread today in southern Europe and the tropics (Nair, 1991; Long and Nair, 1999), but it is not commercially significant in the UK at present. This may be about to change.

Research projects initiated in the mid-1980s, stimulated by the success of silvo-pastoralism in New Zealand (Knowles, 1991) and the emergence of European food surpluses, are now yielding promising results. Crucially, agroforestry offers farmers a flexible way of diversifying and spreading economic risk as they adjust to changing circumstances (Davies *et al.*, 2000; Doyle and Thomas, 2000). It allows new (tree) crops to be established with only minor initial changes to farming practices and yields. For example, in deciduous silvopastoral systems with a tree density of 400 trees/ha, levels of pasture production remain unaffected for 8–10 years and are reduced by less than 15 per cent even after twenty years; the livestock, meanwhile, are initially unaffected and benefit later on from the increasing shelter (McAdam and Sibbald, 2000). At its research centre near Crianlarich, the Scottish Agricultural College aims to integrate sheep production with new woodland, and has planted one-third of a hill sheep farm with birch, pine, willow and alder (Hulbert *et al.*, 1999; Davies *et al.*, 2000). This long-term project aims to show that continuing hill sheep production and woodland establishment are not mutually exclusive. It hopes to encourage diversification on hill farms by demonstrating that woodlands can be an investment for the future without penalising the present.

Silvoarable systems also allow flexibility. They allow a cash-flow from an arable area to be maintained during the early years of tree establishment, thereby minimising the length of time for which land is 'locked up' yielding little or no income prior to the commencement of timber production. Both silvoarable and silvopastoral systems can be highly productive (McAdam *et al.*, 1999a), can increase biodiversity on farms (Bullock *et al.*, 1994; Burgess, 1999), and can have a wide range of non-market benefits for animal welfare, wildlife and landscape (Hislop and Claridge, 2000). Except in areas of very high agricultural productivity, woodlands also enhance the market value of a farm, especially where shooting, conservation and landscape value are high (Sanders, 1999). For all these reasons, agroforestry is increasingly being regarded as a sustainable and realistic land use option, and it can even be argued that it is more sustainable than either agriculture or farm woodlands (McAdam *et al.*, 1999b).

However, despite the apparently bright prospects, some formidable constraints and obstacles stand in the way of an expansion of agroforesty (Long and Nair, 1999). Despite its greater measure of flexibility relative to pure forestry, it nevertheless requires long-term commitments from farmers at a time of uncertainty, hardship and rapid change. Its commercial viability remains doubtful (especially since the collapse of timber prices in the late 1990s), and there remain many unknowns about the interactions between trees, stock and arable crops (Sinclair *et al.*, 2000). It is also more complex to manage than monocultural production. Moreover, no one yet knows whether farmers and the public will adopt and welcome this novel approach. Thus far, although the scientific and forestry communities have shown considerable enthusiasm for silvopastoralism in particular, uptake by hill farmers has been minimal (Hulbert et al., 1999). This is partly because of ignorance and partly because adequate financial incentives are lacking (Doyle and Thomas, 2000).

In practice, then, future uptake of agroforestry in Scotland will depend to a great extent on fluctuations in commodity prices, dissemination of information to farmers, and changes to agricultural and forestry support mechanisms. One of the reasons why agroforestry is so restricted in the UK is the lack of a supportive institutional infrastructure. Although grants specifically for agroforestry have been available since 1991, grant aid has remained primarily orientated either to agriculture or to forestry but not to combinations of the two (Bullock *et al.*, 1994). On the other hand, agroforestry seems to be in tune with moves towards a more diverse, multi-faceted and sustainable agriculture (Sibbald, 1999) and it has the potential to deliver many of the policy objectives of a reformed CAP (McAdam, 1999). Consequently, it may well carve out a niche for itself and become a feature of rural Scotland, especially in areas such as the Borders which are amenable to silvopastoral systems. It cannot, however, be regarded as a panacea for all agricultural and forestry problems.

Policies and incentives to encourage the integration of farming and forestry tend to focus on getting more trees onto agricultural land. However, some also advocate the inverse: utilising existing woodlands for rough grazing, both by sheep and cattle but also perhaps for farmed deer. In New Zealand, grazing in forestry plantations is the most common form of silvopastoralism (Knowles, 1991), and historically such pasture woodlands have long been typical in Scotland (Begg and Watson, 2000; Quelch, 2001) although they are rare today. They confer many benefits both for the stock and for the woodlands, but also also carry risks such as tree and root damage.

11.3 CONCLUSION

Farm forestry and agroforestry represent two ways in which farming and forestry can be integrated at different scales. The first section of this chapter discussed the integration of forestry with deer management. A more holistic and altogether bolder vision, and one which represents one end of the integration continuum, is of woodland integrated with all land uses in a district. This is the concept envisaged by the Cairngorms Working Party (CWP, 1993) in its proposal for the creation of the extensive Forests of Mar and Strathspey, a century-long vision being taken forward by the

Cairngorms Partnership.[4] These two proposed forests, if they come to fruition, will be the antithesis of the wall-to-wall commercial woodlands of the twentieth century. Native woodland regeneration will be the explicit focus, but the forest regions will include a diverse mosaic of habitats, not only forest habitats of various kinds, but also open areas and patches of farmland. Deer will be managed within their natural woodland habitat and not excluded from it, and recreation will be encouraged. The RSPB (1993) recommends a third such forest in the Beauly catchment, and Summers *et al.* (1995) see further potential north and south of the Great Glen. Such forests, broadly defined, therefore have the potential of maintaining diversity of landscape, land use, habitat and employment (CWP, 1993). This holistic vision, in which forestry forms a landscape matrix, chimes in with recent thinking about the development of forest habitat networks (Hampson and Peterken, 1998). The explicit aim of such networks is to counter the long-standing trend of forest fragmentation and to re-integrate woodlands into the landscape.

In summary, the traditionally watertight division between agriculture and forestry is becoming ever more permeable. Progressive integration is set to continue. The provision of EU funds for farm forestry under the Rural Development Regulation (Section 5.2.1.ii) is symbolic of this new intermingling between formerly separate land uses. But although tree planting on farms is now at historically high levels (Burgess *et al.*, 2000), and despite the many potential niches for trees in farming landscapes (Sinclair *et al.*, 2000), farm forestry and agroforestry are unlikely to move centre stage without significant injections of public finance.

NOTES

1. Detailed discussions of the interactions between deer and forestry, together with possible management responses, case studies and census techniques, are provided by Mitchell *et al.* (1977), McIntosh (1987), Ratcliffe and Mayle (1992), Prior (1994, 1995), Mayle *et al.* (1998), Putman (1998) and Stewart and Hester (1998). The impact of deer on woodland biodiversity is discussed by Gill (2000). A special issue of *Forestry* (74(3), 2001) explores the ecological impacts of deer in woodland.
2. Comprehensive reviews of the extensive literature on the types and impacts of deer damage to trees are provided by Gill (1992a, 1992b). Probably the most significant long-term study of deer damage to forestry commenced in 1978 in Glenbranter Forest, Argyll, focusing on Sitka spruce; the results are reported by (*inter alia*) Welch *et al.* (1987, 1991, 1992) and Welch and Scott (1998). Harmer and Gill (2000) discuss deer damage to naturally regenerating broadleaved woodlands, and Putman and Moore (1998) review the impacts of deer on other land uses in lowland areas.
3. For detailed analysis and discussion of agroforestry systems and their development in the UK, see the volumes edited by Jarvis (1991), Burgess *et al.* (1999) and Hislop and Claridge (2000), and the collection of papers in *Scottish Forestry* 53(1), 1999.
4. Responsibility for this will soon pass to the new Cairngorms National Park Authority.

CHAPTER TWELVE

Case studies of conservation conflict

This chapter examines two high-profile conservation controversies which ran simultaneously for much of the 1990s, one concerning a site in the heart of Scotland, one at its western extremity. They are the latest in a long line of conflicts between economic development and conservation, conflicts which have shared common features such as the 'jobs *v.* conservation' tension (Section 8.3.2) and the great difficulty of weighing local feelings and wishes against national and international perspectives. Such apparently incommensurable viewpoints loomed large in the long-running tugs of war considered here. The first was a victory for economic development, the second for conservation.

12.1 The Cairngorms funicular railway

12.1.1 Background: the Cairngorms and downhill skiing

The Cairngorms massif is a magnificent mountain fastness, one of the largest and most unspoilt upland areas in Britain (Fig. 12.1). Its 4,000km^2 subarctic plateau and dramatic mountain landscapes are home to a wide range of flora and fauna, including arctic-alpine plants and rare bird species such as dotterel and snow bunting. It is regarded as the most important area within the EU for subarctic-oceanic mountains, moorland and boreal forest (Conroy *et al.*, 1990). As Crumley's (2000) powerful evocation of the area's special qualities shows, it is an environment of superlatives, whether viewed from an aesthetic or scientific perspective.[1] For all these reasons, the Cairngorms can lay claim to the title of Britain's premier mountain range of international significance (R. Watson, 1990; Curry-Lindahl, 1990). The 'unanimous conviction' of the Cairngorms Working Party was that 'the Cairngorms Area is an outstanding place of world importance which deserves special attention' (CWP, 1993: 2). The ultimate accolade is its status as a candidate World Heritage Site for its natural heritage interest. Yet, as in so many wild areas, these very qualities pose a threat by attracting ever greater numbers of people. The consequence has been a long history of controversy and conflict (Lambert, 2001b).

The need to protect the Cairngorms from inappropriate developments and from being 'loved to death' was recognised long ago. In 1928, for instance, the *Scots Magazine* campaigned for a Cairngorms National Park, and the Cairngorms National Nature Reserve was established in 1954; it now covers 260km^2 of the core montane zone (Fig. 12.2). Government-commissioned studies in 1931, 1945, 1947 and 1990 all

FIGURE 12.1 *Loch Avon and the Cairngorms in winter. Photo: © Patricia & Angus Macdonald/SNH.*

concluded that creating a national park would be the best way of preserving the area's unique characteristics while managing the growing influx of visitors, but the concept was resisted until the very end of the twentieth century (Section 8.3.4). Numerous landscape and nature conservation designations did, however, accrete across the

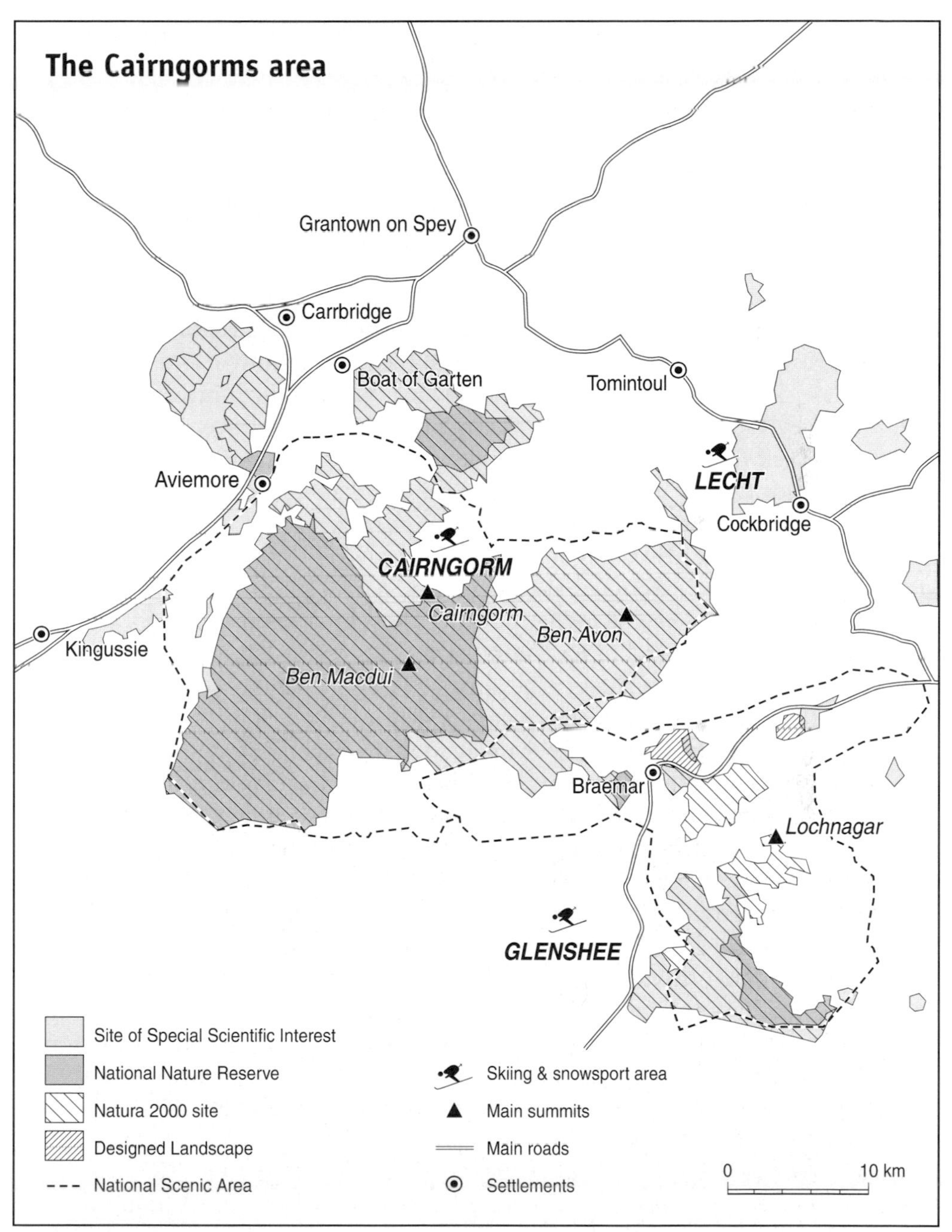

FIGURE 12.2 *The Cairngorms area, showing the location of the downhill ski developments at Cairngorm, Glenshee and the Lecht, and the primary conservation designations. After SNH (2000c).*

Cairngorms (Fig. 12.2), including more than fifty SSSIs. Despite this near-total coverage, it has often been argued that these conservation designations have proved utterly and uniformly ineffective in protecting the environment from the rising tide of threats (Curry-Lindahl, 1990; R. Watson, 1990). These pressures are numerous (A. Watson, 1990; Cairngorms Campaign, 1997). They include over-use of fragile mountain areas, unsuitable commercial developments, pressure from the expanding red deer population, uncontrolled muirburning, inappropriate fencing, and bulldozed tracks in remote mountain core areas (most infamously the track high on Beinn a'Bhuird which the NTS is removing at great expense (Johnston, 2000)).

But over the years, one of the strongest challenges to the preservation of the wild qualities of the Cairngorms has been the desire to develop the potential for downhill skiing. The Scottish Ski Club dates back to 1907, and there are now five skiing areas in the Highlands. Three of them – Glencoe, Glenshee, and Cairngorm – date from the late 1950s and early 1960s. The small area at the Lecht followed, and then in 1989 a major new ski area was opened on Aonach Mor (Nevis Range) near Fort William. The three developments within the Cairngorms area – Glenshee, the Lecht, and Cairngorm itself (Fig. 12.2) – were all built at times when environmental concerns had a low priority, and it shows. The landscape is widely eroded and scarred, the mechanical facilities and buildings are visually intrusive, and the ski roads provide easy access into fragile areas. In contrast, considerable efforts were made at Aonach Mor to minimise the environmental and visual impacts (Smith, 1997).

Despite its relatively long-standing presence, the ski development on Cairngorm has been the focus of several bitter controversies. Increasing environmental concern during the 1970s was brought to a head in 1980 by the proposal for a substantial westwards extension into the northern corries. Ski tows and chairlifts were proposed for Coire an t'Sneachda, Coire an Lochain and Lurcher's Gully, with the possibility of a road across the mouths of the corries. This proposal achieved a first in uniting Scotland's environmental movement in opposition (Smith, 1997). A public inquiry in 1981 supported the conservation case, and the Secretary of State rejected the plan. Nevertheless, planning guidelines for skiing developments published in 1984 identified Cairngorm as a prime area for further development (Brindley, 1990). In 1987 the Cairngorm Chairlift Company (CCC), supported by Highland Regional Council (HRC), put forward a modified proposal for ski development and road access into Lurcher's Gully, prompting another hard-fought battle. The Nature Conservancy Council and the Countryside Commission for Scotland, together with a coalition of environmental groups under the banner of the 'Save the Cairngorms Campaign', strongly opposed the plan, and in 1990 it too was rejected (Mackay, 1995).

12.1.2 The funicular controversy: proposals and chronology

Undaunted by these setbacks, and spurred into action by competition from the Aonach Mor development, the CCC unveiled a new development plan in May 1993. Accepting that expansion into new areas was never going to be permitted, they instead proposed a £16.5 million redevelopment of the existing skiing facilities, presented as an integral part of a strategy to revitalise tourism in the struggling Aviemore tourist area. The scheme involves replacing the chairlift with a 2km funicular (cable-drawn) railway running on a concrete viaduct from the existing Coire Cas car park at 640m to the top station at 1,100m, with the highest section hidden in a 250m-long tunnel to reduce visual impact on the skyline. Also included is the construction of a restaurant and multi-media visitor exhibition at the top station, a short walk from the 1,248m summit of Cairngorm itself. The scheme has the potential to increase the area's maximum capacity from 4,900 to 6,700 skiers and to double the catering capacity. The planning application was submitted in August 1994.

This time round persistence has been rewarded. Despite intense efforts by environmentalists to thwart the plan, HRC unanimously granted planning permission in April 1996. In a bid to present a positive and preferable alternative, a consortium of environmental groups then launched a proposal dubbed 'the GlenMore Gondola'. This envisaged an upgraded and redesigned ski area with new chairlifts which – critically – would only operate in the skiing season. The existing unsightly car parks and access roads would be replaced with a 2km gondola system giving a tree-top ride from Glen More Forest Park to the foot of the existing chairlift. Mass access to the sensitive summit area for non-skiers would thus be avoided, and the consortium argued that the potential benefits to the environment, local economy and tourist trade would all outweigh those offered by the funicular. Though an attractive proposal in many ways, it never got off the ground; the battle lines remained firmly drawn around the funicular concept.

Initially, SNH officially objected to the funicular, but in March 1997, after much agonising, it formally withdrew its objection, a decision that was roundly condemned both in conservation circles and (privately) by SNH staff. The basis on which SNH reluctantly granted approval was the production by CCC of a comprehensive 'closed system' visitor management plan to safeguard the plateau from mass access. The RSPB (which owns the neighbouring estate of Abernethy) and the WWF then took legal action in a last ditch attempt to halt the scheme, but to no avail; the Court of Session rejected their case in October 1998. Around the same time, looming financial problems seemed destined to succeed where conservationists had failed. Funding was eventually secured, however, (two-thirds coming from Highlands and Islands Enterprise (HIE) and the European Regional Development Fund, with stringent conditions attached), and preliminary works began in the summer of 1999. Construction of the funicular itself commenced in the spring of 2000 with completion due in 2002 (Fig. 12.3). Both sides in this protracted battle had strong arguments to make.[2]

12.1.2.i The case in favour of the funicular

No one disagrees that the existing 1960s ski facilities desperately need upgrading if downhill skiing is to continue at Cairngorm. Even by the mid-1980s they were seriously over-stretched (Cramond, 1990). Compared to equipment at other Scottish ski centres such as Aonach Mor, let alone to resorts in the European Alps, the two-stage chairlift is antiquated, slow, unreliable and expensive to maintain. Capacity is insufficient to match peak demand, and whenever the windspeed exceeds 45kph (which it does almost half the time) the chairlifts have to be stopped. The funicular will not be weather-prone, and it will be faster, taking just 4.5 minutes as against the current 17 minute ascent time. It will carry 1,200 skiers per hour, double the present capacity, and 500 (seated) at other times of year. The CCC argues that it will provide a reliable,

FIGURE 12.3 *The Cairngorm funicular nearing completion in late summer 2001.*
a. The upper section of track. The entrance to the tunnel is just visible on the skyline. Photo: © the author.
b. Looking down towards the bottom station, with Loch Morlich and Glen More forest beyond. Photo: © the author.

year-round tourist attraction which could increase the number of non-skiing visitors to 250,000 per annum (a fourfold increase on recent averages), as well as being able to cater for the elderly, the very young, and visitors with special needs (something that existing facilities are patently unable to offer).

All this, in turn, will arguably give a dramatic boost to the local economy, initially during the construction phase, and thereafter through all the tourism knock-ons that should follow. No other single project could bring so much public money into the region. It could create upwards of 200 jobs and undergird 2,500 more, no small contribution in an area as dependent on tourism as Strathspey. The CCC points out that mountain railways are commonplace in other European mountain areas, and argues that none of the other options for redevelopment (a modern chairlift, a cable car, a gondola system or a rack railway) would deliver all the suggested benefits. Moreover, in response to concern about the visual impact, it argues that the redevelopment will actually harmonise and enhance the ski area which has long been an eyesore as a result of its piecemeal development. It is not as if the funicular is being imposed on a pristine environment. Nor does the ski area directly affect a large proportion of the Cairngorms; its 844ha represents less than 1 per cent of the total area, leaving the vast, wild remainder for conservation and recreation (Cramond, 1990). In summary, therefore, the CCC argues that the economic and social benefits of the funicular greatly outweigh the environmental costs.

12.1.2.ii The case against the funicular

Those opposed to the mountain railway could not have disagreed with this conclusion more strongly. They angrily attacked the idea on two main fronts: economics and the environment. The central economic argument was that the scheme was not financially viable because the CCC's costings were based on over-optimistic visitor forecasts. This over-estimation might be as great as 100,000 users per annum, and, if so, there could be an on-going need for public subsidies. While accepting that some jobs would be created and sustained by the funicular, opponents criticised the very high job-creation cost, projected to be £97,000 per job compared to an average figure for the Highlands of £10,000 (Westbrook, 1994). They argued that the funicular could actually undermine other tourist attractions on Speyside. Further, they questioned the fairness and wisdom of investing a disproportionate amount of public money in just one of Scotland's ski centres to the potential detriment of the others. Finally, it was argued that the £16.5 million could be spent in far more effective and environmentally friendly ways on a large number of smaller projects such as the Forestry Commission's Visitor Centre in Glen More Forest Park or the RSPB's osprey-watching facilities at Loch Garten.

The environmental case rests on the international conservation significance of the Cairngorm plateau.[3] The subarctic ecosystem is hardy but fragile; the alpine plant communities, thin soils, and ground-nesting birds are all very susceptible to damage and disturbance. The conservation lobby argued that, far from encouraging more visitors into this sensitive environment, numbers should be reduced, otherwise the very resource that people come to enjoy – the mountain environment – will be degraded.

Winter visitors do little damage because the ground is protected by snow and frozen solid, but in spring and summer, when melting snow patches keep the ground saturated, the plateau is especially vulnerable. These arguments were the rationale behind SNH's insistence on a draconian visitor management plan (VMP) which will prevent people leaving the top station altogether. Conservationists fear a 'Trojan Horse' scenario, however, in which future commercial pressure could force these arrangements to be relaxed (Austin, 1997). This fear was borne out even during construction, with councillors and MSPs calling for unrestricted access onto the plateau. This is an impossible recommendation because the closed system is a legally binding condition of EU financial support. Conservationists advocate the principle of 'the long walk in' (Section 9.4.2) as a better means of safeguarding the core montane zone. Almost unanimous opposition to the stringent controls in the VMP has also come from recreational users such as hillwalkers, climbers and birdwatchers. Many say that they will simply go elsewhere in the Cairngorms, indicating that this 'solution' would merely displace the problem to other, already heavily used areas (Warren, 1997). Landscape arguments against the funicular focused on the aesthetic pollution of the development – the visual intrusiveness of the track and the large top station (Miles, 2001). Finally, some of the funicular's most passionate opponents (e.g. Crumley, 1997) have argued that downhill skiing is fundamentally inappropriate in the Cairngorms, and that the money would be better spent removing the ski development altogether.

12.1.3 Contrasting public perceptions: to ski or not to ski?

The funicular project has provoked powerful reactions at all stages from drawing board to construction. Given that the local area stands to benefit substantially in terms of jobs, increased tourism and multiplier effects, it is not surprising that a majority of locals strongly support the funicular. The CCC believes that it is taking all possible steps to safeguard the environment, and HIE has even described the project as 'a world-class example of environmentally-sensitive tourism development'.[4] Conversely, amongst the public further afield many are horrified. In a letter to *The Times* (27th March 1996), Chris Bonington and others suggested that it 'would do irreversible damage to one of Europe's most important wild areas'. Even SNH's deputy chairman, Chris Smout, writing in a personal capacity to *The Scotsman* (10th May 1996), argued that 'an unsatisfactory proposal of this sort, rushed through in an atmosphere of controversy, is the last thing that should be foisted on one of the most important mountain areas in northern Europe'. This conflict between local and non-local opinion was brought out by the consistently contrasting tone adopted by local and national news media. The former supported the project and bewailed the succession of hitches which threatened to halt it (e.g. 'Cloud of gloom hangs over funicular plans', *Strathspey and Badenoch Herald*, 18th July 1996), whereas national newspapers took a strongly pro-conservation line (e.g. 'Save our granite cathedrals', *The Independent*, 5th February 1996). The latter article, written by Jim Crumley, describes the funicular as 'a hideous prospect in the sanctity of nature's cathedral landscape'. Most damning of all is Sir John Lister-Kaye (2001: 7):

> As the funicular railway progresses steadily toward the summit of the most precious and fragile mountain ecosystem in Britain so it concretes itself into Scottish history as an icon of bad planning and fudged responsibility. . . . In the name of contorted economics and recreation, it stands as a triumph of crass materialism over wildness and the authority of nature.

The controversies which have dogged the Cairngorm ski area for the last twenty years highlight the dilemmas which beset downhill skiing in Scotland (Smith, 1997). On the one hand, it is a sport which has enjoyed rapid growth, is popular, and has brought considerable economic benefits into rural areas. On the other, by its very nature it has to operate in sensitive and vulnerable montane environments. Its commercial viability has always been dependent on unpredictable and erratic snowfall, and current climate change predictions suggest that ski operators may have to contend with warmer, wetter and windier winters in future (Kerr *et al.*, 1999; Scottish Executive, 2000a).[5] At present, however, it is apparent that calls to end downhill skiing will go unheeded. In the Cairngorms, at least, the Cairngorms Partnership (1997: 5) has a policy of facilitating downhill skiing, albeit within tight environmental constraints and restricted 'to the minimum necessary [areas] to secure sustainable use and management'.

The cases for and against the funicular are each strong and coherent, but as in so many environmental controversies, the debate involves trying to solve an equation of incommensurables. How should the value of human livelihoods be weighed alongside the value of wild landscapes? Both are important, yet the scales and the currencies are different. Would that it could be 'both and', but often it has to be 'either or', requiring hard decision making. As Cramond (1990: 28) comments, 'it is easy to make headlines but difficult to devise constructive solutions'. Had downhill skiing not come to Cairngorm in the 1960s there would have been no question about rejecting a funicular proposal in the 1990s; a commercial skiing development in a pristine mountain area of international conservation significance would have been unthinkable. However, given that Cairngorm is well established as Scotland's premier downhill ski area, given that this small part of the Cairngorms is already spoiled beyond short-term redemption, and given too that the economic fortunes of Strathspey are in need of revival, it is at least arguable that the right decision has been taken (though history may judge otherwise). The balance of arguments was rather different, however, in the case of the proposal for a superquarry in the Outer Hebrides.

12.2 ROCKS AND HARD PLACES: CONTROVERSIAL SUPERQUARRIES

12.2.1 Superquarries: the need for aggregates

The idea of pulping a mountain to make motorway foundations sounds bizarre, yet that is effectively what has been proposed for several remote sites around the Scottish coast. Aggregates (crushed rock, sands and gravels) are essential ingredients in most construction projects, especially roads, and millions of tonnes are required every year;

they are a *sine qua non* of modern societies. But in many of Europe's developed and densely populated areas, sufficient aggregates either do not exist because of the basic geology, or have already been quarried. In the UK, demand trebled between 1960 and 1990, and by the early 1990s south-east England could supply only about half of its aggregate requirements (Pearce, 1994). Government forecasts in 1991 predicted a rise in demand of 66 per cent to about 500 million tonnes per annum by 2011 (FoES, 1994), a rise which existing quarries in England would not be able to meet.

Inevitably, therefore, aggregate suppliers began to look further afield to areas which could supply large volumes of crushed rock. The coasts of northern and western Scotland appeared to be prime sites. Government studies had identified several locations with potential for superquarry development (Fig. 12.4), and the concept of developing coastal superquarries had been actively supported in successive government reports and planning guidelines since 1976.[6] Coasts are preferable to inland locations because distant markets can be supplied economically through bulk shipping. In this context, Scotland's coastal geology represents a world-class natural resource. Scotland has 70 per cent of the UK's hard rock resources (Pollock, 1994), and is home to the UK's only superquarry, at Glensanda beside Loch Linnhe (Fig. 12.4). Operating since 1986, Glensanda is currently producing over five million tonnes of granite each year with potential for three times that figure (Larson, 2001). Unlike the crowded south of England, with its tight environmental and planning constraints, parts of Scotland's coastline are sparsely populated, offering the potential for large scale developments with far less human impact and considerable economies of scale. Moreover, superquarries can provide significant economic boosts to rural areas with limited economic options. Glensanda, for example, employs 160 people and brings about £12 million into the Fort William area and the wider Scottish economy each year (Larson, 2001). In the early 1990s it was even being claimed that rock could replace oil as Scotland's great export in the twenty-first century, and that if this opportunity to meet the aggregate demands of industrial Europe was missed, other countries would reap the rewards – notably Norway with its many deep, sheltered fjords. It was in this context of predicted aggregate shortages, combined with a window of opportunity to capitalise on Scotland's rock riches, that a proposal was made for a superquarry at Lingarabay on South Harris.

12.2.2 The Lingarabay controversy

The ideal location for a superquarry is one with deep, sheltered water close in to the shore with enough room to manoeuvre large bulk-carrying vessels. One site that fits the bill is Lingarabay[7] on the sparsely populated south-east coast of Harris (Figs 12.4 & 12.5). It is not a pristine site, having a lengthy history of stop-start, small scale quarrying, but it is located within a National Scenic Area (NSA). In 1991 Redland Aggregates Ltd (now Lafarge Redland Aggregates Ltd (LRAL)) applied for planning permission for a £70 million development: a 4.6km^2, 500m deep superquarry producing 10 million tonnes per annum for sixty years, creating what would become one of the world's largest man-made craters. The proposal initiated a fiercely fought controversy which, ten years later, was still not finally resolved (Table 12.1).[8]

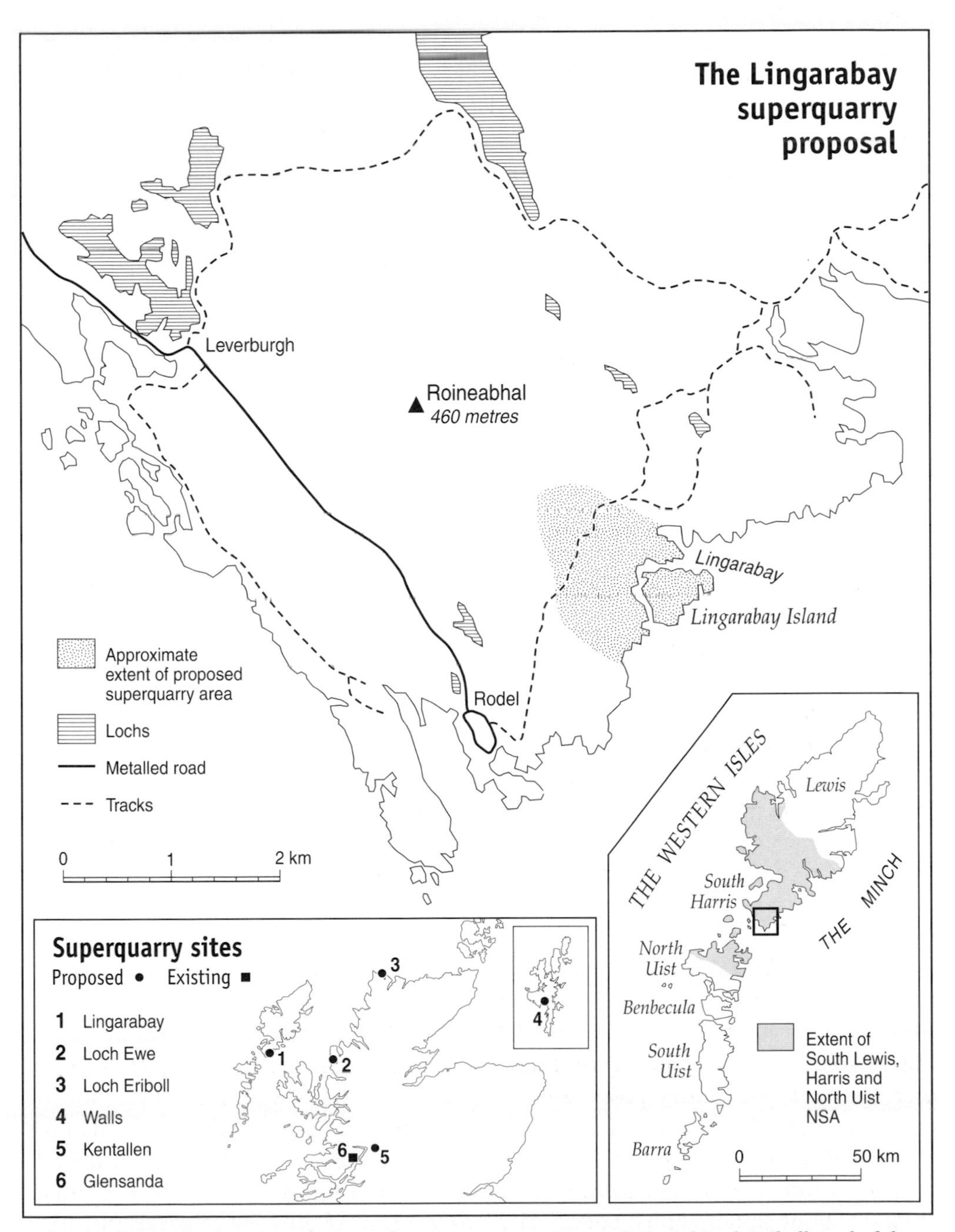

FIGURE 12.4 *The location of proposed superquarries at Lingarabay and Loch Eriboll, and of the Foster Yeoman operation at Glensanda. Other potential superquarry sites identified in government studies are also shown, as is the extent of the National Scenic Area. The detail of the proposal is after Owens and Cowell (1996).*

TABLE 12.1 Chronology of key events in the Lingarabay superquarry controversy. Information collated from sources detailed in Note 7.

March 1991	Redland Aggregates Ltd. applies for planning permission for superquarry
May 1993	Comhairle nan Eilean (the local council) approves the application subject to seventy-eight planning conditions
August 1993	Local referendum on South Harris: 62.1% in favour of superquarry
January 1994	Having received twenty-seven formal objections (including one from SNH), Secretary of State for Scotland calls in proposal for a Public Local Inquiry
October 1994–June 1995	Public Inquiry
May 1995	Local referendum: 68% against superquarry (83% turnout)
March 1996	Coastal Quarry Local Supporters Network formed
April 1999	Public Inquiry Reporter recommends that planning permission should be granted. Final report delivered to Secretary of State for decision. Judged too close to Scottish Parliament elections; decision referred to Scottish Executive
June 2000	Scottish Environment Minister postpones decision further; instructs SNH and JNCC to investigate the possible designation of the Lingarabay site as a Special Area of Conservation
August 2000	Redland (now Lafarge Redland Aggregates Ltd. (LRAL)) launches legal action to force a decision
October 2000	Court of Session rules in favour of LRAL, describing delay as 'scandalous', and orders ministers to make a decision within weeks
November 2000	Scottish Executive rejects superquarry proposal
December 2000	LRAL appeals
March 2001	LRAL begins attempt to resurrect 1965 planning consent for quarrying at Lingarabay

About two basic facts there was no disagreement: the superquarry would permanently transform the landscape, and it would generate jobs and inward investment. Almost every other 'fact' was disputed. Such was the strength of opposition that a Public Local Inquiry was held, running for a record eight months in 1994–5. Given that several other quarry proposals were waiting in the wings, much hung on the outcome. It represented a bifurcation between two radically different futures for Harris and Lewis. More generally, it was of great significance for the UK's minerals extraction industry and for policies for NSAs. It was also a practical test of political rhetoric about sustainability. Specifically, it was a trial of the new minerals planning guidelines, NPPG4 (SOED, 1994), which explicitly set out to provide a sustainable development framework for minerals extraction in Scotland.

In the event, politicians of the *ancien regime* chose to procrastinate and then to pass the buck to the incoming Scottish Parliament in 1999. By then, after a four year wait,

FIGURE 12.5 *Aerial view of south-east Harris showing the settlement of Rodel. The massif on the right is Roineabhal (460m), the right-hand flank of which was the proposed site of the superquarry. Photo: © Patricia & Angus Macdonald/SNH.*

the Inquiry's reporter had concluded in favour of the quarry on the grounds that the socio-economic benefits were of national importance and so outweighed the very considerable environmental concerns. The decision proved to be so hot a political potato, however, that further delays ensued. In 2000, when the Environment Minister asked SNH and JNCC to begin investigating the possible designation of the Lingarabay site as a Special Area of Conservation, LRAL resorted to court action to force a decision. Controversially, the Scottish Executive then overturned the recommendation of the Public Inquiry and rejected the proposal, arguing that the reporter had seriously underestimated the landscape impact on the NSA.

12.2.2.i Pro-superquarry arguments

Operationally and technically the site is ideal, potentially yielding some 600 million tonnes of anorthosite (a uniquely suitable igneous rock) from the rounded massif of Roineabhal (Fig. 12.5). The 24m deep harbour would allow direct loading onto ships of 60,000 tonnes or more, making it economically attractive. Furthermore, the proposal represented a development opportunity which had the potential to reverse the on-going depopulation, unemployment and economic decline that have long blighted Harris and Lewis. Crofting (including fishing) and tourism are the islands' mainstays, but neither have proved secure or substantial enough to reverse the trend. The population of Harris fell by 14.4 per cent in the 1980s, and unemployment stood at 21.5 per cent in 1991 (Black and Conway, 1996b). The steady exodus has continued since,

6.3 per cent of the adult population leaving the island between 1995 and 2000 (Ross, 2000b). LRAL believed that the quarry could halt (and perhaps reverse) this downward spiral by promising 200 or more new jobs, and by acting as a catalyst for an economically sustainable future. Moreover, for every tonne of rock produced, the company offered to pay 1p into a Community Development Fund, potentially bringing in £100,000 per annum to the island economy once full production was underway (though much less in the first decade or more). LRAL pointed out that our developed, Western lifestyles largely depend on the extraction of finite minerals – including oil and hard rock – and that unless the whole of society agrees to change its ways radically, utilisation of non-renewable resources will have to continue. On this basis, they accused environmentalists of the NIMBY (Not In My Back Yard) syndrome.

While accepting that the quarry would transform the landscape locally, LRAL promised that great care would be taken to minimise the environmental impact. It suggested that, in terms of the natural heritage, there was little that was unique or special about the site, and that the integrity of the NSA would remain largely unaffected because the quarry would only occupy 0.19 per cent of the designated area. Facing out to sea, it would be mostly invisible from Harris itself. LRAL even claimed that the artificial 'sea loch' backed by a 'corrie' would be 'a dramatic and beautiful new landscape' (in Mackenzie, 1998: 517) which could eventually become a tourist attraction, just as other large quarries have. This would counterbalance the inevitable losses to existing tourism that would result from industrial operations. Taking a broader perspective, LRAL argued that a single coastal superquarry would obviate the need for numerous smaller onshore quarries, thereby safeguarding other environments closer to major centres of population. Given this catalogue of benefits it is not surprising that HIE supported the proposal, or that Comhairle nan Eilean Siar (the Western Isles Island Council) initially voted 24–3 in favour of it. The night before the end of the Inquiry, however, the Council dramatically reversed its stance. Its minerals policy now excludes Harris.

12.2.2.ii Anti-superquarry arguments

Opposition to the proposal was vigorous. It was led by SNH and the LINK Quarry Group (LQG) (a consortium of environmental groups), and strongly backed by fish-farming and fishing interests. The LQG (1996) marshalled a powerful battery of arguments, illustrated with a shocking photomontage of how the superquarry might look (Fig. 12.6a). Given the site's location within a NSA, the LQG argued that such a development would be deeply inappropriate, replacing a tranquil, small-scale, traditional landscape with an enormous industrial complex. The objectives of the NSA could not fail to be compromised. The quarry would dramatically and permanently alter the environment, creating an eyesore clearly visible from Skye and even from the mainland (due to its large scale and the whiteness of anorthosite); views from five other NSAs would also be affected. Furthermore, it would generate dust, pollution (including light pollution at night) and high noise levels in an unusually quiet area, all of which would affect local residents and wildlife (especially birds). Full production would require the detonation of thirty-six tonnes of explosives per week

FIGURE 12.6 *Contrasting images of a rejected future: photomontages of the Lingarabay superquarry produced by the opposing camps at the time of the Public Inquiry.*
a. A photomontage produced for the LINK Quarry Group. Photomontage by Envision, and used with permission. Aerial photograph reproduced with the permission of Patricia and Angus Macdonald.
b. The proposed superquarry according to the quarry company. Photomontage produced by Hankinson Duckett Associates and SV Modelling, and reproduced with the permission of Lafarge Aggregates Ltd.

(McIntosh, 2001a). The intended end use of the aggregates – road building to facilitate more traffic – only served to rub salt in environmentalists' wounds.

Marine pollution was one of the key issues. Before loading up, incoming ships would dump their ballast waters, collected in industrial ports around the world. This would not only pollute the Grade A waters of the Minch, with knock-ons for plants, sea birds, and the economically important fish and shellfish stocks, but possibly introduce exotic organisms which could destroy native species. Such problems have proved severe in North American and Australian waters. LRAL attempted to quell such fears by guaranteeing that all ships would carry out a triple exchange of ballast waters in the open ocean prior to arrival, but critics were sceptical of their ability to impose such rigorous and novel procedures on ship captains. Another fear was that the increase in shipping in the Minch, potentially by 100 per cent, would greatly increase the risk of collisions with the laden oil tankers which sometimes use this route. Even a small oil spill here would be environmentally disastrous.

Opponents of the superquarry argued that although some jobs would indeed be created, LRAL's estimates were unrealistically high, and that most jobs would anyway go to non-locals. Such an influx could threaten traditional ways of life. For example, quarry operations on Sundays would offend the deeply held sabbatarian views of the Presbyterian residents. LRAL gave assurances that only essential maintenance would be carried out on Sundays, but (as for the funicular) there were fears of a 'Trojan Horse' scenario, whereby economic imperatives would in practice take precedence over religious sensibilities. Environmental damage could actually lead to job losses in the fishing, crofting and tourism sectors. Eventually the island's economy would become critically dependent on a single-company, single-industry development, itself subject to the vagaries of international markets. Critics went on to argue that even if projections of future aggregate demand proved correct, superquarries would simply be unnecessary if the construction industry was less wasteful and used more recycled material. The controversy also stirred up nationalist sentiments: why should a French-owned company satisfy English needs by violating one of Scotland's National Scenic Areas?

12.2.2.iii Evolving local perspectives

During the ten years since LRAL lodged their proposal, local feelings have ebbed and flowed. In a local referendum in 1993, 62 per cent in South Harris supported the development, though those closest to Lingarabay were marginally against. By the end of the Public Inquiry, however, 68 per cent voted against it (Black and Conway, 1996b). Later polls then showed opinion swinging back in favour. Given the divergent and changing views of locals and incomers, it is simplistic to talk of an undifferentiated 'local viewpoint'. Many indigenous residents resented the anti-development views held by the relatively wealthy incomers (referring to them disparagingly as 'white settlers' or 'drawbridge incomers'), and found the outsiders' perception of Harris as a 'wilderness' highly provocative (Barton, 1996; Shucksmith *et al.*, 1996). The long shadow of the Clearances (Section 3.1) also fell across the debate, with opponents of the quarry being accused of conspiring to perpetuate the depeopling of the island and the consequent destruction of the Harris way of life (Hunter, 1995; Mackenzie, 1998).

Because Harris people held divergent views on all the key issues, the dispute proved deeply divisive for the community during the long years that it was left in limbo. It seemed as if the superquarry had come to represent:

> . . . all things to all people: the hope for the future; the death of the island; a testament to the strength of Sabbath observance; a threat to the fishing industry; witness to island apathy; youth employment; or the destruction of the tourist trade. (Shucksmith *et al.*, 1996: 234)

Whereas the national debate focused on environmental and political considerations, local people were primarily concerned about ideological, religious and social issues. Shucksmith *et al.* (1996) found widespread dismay at the likely environmental impact, but a fatalistic, disempowered acceptance of the lack of alternatives. Given a choice between natural heritage damage and the loss of their spiritual and cultural heritage (especially the Gaelic language), most ranked people above landscape. Few welcomed the quarry idea *per se*, but all agreed that development of *some* sort was needed. They saw the quarry as the last tangible option for preserving island life – a price worth paying to restore the economic fortunes of Harris. When the quarry was rejected, the chairman of the Coastal Quarry Local Supporters Network commented bitterly that 'the Scottish Executive has sent a stark message . . . that flora and fauna are more important than people and jobs' (in Ross and Hardie, 2000).

In an apposite comparison, Shucksmith *et al.* (1996) contrast the views of the Harris islanders with the actual experience of people close to the Kishorn oil fabrication yard in Wester Ross. This operated between 1975 and 1986, providing as many as 4,500 jobs. It brought widespread material benefits but also profound social disruption in both the short and long term. A majority now say that they would have preferred small scale, locally based development which obviated the need for substantial in-migration of labourers. As the prospect of the Lingarabay superquarry recedes, the Kishorn experience suggests that the people of Harris have been saved from a future that they would have come to regret.

12.2.3 Retrospective: the debate with hindsight

Quite apart from the key points at issue, the Lingarabay debate was rich in compelling symbolism. Hunter (1995: 17), for example, identifies the irony that although the Gaelic language embodies a deep reverence for mountains, the long-term survival of the Gaelic-speaking Harris community was said to depend on 'the physical removal of just the sort of mountain to which Highlanders have long attached a huge emotional significance'. The dust has yet to settle from the controversy, but some things are already clear.

For a start, there is no doubt that the long delay worked against LRAL in a variety of ways. Given the evidence available at the time of the Public Inquiry, and in the light of existing planning guidelines, government policy, and the apparent lack of alternatives for economic revitalisation of the island, it is arguable that the decision to recommend planning approval was a defensible (if deeply regrettable) one. By the end of 2000, however, much had changed. Demand for aggregate, which was rising rapidly in the early 1990s, subsequently stabilised or fell across much of Europe, and UK predictions for demand for 1997–2007 dropped from 390 to 218 million tonnes (Cowell *et al.*, 1998). The Quarry Products Association has acknowledged that the demand since the 1992 peak has fallen far short of official projections, and it expects little change in demand over the next twenty years. As a result, Glensanda (amongst other quarries) is operating far below its capacity. Given that LRAL's application was tied to the original projections, these downward revisions dramatically weaken a central plank of their case. The LQG now argues that the only justification for permitting the development within a NSA would be if national need was established. Given the latest demand projections this would now be hard to argue. Further, the superquarry's status as the only hope for the island economy has been undermined by the recent success of Harris Development Ltd in creating jobs in electronics, tourism and fish processing. In fact, more jobs have been created for local people than the superquarry would have provided (McIntosh, 2001a).

Further afield, while the wrangling has continued in Scotland, Norway has snatched the initiative by opening up three coastal superquarries, so it is questionable whether there is now room in the European market for a new one in Scotland. Another factor that has gone against LRAL has been the sea-change in the political landscape. The New Labour administration elected in 1997 is committed to sustainable development and recycling, and has dramatically scaled down the road-building programme (both of which constitute a strengthening of existing policies), so demand for primary aggregates could actually decline. If so, the economic viability of superquarries might become marginal. The landfill tax has increased costs for quarry operators, and the aggregates tax of £1.60 per tonne that will apply from 2002 would add £16 million per annum to the quarry's running costs at full production, further threatening its economic viability.

The creation of the Scottish Parliament has also moved the goalposts. LRAL rested its case squarely on national (UK) needs, but since 1999 'national' has increasingly come to refer to Scotland. From a UK-wide viewpoint, meeting aggregate demand from fewer, bigger quarries may cause less environmental damage and affect fewer people, but a superquarry makes no sense in terms of national (Scottish) needs (Raemaekers and Boyack, 1999). This was, in fact, one of the grounds on which the Scottish Executive overturned the Public Inquiry's verdict; the Inquiry Reporter used the DETR's planning guidance on aggregates provision in England (Minerals Planning Guideline 6) as a factor in her decision, whereas the Scottish Executive argued that Scottish planning guidance on minerals (NPPG4) should have been the primary concern. Had the quarry gone ahead, the UK Government could have stood accused of 'exporting unsustainability' from populous England to the remote Scottish periphery (Owens and Cowell, 1996).

At the time of writing, the superquarry saga remains unfinished. LRAL are fighting on, and a proposal for a large quarry at Loch Eriboll (Fig. 12.4) is contained within Highland Council's structure plan. Nevertheless, given the Scottish Executive's refusal of planning permission for Lingarabay in 2000 and the now questionable economics of aggregate superquarries, it is probable that – like 'Ben's beach' in the film *Local Hero* – Lingarabay and other coastal quarry sites are going to be left in peace rather than in pieces.

12.3 RECURRENT THEMES IN ENVIRONMENTAL CONFLICTS

Woven through both these hard-fought sagas are certain common threads, features which have characterised most recent environmental controversies in Scotland.[9] Indeed, these recurrent themes are emblematic of conservation battles worldwide. Some of these – such as the desire to put one's own argument in the best possible light even at the risk of being 'economical with the truth' – are perhaps unavoidable human traits. This often results in selective use of evidence, and a reliance on exaggerated predictions and projections to bolster or demolish arguments (illustrated, for example, by the wildly varying job predictions quoted by opposing parties in both the debates described above). Another striking example of this trait is the radical contrast between the photomontages of Lingarabay produced by the opposing camps (Fig. 12.6), LQG's blinding white, full-frontal eyesore being barely recognisable as the same location as LRAL's muted and pleasingly aesthetic coastal scene.

In addition to such pervasive features, however, there are perhaps five recurrent aspects of high-profile environmental controversies which stand out.

1. **The 'insider' v. 'outsider' friction** (Section 8.3.2). Locals, driven by understandable self-interest, are usually in favour of economic development, while outsiders typically attach greater value to environmental preservation. This sets up considerable frictions, often exacerbated (as in the two examples here) by the prominent involvement of national and international conservation groups with little local representation. As Cairns (1998) comments, 'for many Highlanders, the [funicular] fight was a straight one between badly needed economic development and interfering conservationists'. Local support for the Harris superquarry was, in part, a backlash against the vociferous anti-quarry campaign led by incomers and non-local environmentalists (Shucksmith *et al.*, 1996).
2. **Polarisation.** The positions of 'developers', 'conservationists' and 'locals' typically become highly polarised, and the individuals or groups concerned frequently suffer from an extreme degree of stereotyping. In reality, of course, there is always a 'shades of grey' continuum of views, with considerable overlap and many exceptions to the stereotypical labels. The divergent, evolving views on Harris exemplify this. Polarised stereotyping makes irresistibly good copy for journalists, but it unhelpfully broadens the gulf separating the various 'camps' and makes for much rancour. Worse still, it greatly hinders conflict resolution and minimises the scope for a consensus-building approach.
3. **Contrasting interpretations of sustainable development.** Such is the authority of the sustainable development paradigm that all parties in these controversies appropriated it to their own ends, no one wishing to be tarred with the (politically suicidal) unsustainability brush. Thus conservation and recreation

interests are adamant that the funicular is a profoundly unsustainable project (Gill and Watson, 1996; Perkins, 1998) while the CCC maintains that it is a prime example of sustainable tourism. When economic sustainability and environmental sustainability are incompatible, which should be pursued? Strongly contrasting interpretations of sustainability also emerged during the Lingarabay debate as both LRAL and LQG sought to lay claim to (versions of) the concept in support of their case (Barton, 1996; Owens and Cowell, 1996; Mackenzie, 1998). Working out just what constitutes sustainable use of non-renewable resources such as aggregates is no easy matter (Kellett, 1995; Owens and Cowell, 1996). The chameleon-like quality of sustainable development thus undermines its practical effectiveness (Section 13.3.4). If by sleight of definition everything is sustainable then nothing is unsustainable. Similarly divergent interpretations of the precautionary principle (Section 13.2.2) emerged from these debates, especially concerning when and how it should be applied.

4. **Problems with planning**. Both sagas prompted numerous commentators to criticise the planning system. The system seems to guarantee maximum conflict by pushing the protagonists into entrenched positions as they fight their corner, tempting them to exaggerate their case to give it maximum impact. The middle ground becomes an ill-advised no-man's-land to occupy. Raemaekers and Boyack (1999) sugest that the adversarial inquiry system, modelled on legal practice, works against less powerful parties, and may not best serve the public interest. In a similar vein, Austin (1997) argues that the challenge of reconciling the real needs of local communities with the wider conservation interest requires a proactive approach, rather than simply reacting to developers' proposals. In particular, the situation concerning planning guidance for mineral use has been trenchantly criticised (FoES, 1994). NPPG4 (SOED, 1994) recognises that finite minerals cannot be developed sustainably. Instead, it requires, firstly, that the environmental and social impacts of extraction are minimised, and secondly that demand for aggregates is reduced through recycling and use of secondary aggregates. However, while the planning system can ensure the former it can do little to promote the latter, and it is weakened by its insistence that the level of extraction should be determined by the market (Raemaekers and Boyack, 1999). The Lingarabay Inquiry also put other fundamental questions in the spotlight, such as what kinds of argument and information should be allowed to influence planning decisions. Opponents of the quarry unashamedly drew on spiritual concepts to support their case (McIntosh, 2001a, 2001b), but LRAL refused to accept that 'subjective value judgements' could legitimately challenge 'objective facts' in the planning process (Owens and Cowell, 1996) (c.f. Section 13.1).
5. **The decision making conundrum**. This is perhaps the most intractable problem of all. Who should take decisions such as these? And who chooses the choosers? Given that proposals such as mountain railways and superquarries primarily affect a specific local area, should decisions be taken by local authorities? Or, given the national and international significance of the Cairngorm and Harris

environments, and the amounts of public money involved, should central government have the final say? Where on the spectrum from local council through the Scottish Parliament and the Westminster Government to the EU should such decisions be taken? On the one hand it seems patently unjust for the strong wishes of local communities to be trumped by the views of those living elsewhere. On the other hand, should the self-interest of the people of one locality determine the fate of national assets of international significance?[10] There is no objectively 'right' answer to this dichotomy: how should the value and quality of the lives of human beings (wherever they live) be weighed alongside the value and quality of the natural environment? As Barton (1996: 109, 118) comments:

> The grounds on which [the superquarry proposal] is evaluated, justified or condemned are no more than a manifestation of the environmental values of the evaluators. . . . Should Harris really be sacrificed as an open-cast mine to service England and the continent's unquenchable thirst for more roads, or preserved as an 'unspoilt wilderness' for visitors who spend most of their lives enjoying the convenience of a wealthy city lifestyle?

Even extending the techniques of Cost Benefit Analysis to incorporate intangible and unquantifiable attributes of the environment (Section 13.4) leaves this central dilemma unresolved. These debates about the processes of decision making – about whose views should count, how much, and when – touch on much broader issues. As Harrison and Burgess (2000: 1123) observe, 'reasons for valuing nature reflect, and are reflected in, more extensive commitments to how society should be organised and governed'. Factors like this explain why controversies such as these are so emotive and so difficult to resolve. They suggest that the funicular and superquarry controversies are unlikely to be the last environmental battles to convulse Scotland, and that the much-heralded 'age of consensus' may struggle to be birthed.

NOTES

1. The breadth of scientific interest is well illustrated in the collection of papers edited by McConnell and Conroy (1996) which spans the full range of the earth and biological sciences. The conservation significance of the Cairngorms is, by common consent, as high as in any part of Scotland. See, for example, Thompson *et al.* (1994), Watson (1996), Gordon *et al.* (1998) and Haynes *et al.* (2001) for discussions of its significance, its sensitivity, and of human impacts on the environment. The challenge of conservation management is explored in detail in the volume edited by Conroy *et al.* (1990).
2. The facts and arguments in the following two sections have been collated from various sources. The arguments in favour come primarily from Whittome (1996), CCC publicity material and media coverage. The case against draws on material from Gill and Watson (1996), Austin (1997) and Perkins (1998), together with information from the media and *SCENES*.
3. See note 1 for references on the conservation significance of the Cairngorms.
4. Quoted in *SCENES* 130: 7, 1998.

5. Global climate change could, however, work in *favour* of the Scottish downhill ski industry. If, as some models predict, the North Atlantic Drift (Gulf Stream) were to shift south, Scotland would be plunged into a much colder climate regime.
6. Coastal superquarries are defined in National Planning and Policy Guidelines 4 (NPPG4: Land for Mineral Working (SOED, 1994)) as those capable of producing more than 5 million tonnes (mt) per annum of crushed aggregate with reserves of at least 150 mt, to be transported by sea. See Black and Conway (1996b) for details of the origins and development of the concept.
7. Various spellings for the site exist, including Lingerbay (widely used during the controversy) and Lingerabay, but OS maps show it as Lingarabay.
8. Much has been written about the details of the superquarry controversy and the wide-ranging issues that it raises. See, for example, FoES (1994), Pearce (1994), Pollock (1994), Barton (1996), Black and Conway (1996b), LQG (1996), Owens and Cowell (1996), Dalby and Mackenzie (1997), Mackenzie (1998), Warren (1999d) and McIntosh (2001a, 2001b). This section is a simplified summary collated from these sources, together with media coverage and items in *SCENES*.
9. Another high-profile controversy in which these common themes were writ large was the showdown in the late 1980s between forest interests and nature conservationists over afforestation of the Flow Country. The significance of this episode has been examined in detail by Warren (2000a).
10. For examples of arguments on both sides of this debate in relation to the Cairngorms, contrast Grant (1990) with R. Drennan Watson (1990).

PART FOUR:

Thinking and deciding about the environment

CHAPTER THIRTEEN

Environmental ethics and decision making

13.1 Introduction: why bother with ethics?

Living in a 'green age' of international environmental concern, it is easy to assume that environmental arguments are self-evident and beyond question. Actually they raise a host of moral and ethical issues which lurk (often unrecognised or unacknowledged) just below the surface of environmental decision making. Ethics are commonly seen as being the preserve of philosophers or eminent advisory committees, or perhaps as being restricted to those issues which have obvious ethical dimensions such as genetic engineering (Section 5.3.2). Either way they are commonly perceived as having little to do with the practicalities of every day life. In fact, of course, everyone involved in making decisions about how land should (or should not) be used is consciously or unconsciously making value judgements all the time. By what criteria should such judgements be made? How should we decide that this is right and that is wrong, or (more usually) that this is better than that? Are there any absolutes, or are we adrift in a sea of relativism? Such questions underlie all environmental decisions, yet are too rarely addressed head on.

However obscure and irrelevant environmental philosophy may at first appear, it is actually the unseen air which environmental management breathes. In very practical terms, the values and preferences held within a democratic society affect government policies through the ballot box. Public policy requires public support. Shifts in thinking therefore lead to shifts in policy and in the distribution of funds, and these in turn largely determine land use. For that reason, only a society-wide change of mind, expressed through democratic choices, will ever put effective brakes on environmental degradation. At present, despite the apparently high profile of green issues, the bottom line is that very few people attach a higher priority to environmental quality than to their personal quality of life, a fact which relegates the environment to the second division of the public policy agenda. 'The benefits of abusing the environment appear still to most political actors to outweigh the costs' (Philip, 1998: 273), so the deterioration continues. It is immediately apparent, then, that how people think, what they believe, and the ethical foundations of their choices matter a great deal. Even if our moral intuitions are poorly articulated or barely recognised, they guide much of our behaviour. Constructing a coherant ethical framework for environmental values and decision making is a testing challenge, however. This chapter briefly explores certain important facets of that challenge.[1] Consonant with the hotly contested nature of much of the subject matter, few answers can be offered to the many questions which it raises.

There are two equal and opposite dangers for those concerned with the way that society interacts with the natural world. The first is to act without thinking, and the second is to think without acting. At one end of the spectrum, activists can sometimes be accused of not being 'thought through'. Actions exposed as being counter-productive or based on false premises lose any supposed moral authority or power to effect change. At the other end, academics (in particular) run the risk of getting immobilised in ineffectual 'eco-philosophising' which generates interesting debate but may impinge on the real world hardly at all. Pearce (1998: 17), for example, levels the accusation that 'environmental ethics as an academic discipline is not particularly useful in the practical fight to save natural environments', and that it is often worked out in an intellectual vacuum that takes no account of inconvenient realities like costs, democracy and politics. Avoiding these two extremes, actions grounded in a coherent intellectual framework would clearly be the ideal, but such a framework is strangely elusive. Persuading others of the rightness of any such set of principles proves to be even harder, especially in a postmodern context when so many shades of green philosophy are being vigorously promoted.

The co-existence of multiple conflicting environmental narratives is a leitmotif of our times. Views range from the strongly ecocentric, emphasising the natural, to technocentric viewpoints which stress the historic success of human ingenuity in extricating ourselves from problematic situations. Ecocentrism is cautious, accommodating, egalitarian and holistic, whereas technocentrism is optimistic, managerialist, hierarchist and reductionist (O'Riordan, 1995a). Each of these camps is a very broad church with distinct sub-divisions (Table 13.1). Thus ecocentrics range from deep ecologists who advocate returning to pre-industrial lifestyles (e.g. Naess, 1989), to those who want to lessen the environmental impact of existing economic systems but not overthrow them. Technocentrics see ecocentrics as eccentrics, dismissing their idealism as naivety. This camp also embraces a wide diversity of views, from extreme technocentric optimists or cornucopians such as Beckerman (1994, 1995) to more moderate views (e.g. Pearce *et al.*, 1989). The breadth of this spectrum of views reflects the tensions raised by the fundamental question of how human being should live in this world (Section 13.3.1), tensions which shape environmental ideologies. O'Riordan (1995a) explores four of them. Firstly, there is the tension between the desire for dominance and the reality of our dependence on nature; secondly that between efficiency and equity; thirdly between the demands of the present and those of the immediate or distant future; and fourthly between property rights and responsibilities to one's 'neighbours' (however defined).

Environmental decisions are still usually justified in terms of scientific and economic factors, while the often equally important ethical and emotional motivations are sidelined. For example, in the superquarry debate (Section 12.2), the developer claimed that its 'objective' arguments were superior to the emotional and spiritual

TABLE 13.1 A spectrum of perspectives on human approaches to the environment. Simplified from O'Riordan (1999).

Ecocentrists		*Technocentrists*	
Deep environmentalists	*Soft technologists*	*Accommodators*	*Cornucopians*
Lack faith in technology and its need for elitist expertise, central authority, and inherently undemocratic institutions Believe that materialism for its own sake is wrong, and that economic growth can be geared to provide for the very poor		Believe that economic growth & resource exploitation can continue indefinitely if certain economic and legal conditions are met	Believe that humans can always find a way out of difficulties, either through politics, science or technology Believe that scientific & technological expertise is vital on matters of economic growth and public health & safety
Recognise the intrinsic importance of nature to being fully human	Emphasise small (community) scale in settlement, work and leisure	Accept new project appraisal techniques which promote wider discussions & consensus-building	Accept growth as the legitimate goal of project appraisal & policy formulation
Believe that ecological (and other natural laws) determine morality	Attempt to integrate work & leisure through personal & communal development	Support effective environmental management agencies at national & local level	Are suspicious of attempts to widen participation in project appraisal & policy review
Accept the right of endangered species or unique landscapes to remain unmolested	Stress participation in community affairs and the rights of minorities		Believe that any impediments can be overcome given the will, ingenuity & sufficient resources (which arise from wealth)

arguments of its opponents (Mackenzie, 1998). But as modernism has given way to postmodernism, many of the old certainties have dissolved into debates and discourses. Increasingly, the modernist belief that quantitative criteria are the best foundation for decision making has been challenged by the view that the morals and beliefs of society are crucial aspects to incorporate alongside more dispassionate information. Thus Redclift (1993: 20) argues that 'it is time to redraw the frontiers of knowledge and belief, and to recognise that both have a part to play in avoiding global nemesis'. Acknowledgement of these more subjective dimensions is still often perceived as weakening a case; the word 'love' is rarely found in policy documents. And yet it is love of the natural world which typically motivates people to care and to act. Anderson and Smith (2001: 9) make the point that although feelings and emotions 'make the world as we know it and live it', they remain 'spectacularly unacknowledged' in policy discussions.

This argument that ethics should be considered a 'first division' issue, rather than as an optional also-ran, is now being widely accepted. The following quotations bear this out:

> The attitude formally common in both the scientific and commercial worlds, which regarded ethics as a separate and retrospective matter for society to consider, will no longer suffice. Increasingly, scientists are aware of the need to examine technology and ethics side by side. (Bruce and Bruce, 1998: 275)

> The world that will exist in 100 and 1000 years will, unavoidably, be of human design, whether deliberate or haphazard. The principles that should guide this design must be based on science . . . and on ethics. (Tilman, 2000: 211)

Two of many possible examples of this changing outlook include the acceptance of the emotional, spiritual and moral motivations underlying conservation (Anon., 1994b; Holdgate, 2000b) and the recognition of the ethical dimensions involved in choosing between native and exotic species (Section 10.1). Surely it is much better to make value judgements explicit rather than to pretend that they do not exist. Worrell (1997: 179) argues that being 'up front' in this way promotes tolerance by reminding us that 'we are all somewhere on the same spectrum of views rather than being irrevocably dug into opposing trenches'. Few would now disagree with Brennan's (1993:19) plea that 'environmental decisions should be informed by good science *and* philosophy' (emphasis added), nor with his wry rider that 'putting [this] innocuous resolution into effect may be far from easy'.

13.2 KEY CONCEPTS IN ENVIRONMENTAL THINKING

This section focuses on just two of the contemporary ruling paradigms in environmental thinking – sustainability and the precautionary principle – but woven through these and other concepts are numerous principles and beliefs that guide and motivate environmental action. One important strand, for example, is justice, the belief that there should be equity between generations, nations, sectors of society, and between species. 'An unjust world will not succeed in solving global environmental problems' (Connelly and Smith, 1999: 23). Another is the over-arching concept of Gaia. Named after the Greek goddess of the earth, this is the ancient myth of 'Mother Earth' presented in a quasi-scientific form. Promoted primarily by James Lovelock, it is an holistic model (or perhaps a metaphor) which sees the earth as a living, self-regulating organism (Lovelock, 2000). Thus humans and the whole biosphere exist for the Earth (O'Riordan and Jordan, 2000). Arguably, however, the two concepts discussed below are those of greatest immediate relevance to the management of Scotland's natural heritage.

13.2.1 Sustainability and Sustainable Development

The twin concepts of sustainable development (SD) and sustainability are often used interchangeably, but they are not synonymous: the former is the process to attain the latter. If sustainability is the Holy Grail then the drive for SD is the search for it (Silbergh, 1999). SD is not a new concept, having roots in the ancient philosophy of wise stewardship of the land and in the nineteenth-century silvicultural objective of sustention or maximum sustainable yield (O'Riordan, 1993). But in its contemporary incarnation it has come to dominate international environmental rhetoric, and has joined justice, democracy and liberty as an unassailable ideal.[2] It provides:

> . . . a context within which economic, social and environmental goals can be considered together and debated. . . . It now lies at the heart of European, UK and Scottish policy development. It is likely to provide, for the foreseeable future, the working hypothesis by which a balance is struck . . . between the achievement of environmental and social objectives, and the drive for wealth creation and economic competitiveness. (Maxwell and Cannell, 2000: 32–3)

Much importance is now officially attached to SD in Scotland (Scottish Office, 1999b). This was shown by the Scottish Parliament's resolution in 2000 to place it at the core of its work, and to require all Bills to carry a sustainability statement. The definition of SD adopted by the Parliament is that formulated by the Brundtland Commission: 'development that meets the needs of the present without compromising the ability of future generations to meet their needs' (WCED, 1987: 43). This is by far the most widely accepted definition of SD, but innumerable alternatives exist. SNH (1993:3), for example, suggests that it is about 'finding ways of maintaining human welfare which do not damage the foundations on which it rests'. More evocatively, O'Riordan (2000a: xii) envisages SD as 'the bonding of people to the planet in a placenta of care, equity, justice and progress'.

The essence of SD is the idea of living off interest rather than capital. It requires an anticipatory, integrated and preventive approach in place of the *ad hoc*, case-by-case, reactive mode of operation that has tended to characterise environmental policy-making. SD stresses the inseparability of ecological, social and economic concerns (a trio sometimes dubbed the three-legged stool of sustainability), together with the significance of environmental capital and services. It is future-orientated, an emphasis neatly captured in Holland and Rawles' (1993: 16) paraphrase of the Brundtland definition: 'we must not foul things up for our descendants; it's unfair'. The SD critique of the basic drivers of economic activity constitutes a radical challenge to the *status quo*, requiring a new and much broader conception of 'development'. The belief that scientific and technological advancement is a 'good thing', and the allied notion of progress and development through it, is a fundamental yet largely unrecognised ethical bedrock of much of Western society. SD questions this bedrock.

Despite (or perhaps because of) the endless attempts at definition (Pezzoli, 1997a, 1997b), SD can be interpreted in so many ways that it teeters on the edge of meaninglessness. The spectrum of interpretation varies according to the emphasis placed on its two component words. Thus 'Weak SD' focuses more on development than sustainability; it is technocentric, and accepts the substitution of anthropogenic capital for natural capital. 'Strong SD' focuses on preserving natural capital at all costs and on allowing ecosystems to function naturally. At the ecocentric end of the spectrum is 'Very Strong SD' which rules out use of non-renewable resources altogether. For the foreseeable future, environmental policy in all EU countries will continue to be based on weak sustainability (sometimes dubbed 'ecological modernisation') (Connelly and Smith, 1999). The whole definition problem is deftly sidestepped by Manning (1998) who argues that it is less important to pin down the meaning of SD precisely than to identify what is *not* sustainable. Nevertheless, the breadth of possible interpretations *is* a problem. Although the differences are often obscured at the rhetorical level, the incompatibilities between the various meanings become sharply apparent when real and difficult decisions have to be faced (Owens and Cowell, 1996), as shown in the debates over the Cairngorm funicular and the Harris superquarry (Chapter 12).

Some critics argue that the very concept of SD is fundamentally flawed (Section 13.3.4). Others believe that SD provides a useful framework but that it has been fatally compromised through being co-opted into the *status quo*, coming to mean all things to all people. Ratcliffe (1998: 40) describes it as the 'fashionable but facile concept now firmly in the lexicon of politicians' clap-trap', muzzled by misuse and familiarity, and hijacked by big business. In its 'ecological modernisation' guise, SD loses its radical cutting edge, simply becoming 'business as usual' with a green spin. This is why many environmentalists have been forced to disown the concept that they originally championed. McDowell and McCormick (1999b: 2) wryly observe that the concept provides 'a focus to unite disparate interests in dialogue *if not action*' (emphasis added) and wonder whether it amounts to much more than old whisky in a new bottle. It certainly seems true that while no policy document can now be written without some verbal genuflections to SD, when the levers of power are operated it is often quietly forgotten (Henton and Crofts, 2000). Given that it represents a profound challenge to conventional wisdom, this is hardly surprising. It shows that the chasm between sustainability rhetoric and the adoption of SD policies will take some bridging.

13.2.2 The Precautionary Principle

Uncertainty permeates environmental management (Section 1.4). A major response to this uncomfortable fact has been the adoption of the precautionary principle (PP), an approach which advocates a proactive rather than a reactive approach to risk. Working on the premise that it is better to be roughly right ahead of time than precisely wrong too late (Jordan, 1998), the PP demands that action should be taken before 'proof' of environmental damage or likely damage is available. As the saying goes, an ounce of prevention is worth a pound of cure. It is especially applicable in situations of great complexity or uncertainty, where establishing cause and effect relationships unambiguously is difficult or impossible (O'Riordan and Jordan, 1995). Strictly applied, this principle could paralyse development by halting all projects with uncertain environmental outcomes. In practice, however, it tends to mean that when irreversible or unpredictable consequences are envisaged, all alternatives are given full consideration before action is taken.

The precautionary approach rapidly gained ground in European policy-making during the 1990s, challenging the reactive approach to risk regulation which remains the bedrock on which most systems of health and safety legislation and pollution control are built. A requirement that damage is 'proved' almost inevitably means that serious problems arise prior to controls being put in place, a self-evidently unsatisfactory situation. The damage caused by agrochemicals and the slow, piecemeal development of reactive regulation is a case in point (Section 6.5.3). The PP advocates shutting the stable door *before* the horse has bolted, or even if there is a reasonable chance that the horse *might* bolt. The principle was endorsed in the Rio Declaration, and has now been adopted by the EU and its member states (O'Riordan, 1995b; SNH, 2000f).

However, despite its apparently self-evident wisdom, the PP is controversial for several practical and ethical reasons (Bruce and Bruce, 1998; O'Riordan, 2000c). First and foremost, it alters the balance of power between science and the community, between developers and environmentalists. In erring on the side of caution, it shifts the burden of proof from the disturbed to the disturber, from conservationists to developers; the latter must demonstrate that an action will *not* cause environmental harm. It necessarily relies on probabilities and prediction rather than 'hard facts', something which is hard to swallow for those of a scientific-rationalistic persuasion. Precautionary action has always been emphasised in the EU's environmental regulatory philosophy, whereas 'sound science' has been the preferred foundation for action in the UK (one of many friction-causing philosophical discrepancies (Section 2.2.2)). Ward (1998: 257) observes that 'the UK's commitment to the precautionary principle has been more rhetorical than substantive'. Secondly, there is the danger of erring too far on the side of caution leading to needless restrictions and depriving society of environmentally harmless benefits. Much of the political and commercial opposition to the concept stems from the additional costs and lengthy delays that it can impose. Consequently there is still powerful resistance in some quarters to the idea that a precautionary approach should replace the reactive method of risk assessment as the norm.

In addition to controversy over the approach *per se*, there is also debate about when, where, and for how long it should be applied, and at what price. When is it appropriate to apply the PP? How many data are required and over what time period before a precautionary approach should be relaxed? What constitutes 'sufficient' information and 'reliable' prediction, and who decides? The PP is thus afflicted by similar problems to those which dog sustainable development; though widely applauded as a 'good thing', its meaning is imprecise and it provides few, if any, operable guidelines for policy-makers (O'Riordan and Jordan, 1995). In practice it is also impotent in the face of commercial pressure, as Francis (1996) shows in relation to the development of marine fish farming in western Scotland (Section 6.2.2).

A fourth and final reason why the PP is controversial is that it highlights the fact that Western society holds ambivalent attitudes towards science, seeing it as partly responsible for environmental problems but also as part of the solution. This ambiguous or even contradictory attitude to science is a characteristic of environmental thinking; green campaigners rail against the technocentrism of society and its faith in scientific progress ('scientism'), but at the same time rely on scientific evidence to legitimise their arguments. Porritt (2000: 7) encapsulates this ambivalence:

> I find myself inevitably locating modern science at the very heart of an inherently destructive model of progress, while simultaneously looking to science to provide answers to the otherwise intractable problems we now face.

Dovers and Handmer (1993) dub this tension the 'eco-cultural paradox'.[3] Clearly the PP is applicable in much environmental decision making because of the complexity and uncertainty involved, but although it redresses some of the more obvious failings of the reactive approach, its application raises numerous difficult questions.

13.2.3 SNH's 'Five Commandments'

Working out how these over-arching concepts and ethical principles should be applied in the Scottish context was one of SNH's earliest challenges. It succeeded in developing its own distinctive set of sustainability guidelines, dubbing them 'the five commandments' (SNH, 1993; Crofts, 1995):

1. **Wise use**: non-renewable resources should be used wisely and sparingly so as not to restrict options for future generations.
2. **Carrying capacity**: renewable resources should not be used beyond their maximum sustainable yield.
3. **Environmental quality**: the quality of the natural heritage as a whole should be sustained and improved.
4. **Precautionary principle**: in situations of great complexity or uncertainty, a precautionary approach should be adopted. SNH (2000f: 3) interprets the PP to mean that 'full scientific proof of a possible adverse environmental impact is not required before action is taken to prevent that impact'.
5. **Shared benefits**: there should be equitable distribution of the costs and benefits (material and non-material) of any development. This guideline specifically implements SNH's basic ethical principle of equity (between generations, between species, and within society), and advocates the 'polluter pays' principle.

These are probably as useful and balanced a set of guidelines for sustainable interaction with the natural heritage as one could hope to develop, having the strong advantages of accessibility, simplicity and brevity. However, as SNH (1993) itself acknowledges, a moment's thought reveals that there are practical and philosophical problems with all of them. For example, concerning Guideline 1, any use of non-renewable resources is by definition unsustainable and will necessarily restrict the options of future generations. What, then, constitutes 'wise' and 'sparing' use of such resources? 'Quality', in Guideline 3, is a notoriously slippery concept, incorporating on the one hand elements which can be precisely measured (such as air or water quality) and, on the other, aspects which can only be subjectively assessed (such as aesthetic quality) and about which society has ever-evolving views (Section 8.1.1). Our descendants may have a rather different conception of a 'high quality' environment from our own, just as our views differ dramatically from those of our ancestors. Finally, while sharing costs and benefits is an inherently just aspiration, it is a fraught one to implement because of the conflict between intra-generational and inter-generational equity (Section 13.3.4). Notwithstanding such difficulties, it is surely better to attempt *some* set of ethical guidelines, albeit contested, than to blunder on in a reactive, case-by-case fashion. Sustainable development is the rocky 'road less travelled', so even a fuzzy goal is better than no goal at all.

13.3 Debates in environmental philosophy

13.3.1 'Man and nature': the place of Homo sapiens *within non-human nature*

The question of how we should think of our relationship to (the rest of) nature is one of the most fundamental in environmental ethics (Benson, 2000).[4] What is the place or role of our species in the natural world? Are we part of nature or separate from it? Are we the pinnacle of creation with divinely given responsibility to steward it wisely, or are we unwitting participants in the inscrutable workings of Gaia? Of course, in a strict biological sense we are incontrovertibly part of the biosphere, but there is a pervasive, longstanding tradition in Western thought which views human beings as separate, or at least distinct, from the non-human world. The almost universal adoption of the antitheses between 'artificial' and 'natural', and between 'people' and 'nature', would seem to support the usefulness, if not the logic, of this Cartesian dualism. Others, however, (particularly ecologists), prefer to see human beings as an integral part of nature, and wish to make no distinction between the human and non-human worlds.

The point at issue is whether we perceive a discontinuity of any kind between 'us' and 'the rest' – whether human actions should be seen as 'part of the web of influences on ecological change' or 'external equilibrium-disturbing impacts' (Adams, 1996:89). The way that we answer this question determines the way that we treat our world. There are two closely linked facets to this issue. The first is the question of whether we have any responsibility for caring for the non-human world. The second is whether environments which have been altered by human beings should be regarded as less valuable than pristine wilderness areas.

13.3.1.i A duty of care?

If we are part of nature, then the distinction between artificial and natural breaks down, and it can be argued that we have no greater responsibility for the world than any other species. Consider the provocative observation by Garvin (1953 in Rawles and Holland, 1994: 14) that 'man is the animal for whom it is natural to be artificial'. In common parlance, the less the human influence, the more natural an environment is. But if, as Garvin implies, it is natural for us to modify the world for our own ends – if human actions are seen as being just as natural as any other process in nature – why should any checks be put on our activity? An extreme *laissez faire* approach is thus one possible logical outcome of seeing *Homo sapiens* as simply one species amongst myriads. Conversely, it is often argued that a belief that we are an integral part of nature brings with it an obligation of care. Thus Tapper (1999: 18) argues that 'man has always been part of nature, giving us a right and a responsibility to 'conserve for wise use' '. In similar vein, Smout (2000a: 3) writes:

> For the human species, considering its unparalleled power over the rest of nature, the duty of care should be seen both as an imperative moral responsibility and enlightened self-interest. Nothing less does our full humanity justice.

Prior to the later twentieth century, a belief that the human race is in some sense apart from nature was a commonly shared perspective. This sprang either from theological beliefs (notably the Judeo-Christian teachings of men and women being created in the image of God) or from a recognition of distinctively human attributes such as self-awareness, culture and moral sensibilities which set us apart from other species. After all, ours is the only species to have escaped the confines of the local environment so that 'the whole earth has become our local ecosystem' (Eldredge, 1995: 103). Unfortunately, such a sense of separateness from nature slides all too easily into a belief in superiority over nature. One consequence has been the overweening arrogance which has characterised much of humanity's dealings with nature in recent centuries. The blame for this was placed squarely at Christianity's door by White (1967), who interpreted God's command to humankind to 'have dominion over . . . the earth' (Genesis 1: 28) as amounting to *carte blanche* for environmental destruction. However, White's thesis not only neglects the importance of other belief systems which encourage exploitation (not least Marxist atheism), but it takes one text out of context and turns it into a pretext; taken as a whole, Christian teaching balances rights with responsibilities, making human beings stewards, not masters, of all they survey.[5] On this basis it can be argued that ecological damage has resulted from a loss of humility as we have progressively come to believe that we are owner-occupiers rather than tenants. Interestingly, then, it appears that a belief that human beings have a responsibility to care for the earth can spring either from the view that we are part of the world, or from the perspective that we are distinct from the rest of nature.

13.3.1.ii The value of natural versus *cultural landscapes*

A longstanding tenet of the conservation and environmental movements is that human influence degrades the 'naturalness' and hence the value of non-human nature. The perception that wilderness is sacred and humanity profane results in great value being attached to 'unspoilt nature' (itself a revealing phrase). Even when attempts are made to restore altered nature to its 'natural state', the results are dismissed by some as fakes and artefacts (Brennan, 1993; Budiansky, 1995). This value system was eloquently articulated in the nineteenth century by John Muir who saw humans as the only unclean animals and regarded nature as his church. It is an outlook which has been very influential in British nature conservation (Toogood, 1997) but it is not uncontested. It can be criticised on four counts. Firstly, few parts of the globe are truly pristine; few if any parts of Scotland have remained unaffected by the 9,000 years of human habitation (Section 1.2.2). This means that we are obliged to work with a sliding scale of semi-naturalness, and to accept that we only have 'damaged goods' to handle. Secondly, wilderness itself is far from being separate from humanity. It is, in fact, a wholly human construct (Oelschlaeger, 1991).

> [Wilderness is] entirely the creation of the culture that holds it dear. . . . [It] serves as the unexamined foundation on which so many of the quasi-religious values of modern environmentalism rest. (Cronon, 1996b: 79, 80)

The dualistic paradox that results from sanctifying pristine nature and damning human influence is this: 'if nature dies because we enter it, then the only way to save nature is to kill ourselves' (Cronon, 1996b: 83).

Thirdly, many of the landscapes that are prized for their supposedly natural qualities are in fact cultural landscapes, created and maintained over the millennia by utilitarian human use. Within Scotland, for example, this is certainly true of heather moorlands (Thompson *et al.*, 1995a; Section 7.3.2) and it might apply even to the wilds of the Flow Country (Chambers, 1997; Warren, 2000a). Moreover, landscapes and ecosystems that have seen long human use are valued not only for their cultural and historical meaning, but often for their ecological value as well (Kirby *et al.*, 1997). So perhaps human influence is not universally and inevitably negative. As Quelch (1998: 21) puts it, 'man has not spoiled the Garden of Eden; maybe he was . . . responsible for helping to create that garden in the first place!' A final criticism is that this value system can all too easily give rise to misanthropy – a negative view of people and communities and their desire to improve their lot. People come to be seen as an inconvenient nuisance, a spanner in the finely tuned workings of sublime nature.

These four critiques are neatly combined by Tapper (1999: 18):

> The main problem with the word 'natural' is the implication that anything 'unnatural' is undesirable. Man has affected the environment for so long that such a distinction can be as unhelpful as it is illogical.

Two further quotations develop and reinforce this thinking:

> The idea that the wilderness should be free of people living and working there has less to do with successful ecosystem functioning than it has with the romantic ideal of a primaeval nature independent of 'man'. Palaeoecology shows that ideal to be a myth. (Roberts, 1998: 251)

> A romantic urban-based wilderness fantasy has no place for human beings to make a living. It is dangerous because it leaves no place for discovering what a

> sustainable human place in nature might be. Furthermore, it sets an arbitrarily high standard against which cultural and productive landscapes are regarded as inferior. (Brown, 1997: 196)

Aspects of this debate about the desirability or otherwise of a human presence in the sparsely populated (or depopulated) parts of Scotland have been discussed earlier (Section 1.3; 8.3.1.iii). It is a debate characterised by deeply held but diametrically contrasting views of 'naturalness' and of the appropriate place of *Homo sapiens* in the environment. It is unlikely to be resolved any time soon.

A distinction between natural and artificial has always been hard to pin down (Coates, 1998; Phillips and Mighall, 2000), and the advent of genetic engineering now blurs the boundaries still further. Nevertheless, Coates (1998: 191) concludes that 'the polarity of nature and culture will endure a good deal longer'. That the two poles are connected by a complex continuum is neatly summed up by Kemal and Gaskell (1993: 3): 'human beings are a fragment of nature, and nature is a figment of humanity.'

13.3.1.iii Rights: what species have them?

A final – and particularly fraught – aspect of the relationship between humankind and the non-human world is the issue of rights. Until the late twentieth century it was generally assumed that rights issues were exclusively human, but the rise of the animal rights movement has now initiated a host of complex and emotive debates that cannot be addressed here (but see Bruce and Bruce, 1998). Do other species, both faunal and floral, have rights? Do they all have the same rights? Is it as wrong to swat a wasp as to kill a golden eagle? Does it make sense to 'award' animals rights when these cannot be balanced by reciprocal responsibilities? Do inanimate parts of nature such as rivers and mountains also have rights of some kind? Despite the explosion in the use of rights language in recent years, some philosophers believe that the whole concept of rights is legal not moral, so to talk of species (*including* humans) as having natural rights is incoherant (Ellis, 1990). They point out that attributing rights to species goes against the trend of evolution, in which no creatures appear to have rights, either as individuals or as species, not even the right to exist (Ratcliffe, 1989).

Although there is a cold logic to such arguments, many people share an instinctive feeling that human beings and some other species *do* have rights, or at least that they should be accorded a measure of respect and dignity. If it is accepted that certain rights do exist, a final question is when (if ever) it is allowable to infringe these rights. To what extent is human need an acceptable imperative for utilising the non-human world, for example for food, medicines and resources? Should a cure for cancer that depends on an endangered species be adopted or rejected? Whether or not non-human species and habitats have rights, most would accept that human beings have a duty to use our unrivalled technological powers in an ethical fashion. What is defined and accepted as 'ethical', of course, will depend on philosophical and spiritual persuasions as well as socio-economic considerations. A final word on rights issues goes to Rivers (1997: 3, 4):

> Rights carry an inherent bias favouring individualism over collectivism . . . and conflict over consensus. . . . [Therefore] society will only function well when individuals are prepared to forgo their rights.

If so, the focus needs to shift from insisting on rights to embracing responsibilities. Humility and selflessness move centre stage.

13.3.2 Environmental values: which moral theory?

What is the appropriate moral theory (or theories) to rely on when assessing environmental issues? On what moral foundation do we or should we construct out environmental values? Do existing moral traditions provide the resources we need, or does the new thinking about environmental values require new moral principles? Much has been written on these questions in recent years,[6] so the function of this brief section is not to survey the field but simply to highlight the contested nature of this fundamental issue, and to challenge the coveted notion of objective decision making. There are, of course, numerous ethical systems from which to choose, amongst the most prevalent being utilitarianism, rights-based theories, natural law, ecological holism and the social contract. In addition, despite the decline of ecclesiastical Christianity in Europe, our moral values are still largely shaped by a Christian ethic which 'permeates the fabric of our moral life and history' (Connelly and Smith, 1999: 11). All these ethical positions jostle for space in the moral market place. An obvious problem with such moral pluralism is that we become the judge of morals, rather than *vice versa*. This leads to a 'pick-'n-mix' morality and the predicament of becoming mired in a moral morass. It is hard to avoid the charge that moral pluralism is merely '*ad hocery*' under a fancy label (Brennan, 1990).

In practice, most ethical decisions flow from one of two moral positions, an intrinsic stance or a consequentialist approach. In the first of these, certain choices are regarded as being inherently right or wrong in themselves (often arising out of spiritual convictions, religious or otherwise). A consequentialist approach, by contrast, makes judgements based on the consequences of adopting a particular course of action. For example, some oppose animal experimentation in principle, regardless of the benefits which might flow from it, whereas a consequentialist might conclude that the ends justify the means. The relative merits of these positions have been the stuff of philosophical debate for centuries. In the environmental sphere, debate hinges on whether elements of nature – trees, animals, rivers – have intrinsic value which transcends human interests, or whether our motivation for caring for them should spring from the many ways that they are valuable *for us*. Should we adopt an anthropocentric or a biocentric approach (Section 13.1)?

Given that intrinsic arguments imply moral absolutes, they were regarded as suspect for much of the twentieth century and were given short shrift in the dominantly rationalistic, scientific decision making processes of Western society. In recent years, however, the situation has become complex, with contradictory influences pulling in diverse directions within society (Section 13.1). On the one hand, the increasing interest in spirituality has given a new currency to intrinsic arguments stemming, for example, from religious or deep ecology perspectives. On the other, postmodernism questions the authority of all 'meta-narratives' and truth claims, including those of the scientific worldview. In a plural, postmodern society, intrinsic arguments are respected as personal viewpoints but are given no privileged status (notwithstanding the inherent contradiction of postmodernity's absolute refusal to countenance absolutes). In the practical business of environmental management, of course, intrinsic and consequentialist approaches form two sides of the same coin. Thus intrinsic value principles lie at the heart of environmental legislation but the application of such policies demonstrates a utilitarian or more generally consequentialist philosophy (Spash, 1997).

It is one of the most pervasive myths in liberal democracies to imagine that adopting one of the non-intrinsic stances, whether rationalist, scientific or utilitarian, somehow lifts decision making above or beyond value-laden judgements. The reality, however, is that 'rational, value free policy processes do not and cannot exist' (Dovers *et al.*, 1996: 1150). The highly prized notion of dispassionate objectivity is, at best, extremely relative; scientific and technological perspectives are arguably as value-laden and ideological as any other (Pepper, 1996). As Bruce and Bruce (1998: 82) observe:

> Deep-seated values and belief systems . . . underlie all our lives, whether consciously or not. Intrinsic ethical arguments are by no means limited to religious believers or to environmental groups. Indeed, . . . most of us hold certain fundamental and often unverifiable beliefs about the cosmos and human life.

If ethical neutrality is unobtainable, it is illogical to dismiss intrinsic arguments simply because they are intrinsic. Rather than trying to cling to the wreckage of the objectivity myth, it is important to acknowledge the moral and ethical positions that underpin our value systems and which guide our decision making.

13.3.3 Finding ethical foundations for conservation

Identifying and articulating a foundational rationale for conservation is a difficult task. This is partly because a multitude of biological and cultural arguments are used to promote conservation, and partly because conservationists are motivated by diverse philosophies. Historically, conservation has undertaken a long ethical journey. In its earliest incarnation in Victorian times it initially relied on theological and utilitarian arguments, but these have progressively been abandoned as conservationists have come to believe in the intrinsic value and perceived rights of non-human nature (Thomas, 1983). However, this journey is neither complete nor universally applauded. Debates continue between those advocating theocentric, anthropocentric and biocentric viewpoints. Should conservation be primarily motivated by the instrumental value to society of environmental 'goods' (whether biological or cultural), or by their intrinsic value independent of human beings? Are the needs of people paramount? If not, why and for whom (or what) should we protect the earth? Should nature be conserved for its own sake or for ours? Such questions generate much controversy (Huxham, 2000; Phillips and Mighall, 2000).

In thoughtful explorations of these issues, Holland and Rawles (1993) and Rawles and Holland (1994) identify three guiding principles which capture important aspects of conservation's ethical basis, namely land health, sustainability and nature.

1. **Land health**. This is based on the 'land ethic' developed by Aldo Leopold. One of the foremost ecophilosophers of the twentieth century, Leopold was the first to call for the incorporation of ethics into land use decision making. His land ethic widens the ethical circle to encompass the entire ecological community of which mankind is a part.

   > When we see land as a community to which we belong, we may begin to use it with love and respect. There is no other way for land to survive the impact of mechanized man. . . . The land ethic simply enlarges the boundaries of the community to include soils, waters, plants and animals, or collectively the land. . . . [It] changes the role of *Homo sapiens* from conqueror of the land-community to plain member and citizen of it. It implies respect for his fellow-members, and also respect for the community as such. . . . A thing is right when it tends to preserve the integrity, stability, and beauty of the biotic community. It is wrong when it tends otherwise. (Leopold, 1949: vii; 204; 224–5)

   Importantly, Leopold's compelling synthesis of science and aesthetics was not protectionist; he advocated the preservation of both the beauty *and* the utility of the landscape (Knight, 1996). For many years after he conceived it, his land ethic was confined to the green fringe, but more recently it has emerged into the mainstream in the form of the 'ecosystem management' paradigm (Callicott, 2000). Thus, for example, foresters have started asking whether their codes of professional ethics should be extended beyond duties towards other people to incorporate duties towards the forest itself (Worrell, 1997), and the US Forest Service has abandoned Gifford Pinchot's utilitarian conservation philosophy in favour of Leopold's land ethic (McQuillan, 1993). The virtues of Leopold's vision of humans as citizens rather than conquerors are extolled by Boyd (1993: 163), but he asks despondently, 'Oh, where can we find humility enough for that citizenship!' The need for humility in our dealings with the environment is a theme emerging out of diverse strands of thought. From whatever source this humble attitude is derived – whether from Christian convictions, from Gaian principles, from the proto-ecological vision of John Muir, or from Leopold's land ethic – it is urgently required.
2. **Sustainability**. As commonly formulated (Section 13.2.1), sustainability is couched in terms of human needs and social justice. However, a policy of maximising human welfare is unlikely to achieve minimum conservation standards, even if an extended timeframe is adopted, so by itself sustainability is an inadequate foundation for conservation. It is nevertheless an important component.

3. **Nature.** Holland and Rawles (1993: 17) suggest that a conservationist might be described as:

> . . . someone for whom there is a presumption in favour of the natural, to be trumped only by some serious human interest or the serious interest of some other sentient being.

In the light of the foregoing discussions (Section 13.3.1), this suggestion clearly raises difficult issues about the naturalness or otherwise of human beings, and the value of semi-natural (or humanly modified) landscapes. It is also far from clear cut because some natural things (such as infectious viruses) are damaging, so a general presumption is unjustified. Nevertheless, a biocentric approach has come to underpin and justify conservation philosophy. This does not mean, though, that conservation is entirely altruistic, motivated by the interests of other species. Part of conservation is about preserving the natural world for human enjoyment, and in that sense there is a strong utilitarian element involved (Worrell, 1997). In fact, 'biocentric *v.* anthropocentric' turns out to be a false dichotomy. Given the extent to which humans are dependent on the environment, the idea of a *purely* human interest is perhaps an illusion. If so, 'it is not in the interests of human beings to be anthropocentric!' (Rawles and Holland, 1994: 13). Again, we are left facing the need to adopt a humble duty of care for *all* of nature (humans included).

13.3.4 Sustainable development: a flawed credo?

Sustainable development (SD) is a concept surrounded by a swirling mass of controversy and debate (Section 13.2.1). Despite its high profile in much current thinking and policy-making, there are those who argue that it is a fundamentally false path to follow – that it is a flawed credo. Of the various lines of attack that are adopted, perhaps three stand out. Firstly, it is argued that SD does not consitute a moral imperative. Advocates of SD believe that the sustainable path is self-evidently the best option and so should be followed *simply because it is the sustainable path.* Opponents argue that SD is a purely technical concept, and that the moral choice of whether or not it *should* be followed is entirely separate matter (Beckerman, 1994). (As Price (2000) points out, logically sustainability applies just as much to continuing flows of bad things as good things, but we do not recommend land uses offering sustainable deprivation!) A sustainable path is not always optimal economically. After all, if unsustainability were a reason for rejecting a project, there would be no use of non-renewable resources, no mining or industry, and the world would be a primitive place indeed. In a world of acute human need, a strong case can be made that it is morally wrong not to use all available resources to improve human welfare. Thus Beckerman (1994: 195) regards SD as 'morally repugnant and totally impracticable.'

Secondly, it is argued that SD is unworkable because future needs are unknowable. The Brundtland definition (Section 13.2.1) is based on 'needs', an inherently subjective and elastic concept. Given that we cannot predict the needs of future generations, how can we know what to preserve for them now? Making needless sacrifices benefits no one, and so the argument here is that 'posterity can look after itself'. Beckerman (1994, 1995) argues that SD is a muddling distraction; instead our goal should be to maximise human welfare *in the long term*. Given that human welfare depends on the natural environment, concern for ourselves and our descendants should safeguard critical natural capital. However, given human selfishness and the short-termism built into most democratic political systems, environmentalists believe that such an approach would be a wholly inadequate brake on the destructive forces of the globalised, capitalist economy.

Thirdly, it has been shown by many commentators that SD is riddled with contradictions. Dovers and Handmer (1993) identify eight of these, but three examples will suffice. Firstly, there is the conflict between inter-generational and intra-generational equity. Taking a global perspective, if we set out to satisfy the basic needs of all those now alive, the options of future generations would probably be severely limited. Conversely, if resources are going to be 'saved' now for future generations, how can justice be achieved for the present needy generation? It is clearly unethical to focus on equity across time while ignoring equity across space. Secondly, there is the tension between individual and collective interests, the so-called 'tragedy of the commons' (Hardin, 1968). Individual rights are a basic tenet of Western culture (Section 13.3.1.iii), but environmental issues are collective problems. This tension applies at all scales, from the local community right up to the international level where the insistence of individual nations on their rights often hinders the implementation of solutions to global environmental problems. Climate change negotiations are a classic of the genre. Thirdly, it is argued that SD is actually a contradiction in terms. As currently understood, 'development' is a material concept which, in the *long* term, is unavoidably limited by the finite nature of the earth and so cannot be sustainable.

SD is thus a contested, contradictory and paradoxical phrase, but this has not stopped it being adopted around the world as the as the 'green narrative' of our time (McDowell and McCormick, 1999b). Dovers and Handmer (1993) conclude their critique with the ironic observation that since many aspects of the modern world are riddled with paradoxes and logical inconsistencies, perhaps our ability to live with paradox will be our salvation!

13.3.5 The Voluntary Principle: good or bad?

Governments have many tools at their disposal with which to achieve their policy objectives in the countryside. These include regulation, economic instruments (green taxes, grants and other financial incentives), legal liability, rights-based mechanisms (tradeable permits and quotas) and voluntary persuasion.[7] Hanley and MacMillan (2000) argue that a mixture of all of these should be used. However, particular controversy has surrounded the last option, the voluntary principle. It has long been the foundation on which British nature conservation rests, and has been enshrined in legislation since the 1981 Wildlife and Countryside Act. Alongside sustainable development and the precautionary principle, it has become a foundational principle. For example, the Cairngorms Partnership (1997) bases its integrated management strategy on this trio of principles. Unsurprisingly, landowners and managers are traditionally in favour of a voluntary approach. Many of them feel that rural life is already hedged about with quite enough rules and regulations, and that some room for manoeuvre and private choice must be maintained.

Conservationists by contrast have long argued that the voluntary approach is weak, confusing and ineffective (Francis, 1994; SWCL, 1997), and that the government should use more stick and less carrot.

> The voluntary approach . . . is not ethical as it is presently conceived and translated in practice. As reflected in existing primary legislation, it represents a triumph of individual control and decision making over the common good. It is no longer defensible. (Francis, 2000: 69)

Equally critical is Ratcliffe (1995: 72) who dismisses the voluntary principle as 'no more than a crude slogan to justify shoring up private privilege at public expense'. The principle has meant that unless owners and occupiers adopt conservation measures voluntarily, the only options available to SNH have been court action, compulsory purchase, or to offer payments for things not to be done. None of these are efficient, sustainable, or likely to deliver positive conservation (Crofts, 1995). With an overhaul of the designations process underway (Section 8.3.3), things are now set to improve.

The voluntary principle applies much more widely than simply to nature conservation, being at the heart of debates over public access (Section 9.4) and wildlife management (notably the control of the burgeoning red deer population (Section 7.6.1)). Indeed, a predilection for policies based on persuasion, consent and codes of practice has long been a characteristic of British governments, in contrast to most other European countries where legal enforcement is the tool of choice (Section 2.2.2). Carter and Lowe (1998) argue that inherent within this British tradition of voluntarism is a bias in favour of corporate interests, a bias which may hinder the resolution of environmental problems. The debate over the voluntary principle is a microcosm of that in political philosophy between those on the right who advocate *laisser-faire* economics and the rights of individual citizens, and those on the left who see a role for greater intervention by the state to ensure that private decisions do not compromise the public good. The pendulum is bound to keep swinging.

13.4 ENVIRONMENTAL DECISION MAKING: VALUATION BEYOND PRICE

This section focuses on one facet of the large field of environmental decision making, namely the challenge of incorporating intangibles. Some costs and benefits *can* be measured but many aspects of the environment which matter to people are hard or impossible to measure. A range of ingenious techniques have been developed which endeavour to put prices on these intangibles so that they can carry weight in conventional decision making processes. The aim here is to introduce these techniques and to evaluate their strengths and weaknesses.

13.4.1 The problem of intangibles

Decisions can be made on many grounds: economic, scientific, moral, aesthetic, spiritual, gut instinct, and combinations of these. In our society, however, 'economics has become the language of default' (More *et al.*, 1996: 407). Monetary values override all others. It is argued by some that this has led to an extension of economics into inappropriate areas, because not everything that we value has a *monetary* value. The overall value of the environment to society and individuals consists of many diverse elements. Some of these, such as timber, species diversity or water quality, can be objectively assessed and/or quantified. Others, such as beauty, peacefulness or wildness cannot. Some have monetary value (directly or indirectly), but many do not. Because market forces can only work with cash valuations, the latter are usually described as externalities (external to the market), or as non-market benefits and costs. But simply recognising that these important aspects of the environment lie outwith the market does not alter the fact that decisions have to be made.

Environmental managers are therefore faced with a stark choice between two problematic pathways. The first option is to ignore externalities altogether, while the second is to endeavour to build them in to decision making somehow. The consequence of the first is that decisions are based purely on existing markets. This is a pragmatic approach which acknowledges that, for better or worse, 'money makes the world go round'. It was the route by which decisions were arrived at for most of the twentieth century. Indeed, policy-makers and environmental managers prided themselves on fact-based decision making which delivered 'value for money'. Partial though this approach may be, it has long been argued that it has the salient advantage of avoiding the moral maze by sticking to verifiable, objective, quantitative criteria.

Unquestionably this first option is the simpler of the two, but it is clearly unsatisfactory. Of the wide range of criticisms which can be levelled at it, perhaps four stand out:

1. It implicitly puts a zero price on all externalities, yet these are often the very things to which society now attaches a high value – the things which people really care about. The market cannot incorporate the ancient truth that 'man does not live on bread alone' (Deuteronomy 8:3) or respond to the fact that 'civilisation brings with it . . . a search for fulfilment of the spirit as well as the body' (Mowle, 1997: 143). People assign values using currencies other than cash.
2. Because this approach is exclusive and excluding, reserving decision making for a technocratic minority, it can be accused of being anti-democratic. Too great a reliance on science alienates ordinary people (Section 8.2.2). The emotional impacts and costs for those whose environment will be forever changed are not considered because the privately emotional and the rationally public are not allowed to mix (Anderson and Smith, 2001).
3. It can lead to long delays. Scientists seek all the facts and a full understanding before taking a decision, but as Boehmer-Christiansen (1994: 73) points out, 'management or policy made entirely on the basis of hard facts tends to lead to 'paralysis by analysis' '. Waiting for proof can mean waiting too long.

4. Perhaps most fundamentally, this approach does not in fact offer a means of making objective, dispassionate, value-free decisions – the very thing which it claims to provide. However 'hard' the facts are, the judgement of where the balance of benefits and costs lies, or of what level of impact or human presence is acceptable, is a value judgement (Boehmer-Christiansen, 1994). Long ago the philosopher David Hume pointed out that it is impossible to derive 'ought' statements (values) from 'is' statements (facts), so facts themselves cannot be the last (or only) word on which way a decision should go. As Taylor (1996: 13) succinctly puts it, 'science can inform but it cannot decide'. Nor can it provide moral guidance. Lovelock (2000: 27) observes that 'having supplanted religion, science has left the world in a moral vacuum'. Budiansky (1995: 189, 23) concurs:

> Science is a tool. It is not a moral imperative. . . . Whether it's more important to use our forests as a source of lumber or as a source of inspiration is a question about which science offers no answers.

Consequently, to make sound decisions we need a thorough knowledge not only of the relevant *facts* but also of the relevant *values* (More *et al.*, 1996; Worrell, 1997). This, of course, begs the questions 'whose values are relevant?' and 'who decides what the relevant facts are?' These problematic questions are briefly addressed in Section 12.3. Recognition of this rather obvious but inconvenient reality has increasingly focused attention on the need to improve and augment our decision making tools. This involves adopting the second of the two options above by trying to build a 'value dimension' into Cost Benefit Analysis (the technique which assesses the economic efficiency of proposed developments). The challenge is to find ways of attaching monetary value to things beyond price – of putting hard figures on soft emotions.

13.4.2 Costing the intangible

Conventionally, the economic techniques of Cost Benefit Analysis (CBA) have been limited to balancing only those aggregated costs and benefits to which prices can be attached. Externalities have either been ignored altogether, or, at best, have been addressed by appending a qualitative commentary. This is because CBA works well for traded goods but is unsatisfactory when faced with 'goods' for which no market exists (such as clean air, rare species or landscape beauty). Consequently, the problems and limitations identified above have driven environmental economists to extend CBA by devising a suite of surrogate valuation techniques. That there is a pressing practical need to do this can be illustrated using a single example from Scottish forestry. The economic case for public sector support for native woodland restoration rests almost entirely on the provision of non-market benefits related to wildlife, landscape and recreation (MacMillan and Duff, 1998), but unless or until a quantitative assessment can be made of how valuable these benefits are, it is impossible to know what the appropriate level of public support should be.

One of the leading exponents of the efficacy of using economic tools to protect and manage the environment is David Pearce. He argues that while campaigns, protests and lobbying can all make a difference, nothing would help to save the world's environments more effectively than a reformulation of global economic policies. He explains the essence of his logic concisely (Pearce, 1998: 2, 3, 39):

> In many situations there is . . . an inevitable trade-off between human well-being and environmental conservation. It is the economic perspective that promises to help us make better and informed choices about the nature of that trade-off. To date, conservation has not stood much of a chance. The fact that so much of the value of conserved resources is not marketed means that, when faced with competition from marketed products, conservation loses out. By imparting economic value to the natural world, we can ensure a more 'level playing field' between conservation and development. . . . All economic decisions affect the environment, and so the way to improve the environment is to change the way in which those economic decisions are made. . . . The only way to get the environment onto the economic agenda is to demonstrate that the environment matters to the economy.

Pearce is dismissive of those who advocate a radical moral revolution as the *only* means of saving the environment. Money is a currency that everyone understands, and however much one believes in the intrinsic value of nature, the unpalatable bottom line is that 'we will get the kind of countryside that we are prepared to pay for' (Edwards and Smout, 2000: 24).

The first challenge in costing intangibles is to think clearly about the types of value which people place on the environment. Though 'value' is a multi-faceted word, values can helpfully be divided into two categories – use values and passive use (or non-use) values – and then further subdivided along a continuum of increasingly distant interest (Table 13.2). With the exception of Q-Altruism which is ecocentric, all these types of value are tied to human satisfaction. Categorising non-market benefits in this way is relatively easy, but putting values on them is hard (Henderson-Howat, 1993). This is the problem to which environmental economists have endeavoured to provide solutions. The techniques that they have devised attempt to derive economic values for externalities as if a market for them existed. This then transforms everything into a common currency (money) and makes comparisons possible. The three best established tools of extended CBA are the travel cost method, hedonic pricing and contingent valuation.[8] The first two of these recognise that some resources, though themselves unpriced, have an economic value because people are prepared to pay to benefit from them.

13.4.2.i The travel cost method and hedonic pricing

The travel cost method (TCM), developed in the late 1950s, has been widely used for valuing outdoor recreational resources such as forests, lakes and wetlands. Where a site has no entrance charge, the price of access is the cost of getting there, so the value of places can be estimated by the costs people incur in visiting them. Clearly there are problems that need to be taken into account. For example, local people may value a

TABLE 13.2 A typology of environmental values.

Type of value	*Description*	*Example*
Use values		
Direct use value	Directly used or consumed in some way	Water, food, medicines, recreation, aesthetic/spiritual benefits
Indirect use value	Benefits arising from environmental systems	Drainage, coastal protection
Option value	Not currently used, but valued for potential use	Wilderness areas
Passive use (or non-use) values		
Existence value	Valued because it exists, even if never seen or used	'Save the Whale'
Vicarious use value (or bequest value)	Valued for the benefit of the next generation; the desire to preserve option value for one's children	Clean air or water; unspoilt nature
Altruism	Protection for other people's benefit	Any of the above, but for others
Q-Altruism	Belief in the intrinsic worth of nature, regardless of any human benefit or cost; protecting it for its own sake	Permanent protection for Antarctica

site highly but pay little or nothing to reach it. Some may visit a distant site as part of a multi-purpose holiday, so there is a need to work out how much of their total outlay applies to the site in question. Similarly, how should the cost of the trip be divided amongst the different attributes of the site (landscape, wildlife, cultural interest)? Nevertheless, the TCM can provide a useful indication of the values attached to parts of the natural heritage.

Hedonic pricing followed in the late 1960s. It assesses value *via* the 'quality of life' premium that people will pay in relation to some environmental benefit. For example, many people will pay extra for a house with a beautiful view, or one that is not affected by traffic noise or air pollution. Separating out that part of the total cost of an asset that is due to such environmental benefits is not straightforward, especially given that few markets function freely, but this indirect approach can yield valuable results. TCM and hedonic pricing have both proved useful for measuring use values, but obviously neither can be applied to non-use values. This was the primary task for which the contingent valuation method (CVM) was developed, and this is proving much more controversial.

13.4.2.ii Contingent valuation: the method and its validity

CVM puts a price on environmental costs and benefits based on the economic strength of people's preferences and values. It reveals these preferences by employing sophisticated questionnaire surveys which ask people what monetary value they attach to elements or attributes of the environment. Two avenues can be adopted, both of which rely on creating realistic, though hypothetical, markets. The first is to ask people how much they would be willing to pay (WTP) to preserve or to continue using/enjoying the thing in question (the wood, the moorland, the species). The second is to ask them how much they would be willing to accept (WTA) as adequate compensation if that thing was destroyed. The prices obtained are then incorporated within CBA.

At first sight, CVM solves the problem of ignoring intangibles. It has been promoted as an objective, value-neutral solution, a matrix which can potentially capture all the important market and non-market dimensions of environmental value, including the elusive non-use values. During the 1990s, policy-makers in the UK were increasingly persuaded of the benefits of CVM (Grove-White, 1997), not least by the influential report *Blueprint for a Green Economy* (Pearce *et al.*, 1989). The technique has now become dominant to the extent that the government uses it in setting the level of green taxes (such as the Landfill Tax) and sometimes employs it as a guide in the allocation of public funds. Within Scotland, CVM and other surrogate valuation methods have been widely applied in land use and environmental decision making (Table 13.3). Given the growing influence of CVM in the policy world, it is important to assess its validity.

The key contribution that CVM makes to environmental decision making is its emphasis on the (often considerable) non-monetary values that people attach to nature. For the first time, these can be built into the decision making process rather than having to be awkwardly bolted on, often as an afterthought and frequently carrying little weight. Most would agree that CVM is in pursuit of a highly desirable goal. What its critics argue, however, is that it fails to deliver, both for methodological reasons and because it is flawed in principle. The methodological criticisms are numerous and detailed,[9] including aspects such as the aggregation of individual preferences and the discounting of future values,[10] but three examples will suffice. The monetary values obtained using WTP are typically between two and five times lower than those yielded by WTA (Garrod and Willis, 1999); how are we to know which represent the 'true' value? Secondly, big differences also exist between what people *say* they are willing to pay and what they are actually prepared to pay (partly affected, of course, by ability to pay). Should we therefore take people's stated WTP with a pinch of salt, and if so, how big a pinch? Thirdly, results differ according to how much information is provided to the respondents, and whether the questions are open-ended or specific ('how much would you be willing to pay to save the capercaillie?' *versus* 'would you be willing to pay £10?'). Which results are the more reliable?

These technical difficulties can be addressed in a variety of ways by refining the methods employed, but there are also a range of criticisms that strike more profoundly at the very concept of CVM:

1. One of the most fundamental critiques is that environmental values and monetary values are incommensurable – that it is impossible to transform

the former into the latter in any meaningful way (Sagoff, 1998). The argument is that using preferences to assess values is a 'category mistake' because the two terms are not synonymous. The plurality of values that humans place on the environment simply cannot be represented by a single monetary measure, however convenient this is.

TABLE 13.3 Some examples of the application of environmental economics to aspects of environmental management in Scotland. All studies used the contingent valuation method unless otherwise stated.

Sector	*Description*	*Reference*
Agriculture	Assesses the public benefits of landscape and environmental change resulting from grazing extensification in the Southern Uplands	Bullock (1995); Bullock and Kay (1997)
Agriculture	Compares CVM and Choice Experiments for valuing the benefits of ESAs (Section 5.2.2.i)	Hanley *et al.* (1998)
Agriculture	Examines the benefits of agri-environmental policy in Breadalbane and Machair ESAs	Alvarez-Farizo *et al.* (1999)
Deer management	Employs Choice Experiments to assess values of actual & hypothetical attributes of deer stalking experience in different environments	Bullock (2001)
Forestry	Evaluates the non-market benefits & costs of the proposed native pinewoods in Strathspey and Glen Affric	MacMillan (1995, 1999); MacMillan and Duff (1998)
Nature conservation	Measures the public's WTP to prevent afforestation of the Flow Country	Hanley and Craig (1991)
Nature conservation	At sites in southern Scotland, compares ecological value with the public's WTP for environmental attributes	Edwards-Jones *et al.* (1995)
Nature conservation	Assesses non-market environmental costs and benefits of biodiversity projects in Affric and Strathspey	MacMillan *et al.* (2001b)
Recreation	Valuation of environmental preferences in countryside recreation in Grampian Region	Christie (1999); Christie *et al.* (2000)
Water resources	Assesses public's WTP to maintain river flows at natural levels on the River Almond	Edwards-Jones *et al.* (1997)
Water resources	Assesses the economic value of improvements to water quality	Hanley (1997)
Wildlife management	Uses CVM and Choice Experiments to explore the conflict between goose conservation and agriculture (Section 7.7.3)	MacMillan *et al.* (2001a)

2. A reliance on people's preferences inevitably builds in a bias against the less visible, the long term, and the less attractive. The 'cute furry animals' factor means that obscure, unattractive species are valued little, however crucial their ecological function (Price, 1997; White *et al.*, 2001).
3. Ecocentrics criticise CVM for its utilitarian, anthropocentric approach – its stress on the value *to human beings*. It cannot assess the intrinsic worth of other species or of other parts of the environment. The danger is that once nature is given a price tag, other non-economic arguments tend to be lost. Some participants in CVM surveys actually feel that pricing nature is immoral (Burgess, 1998). 'Must economists reduce all of nature to the metric of unrighteous mammon?' (Bishop, 1999: 19). Another side of this critique is that the trade-offs which are implicit in CVM only make sense within the utilitarian schema of CBA; for those who hold rights-based beliefs, asking for a trade-off is meaningless because the environment *should* be protected, come what may (Hanley and Milne, 1996). Infinite values cannot be traded.
4. CVM techniques are not as objective as they appear because preferences are affected by levels of education, breadth of experience, contrasting world views, and a host of other factors. Thus values revealed by WTP or WTA are highly contextual (Spash, 1997).
5. Contingent valuation reduces humans to bundles of preferences and so is ethically blind, taking no account of whether expressed preferences are good or bad. Moreover, 'placing a monetary value on nature confuses the values people have as consumers with those they hold as citizens' (Adams, 1996: 108).
6. Because CVM cannot incorporate the preferences of future generations, it runs the risk of perpetrating intergenerational injustices.
7. Although it claims to dispense with value-laden decision making, actually CVM often serves to mask rather than to reveal the value judgements which are an unavoidable facet of environmental issues. One of the salient attractions of extended CBA is that it reduces highly complex problems to manageable proportions, but critics turn this around and argue that simply too much of importance is lost in the process of reduction, so that it becomes a case of *reductio ad absurdum*.

Clearly, then, CVM is beset by problems on all sides, with critics arguing that far from incorporating values into decision making these techniques actually *devalue* nature – that they find a price for everything and the value of nothing. Diamond and Hausman (1994: 46), for example, believe that CVM does not measure the preferences that it purports to measure, and that reliance on the technique is 'basically misguided'. More generally, Foster (1997b: 2) wonders whether economic valuation of the environment is anything more than 'a new eco-friendly jargon for licensing our aspirations to technological management and control of nature'. But is it better at least to *try* to build environmental values into decision making, even in a flawed way, than to ignore externalities altogether? Is some number better than no number (Diamond and Hausman, 1994)? No one wants to ignore the importance of non-market values, but if the prices derived to represent those values are of questionable validity, how much further forward does CVM take us?

The choice, however, is not as black and white as such questions would make it appear, because no one is proposing that extended CBA tools should be the sole basis of decision making. Their advocates merely argue that they reveal important information which might otherwise remain hidden, ignored or under-valued. These tools are designed to assist, not to supplant, other approaches to decision making (White *et al.*, 2001). The critics and proponents of CVM are both seeking the same ends, namely greater protection for the environment. Even some of those who are highly critical of CVM still advocate its use and development, arguing that as long as its limitations are clearly understood it can at least indicate the upper and lower bounds of value (Hanley and Spash, 1993; Price, 1997, 2000). Price (1997: 142) argues that CVM and other surrogate valuation methods must help to improve the current situation given that, at present, decisions are made 'by default using seat-of-pants techniques or inscrutable political intuition'.

It is probably fair to conclude that although the various tools of environmental economics can certainly be useful, especially in making a case for the environment in the corridors of power, they 'fail to capture the full range of reasons why nature is important' (Adams, 1996: 8; Grove-White, 1999). Unless or until economic values cease to have the power of trump cards in official decision making, there will be a pragmatic need for techniques which give monetary weight to environmental issues. Pearce (1999: 6) explains why:

> . . . the battle over land use is a battle over real economic values. It is about who makes most profit from the different ways land might be used. As much as we might like the battleground to be a different one, for example an issue of morality, of ethical norms, it is the conflict of economic values that defines the real world.

This reality is unlikely to change any time soon. Since at present there are no obvious alternative ways of obtaining the kinds of information that extended CBA methods provide, for the foreseeable future it seems likely that they will continue to make a useful but imperfect (and fiercely contested) contribution to the complex business of managing our environment.

NOTES

1. For in-depth discussions of the huge range of complex ethical and philosophical issues emerging out of environmental thinking see, for example, Cronon (1996a), Pepper (1996), O'Neill *et al.* (1997), Attfield (1999) and Pratt *et al.* (2000).
2. The volume of literature on sustainable development is vast and continues to expand exponentially. For useful general introductions, see Mitchell (1997), Connelly and Smith (1999) and O'Riordan (2000b). Recent discussions of sustainability issues in Scotland can be found in McDowell and McCormick (1999) and Holmes and Crofts (2000).

3. For detailed discussions of the role of science in environmental decision making, see Porritt (2000) and the volume edited by Huxham and Sumner (2000).
4. The 'man and nature' debate has a long and literature-rich history. For detailed discussions, see Glacken (1967), Black (1970), Passmore (1974), Oelschlaeger (1991), Eldredge (1995), Cronon (1996a), Macnaghten and Urry (1998) and Cooper (2000).
5. Gottlieb (1996) discusses this specific controversy, together with wider issues concerning religion and environment. See also Elsdon (1992), C. Russell (1994), Kinsley (1994) and Northcott (1996).
6. See, for example, Ellis (1990), Brennan (1990), O'Neill *et al.* (1997), Benson (2000) and Pratt *et al.* (2000).
7. For a helpful introduction to and discussion of these various approaches, see Connelly and Smith (1999).
8. Several excellent texts on the techniques and problems of environmental economics exist, and what follows is just the briefest of summaries, focusing on concepts rather than methods. For details of the techniques, and discussions of the field, see Hanley and Spash (1993), Pearce (1998), Garrod and Willis (1999), Perman *et al.* (1999), Adger (2000) and Hanley *et al.* (2001).
9. Numerous authors have provided critiques of the theory and methods of surrogate valuation techniques, notably in the volume edited by Foster (1997a). See also Adams (1993), Redclift (1993), Rawles and Holland (1994), More *et al.* (1996), Price (1997, 2000), Grove-White (1999) and the special issue of *Environmental Values* 9(4), 2000.
10. The practice of discounting future benefits to achieve a net present value is especially controversial because a high discount rate effectively renders the distant future almost valueless. People certainly value certain material benefits more highly if they are accessible immediately rather than in the future, and consequently a high discount rate is often appropriate over short time periods when choosing privately over relatively trivial things. However, when it comes to public choices concerning important, long term issues (such as landscape or biodiversity change), future benefits may be regarded as being of equal value to present benefits, so discounting seems inappropriate. For discussions, see Hanley and Spash (1993) and Pearce (1998).

PART FIVE:

Conclusion

CHAPTER FOURTEEN

Environmental management in twenty-first-century Scotland

14.1 Common threads in the emerging tapestry

Identifying coherent themes within the ever-changing kaleidoscope of environmental management is a subjective exercise. During the last decade, however, a few salient trends have emerged which are very likely to characterise the immediate (and probably the medium term) future. The first three of these – integration, partnerships, and community involvement – are all closely related. They are an expression of a general trend whereby government and management are becoming more open, accountable, inclusive and democratic.

14.1.1 Policy integration and multiple objectives

Through much of the twentieth century, environmental management in Scotland was bedevilled by a sectoral approach, the sole exception being the traditional private estates on which integrated land use and multiple objectives have long been the norm (Section 2.4.3). The concept of managing land and water resources within an integrated, holistic framework is hardly new (c.f. Marsh, 1864; Leopold, 1949), but for decades commentators called in vain for greater integration of rural policy. Only during the 1990s did things begin to change significantly as the self-evident logic of such an approach penetrated the corridors of power. SNH now believes that integration is a *sine qua non* of achieving sustainable development (Crofts, 2000). This change reflects a much wider trend. For example, the EU's Fifth Environmental Action Plan (entitled 'Towards Sustainability' and covering 1993–2000) adopted the view that integrating the environment into the development and implementation of all policies is crucial if sustainable development is the desired objective (Connelly and Smith, 1999). Carter and Lowe (1998) identify four significant developments at European and UK level which have contributed towards this move away from the piecemeal, reactive approach of traditional policy making towards a more strategic, integrated and anticipatory stance. These are the politicisation of environmental issues, a growing openness in the policy process, the gathering momentum of European political integration, and the emergence of the UK Government's sustainable development strategy.

An inevitable corollary of the 'joined up thinking' required for integrated policy making is the abandonment of single purpose management and its replacement with the pursuit of multiple objectives. During the last ten years a remarkably wide range of interests have abandoned one-track policies and embraced the brave new world of multi-purpose management. This sea-change in approach has been apparent across the board, affecting public, private and voluntary sectors alike. Obvious examples include the new emphasis on integrated catchment management (Section 6.4.2), the adoption of multi-benefit forestry (Section 4.5.1), and the new inclusive approach of 'sustainable conservation' (Section 8.2.2). Even in that most conservative of worlds, the deer management fraternity, the single-minded pursuit of sporting objectives is being replaced with a broader ethos that takes account of forestry, agriculture, the natural heritage, and welfare and safety issues, all under the umbrella of sustainability (DCS, 1999).

One interesting symptom of this new era of multi-purpose, integrated policies has been the broadening visions of voluntary organisations. Single-issue campaigning is beginning to fall out of fashion. The RSPB, for example, has progressively transformed itself, leaving behind its exclusive, narrow focus on birds to become an organisation which is a widely respected commentator on (and partner in) many aspects of Scotland's rural scene. On its own properties, such as Abernethy in the Cairngorms and Forsinard in the Flow Country, it has confounded its critics by successfully putting much of its thinking into practice (Beaumont *et al.*, 1995) and is researching solutions to persistent problems such as bird strikes on forest fences (Summers and Dugan, 2001). The RSPB's director goes so far as to argue that conservation organisations and land use interests should stop bickering over detailed issues like raptors (Section 7.6.2) and should instead join forces in lobbying to get the 'big picture' issues right (Housden, 2001). Another facet of integrated thinking is the realisation of the importance of managing the physical environment within natural rather than political (or other artificial) boundaries. Newson (1999: 12) nicely summarises this logic:

> The most appropriate spatial scale for all environmental management is surely that at which the natural system concerned can be holistically conceptualised and practically reconciled with human systems of exploitation and conservation.

Prime examples of this thinking being applied in practice include the nationwide formation of deer management groups (Section 7.5.2) and the moves towards integrated catchment planning (Section 6.4.2).

While positive signs of increasing integration are emerging in and between many sectors, it is important to remember that, for all its faults, a sectoral approach does have strengths as well as weaknesses. It has the advantage of being focused and clearly defined, and of delivering results within specified parameters. D. W. Mackay (1994), for example, argues that greater success in tackling river pollution was achieved in Scotland than in England and Wales because the River Purification Boards in the former had a narrower focus than the multi-functional approach of the National Rivers Authority in the latter. A danger of integration is thinking too broadly and incorporating too much so that managers drown in information and fail to make progress towards any of their multi-purpose goals. Boundaries still have to be drawn somewhere. It is also the case that fine words about integrated policies come far easier than integrated actions.

> There is sometimes a tendency to create the impression that the mere utterance of words like 'integration', 'sustainability', 'holistic management' will somehow bring to an end all problems in resource conservation. Clearly that is far from being the case. (Boon, 1994: 572)

But although the gap between rhetoric and reality may take many years to close, even the fact that integration is now a widely shared aspiration represents heartening progress.

14.1.2 Partnerships and participatory decision making

14.1.2.i The rise of partnership working

One increasingly fashionable method of attempting to implement the ideal of integrated management has been the formation of partnerships. In recent years, many of the traditional faultlines that have divided sectors and interest groups have begun to be bridged. Polarised trench warfare has largely ended as the warring parties have ventured cautiously into no-man's-land and found surprising amounts of common ground. For example, no one would have believed that the Ramblers Association and private landowners (amongst others) could negotiate and reach agreement on an Access Concordat as happened in 1996 (Section 9.3.1.ii). The on-going work of the Access Forum in building a consensus in favour of the new access arrangements, and in hammering out the details of the Scottish Outdoor Access Code (Section 9.4.3), bears testimony to the effectiveness of participatory involvement that goes beyond mere consultation (Davison, 2001).

During the 1990s, partnerships sprang up right across the land, both formal and informal, and both spontaneous (bottom up) and artificial (top down). They now have official blessing (SODD, 1999b), and public bodies are being strongly encouraged to work with the private and voluntary sectors and with local communities. SNH (2000g) is now funding over 150 partnerships of diverse kinds, and SERAD (2000a: 72) states unequivocally that 'partnership is one of the keys to the successful delivery of rural development'. Cynics dismiss partnerships as talking shops for the spin doctors, and there are those who suggest that participatory decision making is less about real empowerment than about public relations. Nevertheless, real and positive results have been flowing from many of these initiatives.

Again, there are innumerable examples at all levels and across all sectors (Table 14.1). One of the most pertinent and high-profile examples is the Cairngorms Partnership which has been described as 'the best attempt, so far, to synthesise local and national interests into a common strategy' (Austin, 1997: 32). In conservation particularly, participatory decision making has increasingly become *de rigueur* (Section 8.2.2.ii) (Goodwin, 1998a; Warburton, 1998). Notably, 'the breadth of the partnerships involved in implementing the [biodiversity] action plans in Scotland is without precedent'

(Usher, 2000: 3), and the Scottish Biodiversity Group is one of the largest partnerships that Scotland has ever seen.

TABLE 14.1 Examples of partnership working in Scottish environmental management.

Sector	*Example*	*Section/reference*
Deer management	Deer management groups	7.5.2
	The Deer Management Round Table	7.5.2
Forestry	The Central Scotland Forest	4.5.3
	The Millennium Forest for Scotland Trust	4.5.5
Nature conservation	BAPs & LBAPs	8.2.3
	Local Agenda 21	8.2.2.ii
Planning	Formulation of Indicative Forest Strategies	4.5.2; SODD, 1999c
	The Community Planning process	SODD, 1999b
Recreation	The Paths for All Partnership	9.3.1.i
Water resources	The River Valleys Project	6.4.2
	The Tweed Foundation	6.4.2

14.1.2.ii The problems of partnership working

Initiatives like these have gained broad support, and demonstrate that co-ordinated, voluntary participation amongst all the stakeholders can yield progressive, practical results. In the light of all this, there can be no doubt that the ambition set out in the Rio principles of engaging people at all stages of decision making is being extensively realised. However, the above should not be taken to imply that all is now sweetness and light, nor that participatory decision making is unproblematic. Getting people around a table does not magically replace conflict with consensus. Partnerships are challenging to operate, taking much time, patience and skill. The acclaimed Access Concordat (Section 9.3.1.ii), for example, emerged from a lot of hard talking and no less than eight drafts. For partnerships to work effectively and 'deliver the goods', there is a need for people to reconsider their views, recognise and reject their prejudices, and accept compromises (Ratcliffe, 1998). Each of those stages is easy to recommend but humbling and hard to achieve. Fragile consensus can easily fall apart as people dig in their heels and insist that *others* do the compromising. It is much easier to sabotage agreements than to achieve them. Some down-to-earth realism is offered by Price (2000: 191) who comments that 'the idealised sweet reasonableness of participatory discussion is not always found in real-world debate, where decisions may favour not the most deserving, but the most obstinate'.

Moreover, as Owens and Cowell (1996: 56) point out, the concept of participation itself is not unproblematic:

> The different interest groups have unequal access to the policy process; local opinion needs to be reconciled with the necessity for strategic direction in policy;

and considerations of intra- and inter-generational equity raise difficult questions about the nature of 'democratic' input.

Partnership management also raises awkward questions about the place of science in environmental decision making (Section 13.2.2). Participatory or stakeholder consultation is described by O'Riordan (2000a: xi) as the 'soft' end of environmental science. He goes on to warn that 'it cannot succeed unless the "hard" end of testing for fundamental laws and of creating plausible models runs sympathetically and in parallel, so as to provide valuable contextual information'. In other words, the necessary swing away from the old style 'expertocracy' – widely criticised as arrogant and alienating – must not be allowed to go so far that the new participatory style is pursued without adequate empirical foundations. For these and other sociological and psychological reasons, there is no guarantee that decisions arrived at in a participatory fashion will automatically be good ones.

A final challenging dimension of the rise and rise of partnership management is the onus it places on individuals to master a far wider range of skills than practitioners of earlier generations. Foresters no longer have the luxury of concerning themselves just with growing trees, nor deer managers only with deer. In the new political and social climate, generic skills such as communication, mediation, political awareness (in the broad sense), and an understanding of human and organisational behaviour have become increasingly important (Bills, 1999). Depth and breadth are now both required. Given the rate at which specialised knowledge is growing, and the increasing complexity of the national and international scene, this is a daunting challenge. In their discussions of the pros and cons of the partnership approach, both Gemmel (1996) and Mitchell (1997) recognise that it is a process dogged by many problems. Nevertheless, both conclude that it is worthwhile (especially in minimising conflict) and that it can often be the only viable path to take. 'Jaw-jaw' must be preferable to 'war-war'.

14.1.3 Community involvement in decision making

For much of recent history, rural communities have been at the mercy of forces outwith their control, whether laws passed at Westminster, the decisions of private landowners, or the policies of government agencies. The empowerment of communities (whether 'communities of neighbourhood' or 'communities of interest') has been a striking development during the last decade. The highest profile and most profound expression of this has seen local people banding together to buy their home ground (Section 3.3.3), but communities do not need to be owners to have a say in decisions which affect them. Increasingly, local people have demanded to be a part of the decision making process, and often of the implementation of the resulting decisions as well. This is a message that government agencies and the voluntary sector have heard loud and clear, many of them publishing statements of how they intend to incorporate communities' views in decision making.

The Forestry Commission took a lead in this area (FC, 1996), and the government gave the concept of community involvement its blessing in 1997 in arguing that rural development should increasingly be driven by the priorities of local people (Scottish Office, 1998d). In 1999 SNH launched an important policy statement concerning its natural heritage work with communities (SNH, 1999b). Its Community Grant Scheme, by providing funds for environmentally orientated community projects, encourages people to take a more active role in enjoying and managing the natural heritage. It seems that SNH is increasingly prepared to abide by local wishes. For example, in the consultation concerning the proposed reintroduction of the beaver to Knapdale (Section 10.4.2), local people were not just consulted but were given a right of veto; reintroduction was made conditional on local acceptance of the idea.[1]

As is the case with partnerships, the wider the net is cast, the longer and more complex decision making becomes. However frustrating this can be, it is often the case that better decisions emerge and that – crucially – they have majority support amongst the interested parties (Bills, 1999). This minimises controversy and reduces the need to reverse decisions and go back to the drawing board. In areas which are much in the public eye (the Cairngorms and Loch Lomond spring to mind) the frequency of consultations runs the risk of inducing 'consultation fatigue' amongst local people, but at least they can no longer complain that they were kept in the dark while decisions were made (and then foisted on them) by faceless bureaucrats in Edinburgh or beyond. The opportunity to play a part in decision making is now routinely on offer.

14.1.4 Environmental education

Practitioners involved in all kinds of environmental action and management regularly bewail the ignorance of the general public and call for better education. Lister-Kaye (2001), for example, sees the current state of environmental education as woefully inadequate, describing it as the string missing from the conservation violin. The Cairngorms Working Party (CWP, 1993: 54) makes a strong case for the promotion of environmental education 'for people of all ages and through all available means', sentiments echoed by Crofts and Holmes (2000). Certainly the public are more likely to respect and care for a particular resource, whether freshwaters, woodlands, or rare species, if they understand its value, but education is a two-way process. Managers and activists also need to take time to hear and understand the views and desires of the public. In a woodland context, for example, 'foresters need to hear what woods mean to people as well as tell people what they mean to foresters' (HGTAC, 1998: 7). As the above sections show, the wishes of the public in general and of rural communities in particular are increasingly being heard and acted upon. The reciprocal flow of information, however, is proving harder to implement.

This is not for lack of trying. Indeed, there has been a surge of activity since the 1992 Earth Summit. Scotland now leads the UK in environmental education, with well developed strategies at national and regional levels (Kerr and Bain, 1997). There are numerous initiatives, large and small, that can broadly be described as environmental education, both in the state and voluntary sectors. Since its inception, SNH has seen education as a vital part of its remit, and it now directs around £600,000 each year towards projects with an environmental education theme (SNH, 2000g). These include training workshops, teaching packs and travel schemes for schools, interpretation facilities, travelling 'roadshow' displays, and the funding of environmental education officers. The Forest Education Initiative is another example (HGTAC, 1998). It recognises that many people are unaware of the range of benefits that forestry can offer because the deforested condition of the country has prevented the development of a forest culture such as that which prevails in Scandinavia. In the voluntary sector initiatives include SWT's 'Web of Life', Plantlife's 'Flowers of the Forest' and WWF's 'Wild Rivers' programme (Kerr and Bain, 1997).

But despite the resources which are now devoted to environmental education, there is no doubt that much more needs to be done. All facets of environmental management have been radically transformed over the last two decades, but the public at large remain blissfully ignorant of the fact, still entertaining notions and perceptions that relate to a bygone era. If the new approaches and aspirations are to penetrate public consciousness and bear fruit, environmental education is urgently required. An important illustration of this relates to the new access arrangements (Section 9.4.3). Public education is going to be critically necessary if these are to have a chance of working well on the ground (Mackay, 2001). This is brought home by the fact that, two years after the launch of the Access Concordat, even mountain users were mostly ignorant of it; half had never heard of it, and only 9 per cent had actually read it (Taylor and MacGregor, 1999). Amongst the general public – the constituency that the new arrangements are aimed at – one can reasonably presume almost total ignorance of the Concordat. This failure of communication cannot afford to be repeated once the new rights of access become law, but education on the appropriate scale will be neither easy nor cheap.

Nature conservation (especially of biodiversity) is another arena about which the general public are still surprisingly ill-informed. Many conservationists argue that the on-going damage to our natural heritage will only be halted by a change in society's priorities, a change which will require education. As Adams (1996: 113) says, 'if we do not tackle the issue of our cultural distance from nature, no amount of tinkering with protected area systems will be of much use'. Conservation needs to be 'owned' by society at large, not just by the green few. Only education on a grand scale will bring this about. It is for reasons such as this that Holdgate (2000a) gives environmental education top priority in his 'Agenda for the Future'. One recent trend which should considerably help the cause of environmental education is the on-going revolution in information technology (IT).

14.1.5 Information technology

Rapid developments in IT are affecting all areas of life, and environmental management is no exception. Of the many examples, perhaps three are particularly noteworthy.

1. **Information provision**. Good management, especially integrated decision making, depends on the availability of high quality information. In the past,

obtaining it was frequently a protracted, frustrating and arduous task, and the data were rarely current. Computer databases, email and the internet changed all this in the space of just a few hectic years in the late 1990s. Prior to 1997 very few players in the Scottish environmental scene had a presence on the web, but by 2000 almost all had websites. Government documents, debates in the Parliament, agency reports, consultation documents, NGO policy objectives, research papers and even some private sector management plans can all now be rapidly accessed and downloaded from the web. This has largely demolished the age-old problem of inaccessible, insufficient and outdated information. It is exacerbating, however, the timeless problem of insufficient time.

2. **Decision Support Systems**. These software packages are tools which can effectively integrate and process a vastly wider range of data than any human manager can, and which can allow value judgements to be brought into the open. They also allow managers to ask 'what if?' questions – to explore the long-term consequences of different management strategies. They have been developed, for example, for assessing the vegetation effects of different grazing and muirburning regimes in the uplands ('HillPlan') (Sibbald, 2001), for exploring deer management strategies ('HillDeer') (Buckland *et al.*, 1998; Gordon and Hope, 1998) and for evaluating the conservation value of woodlands (Carlyle and Edwards-Jones, 1995). Although they can only ever *support* decision making, and are not without their own attendant problems (Carlyle and Edwards-Jones, 1995), they are nevertheless useful virtual tools which are enabling various types of management to become less haphazard.
3. **Geographical Information Systems**. These sophisticated systems have now come of age and are utilised in innumerable management contexts as aids in visualising and analysing spatial data. To give just one current example, a GIS-based Native Woodland Model has been developed by MLURI which predicts site suitability for native woodland (Towers *et al.*, 2001).

The IT revolution has already changed the face of practical environmental management and will doubtless continue to do so as its effects permeate ever further.

14.2 THE TWENTY-FIRST CENTURY: THE DAWN OF CONSENSUS?

In the light of the trends identified here, it does seem that a brave new age of positive collaboration and partnership is struggling to be birthed. In recent years the avowed aim in many quarters has been to adopt measures which institutionalise peaceful coexistence, and these efforts are certainly bearing fruit. Smout (2000a: 63) perceives 'the contest lessening, use and delight beginning to become reconciled, a peace process well begun'. These sentiments refer specifically to the forestry scene, but they resonate much more widely. So are we truly entering what Peterken (1996: 466) calls 'the age of consensus'? Probably not. The reason for this unwelcome verdict is that most environmental debates in the countryside derive from a set of intractable dilemmas, several of which are discussed in Section 12.3.

> Underlying the whole debate about public policy and rural land use are unanswered questions about where the greatest public benefits lie. Is the countryside best used for primary production, for the maintenance of biodiversity or for public recreation? What constitutes good land management? There are many detailed codes of good practice but no consensus on the overarching context. Unless and until we develop accepted standards of management, much debate will continue to generate more heat than light. (A. Raven, 1999: 134)

We thus end where we began (Section 1.4) with the inescapable conclusion that the management of the environment is essentially the management of competing interests, both in time and space, interests which include both human and non-human 'players'. It involves reconciling disparate contemporary aspirations at local, regional, national and international levels, and balancing the needs and wishes of this generation with those of future generations. Further, it involves striking a balance between, on the one hand, the needs, wants and rights of human society with, on the other, the protection, enhancement (and rights?) of the natural environment. It is most unlikely that these balancing acts will ever become easy. As Smout (2000a: 170) acutely observes:

> . . . the quarrel over the countryside is an argument over the limits and rights of property. . . . It can become so bitter because it goes to the root of the question – what does it mean to own and use land? For whom is the benefit and how is it shared?

There are few, if any, absolute rights and wrongs in this arena. Whether a particular strategy or policy is judged to be good or bad will depend on the different priorities attached to the various dimensions of the decision making matrix (Fig. 14.1). Depending on one's perspective, different weight is likely to be attached to present concerns *versus* those of our descendants, to local *versus* international perspectives, and to the importance of people *versus* the value of non-human nature. For example, to risk adopting stereotypes, local people living close to the land might give high priority to the present concerns of local communities (Point A in Fig. 14.1), while members of international conservation organisations might find themselves at the other end of all three axes (Point B), stressing the long-term significance of natural systems from a global perspective. Judgements about these priorities are themselves formed in diverse and contested ways, depending on people's beliefs and value systems, their political outlook, and, for example, the importance they attach to scientific approaches as opposed to other bases of knowledge and decision making. 'Ultimately, it is the balance that Scottish society chooses between achieving its economic, social equity and environmental goals that will determine the environment and land use of the future' (Maxwell and Cannell, 2000: 48).

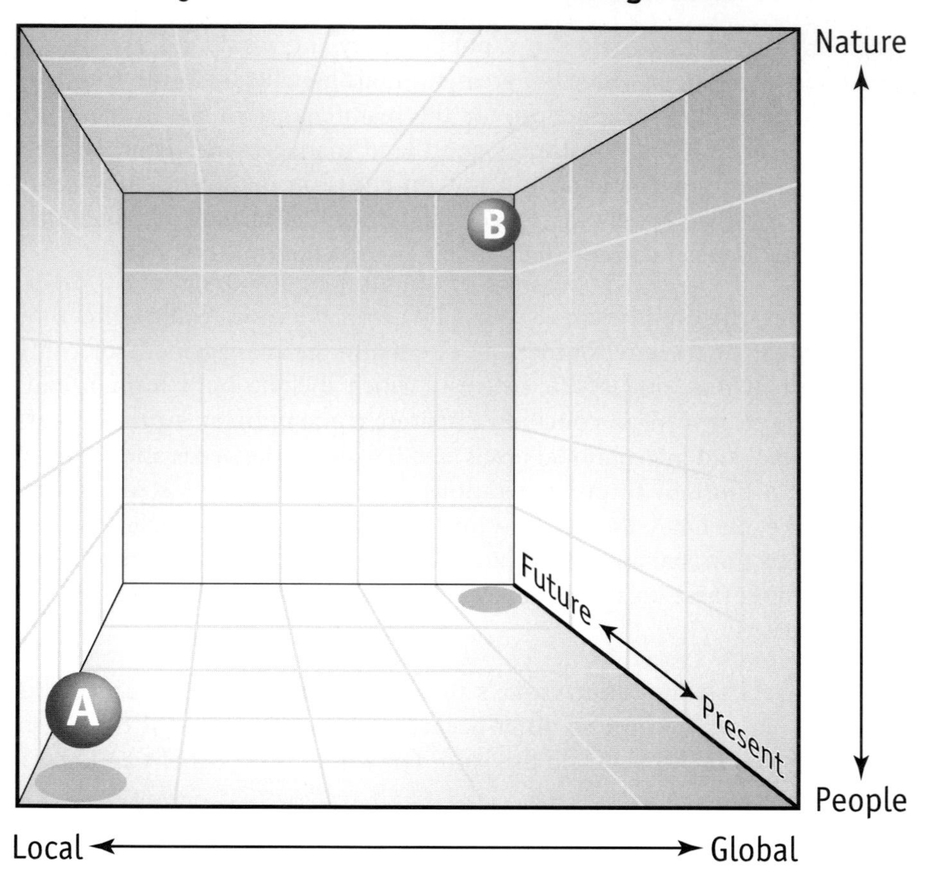

FIGURE 14.1 *Priority axes in environmental management. See text for explanation.*

Despite all the encouraging moves towards consensus, views on where that balance should lie – on what constitutes appropriate and desirable development – still diverge diametrically. Many environmentalists see modern economic development as the cause of the 'environmental problem' whereas members of struggling local communities sometimes identify *lack* of development as the problem (Barton, 1996). Clearly then, even though the formulation of individual policies or decisions may now cause less grief than it once did, there is no society-wide consensus about the ultimate goal that we should be aiming for, or the best way there. On the surface, it seems that sustainability is the avowed goal, and that sustainable development is the agreed means of achieving it. As we have seen, however (Sections 12.3; 13.3.4), interpretations of what these terms actually mean in practice vary so widely that they fail to provide an unambiguous position around which a clear consensus can be built. They nevertheless represent the best hope yet – a fuzzy goal, but a goal nonetheless.

Environmental management was always difficult, even in the era of relative certainty when sectors were kept largely separate, and when neither the goal of economic development nor the techno-scientific, 'expertocratic' model for achieving it were seriously questioned. Today's heady mix of cross-sectoral integrated management, sustainable development and participatory democracy, spiced with diverse belief systems, contested ethical foundations and often mutually exclusive objectives, will continue to make the management of Scotland's natural environment an 'interesting challenge' for the foreseeable future.

NOTE

1. In the event, 65 per cent voted in favour (*SCENES* 158: 2, 2001).

References

Abbott, F. 1999. The role of horses in forestry, based on case studies from the UK and Sweden. *Quarterly Journal of Forestry* 93(4): 299–306.

Abernethy, K. 1994a. *The introduction of sika deer* (Cervus nippon nippon) *to Scotland*. Unpublished PhD thesis, University of Edinburgh.

Abernethy, K. 1994b. The establishment of a hybrid zone between red and sika deer (genus *Cervus*). *Molecular Biology* 3: 551–62.

Abernethy, V. 2000. Local BAPs in Scotland – what difference have they made? *ECOS* 21(2): 21–3.

Access Forum. 1998. *Access to the countryside: the Access Forum's advice*. The Access Forum.

Adams, J. 1993. The Emperor's old clothes: the curious comeback of cost-benefit analysis. *Environmental Values* 2: 247–60.

Adams, W. M. 1996. *Future Nature: a vision for conservation*. Earthscan, London.

Adams, W. M. 2000. Looking forwards. In: *Nature in Transition*. Proceedings of a National Trust/ British Association of Nature Conservationists Conference, Lancaster, 13–14 July, 1999. Available at: www.nationaltrust.org.uk/environment

Adams, W. M. 2001. Joined–up conservation. *ECOS* 22(1): 22–7.

Adger, W. N. 2000. Environmental and ecological economics. In: O'Riordan, T. (ed.), *Environmental Science for Environmental Management*. Second Edition. Pearson, Harlow. pp. 93–118.

ADMG. 1997. *Association of Deer Management Groups Newsletter* No. 9.

ADMG. 2001. *Association of Deer Management Groups Newsletter No. 12*.

Aitken, R. 1997. A vision for Scotland's finest landscapes. In: *Protecting Scotland's Finest Landscapes: Time for action on National Parks*. Proceedings of a conference, 17th September 1997. Scottish Wildlife and Countryside Link, Perth. pp. 9–14.

Aitken, R. 2000. Personal communication, 3rd September 2000.

Aldhous, J. R. (ed.) 1995. *Our Pinewood Heritage*. Forestry Commission/RSPB/SNH.

Aldhous, J. R. 1997. British forestry: 70 years of achievement. *Forestry* 70(4): 283–91.

Alexandra, E. 2000. The Scottish Parliament and the Environment: a first year audit. *Scottish Forestry* 54(3): 130–2.

Allanson, P. and Whitby, M. 1996. Prologue: Rural Policy and the British Countryside. In: Allanson, P. and Whitby, M. (eds), *The Rural Economy and the British Countryside*. Earthscan, London. pp. 1–18.

Allcock, R. and Buchanan, D. 1994. Agriculture and fish farming. In: Maitland, P. S., Boon, P. J. and McLuskey, D. S. (eds), *The freshwaters of Scotland: a national resource of international significance*. Wiley, Chichester. pp. 365–84.

Alley, R. B., Mayewski, P. A. and Saltzman, E. S. 1999. Increasing North Atlantic climate variability recorded in a central Greenland ice core. *Polar Geography* 23(2): 119–31.

Alvarez-Farizo, B., Hanley, N., Wright, R. E. and MacMillan, D. 1999. Estimating the benefits of agri-environmental policy: econometric issues in open-ended contingent valuation studies. *Journal of Environmental Planning and Management* 42(1): 23–43.

Anderson, K. and Smith, S. J. 2001. Emotional geographies. *Transactions of the Institute of British Geographers* 26(1): 7–10.

Anderson, M. L. 1967. *A History of Scottish Forestry*. Thomas Nelson, Edinburgh.

Anon. 1994a. *Sustainable Forestry: the UK Programme*. Command Paper 2429. HMSO, London.
Anon. 1994b. *Biodiversity: the UK Action Plan*. Command Paper 2428. HMSO, London.
Armstrong, A. 1999. Survival is the aim of the game. *Landowning in Scotland* 255: 6–8.
Arnott, J. 1997. The protection of the land. In: Magnusson, M. and White, G. (eds), *The Nature of Scotland: Landscape, Wildlife and People*. Canongate, Edinburgh. pp. 211–25.
Ashmole, P. 2000. Carrifran Wildwood and the Borders Forest Trust – using the past? *Scottish Woodland History Discussion Group: Notes IV* (Proceedings of a meeting at Battleby, Perth, 23rd November 1999). pp. 36–9.
Attfield, R. 1999. *The Ethics of the Global Environment*. Edinburgh University Press, Edinburgh.
Austin, L. 1997. Cairn Gorm: the challenge of sustainability. *ECOS* 18(3/4): 27–33.
Avery, M. and Leslie, R. 1990. *Birds and Forestry*. Poyser, London.
Azeez, G. 2000. *The biodiversity benefits of organic farming*. Soil Association, Bristol.
Bachell, A. 2000. West Affric: learning from the past – new woods in Affric. *Scottish Woodland History Discussion Group: Notes IV* (Proceedings of a meeting at Battleby, Perth, 23rd November 1999). pp. 22–7.
Badger, R. 1996. *Wildlife and Agriculture in Scotland: a secure future*. RSPB, Edinburgh.
Bailey-Watts, A. E. 1994. Eutrophication. In: Maitland, P. S., Boon, P. J. and McLuskey, D. S. (eds), *The freshwaters of Scotland: a national resource of international significance*. Wiley, Chichester. pp. 385–412.
Bainbridge, I. P., Minns, D. W., Housden, S. D. and Lance, A. N. 1987. Forestry in the Flows of Caithness and Sutherland. *Royal Society for the Protection of Birds Conservation Topic Paper* 18.
Baines, D. and Summers, R. W. 1997. Assessment of bird collisions with deer fences in Scottish forests. *Journal of Applied Ecology* 34: 941–8.
Baldock, D. and Lowe, P. 1996. The Development of European Agri-environmental Policy. In: Whitby, M. (ed.), *The European Environment and CAP Reform: Policies and Prospects for Conservation*. CAB International, Wallingford. pp. 8–25.
Balfour, J. 1994. Recreation as a land use in Scotland. In: Fenton, A. and Gillmor, D. A. (eds), *Rural land use on the Atlantic periphery of Europe: Scotland and Ireland*. Royal Irish Academy, Dublin. pp. 143–52.
Balfour, J. 1998. Scotland's forest. *Scottish Forestry* 52(2): 66–7.
Balfour, R. 1998. Landownership and reform of the feudal system. In: *Forestry and Land Use – Forestry in a Changing Scotland*. Proceedings of a Timber Growers Association Conference, Gleneagles, 20th May 1998: 31–6.
Balharry, E., Staines, B. W., Marquiss, M. and Kruuk, H. 1994. *Hybridisation in British Mammals*. JNCC Report 154. Joint Nature Conservation Committee, Peterborough.
Balharry, R. 1990. Introduction. In: Conroy, J. W. H., Watson, A. and Gunson, A. R. (eds), *Caring for the high mountains – conservation of the Cairngorms*. Centre for Scottish Studies, Aberdeen. pp. 11–14.
Ballantyne, C. K. 2002. After the ice: paraglacial and postglacial evolution of the physical environment of Scotland, 20 ka to 5 ka BP. In: Saville, A. (ed.), *Mesolithic Scotland: The Early Holocene Prehistory of Scotland and its European Context*. Society of Antiquaries of Scotland Monograph, Edinburgh, *in press*.
Ballantyne, C. K. and Dawson, A. G. 1997. Geomorphology and landscape change. In: Edwards, K. J. and Ralston, I. B. M. (eds), *Scotland: Environment and Archaeology, 8000 BC–AD 1000*. Wiley, Chichester. pp. 23–44.
Balls, P. W., Macdonald, A., Pugh, K. and Edwards, A. C. 1995. Long-term nutrient enrichment of an estuarine system: Ythan, Scotland (1958–1993). *Environmental Pollution* 90: 311–21.
Bannan, M., Adams, C. E. and Pirie, D. 2000. Hydrocarbon emissions from boat engines: evidence of recreational boating impact on Loch Lomond. *Scottish Geographical Journal* 116(3): 245–56.
Barraclough, F. 1997. Potential implications for land management. In: *Scottish Landownership within Europe*. Proceedings of the Scottish Landowners Federation Conference, Battleby, 23rd April 1997.

Barrow, C. J. 1999. *Environmental Management: Principles and Practice*. Routledge, London.

Barton, H. 1996. The Isle of Harris Superquarry: concepts of the environment and sustainability. *Environmental Values* 5(2): 97–122.

BASC. 2000. *Green Shoots – the contribution of shooting to biodiversity in the UK: an action plan*. British Association for Shooting and Conservation, Wrexham.

Baxter, C. and Thompson, D. 1995. *Scotland – Land of Mountains*. Colin Baxter Photography, Grantown-on-Spey.

Bayfield, N. G., Fraser, N. M. and Calle, Z. 1998. High altitude colonisation of the Northern Corries of Cairn Gorm by Scots Pine (*Pinus sylvestris*). *Scottish Geographical Magazine* 114(3): 172–9.

Beaumont, D., Dugan, D., Evans, G. and Taylor, S. 1995. Deer management and tree regeneration in the RSPB Reserve at Abernethy Forest. *Scottish Forestry* 49(3): 155–61.

Beckerman, W. 1994. 'Sustainable development': is it a useful concept? *Environmental Values* 3: 191–209.

Beckerman, W. 1995. *Small is Stupid: Blowing the Whistle on the Greens*. Duckworth, London.

Beedell, J. D. C. and Rehman, T. 1999. Explaining farmers' conservation behaviour: why do farmers behave the way they do? *Journal of Environmental Management* 57: 165–76.

Begg, J. and Watson, F. 2000. Glen Finglas: native woodland restoration and documentary records. *Scottish Woodland History Discussion Group: Notes IV* (Proceedings of a meeting at Battleby, Perth, 23rd November 1999). pp. 30–5.

Bell, S. 1997. Non-native trees in the rural landscape: why and where. In: Ratcliffe, P. R. (ed.), *Native and Non-Native in British Forestry*. Proceedings of a discussion meeting, University of Warwick, 31 March–2 April, 1995. Institute of Chartered Foresters, Edinburgh. pp. 142–7.

Bell, S. 1999. Plantation management for landscapes in Britain. *International Forestry Review* 1(3): 177–81.

Bennett, K. D. 1995. Post-glacial dynamics of pine (*Pinus sylvestris*) and pinewoods in Scotland. In: Aldhous, J. R. (ed.), *Our Pinewood Heritage*. Forestry Commission/ RSPB/SNH. pp. 22–39.

Benson, J. 2000. *Environmental Ethics: an introduction with readings*. Routledge, London.

Best, G. A. 1994. Afforestation and forestry practice. In: Maitland, P. S., Boon, P. J. and McLuskey, D. S. (eds), *The freshwaters of Scotland: a national resource of international significance*. Wiley, Chichester. pp. 413–33.

Bills, D. 1999. Looking back. Guest editorial. *Scottish Forestry* 53(4): 210–13.

Bishop, K. 1997. The challenge of convergence: countryside conservation and enjoyment in Scotland and Wales. In: Macdonald, R. and Thomas, H. (eds), *Nationality and Planning in Scotland and Wales*. University of Wales Press, Cardiff. pp. 243–63.

Bishop, K. 2000. Gambling on 2020 vision (Editorial). *ECOS* 20(3/4): 1.

Bishop, K., Phillips, A. and Warren, L. 1995. Protected for ever? Factors shaping the future of protected areas policy. *Land Use Policy* 12(4): 291–305.

Bishop, K., Phillips, A. and Warren, L. 1997. Protected areas for the future: models from the past. *Journal of Environmental Planning and Management* 40(1): 81–110.

Bishop, K., Tewdwr-Jones, M. and Wilkinson, D. 2000a. From spatial to local: the impact of the European Union on local authority planning in the UK. *Journal of Environmental Planning and Management* 43(3): 309–34.

Bishop, K., Norton, A. and Phillips, A. 2000b. The impacts of the National Lottery on the countryside. *ECOS* 20(3/4): 11–19.

Bishop, K., Norton, A. and Phillips, A. 2000c. He who pays the piper – the impact of the National Lottery on countryside conservation policy. *ECOS* 20(3/4): 20–8.

Bishop, R. C. 1999. On assessing the validity of market and non–market economic values of forest resources. In: Roper, C. S. and Park, A. (eds), *The Living Forest: non-market benefits of forestry*. Proceedings of an international symposium on non-market benefits of forestry, Edinburgh, 24–28 June, 1996. Stationery Office, London. pp. 17–30.

Bissett, N., Grant, A. T. and Adams, C. E. 2000. Long-term changes in recreational craft utilisation on Loch Lomond, Scotland. *Scottish Geographical Journal* 116(3): 257–66.

Black, A. and Burns, J. 2002. Re-assessing the flood risk in Scotland. *Science of the Total Environment, in press.*

Black, J. 1970. *The Dominion of Man: the search for ecological responsibility*. Edinburgh University Press, Edinburgh.

Black, J. S. and Conway, E. 1996a. The European Community's LEADER programme in the Highlands and Islands. *Scottish Geographical Magazine* 112(2): 101–6.

Black, J. S. and Conway, E. 1996b. Coastal superquarries in Scotland: planning issues for sustainable development. *Journal of Environmental Planning and Management* 39(2): 285–94.

Bloomfield, A. 1999. RSFS response: Forests for Scotland – consultation towards a Scottish Forest Strategy. *Scottish Forestry* 53(4): 262.

Boehmer-Christiansen, S. 1994. Politics and environmental management. *Journal of Environmental Planning and Management* 37(1): 69–85.

Bonner, J. 1996. Red or dead? *New Scientist* 2013 (20th January 1996): 29–31.

Boon, P. J. 1994. Nature conservation. In: Maitland, P. S., Boon, P. J. and McLuskey, D. S. (eds), *The freshwaters of Scotland: a national resource of international significance*. John Wiley, Chichester. pp. 555–76.

Boon, P. J. and Howell, D. L. (eds) 1997. *Freshwater Quality: defining the indefinable?* Scottish Natural Heritage, Edinburgh.

Boscawen, J. T. 1987. Estate implications. In: McIntosh, R. (ed.), *Deer and Forestry*. Proceedings of an ICF Conference, Glasgow. Institute of Chartered Foresters, Edinburgh. pp. 38–47.

Boyack, S. 1999. Planning and the Parliament: challenges and opportunities. In: McDowell, E. and McCormick, J. (eds), *Environment Scotland: Prospects for Sustainability*. Ashgate, Aldershot. pp. 109–26.

Boyd, G. 1999. Land reform and civil society. *Reforesting Scotland* 22: 10–12.

Boyd, G. and Reid, D. (eds) 2000. *Social Land Ownership, Volume Two: eight more case studies from the Highlands and Islands of Scotland*. Community Learning Scotland, Inverness.

Boyd, J. M. 1993. Case study: nature conservation – a Scottish memoir. In: Berry, R. J. (ed.), *Environmental Dilemmas: Ethics and Decisions*. Chapman and Hall, London. pp. 150–63.

Bragg, O. M. 2002. Hydrology and natural heritage: peat-forming wetlands. *Science of the Total Environment, in press.*

Breeze, D. J. 1997. The great myth of Caledon. In: Smout, T. C. (ed.), *Scottish Woodland History*. Scottish Cultural Press, Edinburgh. pp. 47–51.

Brennan, A. A. 1990. *Environmental Philosophy: an introductory survey*. Centre for Philosophy and Public Affairs, University of St. Andrews/Nature Conservancy Council. CPPA, St. Andrews.

Brennan, A. A. 1993. Environmental decision-making. In: Berry, R. J. (ed.), *Environmental Dilemmas: Ethics and Decisions*. Chapman and Hall, London. pp. 1–19.

Brindley, H. 1990. Planning for downhill skiing at Cairn Gorm. In: Conroy, J. W. H., Watson, A. and Gunson, A. R. (eds), *Caring for the high mountains – conservation of the Cairngorms*. Centre for Scottish Studies, Aberdeen. pp. 132–45.

Brogan, J. and Soulsby, C. 1996. Managing riparian tree cover in the Scottish Highlands: a study of Glen Tanar Estate. *Scottish Forestry* 50(3): 133–44.

Brouwer, F. and Lowe, P. (eds) 2000. *CAP Regimes and the European Countryside: prospects for integration between agricultural, regional and environmental policies*. CAB International, Wallingford.

Brown, A. F. and Bainbridge, I. P. 1995. Grouse moors and upland breeding birds. In: Thompson, D. B. A., Hester, A. J., and Usher, M. B. (eds), *Heaths and Moorland: Cultural Landscapes*. HMSO, Edinburgh. pp. 51–66.

Brown, N. 1997. Re-defining native woodland. *Forestry* 70: 191–8.

Bruce, D. and Bruce, A. (eds) 1998. *Engineering Genesis: the ethics of genetic engineering in non-human species*. Working Group of the Church of Scotland 'Society, Religion and Technology Project'. Earthscan, London.

Bryce, J. 1997. Changes in the distributions of Red and Grey Squirrels in Scotland. *Mammal Review* 27(4): 171–6.

Bryce, J. and Balharry, D. 1995. Red squirrel *Sciurus vulgaris* conservation in Scotland: a position statement. *Scottish Natural Heritage Review* No. 59.

Bryden, J. and Mather, A. S. 1996. The 'rural' White Paper – Rural Scotland: People, Prosperity and Partnership. *Scottish Geographical Magazine* 112(2): 114–16.

Bryden, J. and Hart, K. 2000. Land reform, planning and people: an issue of stewardship? In: Holmes, G. and Crofts, R. (eds), *Scotland's Environment: the future*. Tuckwell Press, East Linton. pp. 104–18.

Bryden, J. M. 1997. The land question in Scotland. *Scottish Association of Geography Teachers Journal* 26: 18–24.

Buccleuch, the Duke of. 1996. Consensus forestry. *Scottish Forestry* 50(2): 66–7.

Buckland, S. T., Trenkel, V. M., Elston, D. A., Partridge, L. W. and Gordon, I. J. 1998. A decision support system for red deer managers in Scotland. In: Goldspink, C. R., King, S. and Putman, R. J. (eds), *Population Ecology, Management and Welfare of Deer*. Manchester Metropolitan University. pp. 82–7.

Budiansky, S. 1995. *Nature's Keepers: the new science of nature management*. Weidenfeld and Nicolson, London.

Bullock, C. H. 1995. Measuring the public benefits of landscape and environmental change: a case of upland grazing extensification. In: Thompson, D. B. A., Hester, A. J., and Usher, M. B. (eds), *Heaths and Moorland: Cultural Landscapes*. HMSO, Edinburgh. pp. 277–81.

Bullock, C. H. 1999. Environmental and strategic uncertainty in common property management: the case of Scottish red deer. *Journal of Environmental Planning and Management* 42(2): 235–52.

Bullock, C. H. 2001. Red deer culls, Scots pine and the stalking client. In: Phillips, J., Thompson, D. B. A. and Gruellich, W. H. (eds), *Integrated Upland Management for Wildlife, Field Sports, Agriculture and Public Enjoyment*. The Heather Trust/Bidwells/SNH, Battleby. pp. 107–14.

Bullock, C. H. and Kay, J. 1997. Preservation and change in the upland landscape: the public benefits of grazing management. *Journal of Environmental Planning and Management* 40(3): 315–34.

Bullock, C. H., MacMillan, D. C. and Crabtree, J. R. 1994. New perspectives on agroforestry in lowland Britain. *Land Use Policy* 11(3): 222–33.

Bunce, R. and Jeffers, J. (eds) 1977. *Native pinewoods of Scotland: proceedings of the Aviemore symposium*. Institute of Terrestrial Ecology, Cambridge.

Burgess, J. 1998. *Putting a Price on Nature*. BBC 2, 26th January 1998.

Burgess, P. J. 1999. Effects of agroforestry on farm biodiversity in the UK. *Scottish Forestry* 53(1): 24–7.

Burgess, P. J., Brierley, E. D. R., Morris, J. and Evans, J. (eds) 1999. *Farm woodlands for the future*. BIOS Scientific, Oxford.

Burgess, P. J., Brierley, E. D. R. and Graves, A. 2000. Report on farm woodlands for the future conference. *Quarterly Journal of Forestry* 94(1): 65–9.

Butt, J. and Twidell, J. 1997. The power of Scotland. In: Magnusson, M. and White, G. (eds), *The Nature of Scotland: Landscape, Wildlife and People*. Canongate, Edinburgh. pp. 169–82.

Cairngorms Campaign. 1997. *The Cairngorms – Stepping Forward: a manifesto from the Cairngorms Campaign*. The Cairngorms Campaign, Inverness.

Cairngorms Partnership. 1997. *Managing the Cairngorms: the Cairngorms Partnership Management Strategy*. The Cairngorms Partnership, Grantown-on-Spey.

Cairngorms Partnership. 1999. *From Preparation to Implementation: the Cairngorms Partnership Work Plan 1998–2000*. The Cairngorms Partnership, Grantown-on-Spey.

Cairngorms Partnership. 2000. *Cairngorms Estates: a survey of landowners in the Cairngorms Partnership Area*. The Cairngorms Partnership, Grantown-on-Spey.

Cairns, C. 1998. Tracks of Cairn Gorm's tears. *The Scotsman* 29th October 1998.

Cairns, C. 1999a. Islanders prepare for war on mink invaders. *The Scotsman* 31st August 1999.

Cairns, C. 1999b. New alliance angles for better fishing. *The Scotsman* 15th April 1999.

Cairns, C. 2000. Human needs must win out over call of the wild. *The Scotsman* 7th January 2000.
Calder, I. R. 1993. The Balquhidder Catchment water balance and process experiment results in context: what do they reveal? *Journal of Hydrology* 145(3–4): 467–77.
Callander, R. F. 1987. *A Pattern of Land Ownership in Scotland*. Haughend Publications, Finzean.
Callander, R. F. 1995. Native pinewoods – the last 20 years (1975–94). In: Aldhous, J. R. (ed.), *Our Pinewood Heritage*. Forestry Commission/RSPB/SNH. pp. 40–51.
Callander, R. F. 1998. *How Scotland is owned*. Canongate, Edinburgh.
Callander, R. F. and MacKenzie, N. M. 1991. *The management of wild red deer in Scotland*. Rural Forum Scotland, Perth.
Callicott, J. B. 2000. Harmony between men and land: Aldo Leopold and the foundations of ecosystem management. *Journal of Forestry* 98(5): 4–13.
Cameron, A. D. 1986. *Go listen to the crofters: the Napier Commission and crofting a century ago*. Acair, Stornoway.
Cameron, E. A. 2001. 'Unfinished business': The Land Question and the Scottish Parliament. *Contemporary British History* 15(1): 83–114.
Campbell, R. N. and Maitland, P. S. 1999. From land-uses through to fisheries: integrated management for the aquatic environment. In: Cresser, M. and Pugh K. (eds), *Multiple land use and catchment management*. Proceedings of an international conference, 11–13 September 1996. Macaulay Land Use Research Institute, Aberdeen. pp. 89–96.
Campbell, R. N., Maitland, P. S. and Campbell, R. N. B. 1994. Management of fish populations. In: Maitland, P. S., Boon, P. J. and McLuskey, D. S. (eds), *The freshwaters of Scotland: a national resource of international significance*. Wiley, Chichester. pp. 489–513.
Cannell, M. G. R. and Milne, R. 2000. Kyoto, carbon and Scottish forestry. *Scottish Forestry* 54(1): 11–16.
Carley, M. and Christie, I. 1992. *Managing Sustainable Development*. Earthscan, London.
Carlyle, E. E. and Edwards-Jones, G. 1995. A computerised system for the conservation evaluation of woodlands in southern Scotland. *Scottish Forestry* 49(1): 14–21.
Carrell, S. 2000. Scottish salmon farming revolution that has left the seas awash with toxic chemicals. *The Independent* 2nd October 2000.
Carrell, S. 2001. Salmon farming inquiry demanded by heritage agency. *The Independent* 18th June 2001.
Carter, N. and Lowe, P. 1998. Britain: coming to terms with sustainable development? In: Hanf, K. and Jansen, A.-I. (eds), *Governance and Environment in Western Europe: Politics, Policy and Administration*. Longman, Harlow. pp. 17–39.
Cayford, J. 1993. Black grouse and forestry: habitat requirements and management. *Forestry Commission Technical Paper* 1. Forestry Commission, Edinburgh.
CCS. 1986. *Forestry in Scotland: a policy paper*. Countryside Commission for Scotland, Battleby.
CCS. 1990. The mountain areas of Scotland: conservation and management. *Countryside Commission for Scotland Report*. CCS, Battleby.
CCS. 1991. The mountain areas of Scotland: a report on public consultation. *Countryside Commission for Scotland Report*. CCS, Battleby.
Chadwick, A. H., Ratcliffe, P. R. and Abernethy, K. 1996. Sika Deer in Scotland: density, population size, habitat use, and fertility – some comparisons with red deer. *Scottish Forestry* 50(1): 8–16.
Chambers, F. M. 1997. Bogs as treeless wastes: the myth and the implications for conservation. In: Parkyn, L., Stoneman, R. E. and Ingram, H. A. P. (eds), *Conserving Peatlands*. CAB International, Wallingford. pp. 168–75.
Charman, D. J. 1994. Late-glacial and Holocene vegetation history of the Flow Country, northern Scotland. *New Phytologist* 127: 155–68.
Christie, M. 1999. An economic assessment of the economic effectiveness of recreation policy using contingent valuation. *Journal of Environmental Plannning and Management* 42(4): 547–654.

Christie, M., Crabtree, J. R. and Slee, W. 2000. An economic assessment of informal recreation policy in the Scottish countryside. *Scottish Geographical Journal* 116(2): 125–42.
Christie-Miller, A. 2000. Access to woodlands: the landowner's viewpoint. *Quarterly Journal of Forestry* 94(3): 207–10.
Clark, G. 1983. Rural land use from *c.* 1870. In: Whittington, G. and Whyte, I. D. (eds), *An Historical Geography of Scotland*. Academic Press, London. pp. 217–38.
Clark, G. M. and Johnson, J. A. 1993. Farm woodlands in the Central Belt of Scotland: A socio-economic critique. *Scottish Forestry* 47(2): 15–24.
Clutton-Brock, T. H., Guinness, F. E. and Albon, S. D. 1982. *Red deer: behaviour and ecology of two sexes*. University of Chicago Press, Chicago.
Clutton-Brock, T. H. and Albon, S. D. 1989. *Red Deer in the Highlands*. BSP Professional Books, Oxford.
Clutton-Brock, T. H. and Albon, S. D. 1992. Trial and error in the Highlands. *Nature* 358: 11–12.
Clutton-Brock, T. H. and Thomson, D. 1998. Manipulation of red deer density on Rum: synthesis 1992–1995. *Scottish Natural Heritage Research, Survey and Monitoring Report* No. 97.
Clutton-Brock, T. H. and McIntyre, N. 1999. *Red Deer*. Colin Baxter Photography, Grantown-on-Spey.
Coates, D. 1999. Sustainable rural development and the role of new farm woodlands in England. In: Burgess, P. J., Brierley, E. D. R., Morris, J. and Evans, J. (eds), *Farm woodlands for the future*. BIOS Scientific, Oxford. pp. 1–10.
Coates, P. 1998. *Nature: Western Attitudes since Ancient Times*. Polity Press, Cambridge.
Coghlan, A. 1999. Reap what you sow . . . *New Scientist* 2194 (10th July 1999): 18–19.
Collen, P. 1995. The reintroduction of the beaver (*Castor fiber* L.) to Scotland: an opportunity to promote the development of suitable habitat. *Scottish Forestry* 49(4): 206–16.
Collins, L. 1996. Recycling and the environmental debate: a question of social conscience or scientific reason? *Journal of Environmental Planning and Management* 39(3): 333–55.
Colman, D. 1994. Comparative evaluation of environmental policies: ESAs in a policy context. In: Whitby, M. (ed.), *Incentives for Countryside Management: the case of Environmentally Sensitive Areas*. CAB International, Wallingford. pp. 219–51.
Connelly, J. and Smith, G. 1999. *Politics and the Environment: from theory to practice*. Routledge, London.
Conroy, J. W. H., Watson, A. and Gunson, A. R. (eds) 1990. *Caring for the high mountains – conservation of the Cairngorms*. Centre for Scottish Studies, Aberdeen.
Conroy, J. W. H., Kitchener, A. C. and Gibson, J. A. 1998. The history of the beaver in Scotland and its future reintroduction. In: Lambert, R. A. (ed.), *Species History in Scotland: introductions and extinctions since the Ice Age*. Scottish Cultural Press, Aberdeen. pp. 107–29.
Cook, H. F. 1998. *The Protection and Conservation of Water Resources: a British Perspective*. Wiley, Chichester.
Cooper, A. B. 1987. Internal design. In: McIntosh, R. (ed.), *Deer and Forestry*. Proceedings of an ICF Conference, Glasgow. Institute of Chartered Foresters, Edinburgh. pp. 77–84.
Cooper, N. S. 2000. How natural is a nature reserve?: an ideological study of British nature conservation landscapes. *Biodiversity and Conservation* 9: 1131–52.
Coppock, J. T. 1994. Scottish land use in the twentieth century. In: Fenton, A. and Gillmor, D. A. (eds), *Rural land use on the Atlantic periphery of Europe: Scotland and Ireland*. Royal Irish Academy, Dublin. pp. 39–53.
COSLA. 1987. *Forestry in Scotland: Planning the Way Ahead*. Convention of Scottish Local Authorities, Glasgow.
Costley, T. 2001. Behaviour associated with access to the countryside. In: Usher, M. B. (ed.), *Enjoyment and Understanding of the Natural Heritage*. The Stationery Office, Edinburgh. pp. 57–66.
Cowell, R., Jehlicka, P., Marlow, P. and Owens, S. 1998. *Aggregates, Trade and the Environment: European Perspectives*. IUCN UK Committee, Cardiff.

Cox, G. 1993. 'Shooting a line'?: Field sports and access struggles in Britain. *Journal of Rural Studies* 9(3): 267–76.

Cox, M., Straker, V., and Taylor, D. (eds) 1995. *Wetlands – Archaeology and Nature Conservation.* Proceedings of an International Conference, Bristol, April 1994. HMSO, London.

Crabtree, J. R. 1991. National Park designation in Scotland. *Land Use Policy* 8(2): 241–52.

Crabtree, R. and Bayfield, N. 1998. Developing sustainability indicators for mountain ecosystems: a study of the Cairngorms, Scotland. *Journal of Environmental Management* 52: 1–14.

Crabtree, J. R., Rowan-Robinson, J., Cameron, A. and Stockdale, A. 1994a. Community woodlands in Scotland. *Scottish Geographical Magazine* 110(2): 121–7.

Crabtree, J. R., Leat, P.M. K., Santarossa, J. and Thomson, K. J. 1994b. *The economic impact of wildlife sites in Scotland.* Journal of Rural Studies 10(1): 61–72.

Crabtree, J. R., Chalmers, N. A. and Appleton, Z. E. D. 1994c. The costs to farmers and estate owners of public access to the countryside. *Journal of Environmental Planning and Management* 37(4): 415–29.

Crabtree, J. R., Bayfield, N. G., Wood, A. M., Macmillan, D. C. and Chalmers, N. A. 1997. Evaluating the benefits from farm woodland planting. *Scottish Forestry* 51(2): 84–92.

Cramb, A. 1996. *Who owns Scotland now? The use and abuse of private land.* Mainstream, Edinburgh.

Cramb, A. 1998. *Fragile Land: Scotland's Environment.* Polygon, Edinburgh.

Cramond, R. D. 1990. Cairngorm – Conservation and Development: Living Together. In: Conroy, J. W. H., Watson, A. and Gunson, A. R. (eds), *Caring for the high mountains – conservation of the Cairngorms.* Centre for Scottish Studies, Aberdeen: pp.15–29.

Crawford, I. C. 1997. The conservation and management of machair. *Botanical Journal of Scotland* 49(2): 433–39.

CRC. 1997. *Countryside Sports: their economic, social and conservation significance.* A review and survey by Cobham Resource Consultants for the Standing Conference on Countryside Sports, Reading.

Cresser, M. and Pugh K. (eds) 1999. *Multiple land use and catchment management.* Proceedings of an international conference, 11–13 September 1996. Macaulay Land Use Research Institute, Aberdeen.

Crofters Commission. 1998. *The Way Forward: the role of crofting in rural communities.* The Crofters Commission, Inverness.

Crofters Commission. 1999. *Annual Report 1998/99.* The Crofters Commission, Inverness.

Crofts, R. 1995. The Environment – Who Cares? *Scottish Natural Heritage Occasional Papers* No. 2.

Crofts, R. 1999. Tailor-made solutions for Scotland. *Scotland's Natural Heritage* 14: 2.

Crofts, R. 2000. Sustainable development and environment: delivering benefits globally, nationally and locally. *Scottish Natural Heritage Occasional Papers* No. 8.

Crofts, R. and Holmes, G. 2000. An Agenda for Action. In: Holmes, G. and Crofts, R. (eds), *Scotland's Environment: the future.* Tuckwell Press, East Linton. pp. 134–40.

Cronon, W. (ed.) 1996a. *Uncommon Ground: rethinking the human place in nature.* W. W. Norton and Co., New York.

Cronon, W. 1996b. The trouble with wilderness; or getting back to the wrong nature. In: Cronon, W. (ed.), *Uncommon Ground: rethinking the human place in nature.* W. W. Norton and Co., New York. pp. 69–90.

Crumley, J. 1996. Scotland's company of wolves. *The Times* 27th January 1996.

Crumley, J. 1997. Fast track to conflict. *The Times* 12th April 1997.

Crumley, J. 2000. *A High and Lonely Place.* Whittles Publishing, Latheronwheel.

Cunningham, I. 1991. Forestry expansion – a study of technical, economic and ecological factors: introduction, summary and conclusions. *Forestry Commission Occasional Papers* No. 33.

Curry, N. and Owen, S. 1996. Introduction: Changing Rural Policy in Britain. In: Curry, N. and Owen, S. (eds), *Changing Rural Policy in Britain: Planning, Administration, Agriculture and the Environment.* Countryside and Community Press, Cheltenham. pp. 1–27.

Curry-Lindahl, K. 1990. The Cairngorms National Nature Reserve (NNR), the foremost British conservation area of international significance. In: Conroy, J. W. H., Watson, A. and Gunson, A. R. (eds), *Caring for the high mountains – conservation of the Cairngorms*. Centre for Scottish Studies, Aberdeen. pp. 108–19.

CWP (Cairngorms Working Party). 1993. *Common Sense and Sustainability: A Partnership for the Cairngorms*. HMSO, Edinburgh.

Dalby, S. and MacKenzie, F. 1997. Reconceptualising local community: environment, identity and threat. *Area* 29(2): 99–108.

Dalton, A. 2000. Lottery funding in Scotland: making a case for the countryside. *ECOS* 20(3/4): 43–6.

Danson, M. 1997. Scotland and Wales in Europe. In: Macdonald, R. and Thomas, H. (eds), *Nationality and Planning in Scotland and Wales*. University of Wales Press, Cardiff. pp.14–31.

Davidson, D. A. 1994. Conservation as a land use in Scotland. In: Fenton, A. and Gillmor, D. A. (eds), *Rural land use on the Atlantic periphery of Europe: Scotland and Ireland*. Royal Irish Academy, Dublin. pp. 173–84.

Davidson, D. A. and Carter, S. P. 1997. Soils and their evolution. In: Edwards, K. J. and Ralston, I. B. M. (eds), *Scotland: Environment and Archaeology, 8000 BC–AD 1000*. Wiley, Chichester. pp. 45–62.

Davies, C. M., Kenyon, W. and Wharmby, C. 2000. Uptake of innovative hill land use systems. *Scottish Forestry* 54(4): 203–8.

Davison, R. 2001. New access legislation in Scotland. In: Usher, M. B. (ed.), *Enjoyment and Understanding of the Natural Heritage*. The Stationery Office, Edinburgh. pp. 177–87.

Dawson, W. M. 1999. The social, environmental and economic value of short rotation coppice willow (*Salix*) as an energy crop. In: Burgess, P. J., Brierley, E. D. R., Morris, J. and Evans, J. (eds), *Farm woodlands for the future*. BIOS Scientific, Oxford. pp. 151–60.

DCS. 1998a. *A Policy for Sika Deer in Scotland*. The Stationery Office, Edinburgh.

DCS. 1998b. *Deer Commission for Scotland Annual Report 1997–1998*. The Stationery Office, Edinburgh.

DCS. 1999. *Deer Commission for Scotland Annual Report 1998–1999*. The Stationery Office, Edinburgh.

DCS. 2000a. *Wild deer in Scotland: a long term vision*. Deer Commission for Scotland, Inverness.

DCS. 2000b. *Deer Commission for Scotland Annual Report 1999–2000*. The Stationery Office, Edinburgh.

Dennis, R. 1998. The reintroduction of birds and mammals to Scotland. In: Lambert, R. A. (ed.), *Species History in Scotland: introductions and extinctions since the Ice Age*. Scottish Cultural Press, Aberdeen. pp. 5–7.

Dennis, R., Broad, R., Brockie, K., Crooke, C. and Duncan, K. 1993. *Ospreys in Scotland*. Royal Society for the Protection of Birds, Sandy.

Devine, T. M. 1999. *The Scottish Nation 1700–2000*. Allen Lane, London.

Dewar, D. 1997. National Park for Loch Lomond and the Trossachs. Scottish Office press release, 15th September 1997.

Dewar, D. 1998. *Land reform for the 21st Century*. Fifth John McEwan Memorial Lecture, Aberdeen, 4th September 1998.

Diamond, P. A. and Hausman, J. A. 1994. Contingent valuation: is some number better than no number? *Journal of Economic Perspectives* 8(4): 45–64.

Dickinson, G. 1991. National Parks – Scottish needs and Spanish experience. *Scottish Geographical Magazine* 107(2): 124–9.

Dickinson, G. 2000a. Recreation at Scottish lochs. *Journal of the Scottish Association of Geography Teachers* 29: 41–51.

Dickinson, G. 2000b. The use of the Loch Lomond area for recreation. *Scottish Geographical Journal* 116(3): 231–44.

Dickson, J. H. 1993. Scottish woodlands: their ancient past and precarious present. *Scottish Forestry* 47(3): 73–8.

Dickson, J. H. 1998. Plant introductions to Scotland. In: Lambert, R.A. (ed.), *Species History in Scotland: introductions and extinctions since the Ice Age*. Scottish Cultural Press, Aberdeen. pp. 38–44.

Dixon, J. 1998. Nature conservation. In: Lowe, P. and Ward, S. (eds), *British Environmental Policy and Europe: Politics and Policy in Transition*. Routledge, London. pp. 214–31.

Dolman, P. 2000. Biodiversity and Ethics. In: O'Riordan, T. (ed.), *Environmental Science for Environmental Management*. Second Edition. Pearson, Harlow. pp. 119–48.

Dovers, S. R. and Handmer, J. W. 1993. Contradictions in sustainability. *Environmental Conservation* 20(3): 217–22.

Dovers, S. R., Norton, T. W. and Handmer, J. W. 1996. Uncertainty, ecology, sustainability and policy. *Biodiversity and Conservation* 5: 1143–67.

Dower, M., Buller, H. and Asamer-Handler, M. 1998. The socio-economic benefits of National Parks: a review prepared for Scottish Natural Heritage. *Scottish Natural Heritage Review* No. 104.

Doyle, C. and Thomas, T. 2000. The social implications of agroforestry. In: Hislop, M. and Claridge, J. (eds), Agroforestry in the UK. *Forestry Commission Bulletin* 122. pp. 99–106.

Dressler, C. 2000. The Isle of Eigg Heritage Trust – the first eighteen months. In: Boyd, G. and Reid, D. (eds), *Social Land Ownership, Volume Two: eight more case studies from the Highlands and Islands of Scotland*. Community Learning Scotland, Inverness. pp. 19–25.

Dugan, D. 1997. The Abernethy experience: deer management in a conservation forest. In: Rose, H. (ed.), *Deer in Forestry – Beyond 2000*. Minutes of a conference in March 1997, Battleby. pp. 7–11.

Dugan, D. 2000. Personal communication, 7th September 2000.

Dunion, K. 1999. Sustainable development in a small country: the global and European agenda. In: McDowell, E. and McCormick, J. (eds), *Environment Scotland: Prospects for Sustainability*. Ashgate, Aldershot. pp. 202–13.

Dunsmore, R. D. 1998. Treeline woodlands and the Woodland Grant Scheme. *Scottish Forestry* 52(3/4): 183–4.

Dwyer, J. C. and Hodge, I. D. 1996. *Countryside in Trust: land management by conservation, recreation and amenity organisations*. Wiley, Chichester.

Edwards, A. C., Cook, Y., Smart, R. and Wade, A. J. 2000. Concentrations of nitrogen and phosphorus in streams draining the mixed land-use Dee Catchment, north-east Scotland. *Journal of Applied Ecology* 37(Suppl. 1): 159–70.

Edwards, I. 1999. The tree planter's guide to Sitka spruce. *Reforesting Scotland* 21: 26–7.

Edwards, K. J. and Ralston, I. B. M. 1997. Environment and people in prehistoric and early historical times: preliminary considerations. In: Edwards, K. J. and Ralston, I. B. M. (eds), *Scotland: Environment and Archaeology, 8000 BC–AD 1000*. Wiley, Chichester. pp. 1–10.

Edwards, K. J. and Whittington, G. 1997. Vegetation change. In: Edwards, K. J. and Ralston, I. B. M. (eds), *Scotland: Environment and Archaeology, 8000 BC–AD 1000*. Wiley, Chichester. pp. 63–82.

Edwards, K. J. and Smout, T. C. 2000. Perspectives on human-environment interaction in prehistoric and historical times. In: Holmes, G. and Crofts, R. (eds), *Scotland's Environment: the future*. Tuckwell Press, East Linton. pp. 3–29.

Edwards, P. and Gemmel, C. 1999. Central Scotland Forest: selling the benefits to decision-makers. In: Roper, C. S. and Park, A. (eds), *The Living Forest: non-market benefits of forestry*. Proceedings of an international symposium on non-market benefits of forestry, Edinburgh, 24–8 June, 1996. Stationery Office, London. pp. 320–5.

Edwards, R. 1996. The shooting party takes aim. *New Scientist* 2020 (9th March 1996): 14–15.

Edwards, R. 1997. At the mercy of mink. *New Scientist* 2110 (29th November 1997): 26.

Edwards, R. 1998. Infested waters. *New Scientist* 2141 (4th July 1998): 23.

Edwards-Jones, E. S. 1997a. The River Valleys Project: using integrated catchment planning to improve the quality of two Scottish rivers. In: Boon, P. J. and Howell, D. L. (eds), *Freshwater Quality: defining the indefinable?* Scottish Natural Heritage, Edinburgh. pp. 506–12.

Edwards-Jones, E. S. 1997b. The River Valleys Project: a participatory approach to integrated catchment planning and management in Scotland. *Journal of Environmental Planning and Management* 40(1): 125–41.

Edwards-Jones, G., Edwards-Jones, E. S. and Mitchell, K. 1995. A comparison of contingent valuation methodology and ecological assessment as techniques for incorporating ecological goods into land-use decisions. *Journal of Environmental Planning and Management* 38(2): 215–30.

Edwards-Jones, G., Sloan, C. and Edwards-Jones, E. S. 1997. Monetary valuation of river flows as an element of the landscape: a case study from the River Almond, Scotland. In: Boon, P. J. and Howell, D. L. (eds), *Freshwater Quality: defining the indefinable?* Scottish Natural Heritage, Edinburgh. pp. 448–53.

Egdell, J. M. 1999. Agriculture and the Environment. *Scottish Environmental Audits No. 2.* Scottish Wildlife and Countryside Link, Perth.

Egdell, J. M. and Badger, R. 1996. A 'New Political Economy' perspective on Scottish agri-environmental policy. *Scottish Agricultural Economics Review* 9: 21–33.

Eldredge, N. 1995. *Dominion.* Henry Holt, New York.

Ellis, A. 1990. *Ethics for Environmentalists.* Centre for Philosophy and Public Affairs, University of St. Andrews/Nature Conservancy Council. CPPA, St. Andrews.

Elsdon, R. 1992. *Greenhouse Theology: Biblical perspectives on caring for Creation.* Monarch, Tunbridge Wells.

Evans, D. 1997. *A history of nature conservation in Britain.* Second Edition. Routledge, London.

Evans, J. 1984. Silviculture of Broadleaved Woodland. *Forestry Commission Bulletin* 62.

Evans, J. 1999. Sustainability of forest plantations: a review of evidence and future prospects. *International Forestry Review* 1(3): 153–62.

Evans, P. 1999a. Community woodlands in the Central Scotland Forest. *Scottish Forestry* 53(2): 100–1.

Evans, P. 1999b. *Saving the Species.* Radio 4, 19th August 1999.

FAI. 2001. *An economic study of Scottish Grouse Moors: an update.* A report by the Fraser of Allander Institute for Research on the Scottish Economy. Game Conservancy Scottish Research Trust, Fordingbridge.

FAPIRA. 1995. *Forests and People in Rural Scotland.* Discussion paper, Forests and People in Rural Areas Initiative, Perth.

FC. 1996. Involving communities in forestry through community participation. *Forestry Practice Guide* 10. Forestry Commission, Edinburgh.

FC. 1998a. *The UK Forestry Standard: the Government's approach to sustainable forestry.* Forestry Commission, Edinburgh.

FC. 1998b. *Forestry Commission Annual Report Highlights.* Forestry Commission, Edinburgh.

FC. 1999a. *Forests for Scotland: consultation towards a Scottish forestry strategy.* Forestry Commission, Edinburgh.

FC. 1999b. *Annual Report and Accounts 1998–1999.* Forestry Commission, Edinburgh.

FC. 1999c. *Public opinion of forestry 1999.* Forestry Commission, Edinburgh.

FC. 2000a. Personal communication from the Forestry Commission, 10th August 2000.

FC. 2000b. *Forests and Water Guidelines.* Third Edition. Forestry Commission, Edinburgh.

FC. 2001. *Forestry Statistics.* Forestry Commission, Edinburgh.

Featherstone, A. 1997. The wild heart of the Highlands. *ECOS* 18(2): 48–61.

Fenton, A. 1997. The farming of the land. In: Magnusson, M. and White, G. (eds), *The Nature of Scotland: Landscape, Wildlife and People.* Canongate, Edinburgh. pp. 111–21.

Fenton, A. and Gillmor, D. A. (eds) 1994. *Rural land use on the Atlantic periphery of Europe: Scotland and Ireland.* Royal Irish Academy, Dublin.

Fenton, J. 1997. Native woods in the Highlands: thoughts and observations. *Scottish Forestry* 51(3): 160–4.

Fenton, J. 1999. Scotland: reviving the wild. *ECOS* 20(2): 67–9.

Fernanda, F. C. M., Buckland, S. T., Goffin, D., Dixon, C. E., Borchers, D. L., Mayle, B. A. and

Peace, A. J. 2001. Estimating deer abundance from line transect surveys of dung: sika deer in southern Scotland. *Journal of Applied Ecology* 38: 349–63.
Ferguson, M. P. 1988. National Parks for Scotland. *Scottish Geographical Magazine* 104(1): 36–40.
Ferrier, R. 2002. Water quality issues for Scotland in the 21st century: an hydrological perspective. *Science of the Total Environment, in press.*
Fledmark, J. M. 1988. The planning framework. In: Selman, P. (ed.), *Countryside Planning in Practice: the Scottish experience.* Stirling University Press, Stirling. pp. 49–67.
Fleming, L. V., Newton, A. C., Vickery, J. A. and Usher, M. B. (eds). 1997. *Biodiversity in Scotland: status, trends and initiatives.* The Stationery Office, Edinburgh.
FoES. 1994. *The superquarry debate.* Briefing Paper. Friends of the Earth Scotland, Edinburgh.
FoES. 1996. *Towards a Sustainable Scotland.* Friends of the Earth Scotland, Edinburgh.
Foster, J. (ed.) 1997a. *Valuing Nature: ethics, economics and the environment.* Routledge, London.
Foster, J. 1997b. Introduction: environmental value and the scope of economics. In: Foster, J. (ed.), *Valuing Nature: ethics, economics and the environment.* Routledge, London. pp. 1–17.
Francis, J. M. 1994. Nature conservation and the Voluntary Principle. *Environmental Values* 3: 267–71.
Francis, J. M. 1996. Nature conservation and the Precautionary Principle. *Environmental Values* 5: 257–64.
Francis, J. M. 2000. Attitudes to nature and the ethics of conservation. In: Holmes, G. and Crofts, R. (eds), *Scotland's Environment: the future.* Tuckwell Press, East Linton. pp. 63–72.
Franklin, T. B. 1952. *A History of Scottish Farming.* Thomas Nelson, Edinburgh.
Gale, M. F. 1998. The view from the timber industry. In: *Forestry and Land Use – Forestry in a Changing Scotland.* Proceedings of a Timber Growers Association Conference, Gleneagles, 20th May 1998: 24–7.
Gardiner, J. 2000. Analysis of data from automatic people counters in the Cairngorms 1996–99. *Scottish Natural Heritage Commissioned Report F99AA615* (Unpublished report).
Garner, R. 1989. National Parks for Scotland? The designations debate. *ECOS* 10(3): 13–16.
Garrod, G. 1996. Valuing environmental goods in the countryside. In: Allanson, P. and Whitby, M. (eds), *The Rural Economy and the British Countryside.* Earthscan, London. pp. 83–98.
Garrod, G. and Willis, K. G. 1999. *Economic Valuation of the Environment: Methods and Case Studies.* Edward Elgar, Cheltenham.
Gauld, M. 2001. Personal communication from Munro Gauld, Reforesting Scotland, 19th March 2001.
Gay, H. 2000. Countryside conservation and the Heritage Lottery Fund: an under-exploited opportunity. *ECOS* 20(3/4): 52–60.
Gemmel, J. C. 1996. The land and the people: problems in partnership. *Scottish Forestry* 50(4): 212–19.
Gibson, A. and Warren, C. R. 1997. Return of the timber chute? An assessment of plastic chutes for extraction of thinnings. *Scottish Forestry* 51(1): 26–30.
Giddens, A. 1999. *Runaway World: how globalisation is reshaping our lives.* Profile Books, London.
Gill, P. and Watson, A. 1996. Cairngorms campaign against funicular railway. *ECOS* 17(2): 87–9.
Gill, R. M. A. 1992a. A review of damage by mammals in North Temperate Forests: 1. Deer. *Forestry* 65(2): 145–69.
Gill, R. M. A. 1992b. A review of damage by mammals in North Temperate Forests: 3. Impacts on trees and forests. *Forestry* 65(2): 363–88.
Gill, R. M. A. 2000. The impact of deer on woodland biodiversity. *Forestry Commission Information Note* 36. Forestry Commission, Edinburgh.
Gill, R. M. A., Webber, J. and Peace, A. 2000. The economic implications of deer damage: a review of current evidence. In: *Deer Commission for Scotland Annual Report 1999–2000.* The Stationery Office, Edinburgh. pp. 48–9.
Gilvear, D. J. 1994. River flow regulation. In: Maitland, P. S., Boon, P. J. and McLuskey, D. S. (eds), *The freshwaters of Scotland: a national resource of international significance.* Wiley, Chichester. pp. 463–88.

Gilvear, D. J., Hanley, N., Maitland, P. S. and Peterken, G. 1995. *Wild Rivers: Phase 1 – Technical Paper*. WWF Scotland, Aberfeldy.

Gilvear, D. J., Heal, K. V. and Steven, A. 2002. Hydrology and the ecological quality of Scottish river systems. *Science of the Total Environment*, in press.

Gimmingham, C. H. 1995. Heaths and moorland: an overview of ecological change. In: Thompson, D. B. A., Hester, A. J., and Usher, M. B. (eds), *Heaths and Moorland: Cultural Landscapes*. HMSO, Edinburgh. pp. 9–19.

Gittins, J. 1999. Recreation and multiple land use catchment management: problems and issues. In: Cresser, M. and Pugh K. (eds), *Multiple land use and catchment management*. Proceedings of an international conference, 11–13 September 1996. Macaulay Land Use Research Institute, Aberdeen. pp. 55–63.

Glacken, C. J. 1967. *Traces on the Rhodian shore: nature and culture in western thought from ancient times to the end of the eighteenth century*. University of California Press, Berkeley.

Goldsmith, F. B. 1991. The selection of protected areas. In: Spellerberg, I. F., Goldsmith, F. B. and Morris, M. G. (eds), *The Scientific Management of Temperate Communities for Conservation*. Blackwell, Oxford. pp. 273–91.

Goldspink, C. R., King, S. and Putman, R. J. (eds) 1998. *Population Ecology, Management and Welfare of Deer*. Manchester Metropolitan University, Manchester.

Goodall, S. 2000. Forest certification and UK Woodland Assurance Scheme. *Quarterly Journal of Forestry* 94(3): 239–44.

Goodstadt, V. J. 1996. Environmental sustainability and the role of Indicative Forest Strategies. *Scottish Forestry* 50(2): 77–84.

Goodwin, P. 1998a. 'Hired hands' or 'local voice': understandings and experience of local participation in conservation. *Transactions of the Institute of British Geographers* 23(4): 481–99.

Goodwin, P. 1998b. Challenging the stories about conservation: understanding local participation in conservation. *ECOS* 19(2): 12–19.

Gordon, J. E. and McKirdy, A. P. 1997. Quaternary landforms and deposits as part of Scotland's Natural Heritage. In: Gordon, J. E. (ed.), *Reflections on the Ice Age in Scotland*. Scottish Association of Geography Teachers/Scottish Natural Heritage, Glasgow. pp. 179–88.

Gordon, I. J. and Hope, I. M. 1998. The future management of red deer in Scotland: aiding decision-making in a complex world. In: Goldspink, C. R., King, S. and Putman, R. J. (eds), *Population Ecology, Management and Welfare of Deer*. Manchester Metropolitan University, Manchester. pp. 79–81.

Gordon, J. E. and Leys, K. F. (eds) 2001. *Earth Science and the Natural Heritage: interactions and integrated management*. Stationery Office, Edinburgh.

Gordon, J. E., Thompson, D. B. A., Haynes, V. M., Brazier, V. and Macdonald, R. 1998. Environmental sensitivity and conservation management in the Cairngorms Mountains, Scotland. *Ambio* 27(4): 335–44.

Gottlieb, R. S. 1996. Religion in an age of environmental crisis. In: Gottlieb, R. S. (ed.), *This Sacred Earth: religion, nature, environment*. Routledge, New York. pp. 3–14.

Gourlay, D. and Slee, B. 1998. Public preferences for landscape features: a case study of two Scottish Environmentally Sensitive Areas. *Journal of Rural Studies* 14(2): 249–63.

Grant, J. 1990. Managing Rothiemurchus Estate. In: Conroy, J. W. H., Watson, A. and Gunson, A. R. (eds), *Caring for the high mountains – conservation of the Cairngorms*. Centre for Scottish Studies, Aberdeen. pp. 50–4.

Green, B. H. 1993. Case study: agricultural plenty – more or less farming for the environment? In: Berry, R. J. (ed.), *Environmental Dilemmas: Ethics and Decisions*. Chapman and Hall, London. pp. 104–17.

Green, J. and Green, R. 1994. Mammals. In: Maitland, P. S., Boon, P. J. and McLuskey, D. S. (eds), *The freshwaters of Scotland: a national resource of international significance*. Wiley, Chichester. pp. 251–60.

Green, R. E. and Etheridge, B. 1999. Breeding success of the hen harrier *Circus cyaneus* in relation to the distribution of grouse moors and the red fox *Vulpes vulpes. Journal of Applied Ecology* 36: 472–83.

Green, R. and Riley, H. 1999. *Corncrakes*. Scottish Natural Heritage, Battleby.

Greene, D. 1996. Leisure day trips to the Scottish countryside and coast, 1987–1992. *Scottish Natural Heritage Research, Survey and Monitoring Report* No. 10.

Greene, D. 1997. Presentation on access issues, University of St Andrews, 26th November 1997.

Grove-White, R. 1997. The environmental valuation controversy: observations on its recent history and significance. In: Foster, J. (ed.), *Valuing Nature: ethics, economics and the environment*. Routledge, London. pp. 21–31.

Grove-White, R. 1999. Are surrogate valuation methods useful? A sceptic's view. In: Roper, C. S. and Park, A. (eds), *The Living Forest: non-market benefits of forestry*. Proceedings of an international symposium on non-market benefits of forestry, Edinburgh, 24–28 June, 1996. Stationery Office, London. pp. 31–6.

Gurnell, J. 1994. *The Red Squirrel*. The Mammal Society, London.

Gurnell, J. and Pepper, H. 1993. A critical look at conserving the British red squirrel, *Sciurus vulgaris. Mammal Review* 23: 127–37.

Gurnell, J. and Lurz, P. (eds) 1997. *The conservation of red squirrels,* Sciurus vulgaris *L.* People's Trust for Endangered Species, London.

Hale, S. E., Quine, C. P. and Suárez, J. C. 1998. Climatic conditions associated with treelines of Scots Pine and birch in Highland Scotland. *Scottish Forestry* 52(2): 70–6.

Halhead, A. V. 1987. *Farm woodlands in Central Scotland: the experience of the Central Scotland Woodlands Project*. Countryside Commission for Scotland, Battleby.

Hambrey, J. 1997. The seas of plenty. In: Magnusson, M. and White, G. (eds), *The Nature of Scotland: Landscape, Wildlife and People*. Canongate, Edinburgh. pp. 33–48.

Hammerton, D. 1994. Domestic and industrial pollution. In: Maitland, P. S., Boon, P. J. and McLuskey, D. S. (eds), *The freshwaters of Scotland: a national resource of international significance*. Wiley, Chichester. pp. 347–64.

Hampson, A. M. 1999. A framework for forest development in the Cairngorms. *SNH Information and Advisory Note* 102. Scottish Natural Heritage, Battleby.

Hampson, A. M. and Peterken, G. F. 1998. Enhancing the biodiversity of Scotland's forest resource through the development of a network of forest habitats. *Biodiversity and Conservation* 7: 179–92.

Handmer, J., Norton, T. W. and Dovers, S. R. 2001. *Ecology, Uncertainty and Policy: managing ecosystems for sustainability*. Pearson, Harlow.

Hanf, K. and Jansen, A.-I. 1998. Environmental policy – the outcome of strategic action and institutional characteristics. In: Hanf, K. and Jansen, A.-I. (eds), *Governance and Environment in Western Europe: Politics, Policy and Administration*. Longman, Harlow. pp. 1–16.

Hankey, M. 1999. The role of field sports. Presentation at The Heather Trust Conference, 'Integrated Upland Management for Wildlife, Field Sports, Agriculture and Public Enjoyment', Battleby, 24–25 September, 1999.

Hanley, N. 1997. Assessing the economic value of freshwaters. In: Boon, P. J. and Howell, D. L. (eds), *Freshwater Quality: defining the indefinable?* Scottish Natural Heritage, Edinburgh. pp. 435–47.

Hanley, N. 1998. Britain and the European policy process. In: Lowe, P. and Ward, S. (eds), *British Environmental Policy and Europe: Politics and Policy in Transition*. Routledge, London. pp. 57–66.

Hanley, N. and Craig, S. 1991. Wilderness development decisions and the Krutilla-Fisher Model: the case of Scotland's 'Flow Country'. *Ecological Economics* 4: 145–64.

Hanley, N. and Spash, C. L. 1993. *Cost-Benefit Analysis and the Environment*. Edward Elgar, Aldershot.

Hanley, N. and Milne, J. 1996. Ethical beliefs and behaviour in contingent valuation surveys. *Journal of Environmental Planning and Management* 39: 255–72.

Hanley, N. and MacMillan, D. 2000. Using economic instruments to improve environmental management. In: Holmes, G. and Crofts, R. (eds), *Scotland's Environment: the future*. Tuckwell Press, East Linton. pp. 92–103.

Hanley, N., Wright, R. E., Bullock, C., Simpson, C., Parsisson, D. and Crabtree, R. 1998. Contingent valuation versus choice experiments: estimating the benefits of environmentally sensitive areas in Scotland. *Journal of Agricultural Economics* 49(4): 1–15.

Hanley, N., Shogren, J. F. and White, B. 2001. *Introduction to Environmental Economics*. Oxford University Press, Oxford.

Hardin, G. 1968. The tragedy of the commons. *Science* 162: 1234–48.

Harmer, R. and Gill, R. M. A. 2000. Natural regeneration in broadleaved woodlands: deer browsing and the establishment of advance regeneration. *Forestry Commission Information Note* 35. Forestry Commission, Edinburgh.

Harris, S., Morris, P., Wray, S. and Yalden, D. 1995. *A Review of British Mammals*. Joint Nature Conservation Committee, Peterborough.

Harrison, C. 1993. Nature conservation, science, and popular values. In: Goldsmith, F. B. and Warren, A. (eds), *Conservation in Progress*. John Wiley, Chichester. pp. 35–49.

Harrison, C. and Burgess, J. 2000. Valuing nature in context: the contribution of common-good approaches. *Biodiversity and Conservation* 9(8): 1115–30.

Harrison, S. J. and Kirkpatrick, A. H. 2001. Climatic change and its potential implications for Scotland's natural heritage. In: Gordon, J. E. and Leys, K. F. (eds), *Earth Science and the Natural Heritage: interactions and integrated management*. Stationery Office, Edinburgh. pp. 296–305.

Hart-Davis, D. 1978. *Monarchs of the Glen: a history of deer stalking in the Scottish Highlands*. Cape, London.

Hart-Davis, D. 2000. Langholm's bleak lessons. *The Field*, July 2000: 55–8.

Haynes, V. M., Grieve, I. C., Gordon, J. E., Price-Thomas, P. and Salt, K. 2001. Assessing geomorphological sensitivity of the Cairngorm high plateaux for conservation purposes. In: Gordon, J. E. and Leys, K. F. (eds), *Earth Science and the Natural Heritage: interactions and integrated management*. Stationery Office, Edinburgh. pp. 120–3.

Heggs, G. 1997. Deer management groups: the way forward for forestry. In: Rose, H. (ed.), *Deer in Forestry – Beyond 2000*. Minutes of a conference in March 1997, Battleby. pp. 19–22.

Helliwell, R. 1999. Continuous cover forestry. *Journal of Practical Ecology and Conservation* 3(1): 59–63.

Helliwell, R. 2000. The value of wildlife in woodland. *Quarterly Journal of Forestry* 94(1): 35–40.

Henderson, R. 2000. Where the draft strategy falls down (Timber Report). *Landowning in Scotland* 257: 35.

Henderson-Howat, D. 1993. The value of forestry. In: Neustein, S. A. (ed.), *The winds of change: a re-evaluation of British forestry*. Proceedings of an ICF Discussion Meeting, York. Institute of Chartered Foresters, Edinburgh. pp. 78–87.

Henderson-Howat, D. 1996. The balance of forestry and other land uses in Scotland. *Scottish Forestry* 50(2): 67.

Henton, T. and Crofts, R. 2000. Key issues and objectives for sustainable development. In: Holmes, G. and Crofts, R. (eds), *Scotland's Environment: the future*. Tuckwell Press, East Linton. pp. 75–91.

Herbert, R., Samuel, S. and Patterson, G. 1999. Using local stock for planting native trees and shrubs. *FC Practice Note* 8. Forestry Commission, Edinburgh.

Herries, J. 1998. Glen Shiel hillwalking survey 1996. *Scottish Natural Heritage Research, Survey and Monitoring Report* No. 106.

Hester, A. J. and Miller, G. R. 1995. Scrub and woodland regeneration: prospects for the future. In: Thompson, D. B. A., Hester, A. J., and Usher, M. B. (eds), *Heaths and Moorland: Cultural Landscapes*. HMSO, Edinburgh. pp. 140–53.

Hester, A. J., Miller, D. R. and Towers, W. 1996. Landscape-scale vegetation change in the Cairngorms, Scotland, 1946–1988: implications for land management. *Biological Conservation* 77: 41–51.

Hettinger, N. 2001. Exotic species, naturalisation and biological nativism. *Environmental Values* 10: 193–224.

HGTAC. 1998. *The Future for Forestry – a framework for forestry in Great Britain: advice to the Forestry Commission from the Home Grown Timber Advisory Committee*. Forestry Commission, Edinburgh.

HIE. 1996. *The economic impacts of hillwalking, mountaineering and associated activities in the Highlands and Islands of Scotland*. Highlands and Islands Enterprise, Inverness.

Hill, D. A. 1993. Factors affecting winter pheasant density in British woodlands. *Journal of Applied Ecology* 30: 459–64.

Hislop, M. and Claridge, J. (eds) 2000. Agroforestry in the UK. *Forestry Commission Bulletin* 122.

Hodge, S. J., Patterson, G. and McIntosh, R. 1998. The approach of the British Forestry Commission to the conservation of forest biodiversity. *Scottish Forestry* 52(1): 30–6.

Holdgate, M. 1997. What future for nature? *Scottish Natural Heritage Occasional Papers* No. 4.

Holdgate, M. 2000a. Summing up. In: Holmes, G. and Crofts, R. (eds), *Scotland's Environment: the future*. Tuckwell Press, East Linton. pp. 126–33.

Holdgate, M. 2000b. Making choices in biodiversity conservation. In: *Nature in Transition*. Proceedings of a National Trust/British Association of Nature Conservationists Conference, Lancaster, 13th–14th July, 1999. Available at: www.nationaltrust.org.uk/environment

Holdgate, M. 2001. Adapting to climate change – new opportunities and lost causes. ECOS 22(1): 19–21.

Holland, A. and Rawles, K. 1993. Values in conservation. *ECOS* 14(1): 14–19.

Holmes, G. and Crofts, R. (eds) 2000. *Scotland's Environment: the future*. Tuckwell Press, East Linton.

Holt, D. 2001. Should wolves be reintroduced to the Highlands of Scotland? *Reforesting Scotland* 26: 38–40.

Hossell, J. E., Briggs, B. and Hepburn, I. R. 2000. *Climate change and UK nature conservation: a review of the impact of climate change on UK species and habitat conservation policy*. Department of the Environment, Transport and the Regions, London.

Housden, S. 2001. Integration of enterprises. In: Phillips, J., Thompson, D. B. A. and Gruellich, W. H. (eds), *Integrated Upland Management for Wildlife, Field Sports, Agriculture and Public Enjoyment*. The Heather Trust/Bidwells/SNH, Battleby. pp. 115–22.

House, S. 2000. Authenticity and administration of the Woodland Grant Scheme. *Scottish Woodland History Discussion Group: Notes IV* (Proceedings of a meeting at Battleby, Perth, 23rd November 1999). pp. 5–9.

Howell, D. L. 1994. Role of environmental agencies. In: Maitland, P. S., Boon, P. J. and McLuskey, D. S. (eds), *The freshwaters of Scotland: a national resource of international significance*. Wiley, Chichester. pp. 577–611.

Howells, O. and Edwards-Jones, G. 1997. A feasibility study of reintroducing Wild Boar *Sus scrofa* to Scotland: are existing woodlands large enough to support minimum viable populations? *Biological Conservation* 81: 77–89.

HRC. 1989. *Caithness and Sutherland HRC Working Party: Summary Report and Land Use Strategy*. Highland Regional Council, Inverness.

Hudson, P. J. 1992. *Grouse in Space and Time: the Population Biology of a Managed Gamebird*. Game Conservancy, Fordingbridge.

Hudson, P. J. 1995. Ecological trends and grouse management in upland Britain. In: Thompson, D. B. A., Hester, A. J., and Usher, M. B. (eds), *Heaths and Moorland: Cultural Landscapes*. HMSO, Edinburgh. pp. 282–93.

Hughes, R. and Buchan, N. 1999. The landscape character assessment of Scotland. In: Usher, M. B. (ed.), *Landscape Character: perspectives on management and change*. The Stationery Office, Edinburgh. pp. 1–12.

Hulbert, I., Waterhouse, T., Gordon, P. and Morgan-Davies, C. 1999. Silvo-pastoralism for the uplands of Scotland – a new approach to an old problem: the integration of farming and forestry. *Scottish Forestry* 53(4): 231–5.

Humphrey, J. W. and Nixon, C. J. 1999. The restoration of upland native oakwoods following removal of conifers: general principles. *Scottish Forestry* 53(2): 68–76.

Humphrey, J. W., Newton, A. C., Peace, A. J. and Holden, E. 2000. The importance of conifer plantations in northern Britain as a habitat for native fungi. *Biological Conservation* 96: 241–52.

Hunt, J. F. 2000. The MFS approach to native woodland restoration. *Scottish Woodland History Discussion Group: Notes IV* (Proceedings of a meeting at Battleby, Perth, 23rd November 1999). pp. 10–12.

Hunt, J. F. 2001a. Personal communications, 23rd March 2001 and 6th April 2001.

Hunt, J. 2001b. How do people enjoy the natural heritage? Patterns, trends and predictions of recreation in the Scottish countryside. In: Usher, M. B. (ed.), *Enjoyment and Understanding of the Natural Heritage*. The Stationery Office, Edinburgh. pp. 21–37.

Hunter, J. 1991. *The claim of crofting: the Scottish Highlands and Islands, 1930–1990*. Mainstream, Edinburgh.

Hunter, J. 1995. *On the other side of sorrow: nature and people in the Scottish Highlands*. Mainstream, Edinburgh.

Hunter, J. 1996. Land reform in Scotland. *The Field* June 1996: 63.

Hunter, J. 1999. Foreword. In: MacAskill, J. *We have won the land*. Acair, Stornoway. pp. 7–11.

Hunter, J. 2000. *The Making of the Crofting Community*. Second Edition. John Donald, Edinburgh.

Huxham, M. 2000. Why conserve wild species? In: Huxham, M. and Sumner, D. (eds), *Science and Environmental Decision Making*. Pearson, Harlow. pp. 142–69.

Huxham, M. and Sumner, D. (eds) 2000. *Science and Environmental Decision Making*. Pearson, Harlow.

Inskipp, C. 1997. *Agriculture and the Environment in Scotland*. WWF Scotland/SNH Data Support Sheet No. 5.

Institute of Hydrology. 1998. Broadleaf Woodlands: the implications for water quantity and quality. *Environment Agency Research and Development Publication* 5.

Isaacson, R. 1999. Chances with wolves. *Geographical Magazine* July 1999: 58–63.

Jaakko Pöyry. 1998. *Executive summary report on the future development prospects for British grown softwood*. Jaakko Pöyry Consulting, London.

Jamieson, A. D. and Sheldon, J. C. 1994. Planning. In: Maitland, P. S., Boon, P. J. and McLuskey, D. S. (eds), *The freshwaters of Scotland: a national resource of international significance*. Wiley, Chichester. pp. 531–54.

Jamieson, D. 1995. Ecosystem health: some preventive medicine. *Environmental Values* 4: 333–44.

Jansen, A.-I., Osland, O. and Hanf, K. 1998. Environmental challenges and institutional changes. An interpretation of the development of environmental policy in Western Europe. In: Hanf, K. and Jansen, A.-I. (eds), *Governance and Environment in Western Europe: Politics, Policy and Administration*. Longman, Harlow. pp. 277–325.

Jardine, I. 1998. A view from Scottish Natural Heritage. In: *Protecting Scotland's Finest Landscapes: National Parks – The Opportunities*. Proceedings of a conference, 16th March 1998. Scottish Wildlife and Countryside Link, Perth. pp. 15–17.

Jarvis, P. G. (ed.) 1991. *Agroforestry: Principles and Practice*. Elsevier, Amsterdam.

Jeffrey, W. G. (ed.) 1994. *State Forestry: the way forward*. Report of Edinburgh Conference, 25th February 1994. Royal Scottish Forestry Society, Edinburgh.

Jeffries, M. J. 1999. *Biodiversity and Conservation*. Routledge, London.

JNCC. 1996a. *UK strategy for red squirrel conservation*. Joint Nature Conservation Committee, Peterborough.

JNCC. 1996b. *An introduction to the Geological Conservation Review*. Geological Conservation Review Series No. 1. Joint Nature Conservation Committee, Peterborough.

JoH. 1993. The Balquhidder Catchment and Process Studies. *Journal of Hydrology* 145 (3–4), Special Issue.

Johnson, F. G. 1994. Hydro-electric generation. In: Maitland, P. S., Boon, P. J. and McLuskey, D. S.

(eds), *The freshwaters of Scotland: a national resource of international significance*. Wiley, Chichester. pp. 297–316.

Johnson, P. C. (ed.) 1995. Effects of upland afforestation on water resources. *Institute of Hydrology Report* 116. Second Edition.

Johnston, J. L. 2000. *Scotland's Nature in Trust: the National Trust for Scotland and its wildlife and crofting management*. Poyser, London.

Jolley, T. 2000. The hydrology and water management of Loch Lomond, Scotland. *Scottish Geographical Journal* 116(3): 197–212.

Jones, A. 1999a. Managing and developing a community woodland at Cumbernauld. *Scottish Forestry* 53(2): 93–5.

Jones, A. T. 1999b. The Caledonian Pinewood Inventory of Scotland's native Scots Pine woodlands. *Scottish Forestry* 53(4): 237–42.

Jordan, A. 1998. The impact on UK environmental administration. In: Lowe, P. and Ward, S. (eds), *British Environmental Policy and Europe: Politics and Policy in Transition*. Routledge, London. pp. 173–94.

Kay, J. J. and Scheider, E. 1994. Embracing complexity: the challenge of the ecosystem approach. *Alternatives* 20(3): 32–9.

Kellett, J. E. 1995. The elements of a sustainable aggregates policy. *Journal of Environmental Planning and Management* 38(4): 569–79.

Kemal, S. and Gaskell, I. (eds) 1993. *Landscape, Natural Beauty and the Arts*. Cambridge University Press, Cambridge.

Kempe, N. 1994. The long walk in: a defence of Shanks pony against myriad threats. In: Mollison, D. (ed.), *Sharing the Land*. John Muir Trust, Musselburgh. pp. 65–72.

Kenrick, J. 1993. Guest editorial. *Reforesting Scotland*. 9:4.

Kerr, A., Shackley, S., Milne, R. and Allen, S. 1999. *Climate Change: Scottish implications scoping study*. Scottish Executive Central Research Unit. The Stationery Office, Edinburgh.

Kerr, A. J. and Bain, C. 1997. Perspectives on current action for biodiversity conservation in Scotland. In: Fleming, L. V., Newton, A. C., Vickery, J. A. and Usher, M. B. (eds), *Biodiversity in Scotland: status, trends and initiatives*. The Stationery Office, Edinburgh. pp. 273–85.

Kerr, G. 1999. The use of silvicultural systems to enhance the biological diversity of plantation forests in Britain. *Forestry* 72(3): 191–205.

Kinsley, D. 1994. *Ecology and Religion*. Prentice-Hall, New Jersey.

Kirby, K. J. 1999. Trees, people and profits – into the next millennium: biodiversity and forestry. *Quarterly Journal of Forestry* 93(3): 221–6.

Kirby, K. J., Latham, J. and Hampson, A. 1997. The case for native trees and woodland for nature conservation and the merits of non-native species. In: Ratcliffe, P. R. (ed.), *Native and Non-Native in British Forestry*. Proceedings of a discussion meeting, University of Warwick, 31 March–2 April, 1995. Institute of Chartered Foresters, Edinburgh. pp. 160–70.

Kirby, K. J., Buckley, G. P. and Good, J. E. G. 1999. Maximising the value of new farm woodland biodiversity at a landscape scale. In: Burgess, P. J., Brierley, E. D. R., Morris, J. and Evans, J. (eds), *Farm woodlands for the future*. BIOS Scientific, Oxford. pp. 45–56.

Kitchener, A. C. 1995. *The Wildcat*. The Mammal Society, London.

Kitchener, A. C. 1997. The world of mammals. In: Magnusson, M. and White, G. (eds), *The Nature of Scotland: Landscape, Wildlife and People*. Canongate, Edinburgh. pp. 63–73.

Kitchener, A. C. 1998. Extinctions, introductions and colonisations of Scottish mammals and birds since the last ice age. In: Lambert, R. A. (ed.), *Species History in Scotland: introductions and extinctions since the Ice Age*. Scottish Cultural Press, Aberdeen. pp. 63–92.

Klaffke, O. 1999. The company of wolves. *New Scientist* 2172 (6th February 1999): 18–19.

Knight, R. L. 1996. Aldo Leopold, the land ethic, and ecosystem management. *Journal of Wildlife Management* 60(3): 471–4.

Knightbridge, R. 2000. The UK BAP – five years on. *ECOS* 21(2): 2–8.

Knowles, R. L. 1991. New Zealand experience with silvopastoral systems: a review. *Forest Ecology and Management* 45: 251–67.

Koch, M. and Grubb, M. 1993. Agenda 21. In: Grubb, M. (ed), *The Earth Summit Agreements: a guide and assessment. An analysis of the Rio '92 UN Conference on Environment and Development*. Earthscan, London. pp. 97–158.

Kortland, K. 2000. Crisis time for capercaillie. *Landowning in Scotland* 258: 14–16.

Krebs, J. R., Wilson, J. D., Bradbury, R. B. and Siriwardena, G. M. 1999. The second Silent Spring? *Nature* 400 (6745): 611–12.

Kruuk, L. E. B., Clutton-Brock, T. H., Albon, S. D., Pemberton, J. M. and Guinness, F. E. 1999. Population density affects sex ratio variation in red deer. *Nature* 399 (6735): 459–61.

Lambert, R. A. (ed.) 1998. *Species History in Scotland: introductions and extinctions since the Ice Age*. Scottish Cultural Press, Aberdeen.

Lambert, R. A. 2001a. Grey seals: to cull or not to cull? *History Today* 51(6): 30–2.

Lambert, R. A. 2001b. *Contested Mountains: Nature, Development and Environment in the Cairngorms Region of Scotland, 1880–1980*. White Horse Press, Cambridge.

Lampkin, N., Padel, S. and Foster, C. 2000. Organic farming. In: Brouwer, F. and Lowe, P. (eds), *CAP Regimes and the European Countryside: prospects for integration between agricultural, regional and environmental policies*. CAB International, Wallingford. pp. 221–38.

Lance, A., Thaxton, R. and Watson, A. 1991. Recent changes in footpath width in the Cairngorms. *Scottish Geographical Magazine* 107: 106–9.

Larson, K. 2001. Personal communication from Kurt Larson of Foster Yeoman, 11th April 2001.

Latham, J., Fairweather, A. and Staines, B. W. 1998. How effective is culling for controlling deer populations in plantation forests? In: Goldspink, C. R., King, S. and Putman, R. J. (eds), *Population Ecology, Management and Welfare of Deer*. Manchester Metropolitan University, Manchester. pp. 73–8.

Lawson, T. 1998. No dough means BAPs fail to rise. *ECOS* 19(2): 82–3.

Leach, M. A., Bauen, A. and Lucas, N. J. D. 1997. A systems approach to materials flow in sustainable cities: a case study of paper. *Journal of Environmental Planning and Management* 40(6): 705–23.

Lean, G. 1996. Where have all the woods gone? *Independent on Sunday* 16th June 1996: 4–7.

Leaper, R., Massei, G., Gorman, M. L. and Aspinall, R. 1999. The feasibility of reintroducing Wild Boar (*Sus scrofa*) to Scotland. *Mammal Review* 29(4): 239–59.

Ledoux, L., Crooks, S., Jordan, A. and Turner, R. K. 2000. Implementing EU biodiversity policy: UK experiences. *Land Use Policy* 17(4): 257–68.

Legg, W. 2000. The environmental effects of reforming agricultural policies. In: Brouwer, F. and Lowe, P. (eds), *CAP Regimes and the European Countryside: prospects for integration between agricultural, regional and environmental policies*. CAB International, Wallingford. pp. 17–30.

Leif, A. 1994. Survival and reproduction of wild and pen-reared pheasants. *Journal of Environmental Management* 58(3): 501–6.

Leighton, E. 1999. Conserving environmental quality: the underpinning to catchment management. In: Cresser, M. and Pugh K. (eds), *Multiple land use and catchment management*. Proceedings of an international conference, 11–13 September 1996. Macaulay Land Use Research Institute, Aberdeen. pp. 49–54.

Leopold, A. 1949. *A Sand County Almanac, and sketches here and there*. Oxford University Press, New York.

Lever, C. 1980. No beavers for Britain. *New Scientist* 87: 471–2.

Leys, K. 2001. The sustainable use of freshwater resources: a case study from the River Spey Site of Special Scientific Interest (SSSI) and candidate Special Area of Conservation. In: Gordon, J. E. and Leys, K. F. (eds), *Earth Science and the Natural Heritage: interactions and integrated management*. Stationery Office, Edinburgh. pp. 169–79.

Lindsay, R. A. 1995. *Bogs: the ecology, classification and conservation of ombrotrophic mires*. SNH, Battleby.

Lindsay, R. A., Charman, D. J., Everingham, F., O'Reilly, R. M., Palmer, M. A., Rowell, T. A. and Stroud, D. A. 1988. *The Flow Country: the peatlands of Caithness and Sutherland*. Nature Conservancy Council, Peterborough.

Linklater, M. 2000. Keep these bird brains out of the Highlands. *The Times* 2nd November 2000.

Lister-Kaye, J. 1994. Ill fares the land: a sustainable land ethic for the sporting estates of the Highlands and Islands of Scotland. *Scottish Natural Heritage Occasional Papers* No. 3.

Lister-Kaye, J. 1995. Native pinewoods and sustainable development of the Highlands. In: Aldhous, J. R. (ed.), *Our Pinewood Heritage*. Forestry Commission/RSPB/SNH. pp. 60–3.

Lister-Kaye, J. 1998. A conservation perspective. In: *Forestry and Land Use – Forestry in a Changing Scotland*. Proceedings of a Timber Growers Association Conference, Gleneagles, 20th May 1998: 28–9.

Lister-Kaye, J. 2001. The enjoyment and understanding of nature and wildness. In: Usher, M. B. (ed.), *Enjoyment and Understanding of the Natural Heritage*. The Stationery Office, Edinburgh. pp. 3–10.

Lloyd, M. G. 1997. Structure and Culture: regional planning and institutional planning in Scotland. In: Macdonald, R. and Thomas, H. (eds), *Nationality and Planning in Scotland and Wales*. University of Wales Press, Cardiff. pp. 113–32.

Lloyd, M. G. 1999. Scottish Environment Protection Agency: making sense of a fragmenting environment. *Scottish Affairs* 29: 28–42.

Lloyd, M. G. and Rowan-Robinson, J. 1992. Review of strategic planning guidance in Scotland. *Journal of Environmental Planning and Management* 35(1): 93–9.

Lloyd, T., Watkins, C. and Williams, D. 1996. Farmers' attitudes to woodland planting grants and the potential effects of new forestry incentives. In: Curry, N. and Owen, S. (eds), *Changing Rural Policy in Britain: Planning, Administration, Agriculture and the Environment*. Countryside and Community Press, Cheltenham. pp. 244–61.

LLTWP. 1993. *The management of Loch Lomond and the Trossachs*. The Report of the Loch Lomond and the Trossachs Working Party to the Secretary of State for Scotland. The Scottish Office, Edinburgh.

Long, A. J. and Nair, P. K. R. 1999. Trees outside forests: agro-, community, and urban forestry. *New Forests* 17: 145–74.

Lovelock, J. 2000. Living planet: interview with James Lovelock. *Geographical Magazine* 72(8): 25–7.

Lowe, P. and Ward, S. 1998a. Britain in Europe: themes and issues in national environmental policy. In: Lowe, P. and Ward, S. (eds), *British Environmental Policy and Europe: Politics and Policy in Transition*. Routledge, London. pp. 3–30.

Lowe, P. and Ward, S. (eds) 1998b. *British Environmental Policy and Europe: Politics and Policy in Transition*. Routledge, London.

Lowe, P. and Baldock, D. 2000. Integration of environmental objectives into agricultural policy making. In: Brouwer, F. and Lowe, P. (eds), *CAP Regimes and the European Countryside: prospects for integration between agricultural, regional and environmental policies*. CAB International, Wallingford. pp. 31–52.

Lowe, P. and Brouwer, F. 2000. Agenda 2000: a wasted opportunity? In: Brouwer, F. and Lowe, P. (eds), *CAP Regimes and the European Countryside: prospects for integration between agricultural, regional and environmental policies*. CAB International, Wallingford. pp. 321–34.

Lowe, P., Cox, G., MacEwen, M., O'Riordan, T. and Winter, M. 1986. *Countryside Conflicts: the politics of farming, forestry and conservation*. Gower Publishing, Aldershot.

LQG (LINK Quarry Group). 1996. *The case against the Harris Superquarry*. Scottish Wildlife and Countryside Link, Perth.

LRPG (Land Reform Policy Group). 1998a. *Identifying the problems*. The Stationery Office, Edinburgh.

LRPG (Land Reform Policy Group). 1998b. *Identifying the solutions*. The Stationery Office, Edinburgh.

LRPG (Land Reform Policy Group). 1999. *Recommendations for action.* The Stationery Office, Edinburgh.

Lurz, P. and Cooper, M. 1997. *Red Squirrels.* Scottish Natural Heritage, Battleby.

Lusby, P. 1998. On the extinct plants of Scotland. In: Lambert, R. A. (ed.), *Species History in Scotland: introductions and extinctions since the Ice Age.* Scottish Cultural Press, Aberdeen. pp. 45–62.

Lyddon, D. 1994. Land use planning and management in Scotland. In: Fenton, A. and Gillmor, D. A. (eds), *Rural land use on the Atlantic periphery of Europe: Scotland and Ireland.* Royal Irish Academy, Dublin. pp. 195–208.

Mabey, R. 1997. *Flora Britannica.* Chatto and Windus, London.

MacAskill, J. 1999. *We have won the land.* Acair, Stornoway.

MacBrayne, C. G. 1982. The case for farm forestry. *Scottish Forestry* 26(2): 123–30.

MacCaig, N. 1990. *Collected Poems.* Chatto and Windus, London.

Macdonald, A. J., Kirkpatrick, A. H., Hester, A. J. and Sydes, C. 1995. Regeneration by natural layering of heather (*Calluna vulgaris*): frequency and characteristics in upland Britain. *Journal of Applied Ecology* 32: 85–99.

Macdonald, D. W. and Johnson, P. J. 2000. Farmers and the custody of the countryside: trends in loss and conservation of non-productive habitats 1981–1998. *Biological Conservation* 94: 221–34.

Macdonald, D. W., Tattersal, F. H., Brown, E. D. and Balharry, D. 1995. Reintroducing the European Beaver to Britain: nostalgic meddling or restoring biodiversity? *Mammal Review* 25(4): 161–200.

MacDonald, F. 1998. Viewing Highland Scotland: ideology, representation and the 'natural heritage'. *Area* 30(3): 237–44.

Macdonald, R. and Thomas, H. (eds) 1997. *Nationality and Planning in Scotland and Wales.* University of Wales Press, Cardiff.

Macdonald, T. D. 1994. Water supply. In: Maitland, P. S., Boon, P. J. and McLuskey, D. S. (eds), *The freshwaters of Scotland: a national resource of international significance.* Wiley, Chichester. pp. 279–96.

MacGarvin, M. 2000. *Scotland's Secret? Aquaculture, nutrient pollution, eutrophication and toxic blooms.* WWF Scotland, Aberfeldy.

MacGregor, B. D. 1988. Owner motivation and land use on landed estates in the North-west Highlands of Scotland. *Journal of Rural Studies* 4(4): 389–404.

MacGregor, B. D. 1993. *Land tenure in Scotland.* First John McEwan Memorial Lecture, Rural Forum, Perth.

MacGregor, B. D. and Stockdale, A. 1994. Land use change on Scottish Highland estates. *Journal of Rural Studies* 10(3): 301–9.

MacGuire, F. A. S. and Childs, M. 1998. Wastepaper management and protection of forest biodiversity. *Journal of Environmental Planning and Management* 41(3): 403–6.

Mackay, D. 1995. *Scotland's rural land use agencies.* Scottish Cultural Press, Aberdeen.

Mackay, D. G. 1994. Forestry as a land use in Scotland. In: Fenton, A. and Gillmor, D. A. (eds), *Rural land use on the Atlantic periphery of Europe: Scotland and Ireland.* Royal Irish Academy, Dublin. pp. 117–30.

Mackay, D. W. 1994. Pollution control. In: Maitland, P. S., Boon, P. J. and McLuskey, D. S. (eds), *The freshwaters of Scotland: a national resource of international significance.* Wiley, Chichester. pp. 517–30.

Mackay, J. 2001. Public perceptions and enjoyment. In: Phillips, J., Thompson, D. B. A. and Gruellich, W. H. (eds), *Integrated Upland Management for Wildlife, Field Sports, Agriculture and Public Enjoyment.* The Heather Trust/Bidwells/SNH, Battleby. pp. 123–9.

MacKenzie, A. F. D. 1998. 'The Cheviot, the Stag . . . and the White, White Rock?': community, identity, and environmental threat on the Isle of Harris. *Environment and Planning D: Society and Space* 16: 509–32.

MacKenzie, D. 1999. Sickly swine. *New Scientist* 2205 (25th September 1999): 12.
MacKenzie, D. 2001. This means war. *New Scientist* 2283 (24th March 2001): 12.
MacKenzie, N. A. 1999. The native woodland resource of Scotland: a review 1993–1998. *Forestry Commission Technical Paper* 30.
MacKenzie, N. A. and Callander, R. F. 1995. The native woodland resource in the Scottish Highlands. *Forestry Commission Technical Paper* 12.
MacKenzie, N. A. and Callander, R. F. 1996. The native woodland resource in the Scottish Lowlands. *Forestry Commission Technical Paper* 17.
Mackey, E. C., Shewry, M. C. and Tudor, G. J. 1998. *Land cover change: Scotland from the 1940s to the 1980s*. The Stationery Office, Edinburgh.
MacLellan, R. 1998. Tourism and the Scottish environment. In: MacLellan, R. and Smith, R. (eds), *Tourism in Scotland*. International Thomson Business Press, London. pp. 112–34.
MacLellan, R. and Smith, R. (eds) 1998. *Tourism in Scotland*. International Thomson Business Press, London.
MacMillan, D. C. 1993. Indicative Forest Strategies – an investment perspective in the Borders region of Scotland. *Scottish Forestry* 47(3): 83–9.
MacMillan, D. C. 1995. Non-market benefits of new native pinewoods. In: Aldhous, J. R. (ed.), *Our Pinewood Heritage*. Forestry Commission/RSPB/SNH. pp. 79–83.
MacMillan, D. C. 1999. Non-market benefits of restoring native woodlands. In: Roper, C. S. and Park, A. (eds), *The Living Forest: non-market benefits of forestry*. Proceedings of an international symposium on non-market benefits of forestry, Edinburgh, 24–28 June, 1996. Stationery Office, London. pp. 189–95.
MacMillan, D. C. 2000. An economic case for land reform. *Land Use Policy* 17: 49–57.
MacMillan, D. C. and Duff, E. I. 1998. Estimating the non-market costs and benefits of native woodland restoration using the contingent valuation method. *Forestry* 71(3): 247–59.
MacMillan, D. C., Daw, M., Daw, D., Phillip, L, Patterson, I., Hanley, N., Gustanski, J.-A. and Wright, R. 2001a. *The costs and benefits of managing wild geese in Scotland*. Scottish Executive Central Research Unit, Edinburgh.
MacMillan, D. C., Duff, E. I. and Elston, D. A. 2001b. Modelling the non-market environmental costs and benefits of biodiversity projects using contingent valuation data. *Environmental and Resource Economics* 18:391–410.
Macnaghten, P. and Urry, J. 1998. *Contested Natures*. SAGE Publications, London.
Magnusson, M. 1993. Foreword. In: Smout, T. C. (ed.), *Scotland since Prehistory: Natural Change and Human Impact*. Scottish Natural Heritage/Scottish Cultural Press, Aberdeen. pp. xi–xii.
Magnusson, M. 1995a. Keynote address – Our pinewood heritage. In: Aldhous, J. R. (ed.), *Our Pinewood Heritage*. Forestry Commission/RSPB/SNH. pp. 1–3.
Magnusson, M. 1995b. Foreword. In: Thompson, D. B. A., Hester, A. J., and Usher, M. B. (eds), *Heaths and Moorland: Cultural Landscapes*. HMSO, Edinburgh. pp. xii–xvi.
Magnusson, M. and White, G. (eds) 1997. *The Nature of Scotland: Landscape, Wildlife and People*. Canongate, Edinburgh.
Maitland, P. S. 1992. Fish and angling in SSSIs in Scotland. *Scottish Natural Heritage Review* No. 16.
Maitland, P. S. 1994. Fish. In: Maitland, P. S., Boon, P. J. and McLuskey, D. S. (eds), *The freshwaters of Scotland: a national resource of international significance*. Wiley, Chichester. pp. 191–208.
Maitland, P. S. 1997a. The streams of life. In: Magnusson, M. and White, G. (eds), *The Nature of Scotland: Landscape, Wildlife and People*. Canongate, Edinburgh. pp. 157–67.
Maitland, P. S. 1997b. Sustainable management for biodiversity: freshwater fisheries. In: Fleming, L. V., Newton, A. C., Vickery, J. A. and Usher, M. B. (eds), *Biodiversity in Scotland: status, trends and initiatives*. The Stationery Office, Edinburgh. pp. 167–78.
Maitland, P. S. and Hamilton, J. D. 1994. History of freshwater science. In: Maitland, P. S., Boon, P. J. and McLuskey, D. S. (eds), *The freshwaters of Scotland: a national resource of international significance*. Wiley, Chichester. pp. 3–16.

Maitland, P. S. and Morgan, N. C. 1997. *Conservation Management of Freshwater Habitats: lakes, rivers and wetlands*. Chapman and Hall, London.

Maitland, P. S., Newson, M. D. and Best, G. E. 1990. *The impact of afforestation and forestry practice on freshwater habitats*. Nature Conservancy Council, Peterborough.

Maitland, P. S., Boon, P. J. and McLuskey, D. S. (eds) 1994. *The freshwaters of Scotland: a national resource of international significance*. Wiley, Chichester.

Maitland, P. S., Adams, C. E. and Mitchell, J. 2000. The natural heritage of Loch Lomond: its importance in a national and international context. *Scottish Geographical Journal* 116(3): 181–96.

Malcolm, D. C. 1991. Afforestation in Britain – a commentary. *Scottish Forestry* 45(4): 259–74.

Manning, A. 1997. Biodiversity conservation in Scotland: personal reflections. In: Fleming, L. V., Newton, A. C., Vickery, J. A. and Usher, M. B. (eds), *Biodiversity in Scotland: status, trends and initiatives*. The Stationery Office, Edinburgh. pp. 286–94.

Manning, A. 1998. Scotland – our environment and the world. *Scottish Natural Heritage Occasional Papers* No. 6.

Marchant, J. 2001. The big cull. *New Scientist* 2274 (20th January 2001): 9.

Marren, P. 1993. The siege of the NCC: nature conservation in the Eighties. In: Goldsmith, F. B. and Warren, A. (eds), *Conservation in Progress*. Wiley, Chichester. pp. 283–99.

Marren, P. 2000. Did the Bittern read the BAP? *ECOS* 21(2): 43–6.

Marsden, M. 2000. Proposed EC Water Framework Directive: technical implementation. Paper presented at *Hydrology in Scotland: Agenda for the 21st Century*, University of Dundee, 21–23 March 2000.

Marsh, G. P. 1864. *Man and Nature: or, Physical Geography as Modified by Human Action*. Scribner, New York.

Mason, W. L., Hardie, D., Quelch, P., Ratcliffe, P. R., Ross, I., Stevenson, A. W. and Soutar, R. 1999. 'Beyond the two solitudes': the use of native species in plantation forests. *Scottish Forestry* 53(3): 135–44.

Mather, A. S. 1988. New private forests in Scotland: characteristics and contrasts. *Area* 20: 135–43.

Mather, A. S. 1991. The changing role of planning in rural land use: the example of afforestation in Scotland. *Journal of Rural Studies* 7(3): 299–309.

Mather, A. S. 1992. Land use, physical sustainability and conservation in Highland Scotland. *Land Use Policy* 9(2): 99–110.

Mather, A. S. 1993a. The environmental impact of sheep farming in the Scottish Highlands. In: Smout, T. C. (ed.), *Scotland since Prehistory: Natural Change and Human Impact*. Scottish Natural Heritage/Scottish Cultural Press, Aberdeen. pp. 79–88.

Mather, A. S. 1993b. Afforestation in Britain. In: Mather, A. S. (ed.), *Afforestation: Policies, Planning and Progress*. Belhaven Press, London. pp. 13–33.

Mather, A. S. 1993c. Protected areas in the periphery: conservation and controversy in Northern Scotland. *Journal of Rural Studies* 9(4): 371–84.

Mather, A. S. 1994. European land use: an overview. In: Fenton, A. and Gillmor, D. A. (eds), *Rural land use on the Atlantic periphery of Europe: Scotland and Ireland*. Royal Irish Academy, Dublin. pp. 5–24.

Mather, A. S. 1995. Rural land occupancy in Scotland: resources for research. *Scottish Geographical Magazine* 111(2): 127–31.

Mather, A. S. 1996a. The inter-relationship of afforestation and agriculture in Scotland. *Scottish Geographical Magazine* 112(2): 83–91.

Mather, A. S. 1996b. Rural land use in Britain: agency re-structuring and policy adaptation. In: Curry, N. and Owen, S. (eds), *Changing Rural Policy in Britain: Planning, Administration, Agriculture and the Environment*. Countryside and Community Press, Cheltenham. pp. 87–106.

Mather, A. S. 1997. Mountain recreation in the East Grampians. *Scottish Geographical Magazine* 113(3): 195–8.

Mather, A. S. 1998. East Grampians and Lochnagar Visitor Survey 1995: overview. *Scottish Natural Heritage Research, Survey and Monitoring Report* No. 104.
Mather, A. S. 2001. Forests of consumption: postproductivism, postmaterialism, and the postindustrial forest. *Environment and Planning* C 19(2): 249–68.
Mather, A. S. and Murray, N. C. 1987. Employment and private-sector afforestation in Scotland. *Journal of Rural Studies* 3(3): 207–18.
Mather, A. S. and Murray, N. C. 1988. The dynamics of rural land use change: the case of private sector afforestation in Scotland. *Land Use Policy* 5: 103–20.
Mather, A. S. and Thomson, K. J. 1995. The effects of afforestation on agriculture in Scotland. *Journal of Rural Studies* 11(2): 187–202.
Maxwell, F. 1998. A crofter in command. *The Scotsman* 13th October 1998.
Maxwell, F. 1999. Farming industry plunged into crisis. *The Scotsman* 29th January 1999.
Maxwell, T. J. and Cannell, M. G. R. 2000. The environment and land use of the future. In: Holmes, G. and Crofts, R. (eds), *Scotland's Environment: the future*. Tuckwell Press, East Linton. pp. 30–51.
Mayer, S. 2000. Genetic engineering in agriculture. In: Huxham, M. and Sumner, D. (eds), *Science and Environmental Decision Making*. Pearson, Harlow. pp. 94–117.
Mayle, B. A. 1996. Progress in predictive management of deer populations in British woodlands. *Forest Ecology and Management* 88: 187–98.
Mayle, B. A., Peace, A. J. and Gill, R. M. A. 1999. *How many deer? A field guide to estimating deer population size*. Field Book 18, Forestry Commission, Edinburgh.
McAdam, J. H. 1999. UK Agroforestry Forum. *Scottish Forestry* 53(1): 4–5.
McAdam, J. H. and Sibbald, A. 2000. Grazing livestock management. In: Hislop, M. and Claridge, J. (eds), Agroforestry in the UK. *Forestry Commission Bulletin* 122. pp. 44–57.
McAdam, J. H., Thomas, T. H. and Willis, R. W. 1999a. The economics of agroforestry systems in the UK and their future prospects. *Scottish Forestry* 53(1): 37–41.
McAdam, J. H., Crowe, S. R. and Sibbald, A. R. 1999b. Agroforestry as a sustainable land use option. In: Burgess, P. J., Brierley, E. D. R., Morris, J. and Evans, J. (eds), *Farm woodlands for the future*. BIOS Scientific, Oxford. pp. 127–38.
McCall, I. 1998. The sporting perspective on farm woodland and forestry. In: *Forestry and Land Use – Forestry in a Changing Scotland*. Proceedings of a Timber Growers Association Conference, Gleneagles, 20th May 1998: 37–8.
McCarthy, J. 1998. *An Inhabited Solitude: Scotland, Land and People*. Luath Press, Edinburgh.
McConnell, J. and Conroy, J. W. H. (eds) 1996. Environmental history of the Cairngorms. *Botanical Journal of Scotland* 48(1).
McCormick, F. and Buckland, P. C. 1997. Faunal change: the vertebrate fauna. In: Edwards, K. J. and Ralston, I. B. M. (eds), *Scotland: Environment and Archaeology, 8000 BC–AD 1000*. Wiley, Chichester. pp. 83–103.
McDermott, W. 1997. The potential benefits of National Parks in Scotland. In: *Protecting Scotland's Finest Landscapes: Time for action on National Parks*. Proceedings of a conference, 17th September 1997. Scottish Wildlife and Countryside Link, Perth. pp. 22–6.
McDowell, E. and McCormick, J. (eds) 1999a. *Environment Scotland: Prospects for Sustainability*. Ashgate, Aldershot.
McDowell, E. and McCormick, J. 1999b. Environment Scotland: an overview. In: McDowell, E. and McCormick, J. (eds), *Environment Scotland: Prospects for Sustainability*. Ashgate, Aldershot. pp. 1–15.
McEwan, L. J. 1997. Geomorphological change and fluvial landscape evolution in Scotland during the Holocene. In: Gordon, J. E. (ed.), *Reflections on the Ice Age in Scotland: an update on Quaternary Studies*. Scottish Association of Geography Teachers/Scottish Natural Heritage, Glasgow. pp. 116–29.
McEwen, J. 1977. *Who owns Scotland: a study in land ownership*. Edinburgh University Student Publication Board, Edinburgh.

McGillivary, N. and McIntyre, L. 2001. Millennium Forests for Scotland. *Scottish Forestry* 55(2): 105–8.

McGilvray, J. and Perman, R. 1992. Grouse shooting in Scotland: analysis of its importance to the economy and the environment. In: Hudson, P. J. *Grouse in Space and Time: the Population Biology of a Managed Gamebird*. Game Conservancy, Fordingbridge. pp. 215–17.

McHenry, H. 1996. Understanding farmers' perceptions of changing agriculture: some implications for agri-environmental schemes. In: Curry, N. and Owen, S. (eds), *Changing Rural Policy in Britain: Planning, Administration, Agriculture and the Environment*. Countryside and Community Press, Cheltenham. pp. 225–43.

McHenry, H. 1998. Wild flowers in the wrong fields are weeds! Examining farmers' constructions of conservation. *Environment and Planning A* 30(6): 1039–53.

McIntosh, A. 1997. The French Revolution on Eigg. *ECOS* 18(2): 90–2.

McIntosh, A. 2001a. Sabbath and the corporate mammon: concluding the Harris superquarry debate. *ECOS* 22(1): 46–52.

McIntosh, A. 2001b. *Soil and Soul: people* versus *corporate power*. Aurum Press, London.

McIntosh, A., Wightman, A. and Morgan, D. 1994. Reclaiming the Scottish Highlands: clearance, conflict and crofting. *The Ecologist* 24(2): 64–70.

McIntosh, R. (ed.) 1987. *Deer and Forestry*. Proceedings of an ICF Conference, Glasgow. Institute of Chartered Foresters, Edinburgh.

McIntyre, J. A. R. 2001. Community forestry – the Sunart experience. *Scottish Forestry* 55(2): 99–103.

McLavin, D. 2000. Designations – unwarranted interference or crucial conservations? *Landowning in Scotland* 259: 26–7.

McLean, C. 1993. *Sika Deer Control: a report on a three-year project at Shin Forest, Sutherland*. Red Deer Commission, Inverness.

McLean, C. 1999. The effect of deer culling on tree regeneration on Scaniport Estate, Inverness-shire: a study by the Deer Commission for Scotland. *Scottish Forestry* 53(4): 225–9.

McLean, C. 2000. Personal communication, 28th February 2000.

McLean, C. 2001a. Personal communication, 23rd July 2001.

McLean, C. 2001b. Developments in red deer management. In: Phillips, J., Thompson, D. B. A. and Gruellich, W. H. (eds), *Integrated Upland Management for Wildlife, Field Sports, Agriculture and Public Enjoyment*. The Heather Trust/Bidwells/SNH, Battleby. pp. 69–74.

McLean, C. 2001c. Costs of sika control in native woodland: experience at Scaniport Estate. *Scottish Forestry* 55(2): 109–111.

McNeely, J. A. (ed.) 2001. *The Great Reshuffling: human dimensions of invasive alien species*. IUCN, Gland, Switzerland and Cambridge, UK.

McNeely, J. A., Mooney, H. A., Neville, L. E., Schei, P. J. and Waage, J. K. (eds) 2001. *A Global Strategy on Invasive Alien Species*. IUCN, Gland, Switzerland and Cambridge, UK.

McQuillan, A. G. 1993. Cabbages and kings: the ethics and aesthetics of *new forestry*. *Environmental Values* 2: 191–223.

Meek, D. 1987. The Land Question answered from the Bible: the land issue and the development of a Highland theology of liberation. *Scottish Geographical Magazine* 103(2): 84–9.

Miles, A. 1999. *Silva: the Tree in Britain*. Ebury Press, London.

Miles, S. 2001. A mountain of conflict. *Landscape Design* 300: 28–9.

Miller, G. R., Cummins, R. P. and Hester, A. J. 1998. Red deer and woodland regeneration in the Cairngorms. *Scottish Forestry* 52(1): 14–20.

Milne, J. A., Birch, C. P. D., Hester, A. J., Armstrong, H. M. and Robertson, A. 1998. The impact of vertebrate herbivores on the natural heritage of the Scottish uplands. *Scottish Natural Heritage Review* 95.

Milton, K. 1994. Land-use planning and management in Ireland. In: Fenton, A. and Gillmor, D. A. (eds), *Rural land use on the Atlantic periphery of Europe: Scotland and Ireland*. Royal Irish Academy, Dublin. pp. 209–19.

Minay, C. L. W. 1997. Contrasting approaches to rural economic development. In: Macdonald, R. and Thomas, H. (eds), *Nationality and Planning in Scotland and Wales*. University of Wales Press, Cardiff. pp. 181–203.
Minns, D. 1997. The birds and their habitats. In: Magnusson, M. and White, G. (eds), *The Nature of Scotland: Landscape, Wildlife and People*. Canongate, Edinburgh. pp. 75–87.
Mitchell, B., Staines, B. W. and Welch, B. 1977. *Ecology of red deer: a research review relevant to their management in Scotland*. Institute of Terrestrial Ecology, Cambridge.
Mitchell, B. 1997. *Resource and Environmental Management*. Longman, Harlow.
Mitchell, C. 2001. Natural Heritage Zones: planning the sustainable use of Scotland's natural diversity. In: Gordon, J. E. and Leys, K. F. (eds), *Earth Science and the Natural Heritage: interactions and integrated management*. Stationery Office, Edinburgh. pp. 234–8.
Mitchell, I. 1999. *Isles of the West*. Canongate, Edinburgh.
Mitchell, I. 2000. A flight of fancy and little more. *The Times* 13th May 2000.
Mitchell, I. 2001. SNH defeat has wide implications. *Landowning in Scotland* 261: 34–5.
Moffatt, A. J. 1999. Trees, people and profits – into the next millennium: environmental changes. *Quarterly Journal of Forestry* 93(3): 211–20.
Moir, J. 1991. National Parks: North of the Border. *Planning Outlook* 34(2): 61–7.
Moir, J. 1997. The designation of valued landscapes in Scotland. In: Macdonald, R. and Thomas, H. (eds), *Nationality and Planning in Scotland and Wales*. University of Wales Press, Cardiff. pp. 203–42.
More, T. A., Averill, J. R. and Stevens, T. H. 1996. Values and economics in environmental management: a perspective and critique. *Journal of Environmental Management* 48: 397–409.
Morgan, D. 1996. The progress of Scottish land reform. *Reforesting Scotland* 14: 12–14.
Morris, A. and Robinson, G. 1996. Rural Scotland: problems and prospects. *Scottish Geographical Magazine* 112(2): 66–9.
Morris, C. and Potter, C. 1995. Recruiting the new conservationists: farmers' adoption of agri-environmental schemes in the UK. *Journal of Rural Studies* 11(1): 51–63.
Morris, P. A. 1986. An introduction to reintroductions. *Mammal Review* 16(2): 49–52.
Morrison, B. R. S. 1994. Acidification. In: Maitland, P. S., Boon, P. J. and McLuskey, D. S. (eds), *The freshwaters of Scotland: a national resource of international significance*. Wiley, Chichester. pp. 435–62.
Morrison, I. A. 1983. Prehistoric Scotland. In: Whittington, G. and Whyte, I. D. (eds), *An Historical Geography of Scotland*. Academic Press, London. pp. 1–23.
Moss, B. 2000. Biodiversity in fresh waters – an issue of species preservation or system functioning? *Environmental Conservation* 27(1): 1–4.
Moss, R. and Picozzi, N. 1994. Management of forests for Capercaillie in Scotland. *Forestry Commission Bulletin* 113.
Mowle, A. 1988. Integration: holy grail or sacred cow? In: Selman, P. (ed.), *Countryside Planning in Practice: the Scottish experience*. Stirling University Press, Stirling. pp. 247–64.
Mowle, A. 1997. The managing of the land. In: Magnusson, M. and White, G. (eds), *The Nature of Scotland: Landscape, Wildlife and People*. Canongate, Edinburgh. pp. 133–43.
Munro, J. 1997. Landowning and management in Denmark – a comparison with Scotland. In: *Scottish Landownership within Europe*. Proceedings of the Scottish Landowners Federation Conference, Battleby, 23rd April 1997.
Munton, R. 1995. Regulating rural change: property rights, economy and environment – a case study from Cumbria, UK. *Journal of Rural Studies* 11(3): 269–84.
Mutch, W. E. S. 1987. Responsibilities of land users. In: McIntosh, R. (ed.), *Deer and Forestry*. Proceedings of an ICF Conference, Glasgow. Institute of Chartered Foresters, Edinburgh. pp. 21–9.
Mutch, W. E. S. 1994. The history and development of the state forest service. In: Jeffrey, W.G. (ed.), *State Forestry: the way forward*. Report of Edinburgh Conference, 25th February 1994. Royal Scottish Forestry Society, Edinburgh. pp. 10–21.

Naess, A. 1989. *Ecology, Community and Lifestyle: outline of an ecophilosophy.* Cambridge University Press, Cambridge.

Nair, P. K. R. 1991. State-of-the-art of agroforestry systems. *Forest Ecology and Management* 45: 5–29.

Nash, R. 2001. *Wilderness and the American mind.* Fourth Edition. Yale Nota Bene, New Haven.

NCC. 1986. *Nature Conservation and Afforestation in Britain.* Nature Conservancy Council, Peterborough.

NC/SG (Nautilus Consultants in association with Smiths Gore) 1998. An economic assessment of the major land uses in the Scottish uplands. *Scottish Natural Heritage Research, Survey and Monitoring Report* No. 92.

NCSR. 1999. *Leisure Day Visits: report of the 1998 UK Day Visits Survey.* National Centre for Social Research, London.

Newson, M. 1999. Land, water and development: key themes driving international policy on catchment management. In: Cresser, M. and Pugh K. (eds), *Multiple land use and catchment management.* Proceedings of an international conference, 11–13 September 1996. Macaulay Land Use Research Institute, Aberdeen. pp. 11–21.

Newton, A. C. and Humphrey, J. W. 1997. Forest management for biodiversity: perspectives on the policy context and current initiatives. In: Fleming, L. V., Newton, A. C., Vickery, J. A. and Usher, M. B. (eds), *Biodiversity in Scotland: status, trends and initiatives.* The Stationery Office, Edinburgh. pp. 179–98.

Newton, A. C. and Ashmole, P. 1998. How may native woodland be restored in southern Scotland? *Scottish Forestry* 52(3 & 4): 168–71.

NGF. 2000. *Policy report and recommendations of the National Goose Forum.* Scottish Executive, Edinburgh.

Nicol, R. 1997. The Scottish feudal system of land tenure: can it be improved? In: *Scottish Landownership within Europe.* Proceedings of the Scottish Landowners Federation Conference, Battleby, 23rd April 1997.

Nisbet, T. 1999. The sustainability of afforestation within Highland catchments supporting important salmonid fisheries – the upper Halladale River. In: *Forest Research Annual Report and Accounts 1998–1999.* Forestry Commission, Edinburgh. pp. 41–7.

Nisbet, T. 2001. The role of forest management in controlling diffuse pollution in UK forestry. *Forest Ecology and Management* 143(1–2): 215–26.

Northcott, M. S. 1996. *The Environment and Christian Ethics.* Cambridge University Press, Cambridge.

Oelschlaeger, M. 1991. *The Idea of Wilderness: from prehistory to the age of ecology.* Yale University Press, New Haven.

O'Neill, J., Benson, J., Holland, A. and Roxbee-Cox, J. 1997. *Environmental Values.* Routledge, London.

O'Riordan, T. 1993. The politics of sustainability. In: Turner, R. K. (ed.), *Sustainable Environmental Economics and Management: principles and practice.* Belhaven Press, London. pp. 37–69.

O'Riordan, T. 1995a. Frameworks for choice: core beliefs and the environment. *Environment* 37(8): 4–9; 25–9.

O'Riordan, T. 1995b. The application of the precautionary principle in the United Kingdom. *Environment and Planning A* 27(10): 1534–8.

O'Riordan, T. 1999. From environmentalism to sustainability. *Scottish Geographical Journal* 115(2): 151–65.

O'Riordan, T. (ed.) 2000a. *Environmental Science for Environmental Management.* Second Edition. Pearson, Harlow.

O'Riordan, T. 2000b. The sustainability debate. In: O'Riordan, T. (ed.), *Environmental Science for Environmental Management.* Second Edition. Pearson, Harlow. pp. 29–62.

O'Riordan, T. 2000c. Environmental science on the move. In: O'Riordan, T. (ed.), *Environmental Science for Environmental Management.* Second Edition. Pearson, Harlow. pp. 1–27.

O'Riordan, T. and Jordan, A. 1995. The Precautionary Principle in contemporary environmental politics. *Environmental Values* 4: 191–212.

O'Riordan, T. and Jordan, A. 2000. Managing the global commons. In: O'Riordan, T. (ed.), *Environmental Science for Environmental Management*. Second Edition. Pearson, Harlow. pp. 485–511.

Orr, W. 1982. *Deer Forest, Landlords and Crofters: the Western Highlands in Victorian and Edwardian times*. John Donald, Edinburgh.

Owen, L. and Unwin, T. (eds) 1997. *Environmental Management: readings and case studies*. Blackwell, Oxford.

Owens, S. and Cowell, R. 1996. *Rocks and Hard Places: Mineral Resource Planning and Sustainability*. Council for the Protection of Rural England, London.

Owens, S. and Owens, P. L. 1991. *Environment, Resources and Conservation*. Cambridge University Press, Cambridge.

Palmer, S. C. F, Paterson, I. S., Marquiss, M. and Staines, B. W. 1998. The impact of deer browsing on regeneration of Scots Pine: a preliminary study. In: Goldspink, C. R., King, S. and Putman, R. J. (eds), *Population Ecology, Management and Welfare of Deer*. Manchester Metropolitan University, Manchester. pp. 48–53.

Parker, H., Rosell, F. and Holthe, V. 2000. A gross assessment of the suitability of selected Scottish riparian habitats for beaver. *Scottish Forestry* 54(1): 25–31.

Parkyn, L., Stoneman, R. E. and Ingram, H. A. P. (eds) 1997. *Conserving Peatlands*. CAB International, Wallingford.

Parman, S. 1990. *Scottish Crofters: a historical ethnography of a Celtic village*. Holt, Rinehart and Winston, Fort Worth.

Passmore, J. 1974. *Man's responsibility for nature: ecological problems and Western traditions*. Duckworth, London.

Pearce, D. 1998. *Economics and Environment: Essays on Ecological Economics and Sustainable Development*. Edward Elgar, Cheltenham.

Pearce, D. 1999. Can non-market values save the world's forests? In: Roper, C. S. and Park, A. (eds), *The Living Forest: non-market benefits of forestry*. Proceedings of an international symposium on non-market benefits of forestry, Edinburgh, 24–28 June, 1996. Stationery Office, London. pp. 5–16.

Pearce, D., Markandya, A. and Barbier, E. B. 1989. *Blueprint for a Green Economy*. Earthscan, London.

Pearce, F. 1994. Rush for rock in the Highlands. *New Scientist* 1909 (8th January 1994): 11–12.

Pearce, F. 1997. Burn me. *New Scientist* 2109 (22nd November 1997): 30–4.

Pearce, F. 1999a. That sinking feeling. *New Scientist* 2209 (23rd October 1999): 20–1.

Pearce, F. 1999b. Crops without profit. *New Scientist* 2217 (18th December 1999): 10.

Pemberton, J. M., Swanson, G. M. and Goodman, S. J. 1998. Management of Scottish deer in the face of red-sika hybridisation (Abstract). In: Goldspink, C. R., King, S. and Putman, R. J. (eds), *Population Ecology, Management and Welfare of Deer*. Manchester Metropolitan University, Manchester. p. 118.

Pemberton, J. M., Barton, N. H., Goodman, S. J. and Swanson, G. M. 1999. Introgression of Scottish red deer by sika deer (Abstract). In: *Deer Commission for Scotland Annual Report 1998–1999*. The Stationery Office, Edinburgh. pp. 34–5.

Pemberton, J. M., Jones, F. and Goodman, S. J. 2000. Genetic make up of red deer on proposed refugia. In: *Deer Commission for Scotland Annual Report 1999–2000*. The Stationery Office, Edinburgh. pp. 41–5.

Pepper, D. 1996. *Modern Environmentalism*. Routledge, London.

Pepper, H., Bryce, J. and Cartmel, S. 2001. Co-existence of red squirrels and grey squirrels. In: *Forest Research Annual Report and Accounts 1999–2000*. Stationery Office, Edinburgh. pp. 29–32.

Perkins, S. 1998. People, nature and the Cairngorm funicular. *ECOS* 19(3/4): 70–1.

Perman, R., Ma, Y., McGilvray, J. and Common, M. 1999. *Natural Resource and Environmental Economics*. Pearson, Harlow.

Perretti, J. H. 1998. Nativism and nature: rethinking biological invasion. *Environmental Values* 7: 183–92.

Peterken, G. 1996. *Natural Woodland: Ecology and Conservation in Northern Temperate Regions*. Cambridge University Press, Cambridge.

Peterken, G. 1999. Applying natural forestry concepts in an intensively managed landscape. *Global Ecology and Biogeography* 8: 321–8.

Peterken, G. 2000. Historical considerations in future woodland management and restoration. *Scottish Woodland History Discussion Group: Notes IV* (Proceedings of a meeting at Battleby, Perth, 23rd November 1999). pp. 3–4.

Peterken, G. 2001. Ecological effects of introduced tree species in Britain. *Forest Ecology and Management* 141: 31–42.

Peterken, G., Baldock, D. and Hampson, A. 1995. A Forest Habitat Network for Scotland. *SNH Research, Survey and Monitoring Report* 44. Scottish Natural Heritage, Battleby.

Petty, S. J. 2000. *Capercaillie: a review of research needs*. Scottish Executive, Edinburgh.

Pezzoli, K. 1997a. Sustainable development: a transdisciplinary overview of the literature. *Journal of Environmental Planning and Management* 40(5): 549–74.

Pezzoli, K. 1997b. Sustainable development literature: a transdisciplinary bibliography. *Journal of Environmental Planning and Management* 40(5): 575–601.

PFAP. 2000. *The Paths for All Partnership 2000–2003*. Paths for All Partnership, Alloa.

PFAP. 2001. *The Paths for All Partnership Annual Review 1999–2000*. Paths for All Partnership, Alloa.

Philip, A. B. 1998. The European Union: environmental policy and the prospects for sustainable development. In: Hanf, K. and Jansen, A.-I. (eds), *Governance and Environment in Western Europe: Politics, Policy and Administration*. Longman, Harlow. pp. 253–76.

Phillips, J. 2001. Range management for red grouse. In: Phillips, J., Thompson, D. B. A. and Gruellich, W. H. (eds), *Integrated Upland Management for Wildlife, Field Sports, Agriculture and Public Enjoyment*. The Heather Trust/Bidwells/SNH, Battleby. pp. 20–36.

Phillips, J. and Watson, A. 1995. Key requirements for management of heather moorland: now and for the future. In: Thompson, D. B. A., Hester, A. J., and Usher, M. B. (eds), *Heaths and Moorland: Cultural Landscapes*. HMSO, Edinburgh. pp. 344–61.

Phillips, M. and Mighall, T. 2000. *Society and Exploitation through Nature*. Pearson, Harlow.

Pollard, P. 2000. Conservation of the freshwater natural heritage. Paper presented at *Hydrology in Scotland: Agenda for the 21st Century*, University of Dundee, 21–23 March 2000.

Pollock, S. H. A. 1994. Coastal superquarries in Scotland: guest editorial. *Scottish Geographical Magazine* 110(3): 138–9.

Porritt, J. 2000. *Playing Safe: Science and the Environment*. Thames and Hudson, London.

Potter, C. 1996. Environmental reform of the CAP: an analysis of the short and long range opportunities. In: Curry, N. and Owen, S. (eds), *Changing Rural Policy in Britain: Planning, Administration, Agriculture and the Environment*. Countryside and Community Press, Cheltenham. pp. 165–83.

Potter, C. 1998a. *Against the Grain: Agri-Environmental Reform in the United States and the European Union*. CAB International, Wallingford.

Potter, C. 1998b. Agricultural liberalisation: opportunity or threat? *ECOS* 19(2): 38–43.

Potter, C. 2001. Negotiating the transition: rural policy reform and the restructuring of agriculture. *ECOS* 22(2): 25–30.

Potts, R. 2000. The RSPB: manipulating the truth. *The Field* April 2000: 9.

Pratt, V., Howarth, J. and Brady, E. 2000. *Environment and Philosophy*. Routledge, London.

Price, C. 1997. Valuation of biodiversity: of what, by whom and how? *Scottish Forestry* 51(3): 134–42.

Price, C. 2000. Valuation of unpriced products: contingent valuation, cost–benefit analysis and participatory democracy. *Land Use Policy* 17(3): 187–96.

Pringle, D. 1995. *The first 75 years: a brief account of the history of the Forestry Commission, 1919–1994*. Forestry Commission, Edinburgh.

Prior, R. 1994. *Trees and Deer: how to cope with deer in forest, field and garden*. Swan Hill Press, Shrewsbury.
Prior, R. 1995. *The Roe Deer: conservation of a native species*. Swan Hill Press, Shrewsbury.
PSPS (Peter Scott Planning Services). 1991. Countryside access in Europe: a review of access rights, legislation and provision in selected European countries. *Scottish Natural Heritage Review* No. 23.
Puhr, C. B., Donohhue, D. N. M., Stephen, A. B., Tervet, D. J. and Sinclair, C. 2000. Regional patterns of streamwater acidity and catchment afforestation in Galloway, SW Scotland. *Water, Air and Soil Pollution* 120: 47–70.
Putman, R. J. 1996. Ungulates in temperate forest ecosystems: perspectives and recommendations for future research. *Forest Ecology and Management* 88: 205–14.
Putman, R. J. 1998. The potential role of habitat manipulation in reducing deer impact. In: Goldspink, C. R., King, S. and Putman, R. J. (eds), *Population Ecology, Management and Welfare of Deer*. Manchester Metropolitan University, Manchester. pp. 95–101.
Putman, R. J. 2000. *Sika Deer*. The Mammal Society, London.
Putman, R. J. and Moore, N. P. 1998. Impact of deer in lowland Britain on agriculture, forestry and conservation habitats. *Mammal Review* 28(4): 141–64.
Quelch, P. 1998. An open letter to the Scottish Woodland History Discussion Group. *Native Woodlands Discussion Group Newsletter* 23(1): 20–1.
Quelch, P. 2000. Upland pasture woodlands in Scotland. Part I. *Scottish Forestry* 54(4): 209–14.
Quelch, P. 2001. Upland pasture woodlands in Scotland. Part II: possible origins and examples. *Scottish Forestry* 55(2): 85–92.
RA. 2000. Personal communication from the Ramblers Association, 17th November 2000.
Raemaekers, J. and Boyack, S. 1999. Planning and sustainable development. *Scottish Environmental Audits* No. 3. Scottish Environment LINK, Perth.
Ramsay, A. 1999. Study brings little cheer. *Landowning in Scotland* 252: 4–5.
Ramsay, P. 1993. Land-owners and conservation. In: Goldsmith, F. B. and Warren, A. (eds), *Conservation in Progress*. John Wiley, Chichester. pp. 255–69.
Ramsay, P. 1996. *Revival of the Land: Creag Meagaidh National Nature Reserve*. Scottish Natural Heritage, Battleby.
Randall, J. 1997. The Rural White Paper in Scotland. *Journal of Environmental Planning and Management* 40(3): 385–9.
Randall, J. 1998. National Parks – the Scottish Office position. In: *Protecting Scotland's Finest Landscapes: National Parks – The Opportunities*. Proceedings of a conference, 16th March 1998. Scottish Wildlife and Countryside Link, Perth. pp. 6–7.
Ratcliffe, D. A. 1977. *Highland Flora*. Highlands and Islands Development Board, Inverness.
Ratcliffe, D. A. 1989. Conserving wild nature: purpose and ethics. In: Webb, L. J. and Kikkawa, J. (eds), *Australian Tropical Rainforests*. CSIRO, Australia. pp. 142–9.
Ratcliffe, D. A. 1990. *Bird life of mountain and upland*. Cambridge University Press, Cambridge.
Ratcliffe, D. A. 1995. More thoughts on nature conservation and the Voluntary Principle. *Environmental Values* 4: 71–2.
Ratcliffe, D. A. 1998. Labour's conservation performance. *ECOS* 19(3/4): 37–44.
Ratcliffe, P. R. 1985. Glades for deer control in upland forests. *Forestry Commission Leaflet* 86. HMSO, London.
Ratcliffe, P. R. 1987a. The management of red deer in upland forests. *Forestry Commission Bulletin* 71. HMSO, London.
Ratcliffe, P. R. 1987b. Distribution and current status of Sika Deer, *Cervus nippon*, in Great Britain. *Mammal Review* 17(1): 39–58.
Ratcliffe, P. R. 1995a. Ecological diversity in managed forests. In: Ferris-Kahn, R. (ed.), Managing forests for biodiversity. *Forestry Commission Technical Paper* 8: 3–7.
Ratcliffe, P. R. 1995b. The mongrel of the glen. *The Field* June 1995: 66–9.
Ratcliffe, P. R. (ed.) 1997. *Native and Non-Native in British Forestry*. Proceedings of a discussion

meeting, University of Warwick, 31 March–2 April, 1995. Institute of Chartered Foresters, Edinburgh.

Ratcliffe, P. R. 1998. Woodland deer management: integrating the control of their impact with multiple objective forest management in Scotland. In: Goldspink, C. R., King, S. and Putman, R. J. (eds), *Population Ecology, Management and Welfare of Deer*. Manchester Metropolitan University, Manchester. pp. 67–72.

Ratcliffe, P. R. and Mayle, B. A. 1992. Roe Deer Biology and Management. *Forestry Commission Bulletin* 105. HMSO, London.

Raven, A. 1999. Agriculture, forestry and rural land use. In: McDowell, E. and McCormick, J. (eds), *Environment Scotland: Prospects for Sustainability*. Ashgate, Aldershot. pp. 127–38.

Raven, H. 1999. Land reform. In: McDowell, E. and McCormick, J. (eds), *Environment Scotland: Prospects for Sustainability*. Ashgate, Aldershot. pp. 139–53.

Rawles, K. and Holland, A. 1994. *The ethics of conservation*. Report for Countryside Council for Wales. Lancaster University, Lancaster.

RDC. 1996. *Red Deer Commission Annual Report 1995/1996*. The Stationery Office, Edinburgh.

Redclift, M. 1993. Sustainable development: needs, values, rights. *Environmental Values* 2: 3–20.

Redpath, S. M. and Thirgood, S. J. 1997. *Birds of Prey and Red Grouse*. The Stationery Office, London.

Redpath, S. M. and Thirgood, S. J. 1999. Numerical and functional responses in generalist predators: hen harriers and peregrines on Scottish grouse moors. *Journal of Animal Ecology* 68: 879–92.

Reimers, E. 2001. Range management for large mammals. In: Phillips, J., Thompson, D. B. A. and Gruellich, W. H. (eds), *Integrated Upland Management for Wildlife, Field Sports, Agriculture and Public Enjoyment*. The Heather Trust/Bidwells/SNH, Battleby. pp. 8–19.

Rennie, F. 1997. The way of crofting. In: Magnusson, M. and White, G. (eds), *The Nature of Scotland: Landscape, Wildlife and People*. Canongate, Edinburgh. pp. 123–31.

Reynolds, F. 1998. Environmental planning: land use and landscape policy. In: Lowe, P. and Ward, S. (eds), *British Environmental Policy and Europe: Politics and Policy in Transition*. Routledge, London. pp. 232–43.

Reynolds, P. 1995. Red deer management: key issues and prescriptive leading strings. In: Rose, H. (ed.), *Deer, Habitats and Birds*. Proceedings of a BDS/RSPB Joint Conference, October 1995, Inverness. pp. 7–11.

Richards, E. 2000. *The Highland Clearances: People, Landlords and Rural Turmoil*. Birlinn, Edinburgh.

Rickman, R. 1994. Commentary. In: Jeffrey, W. G. (ed.), *State Forestry: the way forward*. Report of Edinburgh Conference, 25th February 1994. Royal Scottish Forestry Society, Edinburgh. pp. 74–5.

Riddell, J. 2000. Progress in flood hazard management in Scotland. Paper presented at *Hydrology in Scotland: Agenda for the 21st Century*, University of Dundee, 21–23 March 2000.

Ritchie, A. and Ritchie, G. 1997. The coming of man. In: Magnusson, M. and White, G. (eds), *The Nature of Scotland: Landscape, Wildlife and People*. Canongate, Edinburgh. pp. 99–109.

Rival, L. (ed.) 1998. *The Social Life of Trees: anthropological perspectives on tree symbolism*. Berg, Oxford.

Rivers, J. 1997. Beyond rights: the morality of rights language. *Cambridge Papers* 6(3): 1–4.

Roberts, N. 1998. *The Holocene: an environmental history*. Second Edition. Blackwell, Oxford.

Robinson, G. M. 1990. *Conflict and Change in the Countryside*. Belhaven Press, London.

Robinson, G. M. 1994a. Agriculture as a land use in Scotland. In: Fenton, A. and Gillmor, D. A. (eds), *Rural land use on the Atlantic periphery of Europe: Scotland and Ireland*. Royal Irish Academy, Dublin. pp. 75–98.

Robinson, G. M. 1994b. The greening of agricultural policy: Scotland's Environmentally Sensitive Areas (ESAs). *Journal of Environmental Planning and Management* 37(2): 215–25.

Rodwell, J. and Patterson, G. 1994. Creating new native woodlands. *Forestry Commission Bulletin* 112. HMSO, London.

Rogers, D. 2000. Proposed EC Water Framework Directive: implementation in Scotland. Paper presented at *Hydrology in Scotland: Agenda for the 21st Century*, University of Dundee, 21–23 March 2000.

Roper, C. S. and Park, A. (eds) 1999. *The Living Forest: non-market benefits of forestry*. Proceedings of an international symposium on non-market benefits of forestry, Edinburgh, 24–28 June, 1996. Stationery Office, London.

Rose, H. 1995. The pure red deer, a vanished breed. *The Field* June 1995: 69.

Ross, A., Rowan-Robinson, J. and Walton, W. 1995. Sustainable development in Scotland: the role of Scottish Natural Heritage. *Land Use Policy* 12(3): 237–52.

Ross, D. 2000a. Clearances land war raging on. *The Herald* 22nd June 2000.

Ross, D. 2000b. Superquarry falls at final hurdle. *The Herald* 4th November 2000.

Ross, J. 2001. Our land reform intentions are good, but are we rambling into trouble? *The Scotsman* 23rd February 2001.

Ross, J. and Hardie, A. 2000. After ten years, quarry battle is over. *The Scotsman* 4th November 2000.

Rowan-Robinson, J. 1997. The organisation and effectiveness of the Scottish planning system. In: Macdonald, R. and Thomas, H. (eds), *Nationality and Planning in Scotland and Wales*. University of Wales Press, Cardiff. pp. 32–53.

RSFS (Royal Scottish Forestry Society) 1993. A forest policy for Scotland. *Scottish Forestry* 47(4): 147–52.

RSFS (Royal Scottish Forestry Society) 1996. The wood-using industries of Scotland. *Scottish Forestry* 50(1): 31–6.

RSPB. 1993. *Time for Pine*. Royal Society for the Protection of Birds, Edinburgh.

RSPB and 8 conservation NGOs. 1998. *A Future for Grouse Moors: a call for action*. Royal Society for the Protection of Birds, Sandy.

RSPB. 2000. Personal communication from the Royal Society for the Protection of Birds, 28th July 2000.

Runte, A. 1997. *National Parks: the American experience*. Third Edition. University of Nebraska Press, Lincoln, Nebraska.

Russell, N. 1994. Issues and options for agri–environmental policy: an introduction. *Land Use Policy* 11(2): 83–7.

Russell, C. A. 1994. *The Earth, Humanity and God*. UCL Press, London.

Rutherford, A. and Hart, K. 2000. The new Rural Development Regulation: fresh hope for farming and England's countryside? *ECOS* 21(1): 69–75.

Sagoff, M. 1998. Aggregation and deliberation in valuing environmental public goods: a look beyond contingent pricing. *Ecological Economics* 24: 213–30.

Samuel, A. 2001. Rum – nature and community in harmony? *ECOS* 22(1): 36–45.

Sanders, N. D. H. 1999. The effect of woodland on farm valuations. In: Burgess, P. J., Brierley, E. D. R., Morris, J. and Evans, J. (eds), *Farm woodlands for the future*. BIOS Scientific, Oxford. pp. 115–22.

SBG. 1997. *Biodiversity in Scotland: the way forward*. Scottish Biodiversity Group, Edinburgh.

Scambler, A. 1989. Farmers' attitudes towards forestry. *Scottish Geographical Magazine* 105: 47–9.

Scott, D. 1998. Impact of red deer on a Scots Pine plantation after removal of deer fencing. *Scottish Forestry* 52(1): 8–13.

Scott, D. and Palmer, S. C. F. 2000. Damage by deer to agriculture and forestry. In: *Deer Commission for Scotland Annual Report 1999–2000*. The Stationery Office, Edinburgh. pp. 50–3.

Scott, D., Welch, D., Thurlow, M. and Elston, D. A. 2000. Regeneration of *Pinus sylvestris* in a natural pinewood in NE Scotland following reduction in grazing by *Cervus elaphus*. *Forest Ecology and Management* 130: 199–211.

Scott, M. 1997. Awareness and education: selling biodiversity. In: Fleming, L. V., Newton, A. C., Vickery, J. A. and Usher, M. B. (eds), *Biodiversity in Scotland: status, trends and initiatives*. The Stationery Office, Edinburgh. pp. 252–6.

Scott, M. 2000. *Montane Scrub*. Scottish Natural Heritage, Battleby.
Scott, M. 2001. Concluding perspective. In: Usher, M. B. (ed.), *Enjoyment and Understanding of the Natural Heritage*. The Stationery Office, Edinburgh, pp. 207–12.
Scottish Executive. 2000a. *Scottish Climate Change Programme Consultation*. The Stationery Office, Edinburgh.
Scottish Executive. 2000b. *Forests for Scotland: the Scottish Forestry Strategy*. The Stationery Office, Edinburgh.
Scottish Executive. 2000c. *National Parks for Scotland: consultation on the National Parks (Scotland) Bill*. The Stationery Office, Edinburgh.
Scottish Executive. 2001a. *The Nature of Scotland: a policy statement*. The Stationery Office, Edinburgh.
Scottish Executive. 2001b. *Land Reform: the Draft Bill*. The Stationery Office, Edinburgh.
Scottish Executive. 2001c. *A Forward Strategy for Scottish Agriculture*. The Stationery Office, Edinburgh.
Scottish Executive. 2001d. *Key Scottish Environment Statistics*. The Stationery Office, Edinburgh.
Scottish Executive. 2001e. *A Draft Scottish Outdoor Access Code*. The Stationery Office, Edinburgh.
Scottish Office. 1992. *Rural Framework*. Scottish Office, Edinburgh.
Scottish Office. 1995. *Rural Scotland: People, Prosperity and Partnership*. HMSO, Edinburgh.
Scottish Office. 1996. *Natural Heritage Designations Review*. HMSO, Edinburgh.
Scottish Office. 1997. *Biodiversity in Scotland: the way forward*. HMSO, Edinburgh.
Scottish Office. 1998a. *Scottish Environment Statistics No. 6*. Stationery Office, Edinburgh.
Scottish Office. 1998b. *People and Nature: a new approach to SSSI designations in Scotland (Consultation Paper)*. The Stationery Office, Edinburgh.
Scottish Office. 1998c. *Natural Heritage Designations in Scotland: a guide*. The Stationery Office, Edinburgh.
Scottish Office. 1998d. *Towards a development strategy for rural Scotland: the framework*. The Stationery Office, Edinburgh.
Scottish Office. 1999a. *Agenda 2000*. The Scottish Office, Edinburgh.
Scottish Office. 1999b. *Down to Earth: a Scottish perspective on sustainable development*. The Scottish Office, Edinburgh.
Scruton, R. 1999. We should settle our land grievances in the long-term interests of our nation, not just for the pleasure of present generations. *The Times* 23rd January 1999.
SDD. 1990. *Indicative Forest Strategies*. Scottish Development Department Circular No. 13/1990. SDD, Edinburgh.
SEERAD. 2001a. *Agriculture Facts and Figures*. Scottish Executive Environment and Rural Affairs Department, Edinburgh.
SEERAD. 2001b. *Economic Report on Scottish Agriculture 2001 Edition*. Scottish Executive Environment and Rural Affairs Department, Edinburgh.
Selman, P. (ed.) 1988a. *Countryside Planning in Practice: the Scottish experience*. Stirling University Press, Stirling.
Selman, P. 1988b. Foreword. In: Selman, P. (ed.), *Countryside Planning in Practice: the Scottish experience*. Stirling University Press, Stirling. pp. xi–xxi.
Selman, P. 1996. The potential for landscape ecological planning in Britain. In: Curry, N. and Owen, S. (eds), *Changing Rural Policy in Britain: Planning, Administration, Agriculture and the Environment*. Countryside and Community Press, Cheltenham. pp. 28–43.
Selman, P. 2000. A sideways look at Local Agenda 21. *Journal of Environmental Policy and Planning* 2(1): 39–53.
Selmes, R. E. 1999. Priorities for the development of the GB softwood market over the next 15 to 20 years. *Scottish Forestry* 53(2): 104–8.
SEPA. 1998. *SEPA's Environmental Strategy*. Scottish Environment Protection Agency, Stirling.
SEPA. 1999. *Improving Scotland's Water Environment: SEPA State of the Environment Report*. Scottish Environment Protection Agency, Stirling.

SERAD. 2000a. *Rural development regulation (EC) No. 1257/1999: plan for Scotland.* Scottish Executive Rural Affairs Department, Edinburgh.

SERAD. 2000b. *A Forward Strategy for Scottish Agriculture: a discussion document.* Scottish Executive Rural Affairs Department, Edinburgh.

SERAD. 2000c. *Economic Report on Scottish Agriculture 2000 Edition.* Scottish Executive Rural Affairs Department, Edinburgh.

SERAD. 2000d. *Protecting and Promoting Scotland's Freshwater Fish and Fisheries: a review.* Scottish Executive Rural Affairs Department, Edinburgh.

SERAD. 2000e. *Statistical Bulletin Fisheries Series: 1999 catches.* Scottish Executive Rural Affairs Department, Edinburgh.

Sewel, J. 1997. Scottish National Parks to promote sustainable development. Scottish Office press release, 17th September 1997.

Sharp, R. 1998. Responding to Europeanisation: a governmental perspective. In: Lowe, P. and Ward, S. (eds), *British Environmental Policy and Europe: Politics and Policy in Transition.* Routledge, London. pp. 33–56.

Sheail, J. 1975. The concept of national parks in Great Britain, 1900–1950. *Transactions of the Institute of British Geographers* 66: 41–56.

Sheail, J. 1998. *Nature conservation in Britain: the formative years.* The Stationery Office, London.

Sheail, J. 1999. The grey squirrel (*Sciurus carolinensis*) – a UK historical perspective on a vertebrate pest species. *Journal of Environmental Management* 55: 145–56.

Shoard, M. 1980. *The Theft of the Countryside.* Temple Smith, London.

Shoard, M. 1997. *This Land is Our Land: the struggle for Britain's countryside.* Gaia Books, London.

Shoard, M. 1999. *A Right to Roam: should we open up Britain's countryside?* Oxford University Press, Oxford.

Shotbolt, L., Anderson, A. R. and Townend, J. 1998. Changes to blanket bog adjoining forest plots at Bad a'Cheo, Rumster, Caithness. Forestry 71: 311–24.

Shucksmith, M. 1992. The effects of European agricultural policies on the communities of upland Scotland: the case of Grampians. *Revue de Géographie Alpine* 80(4): 97–116.

Shucksmith, M., Chapman, P., Clark, G. M., Black, S. and Conway, E. 1996. *Rural Scotland today: the best of both worlds?* Avebury, Aldershot.

Sibbald, A. R. 1999. Agroforestry principles – sustainable productivity? *Scottish Forestry* 53(1): 18–23.

Sibbald, A. R. 2001. Decision support tools to link ecology and land management. In: *The Macaulay Institute Annual Report 2000.* The Macaulay Institute, Aberdeen. pp. 24–7.

Sidaway, R. 1994. Recreation and the natural heritage: a research review. *Scottish Natural Heritage Review* No. 25.

Sidaway, R. 2001. The effects of recreation on the natural heritage: the need to focus on improving management practice. In: Usher, M. B. (ed.), *Enjoyment and Understanding of the Natural Heritage.* The Stationery Office, Edinburgh. pp. 39–56.

Silbergh, D. 1999. Sustainability and sustainable development. In: McDowell, E. and McCormick, J. (eds), *Environment Scotland: Prospects for Sustainability.* Ashgate, Aldershot. pp. 16–41.

Simmons, I. G. 1993. *Interpreting Nature: cultural constructions of the environment.* Routledge, London.

Simmons, I. G. 1996. *Changing the face of the earth: culture, environment, history.* Second Edition. Blackwell, Oxford.

Simmons, I. G. 2001. *Environmental History of Great Britain.* Edinburgh University Press, Edinburgh.

Sinclair, F. L. 1999. The agroforestry concept – managing complexity. *Scottish Forestry* 53(1): 12–17.

Sinclair, F. L., Eason, W. and Hooker, J. 2000. Understanding and management of interactions. In: Hislop, M. and Claridge, J. (eds), Agroforestry in the UK. *Forestry Commission Bulletin* 122. pp. 17–28.

Skelcher, G. 1997. The ecological replacement of red by grey squirrels. In: Gurnell, J. and Lurz, P. (eds), *The conservation of red squirrels,* Sciurus vulgaris *L.* People's Trust for Endangered Species, London. pp. 67–78.

Skerratt, S. and Dent, J. B. 1996. The challenge of agri-environmental subsidies: the case of Breadalbane Environmentally Sensitive Area, Scotland. *Scottish Geographical Magazine* 112(2): 92–100.

Skinner, J. A., Lewis, K. A., Bardon, K. S., Tucker, P., Catt, J. A. and Chambers, B. J. 1997. An overview of the environmental impact of agriculture in the U.K. *Journal of Environmental Management* 50: 111–28.

Slee, B. 1998. Tourism and rural development in Scotland. In: MacLellan, R. and Smith, R. (eds), *Tourism in Scotland.* International Thomson Business Press, London. pp. 93–111.

SLF. 1993. *Access: towards access without acrimony.* Scottish Landowners' Federation, Edinburgh.

SLF. 1995. *The Management of Country Sports in Scotland.* Scottish Landowners' Federation, Edinburgh.

SLF. 1999. *Supporting Scotland's Countryside – the role of forestry.* Scottish Landowners' Federation Discussion Document, Musselburgh.

SLF. 2000. *A Code of Practice for Responsible Land Management.* Scottish Landowners' Federation, Edinburgh.

Smith, A. 2000. Personal communication, 25th July 2000.

Smith, A., Redpath, S. and Campbell, S. 2000. *The influence of moorland management on grouse and their predators.* Department of the Environment, Transport and the Regions, Bristol.

Smith, J. S. 1993. Changing deer numbers in the Scottish Highlands since 1780. In: Smout, T. C. (ed.), *Scotland since Prehistory: Natural Change and Human Impact.* SNH/Scottish Cultural Press, Aberdeen. pp. 89–97.

Smith, R. 1997. The playground of the future. In: Magnusson, M. and White, G. (eds), *The Nature of Scotland: Landscape, Wildlife and People.* Canongate, Edinburgh. pp. 227–42.

Smith, S. M. 1999. The national inventory of woodland and trees: Scotland. *Scottish Forestry* 53(3): 163–8.

Smout, T. C. 1969. *A History of the Scottish People 1560–1830.* Collins, Glasgow.

Smout, T. C. (ed.) 1986a. *Scotland and Europe 1200–1850.* John Donald, Edinburgh.

Smout, T. C. 1986b. *A Century of the Scottish People 1830–1950.* Collins, Glasgow.

Smout, T. C. (ed.) 1993a. *Scotland since Prehistory: Natural Change and Human Impact.* Scottish Natural Heritage/Scottish Cultural Press, Aberdeen.

Smout, T. C. 1993b. Woodland history before 1850. In: Smout, T. C. (ed.), *Scotland since Prehistory: Natural Change and Human Impact.* Scottish Natural Heritage/Scottish Cultural Press, Aberdeen. pp. 40–9.

Smout, T. C. 1993c. The Highlands and the roots of green consciousness, 1750–1990. *Scottish Natural Heritage Occasional Papers* No. 1.

Smout, T. C. (ed.) 1997a. *Scottish Woodland History.* Scottish Cultural Press, Edinburgh.

Smout, T. C. 1997b. Scottish landownership in a European perspective – an historian's view. In: *Scottish Landownership within Europe.* Proceedings of the Scottish Landowners Federation Conference, Battleby, 23rd April 1997.

Smout, T. C. 1997c. Highland land-use before 1850: misconceptions, evidence and realities. In: Smout, T. C. (ed.), *Scottish Woodland History.* Scottish Cultural Press, Edinburgh. pp. 5–23.

Smout, T. C. 1997d. Bogs and People since 1600. In: Parkyn, L., Stoneman, R. E. and Ingram, H. A. P. (eds), *Conserving Peatlands.* CAB International, Wallingford. pp. 162–7.

Smout, T.C. 1999a. Forests for Scotland. *Scottish Forestry* 53(2): 66–7.

Smout, T. C. 1999b. The past and future forest. *Reforesting Scotland* 20: 10–11.

Smout, T. C. 2000a. *Nature Contested: environmental history in Scotland and northern England since 1600.* Edinburgh University Press, Edinburgh.

Smout, T. C. 2000b. Recreating native woodland: a plea for more thinking. *Scottish Woodland History Discussion Group: Notes IV* (Proceedings of a meeting at Battleby, Perth, 23rd

November 1999). pp. 1–2.

SNH. 1992. *Enjoying the outdoors: a consultation paper on access to the countryside for enjoyment and understanding*. Scottish Natural Heritage, Battleby.

SNH. 1993. *Sustainable development and the natural heritage – the SNH approach*. Scottish Natural Heritage, Battleby.

SNH. 1994a. Agriculture and Scotland's natural heritage. *SNH Policy Paper*. Scottish Natural Heritage, Battleby.

SNH. 1994b. Red deer and the natural heritage. *SNH Policy Paper*. Scottish Natural Heritage, Battleby.

SNH. 1994c. Enjoying the outdoors: a programme for action. *SNH Policy Paper*. Scottish Natural Heritage, Battleby.

SNH. 1995a. *The natural heritage of Scotland: an overview*. Scottish Natural Heritage, Edinburgh.

SNH. 1995b. *Boglands*. Scottish Natural Heritage, Battleby.

SNH. 1997a. Hill-walking in Scotland. *Scottish Natural Heritage Information and Advisory Note* 80.

SNH. 1997b. Long-distance routes in Scotland. *SNH Policy Paper*. Scottish Natural Heritage, Battleby.

SNH. 1998a. *Machair*. Scottish Natural Heritage, Battleby.

SNH. 1998b. *National Scenic Areas: a consultation paper*. Scottish Natural Heritage, Battleby.

SNH. 1998c. *Jobs and the natural heritage: the natural heritage in rural development*. Scottish Natural Heritage, Battleby.

SNH. 1998d. *National Parks for Scotland: a consultation paper*. Scottish Natural Heritage, Battleby.

SNH. 1998e. *Access to the Countryside for Open-air recreation: Scottish Natural Heritage's Advice to Government*. Scottish Natural Heritage, Battleby.

SNH. 1998f. *Reintroduction of the European beaver to Scotland: a public consultation*. Scottish Natural Heritage, Battleby.

SNH. 1999a. Impacts of climate change on plants, animals and ecosystems in Scotland. *Scottish Natural Heritage Information and Advisory Note* 111.

SNH. 1999b. *Working with communities: the natural heritage in rural development*. Scottish Natural Heritage, Battleby.

SNH. 1999c. *National Scenic Areas: Scottish Natural Heritage's Advice to Government*. Scottish Natural Heritage, Battleby.

SNH. 1999d. *National Parks for Scotland: Scottish Natural Heritage's Advice to Government*. Scottish Natural Heritage, Battleby.

SNH. 2000a. *Scottish Natural Heritage Facts and Figures 1999/2000*. Scottish Natural Heritage, Battleby.

SNH. 2000b. *A proposal for a Loch Lomond and the Trossachs National Park*. Scottish Natural Heritage, Battleby.

SNH. 2000c. *A proposal for a Cairngorms National Park*. Scottish Natural Heritage, Battleby.

SNH. 2000d. Re-introducing the European Beaver to Scotland. *Scottish Natural Heritage Briefing Paper*.

SNH. 2000e. The effects of mammalian herbivores on natural regeneration of upland, native woodland. *Scottish Natural Heritage Information and Advisory Note* 115.

SNH. 2000f. *Applying the Precautionary Principle to decisions on the natural heritage*. Scottish Natural Heritage, Battleby.

SNH. 2000g. *Scottish Natural Heritage Annual Report 1999/2000*. Scottish Natural Heritage, Battleby.

SNH. 2001. *The report on the proposal for a National Park in the Cairngorms*. Scottish Natural Heritage, Battleby.

Snowdon, P. J. and Slee, R. W. 1998. An appraisal of community-based action in forest management. *Scottish Forestry* 52(3–4): 146–56.

Snowdon, P. J. and Thomson, K. J. 1998. Tourism in the Scottish economy. In: MacLellan, R. and Smith, R. (eds), *Tourism in Scotland*. International Thomson Business Press, London. pp. 70–92.

SNW. 2000. *Restoring and Managing Riparian Woodlands*. Scottish Native Woods, Aberfeldy.

SOAEFD. 1996. *Wild geese and agriculture in Scotland: a discussion paper*. Scottish Office Agriculture Environment and Fisheries Department, Edinburgh.

SOAEFD. 1997. *Report of the Scottish Salmon Strategy Task Force*. Scottish Office Agriculture Environment and Fisheries Department. Stationery Office, Edinburgh.

SODD. 1999a. *National Planning and Policy Guidelines 14: Natural Heritage*. Scottish Office Development Department, Edinburgh.

SODD. 1999b. *National Planning and Policy Guidelines 15: Rural Development*. Scottish Office Development Department, Edinburgh.

SODD. 1999c. *Indicative Forest Strategies*. Scottish Office Development Department Circular No. 9/1999. SODD, Edinburgh.

SOED. 1994. *National Planning and Policy Guidelines 4: Land for Mineral Working*. Scottish Office Environment Department, Edinburgh.

SOED. 1995. *National Planning and Policy Guidelines 7: Planning and Flooding*. Scottish Office Environment Department, Edinburgh.

Sommerville, A. 1997. The conservation of nature. In: Magnusson, M. and White, G. (eds), *The Nature of Scotland: Landscape, Wildlife and People*. Canongate, Edinburgh. pp. 183–95.

South, A., Rushton, S. and Macdonald, D. 2000. Simulating the proposed reintroduction of the European beaver (*Castor fiber*) to Scotland. *Biological Conservation* 93(1): 103–16.

Spash, C. 1997. Environmental management without environmental valuation? In: Foster, J. (ed.), *Valuing Nature: ethics, economics and the environment*. Routledge, London. pp. 170–85.

Spinney, L. 1995. Return to the wild. *New Scientist* 1960 (14th January 1995): 35–8.

SPRM (Scott Porter Reseach and Marketing Ltd.). 1998. Access Consultation: analysis of responses. *SNH Research, Survey and Monitoring Report* No. 134. Scottish Natural Heritage, Battleby.

Staines, B. W. 1995. The impact of red deer on the regeneration of native pinewoods. In: Aldhous, J. R. (ed.), *Our Pinewood Heritage*. Forestry Commission/RSPB/SNH. pp. 107–14.

Staines, B. W. 1999a. Current issues concerning Red deer (*Cervus elaphus*) in Scotland. Part 1. *Deer* 11(1): 36–9.

Staines, B. W. 1999b. Current issues concerning Red deer (*Cervus elaphus*) in Scotland. Part 2. *Deer* 11(2): 102–5.

Staines, B. W. 2000. Wild deer: issues concerned with deer welfare and public safety. In: *Deer Commission for Scotland Annual Report 1999–2000*. The Stationery Office, Edinburgh. pp. 46–7.

Staines, B. W. and Welch, D. 1987. An appraisal of deer damage in conifer plantations. In: McIntosh, R. (ed.), *Deer and Forestry*. Proceedings of an ICF Conference, Glasgow. Institute of Chartered Foresters, Edinburgh. pp. 61–76.

Staines, B. W. and Scott, D. 1994. Recreation and red deer: a preliminary review of the issues. *Scottish Natural Heritage Review* No. 31.

Staines, B. W., Balharry, R. and Welch, D. 1995. The impact of red deer and their management on the natural heritage in the uplands. In: Thompson, D. B. A., Hester, A. J., and Usher, M. B. (eds), *Heaths and Moorland: Cultural Landscapes*. HMSO, Edinburgh. pp. 294–308.

Stephenson, G. 1997. Is there life after subsidies? The New Zealand experience. *ECOS* 18(3/4): 22–6.

Stephenson, T. 1989. *Forbidden Land: the struggle for access to mountain and moorland*. Manchester University Press, Manchester.

Steven, H. M. and Carlisle, A. 1959. *The Native Pinewoods of Scotland*. Oliver and Boyd, Edinburgh.

Stevenson, A. C. and Birks, H. J. B. 1995. Heaths and moorland: long-term ecological changes, and interactions with climate and people. In: Thompson, D. B. A., Hester, A. J., and Usher, M. B. (eds), *Heaths and Moorland: Cultural Landscapes*. HMSO, Edinburgh. pp. 224–39.

Stewart, F. and Hester, A. 1998. Impact of red deer on woodland and heathland dynamics in Scotland. In: Goldspink, C. R., King, S. and Putman, R. J. (eds), *Population Ecology, Management and Welfare of Deer*. Manchester Metropolitan University, Manchester. pp. 54–60.

Stockdale, A. and Jackson, R. E. 1997. The impact of changing legislation and agricultural policies on the letting arrangements of Scottish landowners. *Scottish Geographical Magazine* 113(2): 90–7.

Stockdale, A., Lang, A. J. and Jackson, R. E. 1996. Changing land tenure patterns in Scotland: a time for reform? *Journal of Rural Studies* 12(4): 439–49.

Stoneman, R. and Brooks, S. (eds) 1997. *Conserving Bogs: the management handbook.* The Stationery Office, Edinburgh.

Stroud, D. A., Reed, T. M., Pienkowski, M. W. and Lindsay, R. A. 1987. *Birds, Bogs and Forestry: the peatlands of Caithness and Sutherland.* Nature Conservancy Council, Peterborough.

Strutt and Parker. 2000. *Scottish Estates Review 2000/01.* Strutt and Parker, Edinburgh.

STS (System Three Scotland) 1991. A survey of public attitudes to walking and access. *SNH Research, Survey and Monitoring Report* No. 4. Scottish Natural Heritage, Edinburgh.

STS (System Three Scotland) 1996. Walking in the countryside in Scotland. *SNH Research, Survey and Monitoring Report* No. 11. Scottish Natural Heritage, Edinburgh.

Stuart-Murray, J. 1994. Indicative Forest Strategies – a critique. *Scottish Forestry* 48(1): 16–21.

Stuart-Murray, J. 1999. Review of Sidaway and Turnbull Jeffrey Partnership (1997). *Scottish Forestry* 53(3): 182.

Stuart-Murray, J., Winterbottom, S. J. and Young, J. A. 1999. Evidence for the effectiveness of forestry indicative strategies: a case study in the Scottish Borders. *Scottish Forestry* 53(3): 145–8.

Summers, D. W. 1993. Scottish salmon: the relevance of studies of historical catch data. In: Smout, T. C. (ed.), *Scotland since Prehistory: Natural Change and Human Impact.* Scottish Natural Heritage/Scottish Cultural Press, Aberdeen. pp. 98–112.

Summers, D. W. and Dugan, D. 2001. An assessment of methods used to mark fences to reduce bird collisions in pinewoods. *Scottish Forestry* 55(1): 23–9.

Summers, R. W., Moss, R. and Halliwell, E. C. 1995. Scotland's native pinewoods: the requirements of birds and animals. In: Aldhous, J. R. (ed.), *Our Pinewood Heritage.* Forestry Commission/RSPB/SNH. pp. 222–41.

Swanson, G. M. 1997. *Sika deer in Scotland.* Presentation at the University of St Andrews, March 1997.

Swanson, G. M. 2000. *The genetic and phenotypic consequences of translocations of deer (Genus* Cervus*) in Scotland.* Unpublished Ph.D. thesis, University of Edinburgh.

SWCL. 1994. *At the Watershed: a discussion paper on integrated catchment management for Scotland.* Scottish Wildlife and Countryside Link, Perth.

SWCL. 1997. *Protecting Scotland's Finest Landscapes: a call for action on National Parks in Scotland.* Scottish Wildlife and Countryside Link, Perth.

Swift, J. 2001. Field sports and wildlife conservation. In: Phillips, J., Thompson, D. B. A. and Gruellich, W. H. (eds), *Integrated Upland Management for Wildlife, Field Sports, Agriculture and Public Enjoyment.* The Heather Trust/Bidwells/SNH, Battleby. pp. 52–60.

Tané, H. 1999. The case for integrated river catchment management. In: Cresser, M. and Pugh K. (eds), *Multiple land use and catchment management.* Proceedings of an international conference, 11–13 September 1996. Macaulay Land Use Research Institute, Aberdeen. pp. 5–10.

Tapper, S. (ed.) 1999. *A Question of Balance: Game animals and their role in the British Countryside.* Game Conservancy, Fordingbridge.

Taylor, J. and MacGregor, C. 1999. Cairngorms Mountain Recreation Survey 1997–98. *SNH Research, Survey and Monitoring Report* No. 162. Scottish Natural Heritage, Battleby.

Taylor, P. 1996. Return of the animal spirits. *Reforesting Scotland* 15: 12–15.

Taylor, S. 1995. Pinewood restoration at the RSPB's Abernethy Forest Reserve. In: Aldhous, J. R. (ed.), *Our Pinewood Heritage.* Forestry Commission/RSPB/SNH. pp. 145–54.

Thin, F. 1999. Landscape assessment in the Natural Heritage Zones programme. In: Usher, M. B. (ed.), *Landscape Character: perspectives on management and change.* The Stationery Office, Edinburgh. pp. 23–33.

Thirgood, S., Redpath, S., Haydon, D. T., Rothery, P., Newton, I. and Hudson, P. 2000a. Habitat loss and raptor predation: disentangling long- and short-term causes of red grouse decline. *Proceedings of the Royal Society of London B* 267: 651–6.

Thirgood, S., Redpath, S., Newton, I. and Hudson, P. 2000b. Raptors and red grouse: conservation conflicts and management solutions. *Conservation Biology* 14: 95–104.

Thirgood, S., Redpath, S., Rothery, P. and Aebischer, N. J. 2000c. Raptor predation and population limitation in red grouse. *Journal of Animal Ecology* 69: 504–16.

Thomas, K. 1983. *Man and the natural world: changing attitudes in England 1500–1800.* Allen Lane, London.

Thomas, T. and Willis, R. 2000. The economics of agroforestry in the UK. In: Hislop, M. and Claridge, J. (eds), Agroforestry in the UK. *Forestry Commission Bulletin* 122. pp. 107–25.

Thompson, D. B. A., Horsfield, D., Gordon, J. E. and Brown, A. 1994. The environmental importance of the Cairngorms massif. In: Watson, A. and Conroy, J. (eds), *The Cairngorms – Planning Ahead.* Proceedings of a conference, Ballater, 16th June 1993. Stonehaven. pp.15–23.

Thompson, D. B. A., Hester, A. J., and Usher, M. B. (eds). 1995a. *Heaths and Moorland: Cultural Landscapes.* HMSO, Edinburgh.

Thompson, D. B. A., MacDonald, A. J., Marsden, J. H. and Galbraith, C. A. 1995b. Upland heather moorland in Great Britain: a review of international importance, vegetation change and some objectives for nature conservation. *Biological Conservation* 71: 163–78.

Thompson, D. B. A., Gillings, S. D., Galbraith, C. A., Redpath, S. M. and Drewitt, J. 1997. The contribution of game management to biodiversity: a review of the importance of grouse moors for upland birds. In: Fleming, L. V., Newton, A. C., Vickery, J. A. and Usher, M. B. (eds), *Biodiversity in Scotland: status, trends and initiatives.* The Stationery Office, Edinburgh. pp. 198–212.

Thompson, D. B. A., Gordon, J. E. and Horsfield, D. 2001. Montane landscapes in Scotland: are these natural, artefacts or complex relics? In: Gordon, J. E. and Leys, K. F. (eds), *Earth Science and the Natural Heritage: interactions and integrated management.* Stationery Office, Edinburgh. pp. 105–19.

Thompson, T. D. E. 1999. Interactive elements of catchment ecosystems: agriculture. In: Cresser, M. and Pugh K. (eds), *Multiple land use and catchment management.* Proceedings of an international conference, 11–13 September 1996. Macaulay Land Use Research Institute, Aberdeen. pp. 39–47.

Thomson, J. 2001. Delivering the new countryside recreation agenda in Scotland. In: Usher, M. B. (ed.), *Enjoyment and Understanding of the Natural Hertitage.* The Stationery Office, Edinburgh. pp. 189–97.

Thomson, K. J. (ed.) 1993a. Scottish farm income trends and the natural heritage. *SNH Research, Survey and Monitoring Report* No. 17. Scottish Natural Heritage, Edinburgh.

Thomson, K. J. 1993b. Farming and forestry expansion into the lowlands? In: Neustein, S. A. (ed.) 1993. *The winds of change: a re-evaluation of British forestry.* Proceedings of an ICF Discussion Meeting, York. Institute of Chartered Foresters, Edinburgh. pp. 123–34.

Tickell, O. 1996. Papering over the cracks. *New Scientist* 2018 (24th February 1996): 47.

Tilman, D. 2000. Causes, consequences and ethics of biodiversity. *Nature* 405(6783): 208–11.

Tipping, R. 1993a. A 'History of the Scottish Forests' revisited. *Reforesting Scotland* 8:16–21; 9: 18–21.

Tipping, R., Davies, A. and Tisdall, E. 2000. The West Affric Forest Restoration Initiative: palaeoecological approaches. *Scottish Woodland History Discussion Group: Notes IV* (Proceedings of a meeting at Battleby, Perth, 23rd November 1999). pp. 13–21.

Tipping, R. 2002. Interpretative issues concerning the driving forces of vegetation change in the early Holocene of the British Isles. In: Saville, A. (ed.), *Mesolithic Scotland: The Early Holocene Prehistory of Scotland and its European Context.* Society of Antiquaries of Scotland Monograph, Edinburgh, *in press.*

Tonn, B., English, M. and Travis, C. 2000. A framework for understanding and improving environmental decision making. *Journal of Environmental Planning and Management* 43(2): 163–83.

Toogood, M. 1995. Representing ecology and Highland tradition. *Area* 27(2): 102–9.

Toogood, M. 1997. Semi-natural history. *ECOS* 18(2): 62–8.

Towers, W., Hester, A., Malcolm, A., Stone, D. and Gray, H. 2001. Identifying the potential for native woodland in Scotland. In: *The Macaulay Institute Annual Report 2000*. The Macaulay Institute, Aberdeen. pp. 30–3.

Trenkel, V. M. 2001. Exploring red deer culing strategies using a population-specific calibrated management model. *Journal of Environmental Management* 62(1): 37–53.

Trenkel, V. M., Partridge, L. W., Gordon, I. J., Buckland, S. T., Elston, D. A. and McLean, C. 1998. The management of red deer on Scottish open hills: results of a survey conducted in 1995. *Scottish Geographical Magazine* 114(1): 57–62.

Tudor, G. J., Mackey, E. C. and Underwood, F. M. 1994. *The National Countryside Monitoring Scheme: the changing face of Scotland 1940s to 1970s*. Scottish Natural Heritage, Edinburgh.

Tuley, G. 1995. Small native pinewoods – possible ways of saving them. *Scottish Forestry* 49(1): 22–7.

Turnock, D. 1983. The Highlands: changing approaches to regional development. In: Whittington, G. and Whyte, I. D. (eds), *An Historical Geography of Scotland*. Academic Press, London. pp. 191–216.

Tylden-Wright, R. 2000. The Laggan Forest Partnership. In: Boyd, G. and Reid, D. (eds), *Social Land Ownership, Volume Two: eight more case studies from the Highlands and Islands of Scotland*. Community Learning Scotland, Inverness. pp. 27–35.

UKRWG. 2000. *Report of the UK Raptor Working Group*. DETR/JNCC, Peterborough.

UoA/MLURI. 2001. *Agriculture's contribution to Scottish society, economy and environment*. University of Aberdeen Department of Agriculture and Forestry, and Macaulay Land Use Research Institute. University of Aberdeen, Aberdeen.

Usher, M. B. 1997. Scotland's biodiversity: an overview. In: Fleming, L. V., Newton, A. C., Vickery, J. A. and Usher, M. B. (eds), *Biodiversity in Scotland: status, trends and initiatives*. The Stationery Office, Edinburgh. pp. 5–20.

Usher, M. B. (ed.) 1999a. *Landscape Character: perspectives on management and change*. The Stationery Office, Edinburgh.

Usher, M. B. 1999b. Nativeness or non-nativeness of species. *SNH Information and Advisory Note* 112. Scottish Natural Heritage, Battleby.

Usher, M. B. (ed.) 2000. *Action for Scotland's Biodiversity*. Scottish Biodiversity Group/The Stationery Office, Edinburgh.

Usher, M. B. (ed.) 2001. *Enjoyment and Understanding of the Natural Heritage*. The Stationery Office, Edinburgh.

Usher, M. B. and Thompson, D. B. A. 1993. Variation in the upland heathlands of Great Britain: conservation importance. *Biological Conservation* 66: 69–81.

Usher, M. B. and Balharry, D. 1996. *Biogeographical zonation of Scotland*. Scottish Natural Heritage, Battleby.

Van den Born, R. J. G., Lenders, R. H. J., De Groot, W. T. and Huijsman, E. 2001. The new biophilia: an exploration of visions of nature in Western countries. *Environmental Conservation* 28(1): 65–75.

Walker, M. J. C. and Lowe, J. J. 1997. Vegetation and climate in Scotland, 13,000 to 7,000 radio carbon years ago. In: Gordon, J. E. (ed.), *Reflections on the Ice Age in Scotland: an update on Quaternary Studies*. Scottish Association of Geography Teachers/Scottish Natural Heritage, Glasgow. pp. 105–15.

Walker, S. E. 1994. Tourism and recreation. In: Maitland, P. S., Boon, P. J. and McLuskey, D. S. (eds), *The freshwaters of Scotland: a national resource of international significance*. Wiley, Chichester. pp. 333–46.

Warburton, D. 1998. Participation in conservation: grasping the nettle. *ECOS* 19(2): 2–11.

Ward, N. 1998. Water quality. In: Lowe, P. and Ward, S. (eds), *British Environmental Policy and Europe: Politics and Policy in Transition*. Routledge, London. pp. 244–64.

Warren, C. R. 1997. Conflict in the Cairngorms. *Geography Review* 11(1): 38–9.

Warren, C. R. 1998. Forestry – blight or blessing? *Geography Review* 11(4): 15–17.

Warren, C. R. 1999a. Scottish land reform: time to get lairds a-leaping? *ECOS* 20(1): 2–12.

Warren, C. R. 1999b. National parks: the best way forward for Scotland? *Scottish Forestry* 53(2): 86–92.

Warren, C. R. 1999c. Access to the hills. *Geography Review* 12(5): 28–9.

Warren, C. R. 1999d. Is Scotland the right place for superquarries? *Geography Review* 12(4): 40–1.

Warren, C. R. 2000a. 'Birds, Bogs and Forestry' revisited: the significance of the Flow Country controversy. *Scottish Geographical Journal* 116(4): 315–37.

Warren, C. R. 2000b. Agonising over agriculture. *Geography Review* 13(4): 12–14.

Warren, C. R. and Gibson, A. 2002. Six years after the fences came down: minimal deer damage to commercial forestry at Glenfinnan. *Scottish Forestry, in press.*

Waters, G. R. 1994. Government policies for the countryside. *Land Use Policy* 11(2): 88–93.

Watson, A. 1984. Paths and people in the Cairngorms. *Scottish Geographical Magazine* 100(3): 151–60.

Watson, A. 1990. Human impact on the Cairngorms environment above timberline. In: Conroy, J. W. H., Watson, A. and Gunson, A. R. (eds), *Caring for the high mountains – conservation of the Cairngorms*. Centre for Scottish Studies, Aberdeen. pp. 61–82.

Watson, A. 1991. Increase of people on Cairn Gorm plateau following easier access. *Scottish Geographical Magazine* 107: 99–105.

Watson, A. 1996. Internationally important environmental features of the Cairngorms: research, and main research needs. *Botanical Journal of Scotland* 48(1): 1–12.

Watson, R. D. 1990. A hillman looks at the conservation of the Cairngorms. In: Conroy, J. W. H., Watson, A. and Gunson, A. R. (eds), *Caring for the high mountains – conservation of the Cairngorms*. Centre for Scottish Studies, Aberdeen. pp. 91–107.

Watt, A. D., Carey, P. D. and Eversham, B. C. 1997. Implications of climate change for biodiversity. In: Fleming, L. V., Newton, A. C., Vickery, J. A. and Usher, M. B. (eds), *Biodiversity in Scotland: status, trends and initiatives*. The Stationery Office, Edinburgh. pp. 147–59.

Watt, G. 1999. Trees, people and profits – into the next millennium: future timber markets. *Quarterly Journal of Forestry* 93(3): 204–10.

WCED (World Commission on Environment and Development) 1987. *Our Common Future*. Oxford University Press, Oxford.

Webb, D. A. 1985. What are the criteria for presuming native status? *Watsonia* 15: 231–6.

Welch, D., Staines, B. W., Scott, D., and Catt, D. C. 1987. Bark stripping by red deer in a Sitka spruce forest in western Scotland. I. Incidence. *Forestry* 60(2): 249–62.

Welch, D., Staines, B. W., Scott, D., French, D. D. and Catt, D. C. 1991. Leader browsing by red and roe deer on young Sitka spruce trees in western Scotland. I. Damage rates and the influence of habitat factors. *Forestry* 64(1): 61–82.

Welch, D., Staines, B. W., Scott, D., and French, D. D. 1992. Leader browsing by red and roe deer on young Sitka spruce trees in western Scotland. II. Effects on growth and tree form. *Forestry* 65(3): 309–30.

Welch, D. and Scott, D. 1998. Bark-stripping damage by red deer in a Sitka spruce forest in western Scotland. IV. Survival and performance of wounded trees. *Forestry* 71(3): 225–35.

Welch, D., Carss, D. N., Gornall, J. Manchester, S. J., Marquiss, M., Preston, C. D., Telfer, M. G., Arnold, H. and Holbrook, J. 2001. An audit of alien species in Scotland. *Scottish Natural Heritage Review* No. 139.

Wells, A. 1998. *Glenlivet Estate: a case study in sustainable land use and development*. Crown Estate, Edinburgh. CD ROM.

Werritty, A. 1995. Integrated catchment management: a review and evaluation. *Scottish Natural Heritage Review* No. 58.

Werritty, A. 1997. Enhancing the quality of freshwater resources: the role of integrated catchment management. In: Boon, P. J. and Howell, D. L. (eds), *Freshwater Quality: defining the indefinable?* Scottish Natural Heritage, Edinburgh. pp. 489–505.

Werritty, A. 2002. Living with uncertainty: climate change, river flows and water resource management in Scotland. *Science of the Total Environment*, in press.

Westbrook, S. 1994. *Economic impact of the proposed Cairngorm funicular railway*. Scottish Wildlife and Countryside Link, Perth.

Whitby, M. (ed.) 1994. *Incentives for Countryside Management: the case of Environmentally Sensitive Areas*. CAB International, Wallingford.

Whitby, M. 1996a. The prospect for agri-environmental policies within a reformed CAP. In: Whitby, M. (ed.), *The European Environment and CAP Reform: Policies and Prospects for Conservation*. CAB International, Wallingford. pp. 227–40.

Whitby, M. 1996b. The United Kingdom. In: Whitby, M. (ed.), *The European Environment and CAP Reform: Policies and Prospects for Conservation*. CAB International, Wallingford. pp. 186–205.

White, J. E. 1997. The history of introduced trees in Britain. In: Ratcliffe, P. R. (ed.), *Native and Non-Native in British Forestry*. Proceedings of a Discussion meeting, University of Warwick, 31 March–2 April, 1995. Institute of Chartered Foresters, Edinburgh. pp. 4–8.

White, L. 1967. The historical roots of our ecological crisis. *Science* 155(3767): 1203–7.

White, P. C. L., Bennett, A. C. and Hayes, E. J. V. 2001. The use of willingness-to-pay approaches in mammal conservation. *Mammal Review* 31(2): 151–67.

Whitehead, G. K. 1996. *Half a century of Scottish deer stalking*. Swan Hill Press, Shrewsbury.

Whitney-McIver, H., Blyth, J. F. and Malcolm, D. C. 1992. The application of group selection working in an upland forest in south Scotland. *Scottish Forestry* 46(3): 202–11.

Whittington, G. 1983. Agriculture and society in Lowland Scotland, 1750–1870. In: Whittington, G. and Whyte, I. D. (eds), *An Historical Geography of Scotland*. Academic Press, London. pp. 141–64.

Whittington, G. and Edwards, K. J. 1997a. Climate change. In: Edwards, K. J. and Ralston, I. B. M. (eds), *Scotland: Environment and Archaeology, 8000 BC–AD 1000*. Wiley, Chichester. pp. 11–22.

Whittington, G. and Edwards, K. J. 1997b. Human activity and landscape change during the Holocene. In: Gordon, J. E. (ed.), *Reflections on the Ice Age in Scotland: an update on Quaternary Studies*. Scottish Association of Geography Teachers/Scottish Natural Heritage, Glasgow. pp. 130–5.

Whittome, T. 1996. *The Cairngorm funicular railway*. Paper presented at the Sixth Annual Conference on Advances in Environmental Impact Assessment, London, 9th July 1996.

Whyte, I. D. and Whyte, K. 1991. *The Changing Scottish Landscape 1500–1800*. Routledge, London.

Wigan, M. 1991. *The Scottish Highland Estate: Preserving an Environment*. Swan Hill Press, Shrewsbury.

Wigan, M. 1993. *Stag at Bay: the Scottish red deer crisis*. Swan Hill Press, Shrewsbury.

Wigan, M. 1996a. A dangerous romance. *The Field* June 1996: 60–2.

Wigan, M. 1996b. Sharing the forest with deer. *The Field* February 1996: 30–1.

Wigan, M. 1998. Millennium Scotland: a Brave New World. *The Field* June 1998: 46–9.

Wigan, M. 1999a. Scottish land reform – it's the same old song. *The Field* March 1999: 44–7.

Wigan, M. 1999b. Eat them, don't protect them. *The Field* September 1999: 70–2.

Wigan, M. 1999c. Why we shouldn't go native. *The Field* February 1999: 9.

Wightman, A. (ed.) 1992. *A forest for Scotland – a discussion paper on forest policy*. Scottish Wildlife and Countryside Link, Perth.

Wightman, A. 1996a. *Who owns Scotland?* Canongate, Edinburgh.

Wightman, A. 1996b. *Scotland's mountains: an agenda for sustainable development*. Scottish Wildlife and Countryside Link, Perth.

Wightman, A. 1999. *Scotland: land and power – the agenda for land reform.* Luath Press, Edinburgh.

Wightman, A. 2001a. Land Reform Draft Bill Part II – Community Right-to-Buy. *Caledonia Briefings* 3.

Wightman, A. 2001b. Land Reform: the Draft Bill. *Caledonia Briefings* 4.

Wightman, A. and Higgins, P. 2000. Sporting estates and the recreational economy in the Highlands and Islands of Scotland. *Scottish Affairs* 31: 18–36.

Wightman, A. and Higgins, P. 2001. Sporting estates and outdoor recreation in the Highlands and Islands of Scotland. In: Usher, M. B. (ed.), *Enjoyment and Understanding of the Natural Heritage.* The Stationery Office, Edinburgh. pp. 171–6.

Williamson, R. 2001. Impacts of recreation on land management. In: Usher, M. B. (ed.), *Enjoyment and Understanding of the Natural Heritage.* The Stationery Office, Edinburgh. pp. 67–9.

Williamson, R. B. and Beveridge, M. C. M. 1994. Fisheries and aquaculture. In: Maitland, P. S., Boon, P. J. and McLuskey, D. S. (eds), *The freshwaters of Scotland: a national resource of international significance.* Wiley, Chichester. pp. 317–32.

Wilson, A. 1999. Urban woodland management in Livingston. *Scottish Forestry* 53(2): 96–9.

Wilson, G. A. 1997. Assessing the environmental impact of the Environmentally Sensitive Areas Scheme: a case for using farmers' environmental knowledge? *Landscape Research* 22(3): 303–26.

Wilson, G. A. and Bryant, R. L. 1997. *Environmental Management: new directions for the Twenty-First Century.* UCL Press, London.

Winter, M. 1996. *Rural politics: policies for agriculture, forestry and the environment.* Routledge, London.

Winter, M. 2000. Strong policy or weak policy? The environmental impact of the 1992 reforms to the CAP arable regime in Great Britain. *Journal of Rural Policy* 16: 47–59.

Winter, M. and Gaskell, P. 1998. The Agenda 2000 debate and CAP reform in Great Britain: is the environment being sidelined? *Land Use Policy* 15(2): 217–31.

Wonders, W. C. 1998. Forestry villages in the Scottish Highlands. *Scottish Forestry* 52(3/4): 158–67.

Woods, M. and Moriarty, P. V. 2001. Strangers in a strange land: the problem of exotic species. *Environmental Values* 10: 163–91.

Worrell, R. 1996. The Boreal Forests of Scotland. *Forestry Commission Technical Paper* 14.

Worrell, R. 1997. Can ethical considerations contribute to determining the choice of tree species and woodland types? In: Ratcliffe, P. R. (ed.), *Native and Non-Native in British Forestry.* Proceedings of a discussion meeting, University of Warwick, 31 March–2 April, 1995. Institute of Chartered Foresters, Edinburgh. pp. 171–82.

Wright, P. 1995. Water resources management in Scotland. *Journal of the Chartered Institute of Water and Environmental Managers* 9: 153–63.

Wright, P. 2000. Climate change and Scottish implications. Paper presented at *Hydrology in Scotland: Agenda for the 21st Century*, University of Dundee, 21–23 March 2000.

WWF. 1999. *Europe's Living Rivers: an agenda for action.* World Wide Fund for Nature, Copenhagen.

Yalden, D. W. 1986. Opportunities for reintroducing British mammals. *Mammal Review* 16: 53–63.

Yalden, D. W. 1998. The past, present and future of deer in Britain. In: Goldspink, C. R., King, S. and Putman, R. J. (eds), *Population Ecology, Management and Welfare of Deer.* Manchester Metropolitan University, Manchester. pp. 1–5.

Yalden, D. W. 1999. *The History of British Mammals.* Poyser, London.

Yorke, M. 1995. Continuous cover silvicultural systems in Britain. *Scottish Forestry* 49(3): 162–5.

Yorke, M. 2001. Some misconceptions of continuous cover silviculture. *Scottish Forestry* 55(2): 73–5.

Young, J. 1997. More raptors and more grouse? *The Field* June 1997: 62–5.

Young, R. 2000. All things bright and beautiful? *The Times* 7th October 2000.

Youngson, R. W. 1997. Sika Deer – a growing problem. In: Rose, H. (ed.), *Deer in Forestry – Beyond 2000*. Minutes of a conference in March 1997, Battleby. pp. 5–6.

Youngson, R. W. and Stewart, L. K. 1996. Trends in red deer populations within the Cairngorms core area. *Botanical Journal of Scotland* 48(1): 111–16.

Yull, L. 1998. Opening remarks. In: *Forestry and Land Use – Forestry in a Changing Scotland.* Proceedings of a Timber Growers Association Conference, Gleneagles, 20th May 1998: 12.

Appendix: Student howlers

Statements made in course essays and exams by students taking an Honours course entitled *Environmental Management in Scotland* between 1995 and 2000.

'As human beings, the superior beings, we are surely responsible for the farewell of the ecosystem, environment, and inhabitants.'

'Research on red deer has proved that male morality increases with greater hind numbers.'

'Red deer are a native species of Scotland and have long been cohabiting with Man.'

'The red deer population explosion was triggered not only by the extinction of predators but also by the introduction of refrigerated sheep from Australia and New Zealand in the 1870s.'

'It is important to recognise the interrelationship between different land uses as well as the natural heretic interest.'

'Native woodland is good for the Caper Ceilidh.'

'90 per cent of *Homo sapiens* in Scotland are coniferous. This has occurred due to the ability to grow exotic species faster.'

'Scottish forestry relies on three species: oakwoods, pinewoods and Cytrus spruce.'

'The Cairngorms are a sub-artic plateau of 4,000km^3.'

'Concern has already been lodged against the building of a vernacular railway.'

'Sustainability is about not ruining life for our ancestors.'

'Environmental concerns will ruin supreme over all decisions.'

'If the aggravated benefits outweigh the costs then the action should be allowed.'

'Gaia is the myth of motherhood in scientific form.'

'In the Judo-Christian viewpoint, man is the steward of the earth.'

'Therefore, if the National Park Governing Body wished to purchase, a large amount of land could be purchased in one foul swoop.'

'The relationship between the reduction of heather moorland and grouse populations is a chicken and egg situation.'

'Species such as Harrier Hens and other raptors are perceived as pests to the grouse industry.'

'Greater numbers of predators suggests an increase in morality of grouse, which has a direct impact on the size of the grouse population.'

'The problem of antipathy towards conservation organisations appears to result from a lack of intolerance on both sides.'

'Conservation has a duel purpose.'

Index